Application of the China Meteorological Assimilation Driving Datasets for the SWAT Model (CMADS) in East Asia

Application of the China Meteorological Assimilation Driving Datasets for the SWAT Model (CMADS) in East Asia

Special Issue Editors

Hao Wang
Xianyong Meng

MDPI • Basel • Beijing • Wuhan • Barcelona • Belgrade

MDPI

Special Issue Editors

Hao Wang
State Key Laboratory of Simulation and
Regulation of Water Cycle in River Basin &
China Institute of Water Resources and
Hydropower Research,
China

Xianyong Meng
The University of Hong Kong (HKU),
Hong Kong

Editorial Office
MDPI
St. Alban-Anlage 66
4052 Basel, Switzerland

This is a reprint of articles from the Special Issue published online in the open access journal *Water* (ISSN 2073-4441) from 2017 to 2019 (available at: https://www.mdpi.com/journal/water/special_issues/CMADS)

For citation purposes, cite each article independently as indicated on the article page online and as indicated below:

LastName, A.A.; LastName, B.B.; LastName, C.C. Article Title. *Journal Name* **Year**, *Article Number*, Page Range.

ISBN 978-3-03921-235-4 (Pbk)
ISBN 978-3-03921-236-1 (PDF)

Cover image courtesy of Xianyong Meng.

Contents

About the Special Issue Editors

Hao Wang was born in August 1953. He is an Academician of the Chinese Academy of Engineering, an engineer at professorial level, Ph.D. supervisor in China Institute of Water Resources and Hydropower Research (IWHR) and has the privilege of enjoying the Special Allowance from the State Council of P. R. China since 1994. He is now Director of State Key Laboratory of Simulation and Regulation of Water Cycles in River Basin and Honorary Director of Department of Water Resources, China Institute of Water Resources and Hydropower Research (IWHR) and Director of Key Laboratory of Sustainable Development of Water Environment for East Asia (WEEA), and Visiting Professor at Tsinghua University, Wuhan University, Northwest Agriculture and Forestry University, China University of Geosciences. Dr. Hao Wang also serves concurrently as Secretary-General of Global Water Partnership China (GWP China), a member of National Environmental Advisory Committee, a member of the Science & Technology Committee of the Ministry of Water Resources (MWR), a member of the Science & Technology Committee of National Forestry Administration, Deputy Director of the Research Society of Technological Innovation Methods, Deputy Director of China Society of Natural Resources, Managing Director of China Society of Sustainable Development and Director of the Sub-Committee on Water Issues, Executive Director of Chinese Hydraulic Engineering Society (CHES), and Deputy Director of its Committee on Hydrology, Executive Director of China Society of Forestry, Director of Special Task Force on Water Issues for Chinese Society for Sustainable Development (CSSD).

Xianyong Meng Ph.D. was born in December 1987. He is now Research Associate of Department of Civil Engineering, The University of Hong Kong (HKU) and China Agriculture University (CAU). He has also served as Senior Research Fellow of Research Center of Water Environment and Sustainable Development for East Asia (WEEA). Dr. Xianyong Meng has long been engaged in the study of hydrological model improvement, atmosphere-driven field fusion assimilation, land surface model simulation, parameterization scheme, water resources pollution, and non-point source pollution simulation and remediation. Dr. Xianyong Meng has established CMADS (spatial resolution: 0.33 degrees, 0.25 degrees, 0.125 degrees, and 0.0625 degrees; time resolution: daily), which assimilates the precipitation data of nearly 40,000 regional stations in China, and can provide accurate meteorological data for a variety of hydrological models. Dr. Xianyong Meng also built the double distributed snowmelt runoff model forced by WRF and established a numerical prediction of snowmelt flood simulation platform for the northern slope of Tianshan Mountain, China, which has resulted in the reduction of economic loss for the local people in the order of hundreds of millions of dollars. Dr. Xianyong Meng also assimilated and constructed a 1 km high-precision atmospheric forcing field in Xinjiang, China (XJLDAS), and also participated in the construction of CLDAS2.0 for China Meteorological Administration. In recent years, Dr. Xianyong Meng has hosted or participated in six national science and technology support programs of the National Natural Science Foundation (NSFC) of China. Dr. Xianyong Meng has been published over 70 academic papers and has won many awards for scientific and technological progress from the China government and the China Society of water conservancy.

water | MDPI

Editorial

Significance of the China Meteorological Assimilation Driving Datasets for the SWAT Model (CMADS) of East Asia

Xianyong Meng *,† and Hao Wang *,†

State Key Laboratory of Simulation and Regulation of Water Cycle in River Basin & China Institute of Water Resources and Hydropower Research, Beijing 100038, China
* Correspondence: mxy@iwhr.com (X.M.); wanghao@iwhr.com (H.W.); Tel.: +86-10-68410178 (X.M. & H.W.)
† These authors contributed equally to this work.

Received: 26 August 2017; Accepted: 30 September 2017; Published: 8 October 2017

Abstract: The high degree of spatial variability in climate conditions, and a lack of meteorological data for East Asia, present challenges to conducting surface water research in the context of the hydrological cycle. In addition, East Asia is facing pressure from both water resource scarcity and water pollution. The consequences of water pollution have attracted public concern in recent years. The low frequency and difficulty of monitoring water quality present challenges to understanding the continuous spatial distributions of non-point source pollution mechanisms in East Asia. The China Meteorological Assimilation Driving Datasets for the Soil and Water Assessment Tool (SWAT) model (CMADS) was developed to provide high-resolution, high-quality meteorological data for use by the scientific community. Applying CMADS can significantly reduce the meteorological input uncertainty and improve the performance of non-point source pollution models, since water resources and non-point source pollution can be more accurately localised. In addition, researchers can make use of high-resolution time series data from CMADS to conduct spatial- and temporal-scale analyses of meteorological data. This Special Issue, "Application of the China Meteorological Assimilation Driving Datasets for the SWAT Model (CMADS) in East Asia", provides a platform to introduce recent advances in the modelling of water quality and quantity in watersheds using CMADS and hydrological models, and underscores its application to a wide range of topics.

Keywords: East Asia; CMADS; meteorological input uncertainty; hydrological modelling; SWAT; non-point source pollution models

China and the surrounding region in East Asia are considered to be the birthplace of human civilisation. East Asia experiences the most typical and pronounced monsoon climate in the world, and detailed analyses of the atmospheric hydrological cycle in East Asia can offer a substantial regional contribution to global climate change research.

Travelling back to the 19th century, natural science research in East Asia, and even globally, still followed the paradigm of dividing the whole into smaller components (e.g., dividing systems into elements), and then studying each isolated part individually. In this context, there was no interdisciplinary approach for researching various types of scientific issues simultaneously. By the second half of the 20th century, a highly detailed and complex classification of numerous natural science disciplines had been developed, and scientists were accustomed to dividing scientific fields into a number of sub-fields. This promoted a more professional approach to research and led to the evolution of various cross-disciplines and interdisciplinary disciplines. However, in recent decades, scientists have recognised many problems associated with the single-disciplinary approach of the 19th century. More researchers reconsidered methods for systematically considering and analysing different system elements of multiple disciplines at a more comprehensive level; in particular, they conducted comprehensive integrations based on a high degree of differentiation (i.e., categorising disciplines by specialisations and integrating multiple disciplines). Using research on atmospheric hydrology as an example, in 1962, a number of scientists attempted to use climate data and mathematical models to establish or supplement missing hydrological runoff sequences [1–4]. These studies made pioneering contributions to the field of atmospheric hydrology at a time when there was a lack of hydrological runoff data and meteorology sites were scarce.

With the gradual deepening of scientific research and continuous development of disciplinary theory, atmosphere and hydrology researchers began to conduct in-depth studies of atmospheric or water cycle mechanisms based on their own expertise. Differences in these two fields led to the rise of several subtle distinctions in the research directions of topics such as atmospheric circulation and water cycle research by atmospheric scientists and hydrologists (e.g., in terms of methods, techniques, etc.). For instance, researchers with a background in atmospheric sciences are more likely to focus on macro land-to-air interactions and their macroscopic effects on larger regions and the globe; they generally prefer studying the balance of various fluxes at the surface. For example, when studying atmospheric processes, atmospheric scientists have developed various land–air coupling models to simulate land-to-air interaction processes. From the Bucket model [5] in the mid-1990s to the Community Land Model (CLM) in the 21st century [6], land–air models have undergone a series of complex evolution processes. During this period, meteorologists developed various land models, including BATS [7], Simplified Simple Biosphere (SSIB) [8], A Revised Land Surface Parameterization (SiB2) [9,10], and the Common Land Model (CoLM) [11], etc. Based on the atmosphere-based Bucket model, the above-mentioned models gradually evolved into the more complex and generalised CLM model, by integrating components such as land surface, ocean and sea ice, sulphate aerosols, non-sulphate aerosols, carbon cycle, dynamic vegetation, and atmospheric chemistry [12]. When developing land models, atmospheric researchers are more concerned with improving the accuracy of various elements of the atmospheric forcing field, to reduce its uncertainty as an input in land models. In this process, various assimilation techniques and multi-source data (e.g., observation stations, radar stations, satellite remote sensing data, aerial data, and model data) have been widely used to establish atmospheric reanalysis datasets at various scales. Examples include the National Centers for Environmental Prediction/National Center for Atmospheric Research NCEP/NCAR-(R1) reanalysis dataset and National Centers for Environmental Prediction-Department of Energy (NCEP-DOE)-(R2) reanalysis dataset [13,14], Climate Forecast System Reanalysis (CFSR) by NCEP [15], European Centre for Medium-Range Weather Forecasts (ECMWF) 15-year Re-Analysis (ERA-15) [16], ECMWF Re-Analysis from September 1957 to August 2002 (ERA-40) [17], ECMWF Reanalysis-Interim (ERA-Interim) [18], Japanese 25-year Reanalysis project (JRA-25) [19], and the CMA Land Data Assimilation System (CLDAS) [20]. These reanalysis data sets provide important basic data for global researchers to analyse climate–water cycles. However, in focusing on macro-energy balances, meteorologists do not have sufficient resources to consider the microscale water balance processes of hydrological cycles, which is the main interest of hydrologists. When atmospheric scientists consider the hydrological confluence process, they mostly use simple conceptual methods for calculations (e.g., a vector-based river routing scheme [RAPID]) [21], and such simple conceptual models are not applicable under many conditions (e.g., where there is artificial intervention in a river area or in areas experiencing extreme climate change).

As noted above, hydrologists are primarily concerned with the micro-scale water balance processes of hydrological cycles. In 1959, the development of the Stanford Watershed Model (SWM) [22] set a precedent for the development of hydrological conceptual statistical models. However, it was not until 1979, after the advent of the distributed hydrological model topography-based hydrological model (TOPMODEL) [23], that fully distributed hydrological models for small- and medium-sized watersheds began to be accepted by the scientific community. Representative examples of such models include the Soil and Water Assessment Tool (SWAT) [24] and the Soil and Water Integrated Model (SWIM) [25]. As the development of these hydrological models continued to mature, more physical processes were gradually integrated into the models' calculations, leading to more complete expressions of the physical processes involved therein.

Atmospheric scientists are more inclined to use complex, more accurate atmospheric-driven land models coupled with simple hydrological models (i.e., conceptual models). In contrast, most hydrological scientists use simple atmosphere-driven (e.g., meteorological observatory) conceptual or distributed (complex) hydrological models. This presents two problems. First, the simple hydrological conceptual models currently used by meteorologists cannot reproduce real runoff processes and their related components (e.g., sediment erosion, non-point source pollution, floods, etc.) in areas with complex geological structures and significant artificial influences (or extreme changes in climate). Second, hydrologists, especially researchers in East Asia, cannot easily use the limited number of available meteorological sites to rationally and effectively determine the model parameters of complex distributed hydrological models. Consequently, researchers can only use low-quality meteorological data that are more akin to simulation games than research tools, which are unlikely to yield accurate conclusions.

Owing to the limitations of multiple objective factors, such as economy and geological structure, the overall distribution density of traditional observational meteorological stations (e.g., precipitation, temperature, humidity, wind speed, soil temperature, and soil moisture) in East Asia is low. Atmospheric hydrology studies in various fields over the past few decades in East Asia were not comprehensive, owing to the limited access to meteorological data. Despite the fact that researchers in East Asia can now use existing published reanalysis data (e.g., CFSR, NCEP, etc.) to conduct climate analyses in the region, since the needed assimilation and revision of the above reanalysis products were not conducted by stations in most parts of East Asia [26], the reanalysis results and actual results often differ substantially. For example, the CFSR precipitation data in summer in China are severely overestimated [26]. Recently, scientists in East Asia have started collecting small amounts of meteorological observation data by establishing field monitoring sites, and funding in atmospheric hydrological research in East Asia is being increased in an attempt to reduce the gap between atmospheric hydrology research in East Asia and worldwide. However, distortions of the meteorological input data used in scientific analyses (e.g., acquisition failure, lack of data, presence of outliers, etc.) reduce the reliability of the findings, with differences in input data possibly resulting in entirely different results. A major cause of the emergence of this phenomenon is that meteorological data collection does not follow a standard procedure; data are assimilated from multiple sources, revised based on the large number of stations, and are accessible in the public domain.

East Asia is a part of the largest continent in the world. In addition, it is the world's most densely populated region, with approximately 1.5 billion inhabitants. The underlying geography is complex and highly differentiated, leading to large climate variations. For example, this region contains the Qinghai–Tibet Plateau, the world's highest, which has a unique alpine climate that profoundly influences the climate in East Asian countries and across the globe. Owing to climate change, East Asia's water resources have been facing multiple pressures over recent years, such as uneven distributions of droughts and floods, water pollution, and water shortages. Consistent with the limitations in weather station observations, shortcomings related to economics, terrain, and other objective factors make it difficult to perform large-scale, long-term, high-frequency monitoring studies of water pollution and other related topics (such as floods, droughts, water scarcity, etc.) in East Asia.

To address the many aforementioned difficulties, the China Meteorological Assimilation Driving Datasets for the SWAT model (CMADS) [26,27] was developed by Xianyong Meng using STMAS assimilation techniques [20], as well as big data projection and processing methods (including loop nesting of data, projection of resampling models, and bilinear interpolation). CMADS comprises many variables, including daily average temperature, daily maximum temperature, daily minimum temperature, daily cumulative precipitation (20–20 h), daily average relative humidity, daily average specific humidity, daily average solar radiation, daily average wind, daily average atmospheric pressure, soil temperature, and soil moisture. CMADS was developed to provide high-resolution, high-quality meteorological data for use by the scientific community. Applying CMADS can significantly reduce meteorological input uncertainties and improve the performance of non-point source pollution modelling, since water resources and non-point source pollution can be more accurately localised. In addition, researchers can employ high-resolution time series data from CMADS to perform spatial- and temporal-scale analyses of meteorological data. Over the past few years, the CMADS dataset has received attention from around the world, including researchers in the United States, Germany, Russia, Italy, India, and South Korea, among others. As a developer of CMADS, we have used the CMADS driven SWAT model to simulate the runoff of many watersheds, such as China's Heihe River Basin [26] and Manas River Basin [27], and obtained satisfactory results. We expect researchers around the world to take full advantage of the CMADS owing to its high spatiotemporal resolution, unified procedure (including latitude and longitude, and elevation), and reliable quality. CMADS can be used to carry out studies of various distributed models (e.g., the SWAT and Variable Infiltration Capacity (VIC) models) and high-resolution climate verification and analyses. Given that meteorological data pertaining to East Asia are scarce, the use of CMADS can assist researchers globally to perform more efficient and effective scientific comparisons and in-depth investigations with a standard procedure.

Acknowledgments: This research was financially joint supported by the National Science Foundation of China (51479209, 41701076, 51609260), the National Key Technology R&D Program of China (2016YFA0601602, 2017YFC0404305, 2017YFB0203104), the China Postdoctoral Science Foundation (2017M610950), and National One-Thousand Youth Talent Program of China (122990901606). The authors also served as Guest Editors of this Special Issue and thank the journal editors for their support.

Conflicts of Interest: The authors declare no conflict of interest.

References

1. Fibbing, M.B. On the use of correlation to augment data. *J. Am. Stat. Assoc.* **1962**, *57*, 20–32. [CrossRef]
2. Benoit, B.M.; James, R.W. Noah, Joseph, and operational hydrology. *Water. Resour. Res.* **1968**, *4*, 909–918.
3. Rodríguez-Iturbe, I. Estimation of statistical parameters for annual river flows. *Water. Resour. Res.* **1969**, *5*, 1418–1421. [CrossRef]
4. Stockton, C.W. The Feasibility of Augmenting Hydrologic Records Using Tree-Ring Data. Ph.D. Thesis, The University of Arizona, Tucson, AZ, USA, 1971.
5. Budyko, M.I. The heat balance of the earth's surface. *Sov. Geogr.* **1961**, *2*, 3–13. [CrossRef]
6. Dai, Y.; Zeng, X.; Dickinson, R.E.; Baker, I.; Bonan, G.B.; Bosilovich, M.G.; Scott Denning, A.; Dirmeyer, P.A.; Houser, P.R.; Niu, G.; et al. The common land model. *Bull. Am. Meteorol. Soc.* **2003**, *84*, 1013–1023. [CrossRef]
7. Dickinson, R.E.; Henderson-Sellers, A.; Kennedy, P.J. *Biosphere-Atmosphere Transfer Scheme (BATS) Version 1e as Coupled to the NCAR Community Climate Model*; NCAR Technical Note NCAR/TN-387+STR; National Center for Atmospheric Research: Boulder, CO, USA, 1993.
8. Xue, Y.; Sellers, P.J.; Kinter, J.L.; Shukla, J. A Simplified Biosphere Model for Global Climate Studies. *J. Clim.* **1991**, *4*, 345–364. [CrossRef]
9. Sellers, P.J.; Randall, D.A.; Collatz, G.J.; Berry, J.A.; Field, C.B.; Dazlich, D.A.; Zhang, C.; Collelo, G.D.; Bounoua, L. A Revised Land Surface Parameterization (SiB2) for Atmospheric GCMS. Part I: Model Formulation. *J. Clim.* **1996**, *9*, 676–705. [CrossRef]
10. Sellers, P.J.; Los, S.O.; Tucker, C.J.; Justice, C.O.; Dazlich, D.A.; James Collatz, G.; Randall, D.A. A Revised Land Surface Parameterization (SiB2) for Atmospheric GCMS. Part II: The Generation of Global Fields of Terrestrial Biophysical Parameters from Satellite Data. *J. Clim.* **1996**, *9*, 706–737. [CrossRef]
11. Dai, Y.J.; Zeng, X.B.; Dickinson, R.E. *Common Land Model, Technical Documentation and User's Guide*; Georgia Institute of Technology: Atlanta, GA, USA, 2001; pp. 1–69.
12. Oleson, K.W.; Dai, Y.J.; Bonan, G.; Bosilovich, M.; Dickinson, R.; Dirmeyer, P.; Hoffman, F.; Houser, P.; Levis, S.; Niu, G.-Y.; et al. *Technical Description of the Community Land Model (CLM)*; NCAR Tech Note NCAR/Tn-461+Str; National Center for Atmopheric Research: Boulder, CO, USA, 2004; p. 173.
13. Trenberth, K.E.; Anthes, R.A.; Belward, A.; Brown, O.B.; Habermann, T.; Karl, T.R.; Running, S.; Ryan, B.; Tanner, M.; Wielicki, B. *Challenges of a Sustained Climate Observing System*; Springer: Berlin, Germany, 2013.

14. Kanamitsu, M.; Ebisuzaki, W.; Woollen, J.; Yang, S.-K.; Hnilo, J.J.; Fiorino, M.; Potter, G.L. NCEP-DEO AMIP-II Reanalysis (R-2). *Bull. Am. Meteorol. Soc.* **2002**, *83*, 1631–1643. [CrossRef]
15. Saha, S.; Moorthi, S.; Pan, H.L.; Wu, X.; Wang, J.; Nadiga, S.; Tripp, P.; Kistler, R.; Woollen, J.; Behringer, D.; et al. The NCEP Climate Forecast System Reanalysis. *B. Am. Meteorol. Soc.* **2010**, *91*, 1015–1057. [CrossRef]
16. Gibson, J.K.; Kållberg, P.; Uppala, S.; Nomura, A.; Hernandez, A.; Serrano, E. ERA Description. In *ECMWF ERA-15 Project ReportSeries, No.1*; European Centre for Medium-RangeWeather Forecasts: Shinfield, Reading, UK, 1997; Available online: https://www.ecmwf.int/search/elibrary?authors=Gibson (accessed on 7 October 2017).
17. Uppala, S.M.; Kållberg, P.W.; Simmons, A.J.; Andrae, U.; Da Costa Bechtold, V.; Fiorino, M.; Gibson, J.K.; Haseler, J.; Hernandez, A.; Kelly, G.A.; et al. The ERA-40 re-analysis. *Q. J. R. Meteorol. Soc.* **2005**, *131*, 2961–3012. [CrossRef]
18. Dee, D.P.; Uppala, S.M.; Simmons, A.J.; Berrisford, P.; Poli, P.; Kobayashi, S.; Andrae, U.; Balmaseda, M.A.; Balsamo, G.; Bauer, P.; et al. The ERA-Interim reanalysis: Configuration and performance of the data assimilation system. *Q. J. R. Meteorol. Soc.* **2011**, *137*, 553–597. [CrossRef]
19. Onogi, K.; Tsutsui, J.; Koide, H.; Sakamoto, M.; Kobayashi, S.; Hatsushika, H.; Matsumoto, T.; Yamazaki, N.; Kamahori, H.; Takahashi, K.; et al. The JRA-25 reanalysis. *J. Meteorol. Soc. Jpn.* **2007**, *85*, 369–432. [CrossRef]
20. Meng, X.Y.; Wang, H.; Wu, Y.P.; Long, A.H.; Wang, J.H.; Shi, C.X.; Ji, X.N. Investigating spatiotemporal changes of the land surface processes in Xinjiang using high-resolution CLM3.5 and CLDAS: Soil temperature. *Sci. Rep* **2017**, *7*. [CrossRef]
21. David, C.H.; Habets, F.; Maidment, D.R.; Yang, Z.L. RAPID applied to the SIM-France model. *Hydrol. Process.* **2011**, *25*, 3412–3425. [CrossRef]
22. Crawford, N.H.; Linsley, R.K. *The Synthesis of Continuous Streamflow on a Digital Computer*; Technical Report No. 12; Department of Civil Engineering, Stanford University: Stanford, CA, USA, 1962.
23. Beven, K.J.; Kirkby, M.J. A physically based variable contributing model of basin hydrology. *Hydrol. Sci. Bull.* **1979**, *24*, 43–69. [CrossRef]
24. Neitsch, S.; Arnold, J.; Kiniry, J.; Williams, J. *Soil and Water Assessment Tool Theoretical Documentation Version 2009*; Texas Water Resources Institute Technical Report No. 406: College Station, TX, USA, 2011.
25. Krysanova, V.; Wechsung, F.; Arnold, J.; Srinivasan, R.; Williams, J. *SWIM: Soil and Water Integrated Model*; Potsdam Institute for Climate Impact Research (PIK): Potsdam, Germany, 2000.
26. Meng, X.; Wang, H.; Cai, S.; Zhang, X.; Leng, G.; Lei, X.; Shi, C.; Liu, S.; Shang, Y. The China Meteorological Assimilation Driving Datasets for the SWAT Model (CMADS) Application in China: A Case Study in Heihe River Basin. *Preprints* **2016**. [CrossRef]
27. Meng, X.Y.; Wang, H.; Lei, X.H.; Cai, S.Y.; Wu, H.J.; Ji, X.N.; Wang, J.H. Hydrological Modeling in the Manas River Basin Using Soil and Water Assessment Tool Driven by CMADS. *Teh. Vjesn.* **2017**, *24*, 525–534.

water

MDPI

Editorial

Profound Impacts of the China Meteorological Assimilation Driving Datasets for the SWAT Model (CMADS)

Xianyong Meng [1,2,*], Hao Wang [3,*] and Ji Chen [2,*]

1   College of Resources and Environmental Science, China Agricultural University (CAU),
    Beijing 100094, China
2   Department of Civil Engineering, The University of Hong Kong (HKU), Pokfulam 999077, Hong Kong, China
3   China Institute of Water Resources and Hydropower Research (IWHR), Beijing 100038, China
*   Correspondence: xymeng@cau.edu.cn or xymeng@hku.hk (X.M.); wanghao@iwhr.com (H.W.);
    jichen@hku.hk (J.C.); Tel.: +86-10-68410178 (X.M.)

Received: 5 April 2019; Accepted: 18 April 2019; Published: 19 April 2019

Abstract: As global warming continues to intensify, the problems of climate anomalies and deterioration of the water environment in East Asia are becoming increasingly prominent. In order to assist decision-making to tackle these problems, it is necessary to conduct in-depth research on the water environment and water resources through applying various hydrological and environmental models. To this end, the China Meteorological Assimilation Driving Datasets for the Soil and Water Assessment Tool (SWAT) model (CMADS) has been applied to East Asian regions where environmental issues are obvious, but the stations for monitoring meteorological variables are not uniformly distributed. The dataset contains all of the meteorological variables for SWAT, such as temperature, air pressure, humidity, wind, precipitation, and radiation. In addition, it includes a range of variables relevant to the Earth's surface processes, such as soil temperature, soil moisture, and snowfall. Although the dataset is used mainly to drive the SWAT model, a large number of users worldwide for different models have employed CMADS and it is expected that users will not continue to limit the application of CMADS data to the SWAT model only. We believe that CMADS can assist all the users involved in the meteorological field in all aspects. In this paper, we introduce the research and development background, user group distribution, application area, application direction, and future development of CMADS. All of the articles published in this special issue will be mentioned in the contributions section of this article.

Keywords: CMADS; impact; hydrological modeling; SWAT

---

1. Introduction

The China Meteorological Assimilation Driving Datasets for the Soil and Water Assessment Tool (SWAT) (CMADS) is a product employed before the start of the intensive coupling process of model-driven research of atmospheric science and hydrology [1]. From the perspective of natural processes, the region's sensitivity to climate and its impact on the global climate are important in regard to the geological structure of East Asia. From the perspective of social development, however, the construction of basic meteorological facilities in the region is imperfect due to the economy and various other objective factors. From the perspective of the nature–societal water cycle, research and investment in protecting the underlying surface environment in East Asia has not kept pace with the excessive exploitation and pollution of natural resources in the region. Therefore, because the underlying surface of the region is not well understood, it is improper to use the deviated meteorological data to erroneously analyze the water resources, water environment, and air quality of the region [2] for later decision-making. Such applications will further damage the ecological environment of the region and pose a devastating impact in a vicious circle on the region's economy and ecosystem.

The territory of East Asia is vast and includes many countries, each of which differs significantly in its means of meteorological observation, processing methods, and post-assimilation correction methods for meteorological data. In addition, obstacles exist for the sharing of meteorological data among countries and even departments within the same country. This can result in the problem of repeated data collection, which has further led to the lack of reanalyzed datasets in East Asia that follow uniform procedures with uniform latitude/longitude and resolution that can be corrected by additional sources of reliable observation data [1,3]. In addition, from a cross-disciplinary perspective, atmospheric researchers need to provide meteorological data with higher density and less uncertainty which can be easily accessed and used by others. Such an achievement will effectively match hydrological models that are rapidly being physically modularized [4] and can enhance the accuracy of regional climate models in East Asia [5] to ultimately improve the long-term predictive capability for future East Asian climates [6,7]. The CMADS was created precisely in the above context, and its appearance was originally tailored for the SWAT model. The CMADS can be used efficiently without any adjustments and treatments to the SWAT model, which will save at least 90% of the time spent by SWAT users on meteorological data preparation. From the perspective of data precision, the CMADS has been validated by users worldwide in combination with various international reanalysis data, such as Climate Forecast System Reanalysis (CFSR) and Tropical Rainfall Measuring Mission (TRMM) in various river basins in East Asia, with satisfactory results [8–22]. These verification results demonstrate that the performance of CMADS products in East Asia can be trusted, especially in mountainous and highland areas with high altitudes and large differences of land use and geography, where meteorological stations are sparse.

In April 2016, we released a series of CMADS in the Cold and Arid Regions Science Data Centre at Lanzhou (CARD), China (http://westdc.westgis.ac.cn/). Shortly afterwards, the official website of the SWAT model (https://swat.tamu.edu/software/) included our CMADS. As of 1 January 2019, according to a rough estimation, the official website of CMADS (http://www.cmads.org) has recorded nearly 170,000 visits from the world, and we have received nearly 2630 applications from the teams all over the world (Figure 1). Distribution of CMADS users has rapidly expanded from major scientific research institutes in mainland China to research institutes and governmental agencies in Taiwan (Tamkang University), South Korea (Sungkyunkwan University and Seoul National University), Japan (Kyushu University), Thailand (Khon Kaen University and Naresuan University), the Philippines (University of the Philippines), India (Indian Institute of Technology Kharagpur), Pakistan (The University of Agriculture Peshawar), Russia (Far Eastern Regional Hydrometeorological Research Institute (FERHRI)), Germany (TU Dortmund University and UFZ Helmholtz Centre For Environmental Research), Italy (Polytechnic University of Milan), Canada (Memorial University of Newfoundland), and the United States (Virginia Polytechnic Institute and State University, The University of Nevada, United States Department of Agriculture (USDA), University of Massachusetts, and Massachusetts Institute of Technology).

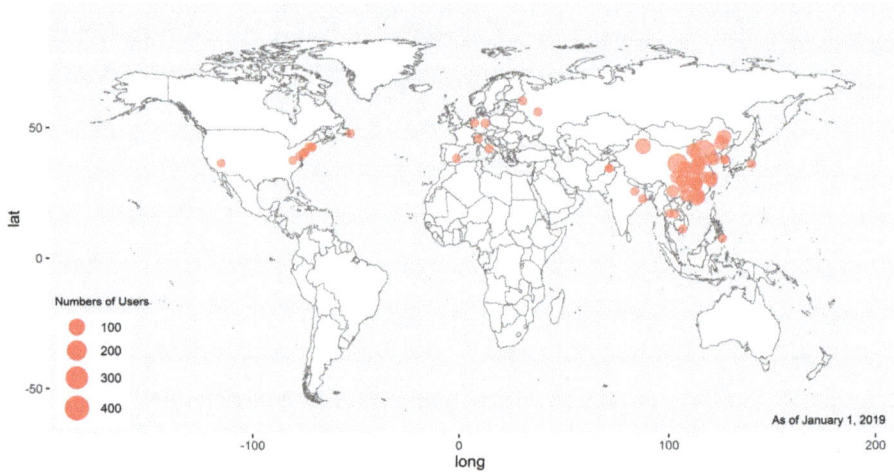

Figure 1. Distribution of the China Meteorological Assimilation Driving Datasets for the Soil and Water Assessment Tool (SWAT) (CMADS) users.

As the founder of CMADS data, our original intention was to make it easier for users in East Asia to obtain high-precision data produced by uniform procedures. According to the data application and usage during the past three years, our original goal has been achieved. We are honoured that a large number of users are in Asia. Through offline emails and online communication of users, we are also excited to learn that CMADS has greatly supported users in East Asia for scientific research. In addition, we are surprised and encouraged to find that many researchers from the regions that are not covered by the CMADS data, such as Europe and North America, have shown great interest.

We evaluated the areas in East Asia to which researchers have applied the CMADS data most often; however, this analysis is a rough estimation because some users did not reveal the application purpose when applying for the data (Figure 2). We found that the CMADS datasets is most widely used in mainland China. The most frequent research area for data application is concentrated in Northwest China, followed by North China, Northeast China, and finally Southwest China, Central China, East China, and South China. Moreover, we found that the number of data applications on the upper left side of the Hu Line is higher than that in the lower right region. Similarly, the number of meteorological stations in China differs significantly on both sides of the Hu Line. Most meteorological stations are located in the lower right region of the Hu Line in Southeast China. This interesting phenomenon may explain why CMADS is used often for filling in missing data owing to a lack of meteorological stations in that region of China. We also noted that the CMADS datasets has been widely used in many countries outside China, including Pakistan, India, Russia, Thailand, South Korea, Japan, and the Philippines (Figure 2).

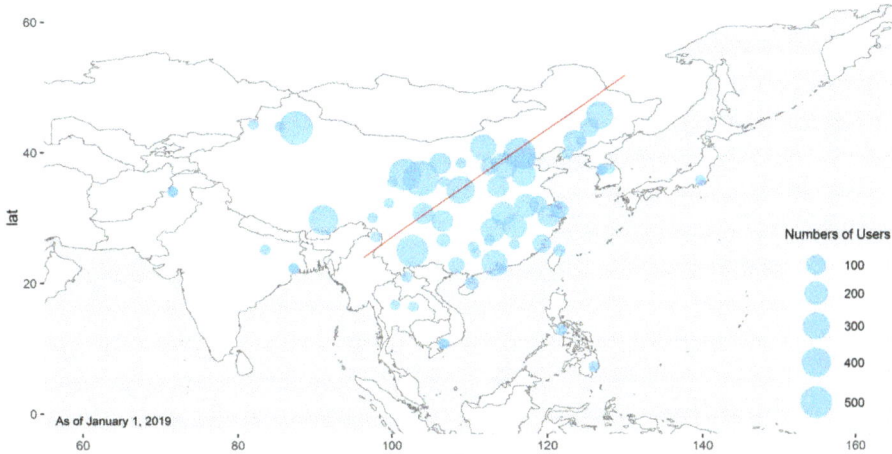

Figure 2. Research hotspot areas of East Asia employing CMADS.

We requested almost all the applicants for the CMADS data to state the purpose of using the data. Figure 3 shows the statistical results of using the data and it can be seen that the hotspot applications are non-point source pollution simulation and water resources modeling, both of which were 23%.

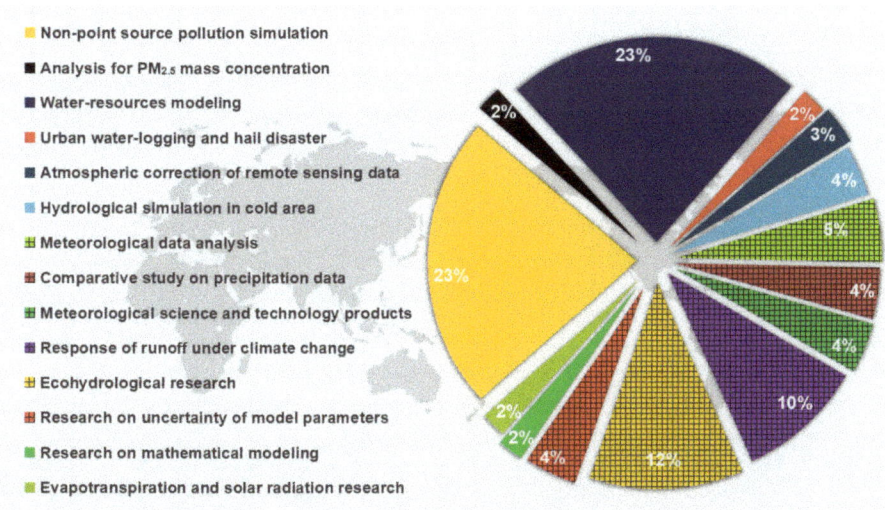

Figure 3. Hotspot application directions of CMADS.

Other research purposes in order of popularity are ecohydrological research (12%), response of runoff under climate change (10%), and meteorological data analysis (5%). There are four purposes with 4%, which are hydrological simulation in cold areas, comparative study on precipitation data, meteorological science and technology products, and research on uncertainty of model parameter. The purpose for atmospheric correction of remote sensing data is 3%. Finally, there are four purposes with 2%, which are analysis for the mass concentration of atmospheric particulate matter (PM) less than 2.5 μm in diameter (PM$_{2.5}$), urban water-logging and hail disasters, research on mathematical modeling, and evapotranspiration (ET) and solar radiation research. We are pleased to report that the CMADS datasets has been applied in different research purposes. Particularly in East Asia, more researchers are using CMADS to focus on non-point source pollution simulation and water resource modeling, which shows that researchers are paying more attention to the current water environment problems in East Asia and the problems of drought and flood caused by uneven spatiotemporal differentiation of water resources. Runoff and ecological hydrology research under climate change is an additional research focus in East Asia. In the future, the CMADS datasets will provide users with historical data, real-time data, and forecast data at multiple resolutions to meet the different needs of users for various research applications.

This special issue provides a platform for researchers by assisting with the use of CMADS to conduct in-depth research on water quality and quantity modeling in East Asia in order to improve the research in the atmosphere, hydrology, and water environment in East Asia. The papers included in this special issue fall into eight broad categories: Meteorological verification and analysis [3,11], non-point source pollution [12], water resource modeling and parameter uncertainty analysis [13–16], comparison of reanalysis products [17–20], optimal operational of reservoirs [21,22], water footprint assessment [23], changes in water resources under climate and land use change [24], hydrological simulation in cold area [25,26], and CMADS-Soil Temperature (ST) application [27]. The following section summarizes the individual research within each application.

2. Contributions

The following is a summary of the papers discussed in the special issue titled "Application of the China Meteorological Assimilation Driving Datasets for the SWAT Model (CMADS) in East Asia".

1. As a CMADS producer, Meng et al. [3] gave a detailed introduction of the CMADS datasets establishment method and the elements provided in this special issue. In addition, we verified the accuracy of the CMADS datasets based on using 2421 automatic weather stations in China.

2. Tian et al. [11] used CMADS to evaluate the potential evapotranspiration (PET) over China with the Penman–Monteith method. Their research compared PET derived from CMADS to that derived from 836 meteorological stations during the period from 2008 to 2016 and analyzed the contribution of different factors to the bias of PET. They concluded that the overall estimation from CMADS agreed well with the observations. In the central and eastern part of China, wind speed and solar radiation were determined to be the major factors influencing the biases in the PET estimation. The wind speed induced biases ranging from −15% to −5% and the solar radiation induced 15% to 50% biases in terms of different locations. Their research discussed the addition of PET elements in CMADS.

3. Qin et al. [12] reported that dam construction changed the watershed nutrient cycle and caused nutrient retention in the reservoir, which led to eutrophication of the surface water. Based on the SWAT model driven by CMADS, the author analyzed the Biliuhe Reservoir Basin in Northeast China and proposed an integrated method to analyse the total nitrogen (TN) accumulation in drinking water reservoirs. Finally, they concluded that fertiliser, atmospheric deposition and soil, and non-point sources accounted for the highest proportions of TN inputs to the Biliuhe Reservoir, at 35.15%, 30.15%, and 27.72%, respectively. Moreover, 19.76% of the TN input was accumulated in the reservoir. Inter-basin water diversion projects also play an important role in the TN accumulation process of the reservoir. Nitrogen pollution of the Biliuhe Reservoir can be alleviated by discharging high nitrogen concentrations of water and sediment through the bottom hole of the dam.

4. Cao et al. [13] used the CMADS-driven SWAT model to validate the runoff in the fan-shaped Lijiang River Basin in China. In addition, the sensitivity and uncertainty of the model parameters were analyzed by using the Sequential Uncertainty Fitting 2 (SUFI-2) method. The authors found that the performance of CMADS-driven SWAT mode was excellent in the calibration and validation periods, with Nash–Sutcliffe model efficiency coefficients (NSEs) of 0.89 and 0.88 and correlation coefficients (R^2) of 0.92 and 0.89, respectively. Based on the accuracy of model results, the temporal and spatial variations of ET, the surface runoff, and the groundwater discharge in the watershed were analyzed. The spatial and temporal variations of surface runoff and groundwater discharge in the Lijiang River Basin were found to be closely related to precipitation and ET was largely controlled by land use type.

In flood and dry years, the contributions of surface runoff, groundwater flow, and lateral flow to the water budget differed significantly.

5. Guo et al. [14] compared the accuracy of three precipitation datasets including CMADS, Tropical Rainfall Measuring Mission (TRMM) 3B42V7, and an interpolation dataset obtained using rainfall gauges. The study area was the Lijiang River Basin, the same as the study area of Cao et al. [14]. The performances of the three types of precipitation datasets were compared through two rainfall runoff models, IHACRES and Sacramento. Compared with the interpolation dataset through rainfall gauges (NSE = 0.83) and TRMM 3B42 V7 (NSE = 0.89), CMADS (NSE = 0.93) performed the best in the rainfall runoff modeling, especially for peak flow modeling. Compared with TRMM 3B42 V7 (CSI = 0.40), CMADS (CSI = 0.61) showed better agreement with the observation from rainfall gauges. The results also showed that both IHACRES and Sacramento performed well in the Lijiang River basin. The uncertainty analysis revealed that IHACRES showed less uncertainty and higher applicability than Sacramento.

6. Zhao et al. [15] considered that, although the SWAT hydrologic model has been commonly used in investigating the hydrological processes at the watershed scale, it can have various uncertainties. It is therefore essential to conduct parameter uncertainty analysis to gain more confidence in the modeling task. They employed high-resolution meteorological driving datasets–CMADS to investigate the parameter uncertainty of SWAT in a semi-arid loess—mountain transitional watershed with an emphasis on filtering the appropriate uncertainty analysis tool. The results showed that the SUFI2 method was more efficient than ParaSol and GLUE, although all three methods can yield good performance for SWAT through using the CMADS. Further analysis showed that the CN2 SOL_K and ALPHA_BF were more sensitive to peak flow, average flow, and low flow simulations, respectively, than others, including the soil evaporation compensation factor (ESCO), channel hydraulic conductivity (CH_K2), and available soil water capacity of the soil layer (SOL_AWC).

7. Zhou et al. [16] reported that the results of runoff simulation are closely related to local climate and catchment conditions. They proposed a three-step framework: (1) Using multiple regression models for parameter sensitivity analysis based on the results of Latin hypercube sampling (LHS-OAT); (2) using the multi-level factorial decomposition method to quantitatively evaluate the impact of individual and interaction parameter effects on the hydrological processes; and (3) analyzing the reasons for changes in dynamic parameters. The authors concluded that the sensitivity of the parameters differed significantly among the periods. Specifically, the interaction effect between the parameters soil bulk density (SOL_BD) and CN2 as well as those between SOL_BD and CH_K2 were obvious, which indicates that SOL_BD affects the surface runoff and the loss of groundwater recharged by the river. Those findings help to provide optimal parameter input for the SWAT model to improve the applicability of the SWAT model.

8. Gao et al. [17] compared CMADS, the National Center for Environment Prediction Climate Forecast System Reanalysis (NCEP-CFSR), the TRMM 3B42 V7, and the Precipitation Estimation from Remotely Sensed Information using Artificial Neural Networks–Climate Data Record (PERSIANN–CDR) with the observed precipitation and evaluated the hydrological application of the datasets in the Xiang River Basin. The results showed that (1) for daily time steps, reanalysis datasets had better linear correlations with gauge observations (>0.55) than satellite-based datasets; (2) CMADS and TRMM 3B42 V7 had better linear correlations with gauge observations than PERSIANN-CDR and NCEP-CFSR, and satellite-based datasets were better than reanalysis datasets in terms of bias for monthly temporal scale; and (3) CMADS and 3B42 V7 simulated streamflow well for both daily and monthly time steps, with NSEs >0.70 and >0.80, respectively. Moreover, CMADS performed slightly better than TRMM 3B42 V7; the performances of NCEP-CFSR and PERSIANN-CDR were not acceptable.

9. Guo et al. [18] evaluated the accuracy and hydrological simulation utility of CMADS, TMPA 3B42 V7, and Integrated Multi-Satellite Retrievals for Global Precipitation Measurement (IMERG-F) products in the Jinsha River, which is a complex terrain area. The results of statistical analysis showed that the three types of datasets had relatively high accuracy on the average grid scale with R^2 values greater

11

than 0.8, at 0.86, 0.81, and 0.88, respectively. In addition, CMADS had the highest success rate of detecting extreme precipitation events. The study analysis showed that all three precipitation products obtained acceptable results when driven the SWAT model. CMADS performed best, followed by TMPA and IMERG with NSEs of 0.55, 0.50, and 0.45, respectively.

10. Liu et al. [19] evaluated the accuracy of five elements from CMADS and CFSR by comparing them at 131 meteorological stations in the Qinghai–Tibetan Plateau. The results indicated that CMADS outperformed CFSR with a higher correlation coefficient and smaller bias. The authors also used different climate data including CMADS, CFSR, and observations to run the SWAT model. The results indicated that CMADS performed best in forcing the SWAT model, with NSEs of 0.78 and 0.68 in calibration and validation periods, respectively. With the help of Geodetector, the authors found that the air temperature, soil moisture, and soil temperature at 1.038 m had a larger impact on snowmelt than other factors. Ultimately, the paper revealed that CMADS is suitable for study regions in the Qinghai–Tibetan Plateau.

11. Vu et al. [20] compared several precipitation products, including PERSIANN, TRMM 3B42 V7, and CMADS, with the observed precipitation. They found that the accuracy of TRMM and CMADS precipitation products was higher than that of PERSIANN and PERSIANN-CDR when compared with traditional observatories through a series of indicator validations. They reported that TRMM and CMADS products can better capture mountainous precipitation. Finally, the authors used all of these precipitation products to drive the SWAT model. After the model was validated in the Han River Basin, Korean Peninsula, they concluded that the model performance driven by gauged rainfall data was the best, at NSE = 0.68, followed by TRMM, CMADS, and PERSIANN with an NSE = 0.49, 0.42, and 0.13, respectively; PERSIANN-CDR had an NSE of 0.16. Because CMADS products have not yet been assimilated in the Korea region, there is a large amount of room for improvement in the expressiveness of CMADS products in that region.

12. Dong et al. [21] indicated the reservoir operation should be incorporated into the hydrological model to quantify its hydrological impact. Accordingly, the authors developed a reservoir module and integrated this module into the Noah Land Surface Model and Hydrology System (LSM-HMS). The module aggregates small reservoirs into one large reservoir by employing a simple statistical approach. The integrated model was applied to the upper Gan River Basin to quantitatively assess the impact of a group of reservoirs on the streamflow. The results indicate that the model can reasonably depict the storage variations of both the large and small reservoirs. With the newly developed module, the performance of the model in simulating streamflow improved at a 0.05 level of significance. The results also indicated that the operation of a group of reservoirs led to an increase in streamflow in dry seasons and a decrease in flood seasons; the impacts of the large and small reservoirs had almost the same order of magnitude.

13. Liu et al. [22] considered that the construction and operation of cascade reservoirs changed the hydrological cycle of the basin and reduced the accuracy of hydrological forecasting. The authors took the Yalong River Basin as the study area and designed eight scenarios using the SWAT model to change the reservoir capacity, operating location, and relative locations of the two reservoirs. After comparing the various scenarios, the authors found that the reservoir decreased and delayed the flood during the flood season and increased the runoff during the dry season. The flood control benefit and the adjustment of the runoff process of the reservoir rose with an increase in the storage capacity. When the reservoir was close to the downstream region, the peak flow of the basin outlet was reduced by 48.9%. The construction of the small reservoir in the upper reaches of the large reservoir resulted in further flood control benefits, with a maximum reduction of 55% in peak flow.

14. Yuan et al. [23] estimated the virtual water in the Bohai Basin, including blue water flow (BWF) and green water flow (GWF) based on the CMADS-driven SWAT model. In addition, they studied the laws of spatiotemporal changes in BWF, GWF and green water coefficient (GWC), and analyzed the sensitivity of GWF and BWF to temperature and precipitation under climate change. Overall, the study showed that the CMADS can be used to detect the observed probability density function of daily

precipitation and temperature. CMADS also performed well in simulations when the relative and absolute deviations of monthly variables of precipitation and temperature were less than 7% and 0.5 °C, respectively. From 2009 to 2016, the BWF increased and the GWF decreased. Ridges of high pressure showed an uneven spatial distribution with a gradual increase from lower altitudes to mountainous areas. However, the spatial distribution of GWF was relatively even. Moreover, the precipitation increased 10% and the BWF increased 20.8%; the watershed-scale GWF increased only 2.5%. When the temperature increased 1.0 °C, the BWF and GWF changed by −3% and 1.7%, respectively. BWF and GWF were more sensitive to precipitation in the areas with lower altitude. Under those conditions, the mountainous water flow was more sensitive to temperature.

15. Shao et al. [24] used meteorological data from traditional observation stations and land use data during the period 1970–2014 of the Hailiutu River Basin combined with the Mann–Kendall (MK) and STARS (Sequential t-Test Analysis of Regime Shift) methods to develop seven land use changes or climate variability scenarios. CMADS was introduced to enhance the representativeness of traditional observation stations during the period 2008–2014. Moreover, the simulation performance was analyzed when CMADS and observed data were used to drive the SWAT model. The results showed that CMADS has an adjustment and optimization effect on the meteorological data of traditional observation stations owing to its high resolution and precision. Therefore, the authors found that in the Hailiutu River Basin, the impact of climate variability on streamflow was more profound than the impact of land use change.

16. Zhang et al. [25] studied the Hunhe River Basin (HRB) in China to evaluate the impact of land-use change on sediment erosion and runoff in alpine regions, using SWAT driven by CMADS. The SWAT model driven by CMADS performed well in the HRB with NSEs of 0.67–0.92 and 0.56–0.95 and R^2 values of 0.69–0.94 and 0.57–0.96 for monthly runoff and monthly sediment, respectively. Forestland played an important role in soil and water conservation, which increased the ET and soil infiltration capacity, decreased the surface runoff (SURQ), and resulted in a reduction in runoff and sediment yield, whereas it decreased the water percolation and increased the runoff in the dry season. The responses of grassland and forestland to runoff and sediment yield were similar, although the former was weaker than the latter in soil and water conservation. Cropland usually increased the SURQ, runoff, and sediment yield. Compared with cropland, when the precipitation is low, urban land might lead to higher sediment yields owing to its higher runoff yield. Additionally, the runoff and sediment yield under different land use scenarios increased along with an increase in average monthly precipitation.

17. Li et al. [26] used CMADS data to drive the SWAT model in the Jingbo River (JBR) Basin in western China to provide scientific elaboration on the region's surface processes. Owing to limitations posed by the local climate and a lack of meteorological stations, in-depth research on surface processes in that region is limited. The authors found that the CMADS-driven SWAT model produced satisfactory results in the JBR Basin, with monthly and daily NSEs of 0.659–0.942 and 0.526–0.815, respectively. Their research also revealed that the soil moisture in the JBR Basin will reach the first peak level in March and April each year as a result of spring snowmelt, whereas the soil moisture after October is constant owing to cold air transit.

18. Zhao et al. [27] used CMADS-ST and soil moisture observation data to study the dynamic characteristics of the near-surface hydrothermal process in a typical frozen soil region near Harbin and analyzed the soil moisture distribution in the black soil region during the freeze–thaw period. The results showed that the shallow soil moisture in the black soil slope farmland had a Gaussian distribution with the freeze–thaw period and that the peak of the soil moisture Gaussian distribution appeared in the early freeze–thaw period in early spring. Under the conditions of shallow melting and deep freezing of the black soil plowing layer, the research results were consistent with the natural phenomenon, such that the snow-melted water infiltrated into the soil in the early spring. Then, for the northeast black soil region, the CMADS-ST was used to explore the change trend of the soil moisture content during the freeze–thaw cycle period. The research would have impacts on decision-making for

the protection of water and soil resources and the environment in northeastern China's seasonal frozen soil zones and the conservation of soil and water of sloping farmland in the black soil zone.

As a summary, Table 1 lists the general information of the above 18 papers.

Table 1. Summary of the contributions published in the special issue.

	Research Focuses	Study Area	Country	Re-Analysis Data	Authors
1	Meteorological verification	China	China	CMADS	Meng et al. [3]
2	PET evaluate	China	China	CMADS	Tian et al. [11]
3	Non-point source pollution	Biliuhe Reservoir Basin	China	CMADS	Qin et al. [12]
4	Sensitivity and uncertainty	Lijiang River Basin	China	CMADS	Cao et al. [13]
5	Comparison of reanalysis products	Lijiang River Basin	China	CMADS, TRMM 3B42 V7	Guo et al. [14]
6	Sensitivity and uncertainty	Jingchuan River Basin	China	CMADS	Zhao et al. [15]
7	Sensitivity and uncertainty	Yellow River	China	CMADS	Zhou et al. [16]
8	Comparison of reanalysis products	Xiang River Basin	China	CMADS, CFSR, TRMM 3B42 V7, PERSIANN-CDR	Gao et al. [17]
9	Comparison of reanalysis products	Jinsha River	China	CMADS, TRMM 3B42 V7, IMERG-F	Guo et al. [18]
10	Comparison of reanalysis products	Qinghai–Tibetan Plateau	China	CMADS, CFSR	Liu et al. [19]
11	Comparison of reanalysis products	Han River Basin	Korean Peninsula	CMADS, TRMM, PERSIANN, PERSIANN-CDR	Vu et al. [20]
12	Reservoir operation	Gan River Basin	China	CMADS	Dong et al. [21]
13	Reservoir operation	Yalong River Basin	China	CMADS	Liu et al. [22]
14	Water footprint assessment	Bohai Basin	China	CMADS	Yuan et al. [23]
15	Changes in water resources under climate and Land Use Change	Hailiutu River Basin	China	CMADS	Shao et al. [24]
16	Hydrological simulation in cold area	Hunhe River Basin	China	CMADS	Zhang et al. [25]
17	Hydrological simulation in cold area	Jing and Bo River Basin	China	CMADS	Li et al. [26]
18	CMADS-ST application	Heilongjiang Province	China	CMADS-ST	Zhao et al. [27]

Funding: This research was financially joint supported by the National Science Foundation of China (41701076) and the National key Technology R & D Program of China (2017YFC0404305,2018YFA0606303).

Acknowledgments: The authors wish to thank the journal editors for their support.

Conflicts of Interest: The authors declare no conflict of interest.

References

1. Meng, X.; Wang, H. Significance of the China Meteorological Assimilation Driving Datasets for the SWAT Model (CMADS) of East Asia. *Water.* **2017**, *9*, 765. [CrossRef]
2. Meng, X.; Wu, Y.; Pan, Z.; Wang, H.; Yin, G.; Zhao, H. Seasonal Characteristics and Particle-size Distributions of Particulate Air Pollutants in Urumqi. *Int. J. Environ. Res. Public Health* **2019**, *16*, 396. [CrossRef] [PubMed]

3. Meng, X.; Wang, H.; Shi, C.; Wu, Y.; Ji, X. Establishment and Evaluation of the China Meteorological Assimilation Driving Datasets for the SWAT Model (CMADS). *Water* **2018**, *10*, 1555. [CrossRef]

4. Meng, X.; Yu, D.; Liu, Z. Energy balance-based SWAT model to simulate the mountain snowmelt and runoff—Taking the application in Juntanghu watershed (China) as an example. *J. Mt. Sci.* **2015**, *12*, 368–381. [CrossRef]

5. Meng, X.; Sun, Z.; Zhao, H.; Ji, X.; Wang, H.; Xue, L.; Wu, H.; Zhu, Y. Spring Flood Forecasting Based on the WRF-TSRM mode. *Teh. Vjesn.* **2018**, *25*, 27–37.

6. Meng, X.; et al. Snowmelt Runoff Analysis Under Generated Climate Change Scenarios for the Juntanghu River Basin in Xinjiang, China. *Tecnología y Ciencias del Agua* **2016**, *7*, 41–54.

7. Xue, L.; Zhu, B.; Yang, C.; Wei, G.; Meng, X. Study on the characteristics of future precipitation in response to external changes over arid and humid basins. *Sci. Rep.* **2017**, *7*, 15148. [CrossRef] [PubMed]

8. Meng, X.; Wang, H.; Long, A.; Wang, J.; Shi, C.; Ji, X. Investigating spatiotemporal changes of the land surface processes in Xinjiang using high-resolution CLM3.5 and CLDAS: Soil temperature. *Sci. Rep.* **2017**, *7*, 13286. [CrossRef] [PubMed]

9. Meng, X.; Wang, H.; Cai, S.; Zhang, X.; Leng, G.; Lei, X.; Shi, C.; Liu, S.; Shang, Y. The China Meteorological Assimilation Driving Datasets for the SWAT Model (CMADS) Application in China: A Case Study in Heihe River Basin. *Preprints.* **2016**, 2016120091.

10. Meng, X.; Wang, H.; Lei, X.H.; Cai, S.Y.; Wu, H.J. Hydrological Modeling in the Manas River Basin Using Soil and Water Assessment Tool Driven by CMADS. *Teh. Vjesn.* **2017**, *24*, 525–534.

11. Tian, Y.; Zhang, K.; Xu, Y.-P.; Gao, X.; Wang, J. Evaluation of Potential Evapo-transpiration Based on CMADS Reanalysis Dataset over China. *Water* **2018**, *10*, 1126. [CrossRef]

12. Qin, G.; Liu, J.; Wang, T.; Xu, S.; Su, G. An Integrated Methodology to Analyze the Total Nitrogen Accumulation in a Drinking Water Reservoir Based on the SWAT Model Driven by CMADS: A Case Study of the Biliuhe Reservoir in Northeast China. *Water* **2018**, *10*, 1535. [CrossRef]

13. Cao, Y.; Zhang, J.; Yang, M. Application of SWAT Model with CMADS Data to Estimate Hydrological Elements and Parameter Uncertainty Based on SUFI-2 Algorithm in the Lijiang River Basin, China. *Water* **2018**, *10*, 742. [CrossRef]

14. Guo, B.; Zhang, J.; Xu, T.; Croke, B.; Jakeman, A.; Song, Y.; Yang, Q.; Lei, X.; Liao, W. Applicability Assessment and Uncertainty Analysis of Multi-Precipitation Datasets for the Simulation of Hydrologic Models. *Water* **2018**, *11*, 1611. [CrossRef]

15. Zhao, F.; Wu, Y. Parameter Uncertainty Analysis of the SWAT Model in a Mountain Loess Transitional Watershed on the Chinese Loess Plateau. *Water* **2018**, *10*, 690. [CrossRef]

16. Zhou, S.; Wang, Y.; Chang, J.; Guo, A.; Li, Z. Investigating the Dynamic Influence of Hydrological Model Parameters on Runoff Simulation Using Sequential Uncertainty Fitting-2-Based Multilevel-Factorial-Analysis Method. *Water* **2018**, *10*, 1177. [CrossRef]

17. Gao, X.; Zhu, Q.; Yang, Z.; Wang, H. Evaluation and Hydrological Application of CMADS against TRMM 3B42V7, PERSIANN-CDR, NCEP-CFSR, and Gauge-Based Datasets in Xiang River Basin of China. *Water* **2018**, *10*, 1225. [CrossRef]

18. Guo, D.; Wang, H.; Zhang, X.; Liu, G. Evaluation and Analysis of Grid Precipitation Fusion Products in Jinsha River Basin Based on China Meteorological Assimilation Datasets for the SWAT Model. *Water* **2019**, *11*, 253. [CrossRef]

19. Liu, J.; Shanguan, D.; Liu, S.; Ding, Y. Evaluation and Hydrological Simulation of CMADS and CFSR Reanalysis Datasets in the Qinghai Tibet Plateau. *Water* **2018**, *10*, 513. [CrossRef]

20. Vu, T.T.; Li, L.; Jun, K.S. Evaluation of Multi Satellite Precipitation Products for Streamflow Simulations: A Case Study for the Han River Basin in the Korean Peninsula, East Asia. *Water* **2018**, *10*, 642. [CrossRef]

21. Dong, N.; Yang, M.; Meng, X.; Liu, X. CMADS-Driven Simulation and Analysis of Reservoir Impacts on the Streamflow with a Simple Statistical Approach. *Water* **2019**, *11*, 178. [CrossRef]

22. Liu, X.; Yang, M.; Meng, X.; Wen, F.; Sun, G. Assessing the Impact of Reservoir Parameters on Runoff in the Yalong River Basin using the SWAT Model. *Water* **2019**, *11*, 643. [CrossRef]

23. Yuan, Z.; Xu, J.; Meng, X.; Wang, Y.; Yan, B.; Hong, X. Impact of Climate Variability on Blue and Green Water Flows in the Erhai Lake Basin of Southwest China. *Water* **2019**, *11*, 424. [CrossRef]

24. Shao, G.; Guan, Y.; Zhang, D.; Yu, B.; Zhu, J. The Impacts of Climate Variability and Land Use Change on Streamflow in the Hailiutu River Basin. *Water* **2018**, *10*, 814. [CrossRef]

25. Zhang, L.; Meng, X.; Wang, H.; Yang, M. Simulated runoff and sediment yield responses to land-use change using SWAT model in Northeast China. *Water* **2019**, in press.
26. Li, Y.; Wang, Y.; Zheng, J.; Yang, M. Investigating Spatial and Temporal Variation of Hydrological Processes in Western China Driven by CMADS. *Water* **2019**, *11*, 435. [CrossRef]
27. Zhao, X.; Xu, S.; Liu, T.; Qiu, P.; Qin, G. Moisture Distribution in Sloping Black Soil Farmland during the Freeze–Thaw Period in Northeastern China. *Water* **2019**, *11*, 536. [CrossRef]

water

MDPI

Article

Establishment and Evaluation of the China Meteorological Assimilation Driving Datasets for the SWAT Model (CMADS)

Xianyong Meng [1,2,*], Hao Wang [3,*], Chunxiang Shi [4], Yiping Wu [5] and Xiaonan Ji [6,*]

[1] College of Resources and Environmental Science, China Agricultural University (CAU), Beijing 100094, China
[2] Department of Civil Engineering, The University of Hong Kong (HKU), Pokfulam 999077, Hong Kong, China
[3] State Key Laboratory of Simulation and Regulation of Water Cycle in River Basin & China Institute of Water Resources and Hydropower Research (IWHR), Beijing 100083, China
[4] National Meteorological Information Center, China Meteorological Administration (CMA), Beijing 100081, China; shicx@cma.gov.cn
[5] Department of Earth & Environmental Science, Xi'an Jiaotong University, Xi'an 710049, China; yipingwu@xjtu.edu.cn
[6] Xinjiang Institute of Ecology and Geography, Chinese Academy of Sciences (CAS), Urumqi 830046, China
* Correspondence: xymeng@cau.edu.cn (X.M.); wanghao@iwhr.com (H.W.); jixiaonan999@163.com (X.J.); Tel.: +86-010-60355970 (X.M.)

Received: 17 September 2018; Accepted: 30 October 2018; Published: 1 November 2018

Abstract: We describe the construction of a very important forcing dataset of average daily surface climate over East Asia—the China Meteorological Assimilation Driving Datasets for the Soil and Water Assessment Tool model (CMADS). This dataset can either drive the SWAT model or other hydrologic models, such as the Variable Infiltration Capacity model (VIC), the Soil and Water Integrated Model (SWIM), etc. It contains several climatological elements—daily maximum temperature (°C), daily average temperature (°C), daily minimum temperature (°C), daily average relative humidity (%), daily average specific humidity (g/kg), daily average wind speed (m/s), daily 24 h cumulative precipitation (mm), daily mean surface pressure (HPa), daily average solar radiation (MJ/m^2), soil temperature (K), and soil moisture (mm^3/mm^3). In order to suit the various resolutions required for research, four versions of the CMADS datasets were created—from CMADS V1.0 to CMADS V1.3. We have validated the source data of the CMADS datasets using 2421 automatic meteorological stations in China to confirm the accuracy of this dataset. We have also formatted the dataset so as to drive the SWAT model conveniently. This dataset may have applications in hydrological modelling, agriculture, coupled hydrological and meteorological modelling, and meteorological analysis.

Keywords: CMADS; SWAT; East Asia; meteorological; hydrological

1. Introduction

Many studies have demonstrated the need for a more realistic distribution of surface climate in meteorological analyses, biogeochemical modelling, and hydrological modelling. Examples of such research include flood and large-scale meteorological studies [1], water balance simulations [2,3], agricultural research [4,5], climate research [6–8], and hydrological modelling [9,10]. It is clear that the resolution and accuracy of the meteorological data may influence the analysis results; indeed, significant amounts of uncertainty exist in coarse-resolution meteorological data. The existing data have been used all over the world, such as in Climate Forecast System Reanalysis (CFSR) [11]; the National Center for Atmospheric Research (NCAR-R1\R2) [12,13]; the ERA-Interim [14],

ERA-15 [15], and EAR-40 [16] products from the European Centre for Medium-Range Weather Forecasts (ECMWF); and the Modern Era Retrospective-Analysis for Research and Applications (MERRA) from the National Aeronautics and Space Administration (NASA). These data are very useful for the water balance analyses and climate change research at the global scale. However, the re-analysis data are too coarse for national or regional scale research. Further, as part of the resolution, it may not be possible to correct deviations in this re-analysis data using local meteorological observation data.

Although China occupies a vast area with complex topographies, meteorological stations are relatively scarce within the country. The existing network of observation stations no longer meets the requirements for large-scale research on hydrological processes, floods, and hydrologic balance. In addition, traditional meteorological stations can only provide data on individual public stations within the country. The use of data obtained from a limited number of meteorological stations clearly does not accurately represent the actual situation on the ground surface at a larger scale. As such, there is an urgent need for a higher resolution dataset that can be used to drive regional hydrological models (such as the SWAT model) to identify the true process occurring in the watershed [17,18]. The Soil and Water Assessment Tool (SWAT) was developed by the United States Department of Agriculture (USDA) Agricultural Research Service (ARS), and designed to predict the impacts of management practices on the quality and quantity of water, sediment, and climate change in large complex watersheds with various soils, land use, and management conditions. SWAT is a physically-based continuous distributed model that operates on a daily time step, and it requires data such as weather, soil properties, topography, vegetation, and land management practices. The SWAT model has been widely applied in simulating soil and water loss and non-point source pollution [19].

This article describes the construction of CMADS over East Asia (0° N–65° N, 60° E–160° E) (Figure 1). The China Meteorological Assimilation Driving Datasets for the SWAT model (CMADS) is a public dataset developed by Dr. Xianyong Meng from China Agricultural University (CAU). CMADS incorporated technologies of Local Analysis and Prediction System/Space-Time Multiscale Analysis System (LAPS/STMAS) [8] and was constructed using multiple technologies and scientific methods, including the loop nesting of data, resampling, and bilinear interpolation.

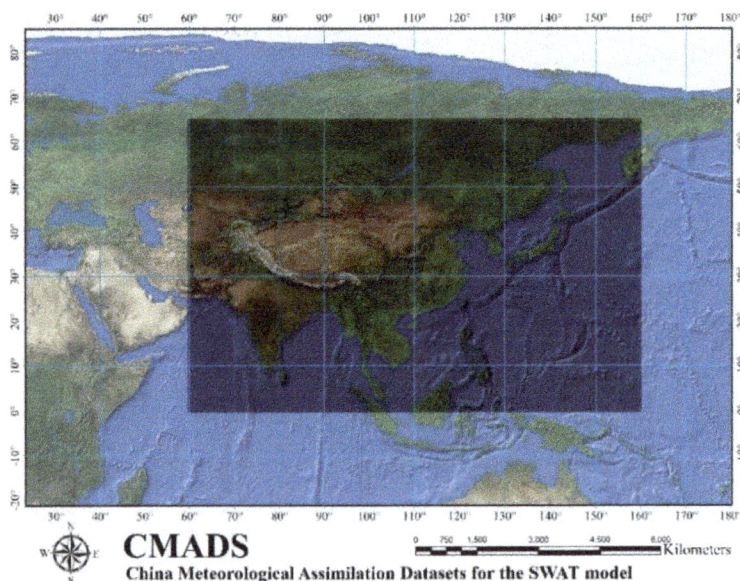

Figure 1. The spatial range of the CMADS.

The CMADS series of datasets can be used to drive various hydrological models, such as SWAT, the Variable Infiltration Capacity (VIC) model, and the Soil and Water Integrated Model (SWIM). It also allows users to conveniently extract a wide range of meteorological elements for detailed climatic analyses. Data sources for the CMADS series include nearly 40,000 regional encrypted stations under China's 2421 national automatic and business assessment centres. This ensures that the CMADS datasets have a wide applicability within the country and that the data accuracy was vastly improved. The CMADS series of datasets has undergone finishing and correction to match the specific format of input and driving data of SWAT models. This reduces the volume of complex work that model builders have to deal with. An index table of the various elements encompassing all of East Asia was also established for SWAT models. This allows the models to utilize the datasets directly, thus eliminating the need for any format conversion or calculations using weather generators. Consequently, significant improvements to the modelling speed and output accuracy of SWAT models were achieved. We used the LAPS/STMAS assimilation method [8] and collected all the relevant meteorological data (e.g., auto observation stations, ECMWF, RADAR, etc.) to construct several versions of datasets for the SWAT model. The CMADS comprises the following variables (Table 1): daily maximum temperature (°C), daily average temperature (°C), daily minimum temperature (°C), daily average relative humidity (%), daily average specific humidity (g/kg), daily average wind speed (m/s), daily 24 h cumulative precipitation (mm), daily mean surface pressure (HPa), daily average solar radiation (MJ/m^2), soil temperature (K), and soil moisture (mm^3/mm^3). Further details on the CMADS datasets will be provided in the following sections: Materials and Methods, Results, Usage Notes, and Conclusion.

Table 1. The information on CMADS.

CMADS Attribute	Records
Variables Provided	daily maximum temperature (°C), daily average temperature (°C), daily minimum temperature (°C), daily average relative humidity (%), daily average specific humidity (g/kg), daily average wind speed (m/s), daily 24 h cumulative precipitation (mm), daily mean surface pressure (HPa), daily average solar radiation (MJ/m^2), soil temperature (K) and soil moisture (mm^3/mm^3)
Spatial range of CMADS	0° N–65° N, 60° E–160° E
Timescale of CMADS	1 January 1980–31 December 2017 (Periodic update)
Spatiotemporal resolution	1/3°, 1/4°, 1/8°, 1/16° (Daily)

2. Materials and Methods

The CMADS have a very strict data assimilation process and have been comprehensively described by Meng et al. [8,20]. First, let us describe the various raw data from the meteorological stations that were incorporated during the process of establishing the CMADS datasets, the assimilation process for the CMADS assimilation field data, and the post-processing of the CMADS data. The raw meteorological data used in this study mainly included the regular raw input data (e.g., regional encrypted stations, national automatic stations, and radar stations), and data from satellites, radars, automatic stations, and background fields of the ECMWF. Several important raw input data for the study are shown in Figure 2, namely data from regional encrypted stations, national automatic stations, and radar stations. The details of the various raw data used to construct the datasets are described below.

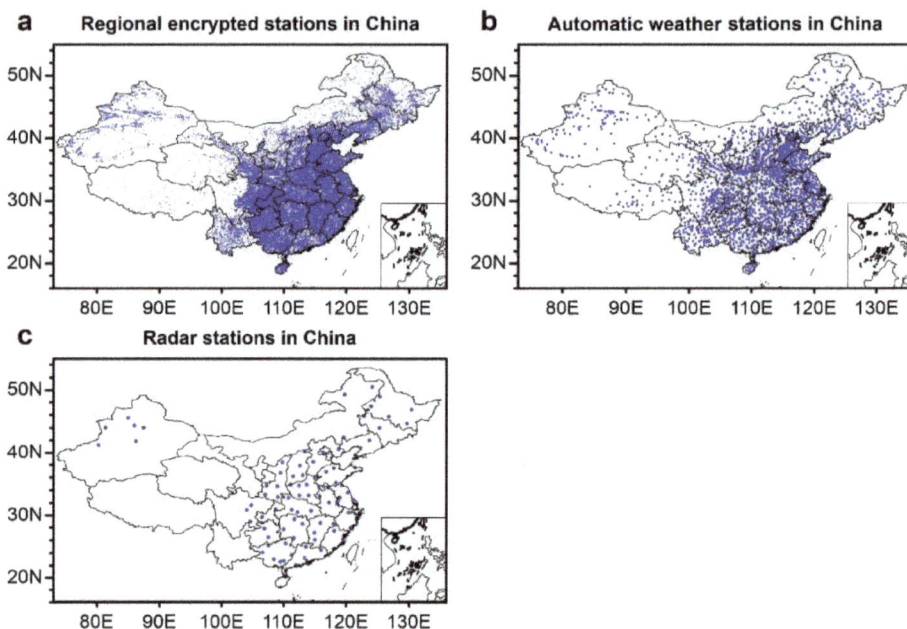

Figure 2. The several important raw input data for the CMADS datasets (**a**) Regional encrypted stations, (**b**) National automatic stations, (**c**) Radar stations.

2.1. Raw Data for the CMADS Datasets

i.   *Data from regional encrypted stations:*

China has nearly 40,000 regional encrypted stations which provide information on various surface meteorological elements. The main information includes the station numbers, coordinates (latitude and longitude) for the location of each station, altitude of each observation field, altitude based on the barometric pressure sensor, and observation data of each station. The last category includes hourly data of 49 atmospheric elements, including wind direction and speed, temperature, relative humidity, dew point, air pressure, hourly precipitation, and ground temperature. All of the aforementioned data have been subjected to dynamic quality controls, ensuring the accuracy and reliability of the meteorological information.

ii.   *Data from national automatic stations:*

There are 2421 automatic stations nationwide. These provide real-time information on multiple elements, including the daily average pressure; maximum and minimum pressure; average, maximum, and minimum temperature; average and minimum relative humidity; average wind speed; maximum and extreme wind speed and direction; sunshine duration; and precipitation. All the data are subjected to stringent quality control checks. In this study, these were used as the initial assimilation data source for the correction of the various elements [8].

iii.   *Data from radar stations:*

Radar data have become an important component of the weather monitoring network in China. There are 131 radar detection stations in China, and these provide photographs, charts, and data of radar return signals for meteorological phenomena (such as regions with sporadic precipitation). These serve as important bases for Chinese meteorological departments to make weather forecasts for the short- and very short-term (0–72 and 0–12 h, respectively), and especially for the forecasting of precipitation. These are also the main tools used by the China Meteorological Administration (CMA), which provides forecasting services for approaching weather (0–2 h). Given the important role of radar data in China's weather detection, these were included as one of the raw input data sources for this study.

All three aforementioned categories of data had been subjected to strict quality control (Include formatting, theoretical limit check, climatic extremum check, factor correlation check, Time consistency check, and horizontal homogeneity check), and only those data marked under quality control as being of Grade 1 accuracy were used. Prior to the assimilation and integration of the observation data and background fields, the numerous uncertainties existing in the metadata were highlighted. Some of the observation stations that did not participate in the assessment were also noted. Other than the various types of traditional observation data, the CMADS also uses the six-hourly reanalysis components of the ERA-Interim datasets released by the European Centre for Medium-range Weather Forecasts (ECMWF) as its basic background fields. These include six-hourly data on the pressure, potential temperature, and vorticity under the regional mode. This data product was jointly released by the ECMWF and the Integrated Forecasting System (IFS) system (established in 2006). The IFS system contains four-dimensional variational (4D-VAR) modules spanning 12-hourly windows for analysis.

2.2. The Assimilation Process for Source Data

The integration of air temperature, air pressure, humidity, and wind speed data was mainly achieved through the LAPS/STMAS system [21]. The LAPS system is a comprehensive analytical system containing data from multiple sources. It has five major functional modules [22] that analyse wind, ground surface, temperature, clouds, and water vapour. The analysis must be carried out in a specific sequence [22] because analytical results from earlier stages are required for subsequent analyses. The analytical results of the five modules can be used for diagnostic analysis to arrive at certain values to support the weather diagnosis. The results can also be entered into numerical models after undergoing balance analysis, thereby realizing the warm boot of the modules.

STMAS is a new-generation integration system that was developed under the LAPS framework. Its algorithm uses a multi-grid sequential variational method, which is different from the traditional LAPS system. Functionally, STMAS' ground surface analytical module replaces the LAPS' ground surface analysis, and its STMAS3D module replaces LAPS' wind and temperature analyses. Input–output analyses and analysis of clouds, water vapour, and energy balance and hydrological balance are still dependent on LAPS. There are plans for the gradual integration of LAPS' non-adiabatic initialization technology with STMAS to form a separate system [21].

2.3. The Integration Process for the Precipitation Data

Precipitation data of CMADS were stitched using CMORPH's global precipitation products [23], the National Meteorological Information Center's data of China (which is based on CMORPH's integrated precipitation products) [24]. The latter contains daily precipitation records observed at 2400 national meteorological stations and the CMORPH satellite's inversion precipitation products. It was developed with a two-step data integration method that combined the probability density function (PDF) matching and optimal interpolation (OI) [25].

After comparison with heavy precipitation events monitored in China, this dataset was found to describe changes in precipitation intensity more accurately, as well as provide greater details on the spatial distribution of precipitation. It has obvious advantages in capturing the small-scale features of precipitation and has the characteristics of a precipitation product with both high resolution and high precision.

2.4. The Assimilation Process for the Radiation Data

The inversion algorithm used for creating CMADS solar radiation at the ground surface makes use of the discrete longitudinal method by Stamnes et al. [26], the same method as used for CLDAS.

This algorithm can be used to calculate the radiance in any direction as it takes into account the anisotropy when the top of the atmosphere reflects solar radiation. First, the radiance of reflected solar radiation at the top of the atmosphere in the direction of the observation of the satellite is calculated. Next, the results are converted to bidirectional visible albedo as observed by the satellite's visible light channel.

The transmission process by which incoming solar radiation at the top of the atmosphere travels through the atmosphere and reaches the ground surface involves a series of physical processes that interact with both the atmosphere and ground surface. The following are considered by the inversion model: (i) ozone absorption, (ii) multiple molecular Rayleigh scattering, (iii) multiple scattering and absorption of cloud droplets, (iv) absorption of water vapour, (v) multiple scattering and absorption of aerosol, and (vi) multiple reflections of the ground surface and atmosphere [27].

The CMADS in grid format (hereafter referred to as CMADS-GRID) was eventually constructed after assimilation of the various types of observation and background data. It serves as the source data for CMADS, but it does not provide the relative humidity component as its output. In addition, the source data for CMADS were processed using LAPS and other means before their format was standardized as NetCDF.

2.5. The Construction Process for the CMADS Datasets

Preparation of the CMADS datasets was completed through the processes of data interpolation and resampling, calculating relative humidity elements (See Section 2.5.2) and format conversion, and elevation extraction. This ensures that the various hydrological models are able to read and access the data.

2.5.1. The Configuration of Spatiotemporal Resolutions

The maximum spatiotemporal resolution of the CMADS-GRID is 1/16°, 1 h. If the weather stations loaded into the ArcSWAT over a certain number, SWAT will refuse to read it [14]. Study areas at various scales also have different requirements in terms of the number of meteorological stations from which to obtain data. For example, if the scale of the study area is small, an atmospheric drive field with a coarse resolution would not be able to reflect the true state of the atmospheric components at the ground surface effectively. Taking into account these two constraints (limiting the number of meteorological stations and research needs), four versions of the CMADS datasets at various resolutions were considered. Specifically, the resolutions for CMADS V1.0, V1.1, V1.2, and V1.3 were 1/3°, 1/4°, 1/8°, and 1/16°, respectively. Currently, the requisite integral timescale used by most SWAT models to drive data is daily steps. However, the atmospheric driving fields of the CMADS-GRID are based on hourly steps. As such, the CMADS-GRID dataset has to be aggregated to daily time steps.

In this study, the daily output of the CMADS-GRID' atmospheric assimilation fields was averaged and cumulated on a daily basis, following which it was screened. For the air temperature element, the CMADS screened the maximum and minimum values from the CMADS-GRID' intra-day data to establish the maximum and minimum daily temperatures. Calculations were also done using the 24-h data from the CMADS datasets to obtain the average daily values for the following elements: temperature (°C), pressure (HPa), specific humidity (g/kg), wind speed (m/s), and solar radiation (MJ/m^2). For precipitation, the summation of the 24-h data gave the cumulative daily precipitation (mm).

As mentioned earlier, datasets with various spatial resolutions were constructed in this study to overcome the model's limitations on the number of stations from which the data were accepted, and the requirements imposed by research areas of different scales. Since the resolution of the source data for the CMADS was 1/16°, two sampling methods were considered when the interpolation calculations were done for the four resolutions. Since the resolution 1/3° is an integer multiple of

1/16°, the bilinear interpolation method was applied (Figure 3a). For the other higher resolutions (1/4°, 1/8°, and 1/16°), the nested assignment was used for data reconstruction (Figure 3b).

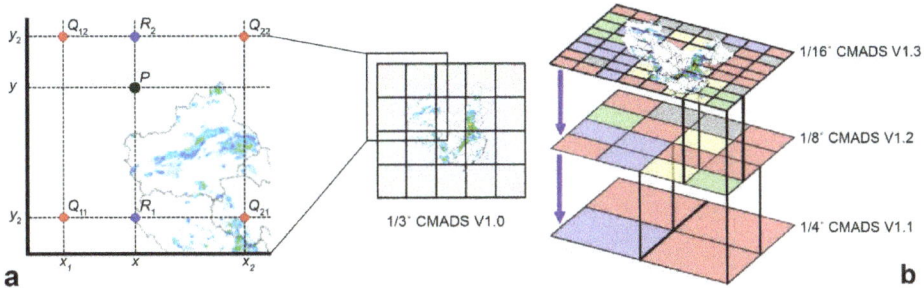

Figure 3. The interpolation methods used in CMADS. (**a**) Bilinear interpolation for CMADS V1.0, (**b**) Nested assignment for CMADSV1.0 to CMADS1.3.

2.5.2. Calculation of the Relative Humidity Element

An important input element for SWAT models is relative humidity data, but this component was not provided by the CMADS-GRID (CMADS-GRID provides: hourly temperature (°C), hourly Specific Humidity (g/kg), hourly wind speed (m/s), hourly precipitation (mm), hourly surface pressure (HPa), hourly solar radiation (MJ/m^2)). Hence, calculation of the relative humidity element was critical. This was achieved using a conversion relationship between the specific and relative humidity.

Equations (1) and (2) were used for calculating the specific humidity (SH) and the corresponding relative humidity (Φ), respectively.

$$SH = \frac{0.622 \times p_{H2O}}{p - 0.387 \times p_{H2O}}, \tag{1}$$

where SH represents the specific humidity, P_{H2O} represents the water vapour pressure and p represents the air pressure.

$$\Phi = \frac{SH \times p}{(0.622 + 0.378 \times SH)p_{H2O}} \tag{2}$$

However, the specific humidity is also defined as the ratio of the mass of water vapour to that of the entire air system (dry air plus water vapour).

2.5.3. Processing of the Data Format

The data in the CMADS datasets were strictly processed to match the format required to be readable and accessible by SWAT models. For SWAT (ArcSWAT and SWAT plus) and other models, the CMADS datasets provide two formats (.dbf and .txt) by which the models could directly access elements of the stations on a daily basis, including maximum and minimum temperature (°C), average wind speed (m/s), average solar radiation (MJ/m^2), cumulative precipitation (24 h), and average surface pressure (HPa).

At the same time, CMADS also provides data in the .txt format for use in other hydrological models, such as VIC and SWIM. This format also facilitates data analysis by climate analysts and researchers. The main elements provided in the txt format on a daily basis include maximum, average and minimum air temperature (°C); specific humidity (g/kg), and relative humidity (%); daily wind speed (m/s), daily 24 h cumulative precipitation (mm), surface pressure (HPa), solar radiation (MJ/m^2), soil temperature (K) and soil moisture (mm^3/mm^3).

2.5.4. Other Elements Provided by the CMADS Datasets

In addition to providing users with data on the various surface meteorological elements recorded by the stations, the CMADS datasets also include the specific latitude, longitude, and elevation for the geographical location of each element. The latitudes and longitudes were projected based on the WGS84 spatial and geographical coordinates, while the altitudinal zones were extracted using the Global 30 Arc-Second Elevation (GTOPO30) [28,29]. The GTOPO30 is a global digital elevation model (DEM) with a horizontal grid spacing of 30 arc seconds (approximately 1 km). After altitudinal extraction, the CMADS datasets would create an index table for all grid points within East Asia based on the requisite format of SWAT models. The index table facilitates direct reading and access by SWAT models and concurrently allows other model users to access information on the various meteorological stations.

2.5.5. CMADS Data Records

CMADS datasets are available at the CMADS official website (http://www.cmads.org/) and the SWAT official website (https://swat.tamu.edu/software/). Table 1 contains a brief summary of the CMADS datasets. Currently, CMADS has been updated until Version 1.1. In CMADS V1.0 (at a spatial resolution of 1/3°), East Asia was spatially divided into 195 × 300 grid cells containing 58,500 grid points. Despite being at the same time resolution as CMADS V1.0, CMADS V1.1 contains more data, with 260 × 400 grid cells containing 104,000 grid points. In the near future, CMADS will release versions 1.2 and 1.3, the 1.2 version will divide into 520 × 800 grid cells containing 416,000 grid points and the 1.3 version will divide into 1040 × 1600 grid cells containing 1,664,000 grid points.

The CMADS datasets provide users with data in both the .txt and .dbf formats. The file naming convention of the SWAT subsets in the CMADS datasets is as follows: element code: R, P, S, T, or W (the first letter of the meteorological variables) + latitude grid number − longitude grid number. The CMADS-ST and CMADS-SM provide the daily average soil temperature and soil moisture of 10 layers (First Layer: 0.007 m, Second Layer: 0.028 m, Third Layer: 0.062 m, Fourth Layer: 0.119 m, Fifth Layer: 0.212 m, Sixth Layer: 0.366 m, Seventh Layer: 0.620 m, Eighth Layer: 1.038 m, Ninth Layer: 1.728 m, Tenth Layer: 2.864 m).

3. Results

3.1. CMADS: Distribution of Related Variables and Verification of Applicability on China

Since prior studies on precipitation and solar radiation exist [24,25,27], further verification of these elements was not performed in this study. This section describes the verification that was performed for the remaining four elements (namely, air temperature, surface pressure, relative humidity, and wind speed) to test their applicability to China. The observation data used for verification were obtained from the national automatic stations. Given the space limitations, only the verification results for 2011–2013 are shown.

In order to verify the applicability of the various CMADS meteorological elements, this study selected three years of CMADS datasets (2011–2013) for China, extracted the elements on a daily basis, and then calculated the annual averages. Next, the bilinear interpolation method was used for data verification. This was achieved via sample matching of the elements (temperature, pressure, humidity, wind speed, precipitation, and radiation) between the CMADS datasets and records by China's national automatic stations (2421 in total). Verification of only the first four elements is demonstrated in this study due to space limitations. The various elements from the observation stations that were selected for the matching process were made to pass strict quality controls (including thresholds for the regional and climatic boundaries, and tests for spatiotemporal consistency), the usability rate of the stations' verification data reached 98.9%. The spatial distributions of the biases and root mean square errors (RMSEs) for the elements of temperature, pressure, relative humidity, and wind speed between the CMADS datasets and national automatic stations are shown in Figures 4–7, respectively.

The verifications showed that the CMADS datasets accurately reflected the spatial characteristics and distribution of various types of surface elements in China.

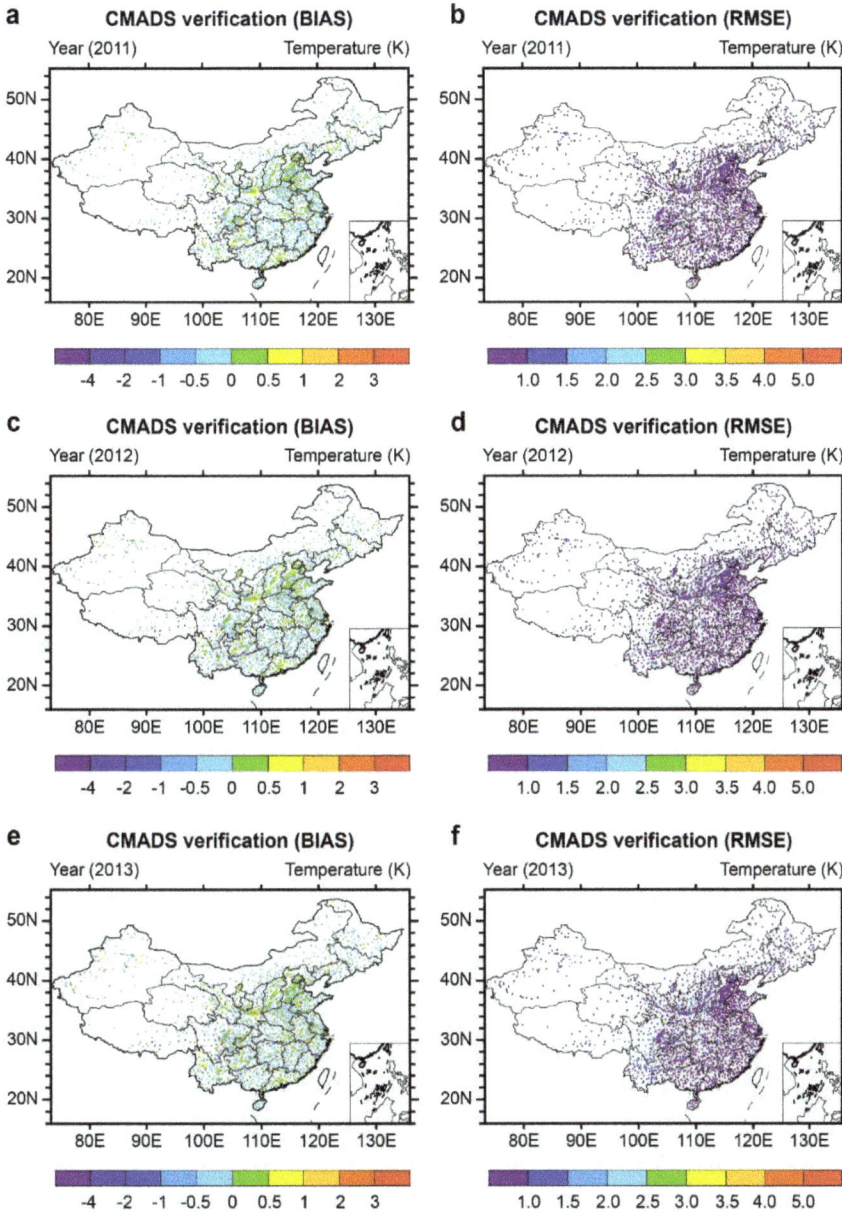

Figure 4. The evaluation of indicators for temperature in China (2011–2013). (**a**) The spatial distributions of the biases for the temperature in year 2011, (**b**) The spatial distributions of the RMSEs for the temperature in year 2011, (**c**) The spatial distributions of the biases for the temperature in year 2012, (**d**) The spatial distributions of the RMSEs for the temperature in year 2012, (**e**) The spatial distributions of the biases for the temperature in year 2013, (**f**) The spatial distributions of the RMSEs for the temperature in year 2013.

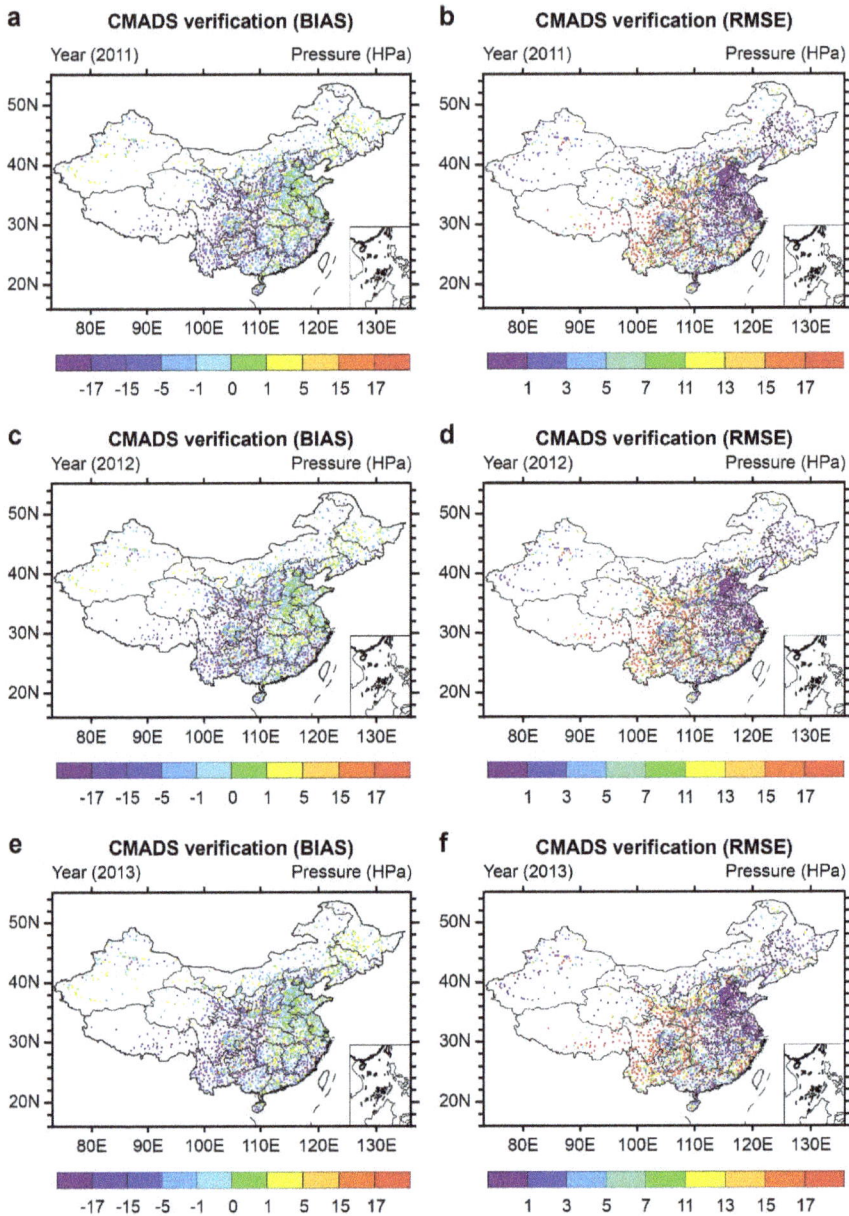

Figure 5. The evaluation of indicators for atmospheric pressure in China (2011–2013). (**a**) The spatial distributions of the biases for the atmospheric pressure in year 2011, (**b**) The spatial distributions of the RMSEs for the atmospheric pressure in year 2011, (**c**) The spatial distributions of the biases for the atmospheric pressure in year 2012, (**d**) The spatial distributions of the RMSEs for the atmospheric pressure in year 2012, (**e**) The spatial distributions of the biases for the atmospheric pressure in year 2013, (**f**) The spatial distributions of the RMSEs for the atmospheric pressure in year 2013.

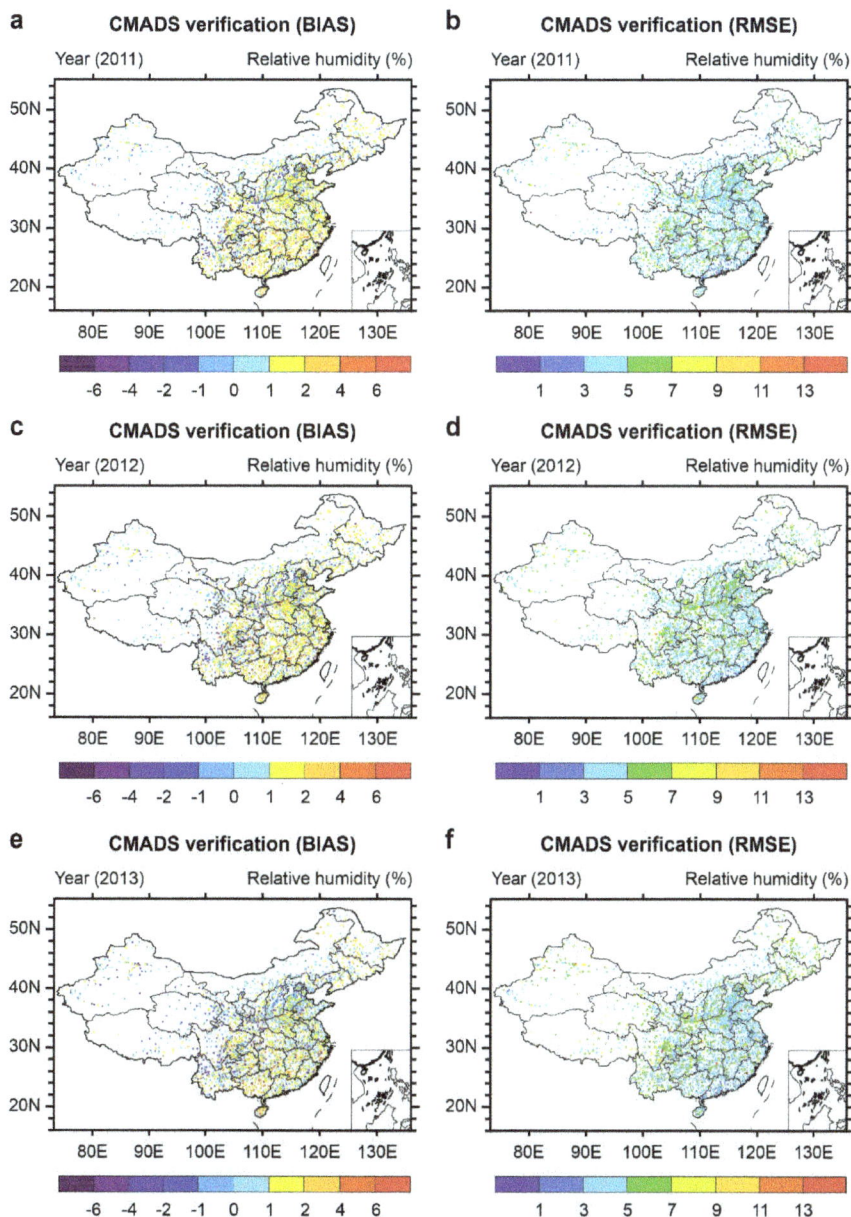

Figure 6. The evaluation of indicators for relative humidity in China (2011–2013). (**a**) The spatial distributions of the biases for the relative humidity in year 2011, (**b**) The spatial distributions of the RMSEs for the relative humidity in year 2011, (**c**) The spatial distributions of the biases for the relative humidity in year 2012, (**d**) The spatial distributions of the RMSEs for the relative humidity in year 2012, (**e**) The spatial distributions of the biases for the relative humidity in year 2013, (**f**) The spatial distributions of the RMSEs for the relative humidity in year 2013.

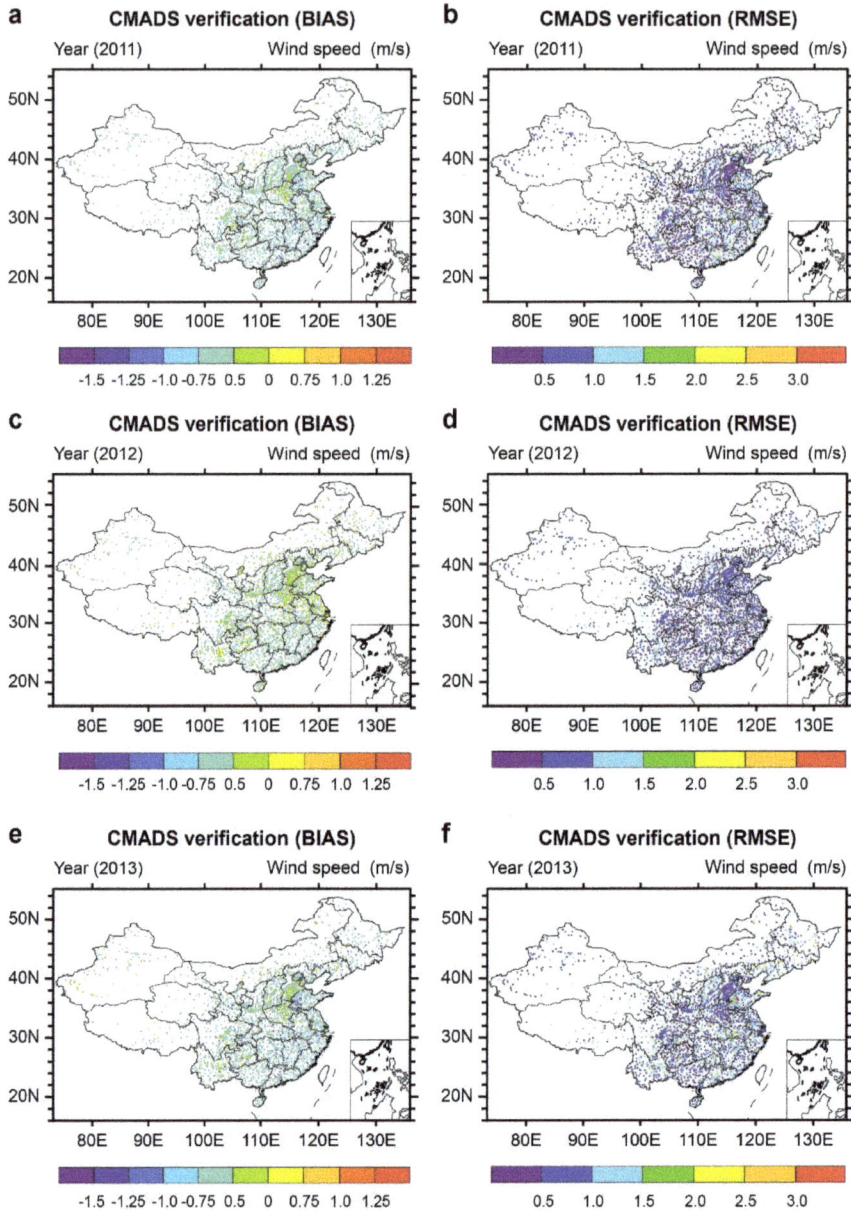

Figure 7. The evaluation of indicators for wind speed in China (2011–2013). (**a**) The spatial distributions of the biases for the wind speed in year 2011, (**b**) The spatial distributions of the RMSEs for the wind speed in year 2011, (**c**) The spatial distributions of the biases for the wind speed in year 2012, (**d**) The spatial distributions of the RMSEs for the wind speed in year 2012, (**e**) The spatial distributions of the biases for the wind speed in year 2013, (**f**) The spatial distributions of the RMSEs for the wind speed in year 2013.

3.2. Distribution and Verification of Indicators for Temperature

The spatial distribution of indicators for temperatures in 2011–2013 and the related verifications are shown in Figure 4. For the temperature element of the CMADS datasets (2011–2013), the spatiotemporal distributions of the biases and RMSEs are shown in Figure 4a–f, respectively. It was found that over the three-year verification period, the temperature element performed very well for the majority of territories within China in terms of bias. In 2011–2013, the biases between the temperature element of CMADS and those of the observation stations were maintained between −0.5 K and 0.5 K. For North China, the northwest region of Northeast China, the southern region of Southwest China, and the southern region of South China, the biases were concentrated at 0.5–1 K. This phenomenon of weak and positive biases in temperature for these four regions/sub-regions was apparent and occurred consistently over the three years. In contrast, the phenomenon of substantial negative biases in temperature was noted for a minority of stations located in Southwest China (including the eastern portion of the Tibet Autonomous Region, the entire western region of the Sichuan Province, the southern part of Gansu Province, and the western part of Yunnan Province). Most of the negative biases were concentrated between −1 K and −4 K. On the whole, the temperature data in the CMADS datasets were acceptable with regards to the bias indicator for most of the country. Overall, the datasets were verified to be good.

In order to analyze the performance of the temperature element in the CMADS datasets more objectively, we further evaluated its RMSE indicator for China (Figure 4b,d,f). For most of the stations in the country, the RMSEs of the temperature element were controlled within 1 K. In the southeastern part of Northwest China (such as the southern part of Gansu Province, the northern part of the Ningxia Hui Autonomous Region, and the western part of Shaanxi Province) and parts of the southern region of Southwest China (such as the southern part of Sichuan province, and the western and southern parts of Yunnan Province), the RMSEs of most of the temperature elements remained within 2.5 K, with a small number of stations being at 3.5 K. This study also found that the RMSEs of the aforementioned regions gradually increased annually during the three-year period. However, this increase in errors applied to stations located in small areas only (such as the western region of Shaanxi Province). In terms of the RMSE indicator, the verification results for the temperature element of the CMADS dataset were good.

3.3. Distribution and Verification of Indicators for Atmospheric Pressure

The spatial distribution of indicators for pressures during 2011–2013 and the related verifications are shown in Figure 5. For the pressure element of the CMADS datasets (2011–2013), the spatiotemporal distributions of the biases and RMSEs are shown in Figure 5a–f, respectively. It can be seen from Figure 5a,c,e that overall, the CMADS' biases for atmospheric pressure were mainly found in the eastern region of West China.

Upon detailed analysis, it was found that these biases were controlled within the range of −1 HPa to 5 HPa for the following regions: the central–southern part of Northeast China (the western part of Liaoning Province, all of Jilin Province, and the southern part of Heilongjiang Province); North China; the northern parts of Central and East China (Zhejiang Province, Jiangsu Province, Anhui Province, and Jiangxi Province); the central part of South China (the western part of Guangdong Province); the northern part of Ningxia Hui Autonomous Region; and the intersection between the Chongqing and Sichuan provinces. In contrast, the biases were controlled between −1 HPa and −17 HPa for the southern part of East China (Fujian Province), most of the areas in Southwest China, and Northwest China. Among these regions, the biases were greater (−5 HPa to −17 HPa) for the majority of stations in the southern part of East China (Fujian Province) and most of the areas in Southwest China.

The RMSEs of the atmospheric pressure element in the CMADS datasets (Figure 5b,d,f) presented a similar situation such as that for biases. Specifically, the RMSEs were greater for most of Southwest China, the eastern part of Northwest China, and the southern part of East China (Fujian Province). The errors were mainly positive and contained at 5–17 HPa. For the remaining regions, the performance of the RMSEs was good, with the errors for the majority of stations being controlled within 3 HPa.

For both indicators, the performance of the atmospheric pressure data in the CMADS datasets was deemed to be reliable when applied to the entire country.

3.4. Distribution and Verification of Indicators for Relative Humidity

The spatial distribution of indicators for relative humidity in 2011–2013 and the related verifications are shown in Figure 6. Figure 6a–f show the spatiotemporal distributions of the biases and RMSEs for the relative humidity element of the CMADS datasets, respectively. Analyses of the biases in relative humidity from an overall perspective revealed that generally, there was a positive bias effect. This effect appeared in Northeast China; North China; East China; Central China; South China; the southeastern parts of South and Southwest China (mainly in the eastern part of the Sichuan Province, Chongqing, Guizhou area, and the southern part of the Yunnan Province); and the southeastern part of Northwest China (mainly in the southern part of the Gansu Province and the central part of the Shaanxi Province). The positive biases for these regions were limited between −1% and 6%.

Slight negative biases (within −1%) were found in Northwest China, the central and northern parts of Southwest China, and Inner Mongolia, and very few stations showed a negative bias between −4 to −1%. Interannual analyses of the relative humidity biases in the CMADS datasets for 2011–2013 indicated a declining trend year-on-year. For some of the stations, positive biases for relative humidity in 2011 became negative biases two years later. The phenomenon of negative biases was the most evident throughout Northwest China, but it also existed generally throughout Southeast China. Analyses of the distribution of RMSEs for relative humidity over China (Figure 6b,d,f) indicated that these were controlled between 3% and 9% for most of the country. As with the biases, the errors were greater (5–9%) in North China; Central China; the eastern part of Northwest China (the southern part of Shaanxi Province); and Southwest China (the southeastern part of Sichuan Province, Yunnan Province, and Guizhou Province). For most of the other regions, the errors were limited to 3–5%. During the period from 2011 to 2013 in this region, a correlation was seen between the trends in the spatial distribution of the RMSE and those of the positive biases in relative humidity. Nevertheless, in terms of overall performance for the entire country, both the biases and RMSEs in the CMADS data on relative humidity were considered to be acceptable.

3.5. Distribution and Verification of Indicators for Wind Speed

The spatial distribution of indicators for wind speed in 2011–2013 and the related verifications are shown in Figure 7. The distribution of biases and RMSEs for the wind speed element of the CMADS datasets are shown in Figure 7a–f, respectively.

Analyses of the biases for the entire country led to the conclusion that its performance was good, with the general range being between −1.0 m/s and 0.75 m/s. For some stations, the verification results showed greater negative biases for 2011 and 2013. Examples included parts of North China (Shandong Province); East China (Jiangxi Province, Zhejiang Province, and Fujian Province); parts of South China (such as Guangdong Province); and parts of Central China (Hunan Province and Hubei Province). A small number of stations in these regions had biases ranging between −1.5 m/s and −1.0 m/s. The bias effect was generally better for 2012 when it was controlled between −0.75 m/s and 0.75 m/s for the whole country. Over the three years, positive biases for some stations were maintained at 0–0.75 m/s for parts of North China (such as Hebei Province and Henan Province); parts of East China (Jiangsu Province); and parts of Southwest China (such as the southern parts of Yunnan, Sichuan, and Guizhou Provinces). The wind speed element mostly contained weak positive biases. In contrast, most of the stations in the other provinces presented the general phenomenon of negative biases. These were between −1.5 m/s and −1.0 m/s for some stations.

Upon analysis, the distribution of RMSEs (Figure 7b,d,f) was found to be similar to the situation for biases. In 2012, the error indicators for most stations in China were within 0.5 m/s. For scattered stations in Inner Mongolia; the southeastern part of Qinghai Province; and parts of Southwest China (such as the southeastern part of the Tibet Autonomous Region, Sichuan Province, and the southeastern part of Yunnan Province), the RMSEs were controlled between 1.0 m/s and 1.5 m/s. Verification of the situations in 2011 and 2013 showed that the RMSEs of the wind speed element were similar. With the exception of parts of North China (such as Shandong Province and the northern part of Shanxi Province); parts of East China (such as Zhejiang Province, Fujian Province, and Guangdong Province); and Central China (the southern part of Hunan Province, and Hubei Province), where its performance was poor, the RMSEs for a minority of stations were at 1.0–1.5 m/s.

Besides the aforementioned regions, the RMSEs of the wind speed element for most of the stations were within 1.0 m/s. The overall verification results proved that the CMADS wind speed element was able to accurately reflect the distribution of wind velocities at the national level over multiple years.

4. Discussion

The uncertainty of hydrological models is greatly influenced by meteorological data. Establishing CMADS is quite useful because the CMADS has defines a unified site location (longitude and latitude), assimilated more data sources, and has been corrected by more observation stations. Importantly, the data is freely available to the public. Further, this set of data can serve climate change analysis, water resources, and water pollution assessment. What needs to be emphasized is: the accuracy of the CMADS data set is achieved by the advanced STMAS method and assimilating/correcting the ECMWF background field using a large number of observations. In areas without observatories, CMADS can still be supported by the corrected background fields to ensure data availability and superiority. However, although we have carried out corrective experiments on the entire East Asia region using observed data in China, more assimilation needs to be followed up to ensure that the CMADS background fields in these regions are close to the real world. Besides, we admitted that the factors controlling CMADS accuracy have not yet been systematically analyzed at the present stage, which will be carried out in our future studies.

The assessment of using CMADS for driving SWAT outside China is acceptable. For example, researchers from Korea used CMADS to drive SWAT in the Han River Basin in the Korean Peninsula with a satisfactory performance [30], and the results were acceptable. In China, scientists used CMADS to drive the hydro-meteorological model for the Qinghai-Tibet Plateau [31], the Yangtze River Basin [32], the Yellow River Basin [33–35], the Pearl River Basin [36], and the inland arid areas in Northwest China [37,38]. The above studies show that CMADS has been widely verified in many regions of East Asia. Furthermore, Researchers from China also used CMADS data and the Penman–Monteith method to calculate potential evapotranspiration (PET) across China with a good performance [39]. All the above studies show that the application of CMADS dataset in East Asia is satisfactory. Although we have gained many advantages in historical simulations, we believe the period the current CMADS covered is still relatively short. Therefore, we plan to improve the data in duration (backward to 1980s) and the time step (hourly). We also plan to produce CMADS forecast data using WRF (CMADS-WRF), and thus the future CMADS would support flood prediction and analysis.

5. Usage Notes

These datasets are more than mere supporting data for the development of SWAT models. These can also be extracted in text format from the for-other-model directory, and then used to drive other models. We recommend using the Notepad++ software which developed by Dr. Don Ho (download from https://notepad-plus-plus.org) to access the data from this directory. If you are accustomed to using a text reader (on the Windows platform), you only need to use the Unix2Dos (developed by Benjamin Lin, download from http://dos2unix.sourceforge.net/) command for execution in this directory layer.

6. Conclusions

This study evaluated the accuracy of the CMADS data for application in China against data recorded by the national automatic stations. The datasets were found to match the actual observation data recorded by the national observation stations very well. This confirmed the applicability of the datasets for the country. It was noted that the distribution of the observation data was scattered, with extremely uneven distributions between the eastern and western regions of China. Thus, the CMADS datasets can more than adequately make up for the lack of traditional meteorological stations in China, especially in West China.

Author Contributions: X.M. and X.J. organized and wrote the manuscript. X.M. collected and analyzed the data. X.M., Y.W., C.S. and H.W. contributed to discussing the results. All authors reviewed the manuscript.

Funding: This research was funded by the National key Technology R & D Program of China (2017YFC0404305) and the National Science Foundation of China (41701076).

Acknowledgments: The authors would like to appreciate the anonymous reviewers for their helpful comments on an earlier draft of this manuscript.

Conflicts of Interest: The authors declare no conflicts of interest.

References

1. Lavers, D.A.; Villarini, G.; Allan, R.P.; Wood, E.F.; Wade, A.J. The detection of atmospheric rivers in atmospheric reanalyses and their links to British winter floods and the large-scale climatic circulation. *J. Geophys. Res.* **2012**, *117*, D20106. [CrossRef]
2. Quadro, M.F.L.; Berbery, E.H.; Dias, M.A.F.S.; Herdies, D.L.; Goncalves, L.G.G. The atmospheric water cycle over South America as seen in the new generation of global reanalyses. *AIP Conf. Proc.* **2013**, *732*, 732–735.
3. Wei, W.; Jilong, C.; Ronghui, H. Water budgets of tropical cyclones: Three case studies. *Adv. Atmos. Sci.* **2013**, *30*, 468–484.
4. Nicholls, N. Increased Australian wheat yield due to recent climate trends. *Nature* **1997**, *387*, 484–485. [CrossRef]
5. Changnon, S.A.; Kunkel, K.E. Rapidly expanding uses of climate data and information in agriculture and water resources: Causes and characteristics of new applications. *Bull. Am. Meteorol. Soc.* **1999**, *80*, 821–830. [CrossRef]
6. Ozturk, T.; Altinsoy, H.; Türkeş, M.; Kurnaz, M. Simulation and spatiotemporal pattern of air temperature and precipitation in Eastern Central Asia using RegCM. *Sci. Rep.* **2018**, *8*, 3639.
7. Hulme, M.; Mitchell, J.; Ingram, W.; Lowe, J.; Johns, T.; New, M.; Viner, D. Climate change scenarios for global impacts studies. *Glob. Environ. Chang.* **1999**, *9*, S3–S19. [CrossRef]
8. Meng, X.; Wang, H.; Wu, Y.; Long, A.; Wang, J.; Shi, C.; Ji, X. Investigating spatiotemporal changes of the land-surface processes in Xinjiang using high-resolution CLM3.5 and CLDAS: Soil temperature. *Sci. Rep.* **2017**, *7*, 13286. [CrossRef] [PubMed]
9. Wang, Y.J.; Meng, X.Y. Snowmelt runoff analysis under generated climate change scenarios for the Juntanghu River basin in Xinjiang, China. *Tecnología y Ciencias del Agua* **2016**, *7*, 41–54.
10. Meng, X.; Sun, Z.; Zhao, H.; Ji, X.; Wang, H.; Xue, L.; Wu, H.; Zhu, Y. Spring Flood Forecasting Based on the WRF-TSRM mode. *Tehnički Vjesnik* **2018**, *25*, 27–37.
11. Saha, S.; Moorthi, S.; Pan, H.-L.; Wu, X.; Wang, J.; Nadiga, S.; Tripp, P.; Kistler, R.; Woollen, J.; Behringer, D.; et al. The NCEP climate forecast system reanalysis. *Bull. Am. Meteorol. Soc.* **2010**, *91*, 1015–1057. [CrossRef]
12. Trenberth, K.E.; Anthes, R.A.; Belward, A.; Brown, O.B.; Habermann, T.; Karl, T.B.; Running, S.; Ryan, B.; Tanner, M.; Wielicki, B. *Climate Science for Serving Society: Research, Modeling and Prediction Priorities*; Hurrell, J.W., Asrar, G., Eds.; Springer: New York, YN, USA, 2013; Chapter 2.
13. Kanamitsu, M.; Ebisuzaki, W.; Woollen, J.; Yang, S.K.; Hnilo, J.J.; Fiorino, M.; Potter, G.L. NCEP-DEO AMIP-II reanalysis (R-2). *Bull. Am. Meteorol. Soc.* **2002**, *83*, 1631–1643. [CrossRef]
14. Dee, D.P.; Uppala, S.M.; Simmons, A.J.; Berrisford, P.; Poli, P.; Kobayashi, S.; Andrae, U.; Balmaseda, M.A.; Balsamo, G.; Bauer, P.; et al. The ERA-Interim reanalysis: Configuration and performance of the data assimilation system. *Q. J. R. Meteorol. Soc.* **2011**, *137*, 553–597. [CrossRef]

15. Gibson, J.K.; Kållberg, P.; Uppala, S.; Hernandez, A.; Nomura, A.; Serrano, E. *ERA Description*; Re-Analysis Project Report Series No. 1; European Centre for Medium-Range Weather Forecasts (ECMWF): Reading, UK, 1997; Available online: www.ecmwf.int/sites/default/files/elibrary/1997/9584-era-description.pdf (accessed on 1 May 1999).

16. Uppala, S.M.; Kållberg, P.W.; Simmons, A.J.; Andrae, U.; Bechtold, V.D.C.; Fiorino, M.; Gibson, J.K.; Haseler, J.; Hernandez, A.; Kelly, G.A.; et al. The ERA-40 re-analysis. *Q. J. R. Meteorol. Soc.* **2005**, *131*, 2961–3012. [CrossRef]

17. Daly, C.; Neilson, R.P.; Phillips, D.L. A statistical-topographic model for mapping climatological precipitation over mountainous terrain. *J. Appl. Meteorol.* **1994**, *33*, 140–158. [CrossRef]

18. Frei, C.; Schar, C. A precipitation climatology of the Alps from high-resolution rain-gauge observations. *Int. J. Climatol.* **1998**, *18*, 873–900. [CrossRef]

19. Neitsch, S.L.; Arnold, J.G.; Kiniry, J.R.; Williams, J.R. *Soil and Water Assessment Tool: Theoretical Documentation—Version 2009*; Texas Water Resources Institute Technical Report No. 406; Agricultural Research Service (USDA) & Texas Agricultural Experiment Station, Texas A&M University: Temple, TX, USA, 2011.

20. Meng, X.; Wang, H. Significance of the China Meteorological Assimilation Driving Datasets for the SWAT Model (CMADS) of East Asia. *Water* **2017**, *9*, 765. [CrossRef]

21. Xie, Y.; Koch, S.; McGinley, J.; Albers, S.; Bieringer, P.E.; Wolfson, M.; Chan, M. A Space–Time Multiscale Analysis System: A Sequential Variational Analysis Approach. *Mon. Weather Rev.* **2011**, *139*, 1224–1240. [CrossRef]

22. Albers, S.C.; Xie, Y.; Raben, V.; Toth, Z.; Holub, K. The Local Analysis and Prediction System (LAPS) Cloud Analysis: Validation with All-sky Imagery and Development of a Variational Cloud Assimilation. In *AGU Fall Meeting Abstracts*; American Geophysical Union: Washington, DC, USA, 2013.

23. Xie, P.; Joyce, R.; Wu, S.; Yoo, S.H.; Yarosh, Y.; Sun, F.; Lin, R. Reprocessed, Bias-Corrected CMORPH Global High-Resolution Precipitation Estimates from 1998. *J. Hydrometeorol.* **2017**, *18*, 1617–1641. [CrossRef]

24. Yang, F.; Lu, H.; Yang, K.; He, J.; Wang, W.; Wright, J.S.; Li, C.; Han, M.; Li, Y. Evaluation of multiple forcing data sets for precipitation and shortwave radiation over major land areas of china. *Hydrol. Earth Syst. Sci.* **2017**, *21*, 1–32. [CrossRef]

25. Shen, Y.; Pan, Y.; Yu, J.J. Application of Probability Density Function-Optimal Interpolation in Hourly Gauge-Satellite Merged Precipitation Analysis over China. Available online: http://www.isac.cnr.it/~ipwg/meetings/saojose-2012/pres/Shen.pdf (accessed on 16 October 2012).

26. Stamnes, K.; Tsay, S.-C.; Wiscombe, W.; Jayaweera, K. Numerically stable algorithm for discrete-ordinate-method radiative transfer in multiple scattering and emitting layered media. *Appl. Opt.* **1988**, *27*, 2502–2509. [CrossRef] [PubMed]

27. Liu, J.J.; Shi, C.X. Retrievals and Evaluation of Downward Surface Solar Radiation Derived from FY-2E. *Remote Sens. Inf.* **2018**, *33*, 104–110.

28. Bliss, N.B.; Olsen, L.M. Development of a 30-arc-second digital elevation model of South America. In Proceedings of the Pecora Thirteen, Human Interactions with the Environment—Perspectives from Space, Sioux Falls, SD, USA, 20–22 August 1996.

29. Danielson, J.J. Delineation of drainage basins from 1 km African digital elevation data. In Proceedings of the Pecora Thirteen, Human Interactions with the Environment—Perspectives from Space, Sioux Falls, SD, USA, 20–22 August 1996.

30. Vu, T.T.; Li, L.; Jun, K.S. Evaluation of Multi-Satellite Precipitation Products for Streamflow Simulations: A Case Study for the Han River Basin in the Korean Peninsula, East Asia. *Water* **2018**, *10*, 642. [CrossRef]

31. Liu, J.; Shanguan, D.; Liu, S.; Ding, Y. Evaluation and Hydrological Simulation of CMADS and CFSR Reanalysis Datasets in the Qinghai-Tibet Plateau. *Water* **2018**, *10*, 513. [CrossRef]

32. Gao, X.; Zhu, Q.; Yang, Z.; Wang, H. Evaluation and Hydrological Application of CMADS against TRMM 3B42V7, PERSIANN-CDR, NCEP-CFSR, and Gauge-Based Datasets in Xiang River Basin of China. *Water* **2018**, *10*, 1225. [CrossRef]

33. Zhao, F.; Wu, Y.; Qiu, L.; Sun, Y.; Sun, L.; Li, Q.; Niu, J.; Wang, G. Parameter Uncertainty Analysis of the SWAT Model in a Mountain-Loess Transitional Watershed on the Chinese Loess Plateau. *Water* **2018**, *10*, 690. [CrossRef]

34. Zhou, S.; Wang, Y.; Chang, J.; Guo, A.; Li, Z. Investigating the Dynamic Influence of Hydrological Model Parameters on Runoff Simulation Using Sequential Uncertainty Fitting-2-Based Multilevel-Factorial-Analysis Method. *Water* **2018**, *10*, 1177. [CrossRef]

35. Shao, G.; Guan, Y.; Zhang, D.; Yu, B.; Zhu, J. The Impacts of Climate Variability and Land Use Change on Streamflow in the Hailiutu River Basin. *Water* **2018**, *10*, 814. [CrossRef]

36. Cao, Y.; Zhang, J.; Yang, M.; Lei, X.; Guo, B.; Yang, L.; Zeng, Z.; Qu, J. Application of SWAT Model with CMADS Data to Estimate Hydrological Elements and Parameter Uncertainty Based on SUFI-2 Algorithm in the Lijiang River Basin, China. *Water* **2018**, *10*, 742. [CrossRef]

37. Meng, X.; Wang, H.; Cai, S.; Zhang, X.; Leng, G.; Lei, X.; Shi, C.; Liu, S.; Shang, Y. The China Meteorological Assimilation Driving Datasets for the SWAT Model (CMADS) Application in China: A Case Study in Heihe River Basin. *PearlRiver* **2016**, *37*, 1–19.

38. Meng, X.Y.; Wang, H. Hydrological Modeling in the Manas River Basin Using Soil and Water Assessment Tool Driven by CMADS. *Tehnički Vjesnik* **2017**, *24*, 525–534.

39. Tian, Y.; Zhang, K.; Xu, Y.-P.; Gao, X.; Wang, J. Evaluation of Potential Evapotranspiration Based on CMADS Reanalysis Dataset over China. *Water* **2018**, *10*, 1126. [CrossRef]

water

MDPI

Article

Evaluation of Potential Evapotranspiration Based on CMADS Reanalysis Dataset over China

Ye Tian [1,*], Kejun Zhang [1], Yue-Ping Xu [2], Xichao Gao [3] and Jie Wang [1]

[1]   School of Hydrology and Water Resources, Nanjing University of Information Science & Technology, Nanjing 210044, China; zhangkj96@163.com (K.Z.); wangjie0775@163.com (J.W.)
[2]   Department of Civil Engineering, Institute of Hydrology and Water Resources, Zhejiang University, Hangzhou 310058, China; yuepingxu@zju.edu.cn
[3]   China Institute of Water Resources and Hydropower Research; Beijing 100038, China; 999gaoxichao@163.com
*   Correspondence: tianye@nuist.edu.cn; Tel.: +86-25-5869-5608

Received: 11 July 2018; Accepted: 21 August 2018; Published: 23 August 2018

Abstract: Potential evapotranspiration (PET) is used in many hydrological models to estimate actual evapotranspiration. The calculation of PET by the Food and Agriculture Organization of the United Nations (FAO) Penman–Monteith method requires data for several meteorological variables that are often unavailable in remote areas. The China Meteorological Assimilation Driving Datasets for the SWAT model (CMADS) reanalysis datasets provide an alternative to the use of observed data. This study evaluates the use of CMADS reanalysis datasets in estimating PET across China by the Penman–Monteith equation. PET estimates from CMADS data (PET_cma) during the period 2008–2016 were compared with those from observed data (PET_obs) from 836 weather stations in China. Results show that despite PET_cma overestimating average annual PET and average seasonal in some areas (in comparison to PET_obs), PET_cma well matches PET_obs overall. Overestimation of average annual PET occurs mainly for western inland China. There are more meteorological stations in southeastern China for which PET_cma is a large overestimate, with percentage bias ranging from 15% to 25% for spring but a larger overestimate in the south and underestimate in the north for the winter. Wind speed and solar radiation are the climate variables that contribute most to the error in PET_cma. Wind speed causes PET to be underestimated with percentage bias in the range −15% to −5% for central and western China whereas solar radiation causes PET to be overestimated with percentage bias in the range 15% to 30%. The underestimation of PET due to wind speed is offset by the overestimation due to solar radiation, resulting in a lower overestimation overall.

Keywords: potential evapotranspiration; Penman-Monteith; CMADS; China

---

1. Introduction

Evapotranspiration (ET) is a fundamental component of the hydrological cycle and a route for energy transfer between the earth's surface and the atmosphere. It is important in activities such as evaluating water resources [1], drought forecasting [2], and managing irrigation [3]. It is also important for an understanding of the behavior of water in soil–vegetation–atmosphere interactions [4] and in rainfall-runoff modeling.

Actual evapotranspiration is usually measured indirectly by eddy covariance (EC) [5], the Bowen ratio method [6], lysimetry [7], and scintillometry [8]; ET cannot be observed at a large scale by these techniques. Eddy towers and the Bowen ratio method observe over only hundreds of meters, depending on wind speed, tower height, and canopy level. The lysimeter only functions on a scale of several meters. The scintillometer measures sensible heat flux on the scale of kilometers [9]. Wang and Dickinson [10] review these observation methods and summarize different methods of modeling

ET. For studies on a catchment, regional, or global scale, ET can be estimated by remote sensing methods [11], hydrological models, and land surface models [12,13]. The use of remote sensing methods may be restricted by a relatively low temporal resolution, and some remote sensing methods are only suitable in the clear sky condition [11]. Hydrological models and land surface models can estimate ET for different spatial and temporal scales, which makes them able to support the management and planning of water resources better than other techniques.

Potential evapotranspiration (PET) has been used in many hydrological models to estimate actual evapotranspiration using a soil moisture extraction function [14]. PET is defined as the amount of water that can potentially be removed from a vegetated surface through the processes of evaporation or transpiration with no forcing other than atmospheric demand [15]. The common methods of estimating PET can be divided into four categories: radiation-based, temperature-based, a combination of these two, and mass transfer [16,17]. These methods differ in their input requirements and their underlying assumptions. Some are developed for a specific climatic region [18]. The Penman–Monteith equation has been incorporated in many hydrological models, such as SWAT [19], VIC [20], and SHE [21], to estimate PET. Many studies have compared different PET estimation methods [18,22]. Kite and Drooger [23] evaluated eight different methods and found the Food and Agriculture Organization of the United Nations Penman–Monteith method (FAO-PM), which combines mass transfer and energy balance with temperature and vegetation conductance, best models PET, and most closely matches field observations. FAO-PM requires observed maximum temperature, minimum temperature, air temperature, wind speed, relative humidity, and solar radiation (or solar duration) as input variables. In China, meteorological stations are not evenly distributed spatially, and observed records of these climate variables are difficult to obtain in some rural areas. Reanalysis datasets, which have high precision and high spatiotemporal resolution, complement this data paucity and they have been extensively used in hydrological modeling.

There are many widely used reanalysis products, which include climate forecast system reanalysis (CFSR) [24], NCEP/DOE [25] and NCEP/NCAR [26] from NCEP, ERA-15 [27], ERA40 [28] and ERA-Interim [29] from ECMWF, JRA-55 from the Japanese meteorological agency [30], and MERRA from NASA [31]. They have provided accurate meteorological data and have shown they can overcome the disadvantages of thinly distributed observation networks. There have been extensive evaluations of the performance of these reanalysis products over different regions of the world [32–34]. PET estimates derived from reanalysis products have been compared. Weiland et al. used CFSR and compared global PET estimates from six different methods [35]. Srivastava et al. evaluated PET calculated from NCEP and ECMWF ERA-Interim data in England [36]. Trambauer et al. compared observed evaporation for Africa using a continental version of the global hydrological model PCR-GLOBWB with PET estimates given by the ECMWF reanalysis products ERA-Interim and ERA-Land, and satellite-based products MOD16 and GLEAM [37]. These products provide global climate datasets having a relatively coarse spatial resolution.

The newly developed China Meteorological Assimilation Driving Datasets for the SWAT model (CMADS) covers East Asia (60°–160° E, 0°–65° N). CMADS was developed to provide meteorological data with fine resolution. Temperature, pressure, and wind speed data were derived from hourly observations from 2421 national weather stations and 29,452 regional weather stations. The data were combined using the Space and Time Multiscale Analysis System (STMAS) with European Centre for Medium-Range Weather Forecasts (ECMWF) ambient field. Precipitation data were incorporated by combining observed data from weather stations with precipitation reanalysis data from the NOAA with CPC MORPHing technique (CMORPH). Solar radiation data obtained from radiance data of the International Satellite Cloud Climatology Project (ISCCP) were combined with data retrieved from the FY-2E satellite using the discrete-ordinate radiative transfer (DISTORT) model. Big data projection and processing methods, such as loop nesting of data, projection of resampling models, and bilinear interpolation, were used to build the datasets [38–40]. There have been many similar applications in different river basins in East Asia that have achieved satisfactory results in rainfall–runoff simulation [41–49].

The objectives of this study are: (1) to evaluate the accuracy of PET estimated by the Penman-Monteith equation (PM) using CMADS reanalysis data by comparing it with PET estimated by PM using observed data provided by 836 weather stations; (2) to analyze the contribution of each reanalysis variable from the CMADS data to error in PET by controlling the variables. Several statistical measures were chosen to evaluate the comparison between CMADS-derived PET and observed PET.

2. Materials and Methods

2.1. Data and Study Area

Meteorological data for maximum temperature, minimum temperature, wind speed at 10 m height, relative humidity, and solar duration from 836 weather stations were collected to estimate PET. The spatial distribution of the 836 weather stations across China is shown in Figure 1. PM is one of the most widely used methods of calculating PET [50–54]. Details of estimating PET using PM are given in the following paragraph. In the rest of this paper, PET values derived from the weather station observations are referred to as PET_obs. They are taken to be the real values against which other predicted PET values can be compared.

PET over China was also estimated using CMADS (version 1.1). The CMADS dataset spans the period 2008–2016 at a spatial resolution of 0.25°; it covers East Asia (60°–160° E, 0°–65° N) and provides daily data for the meteorological variables. The datasets were developed by Xianyong Meng and are available from the website: http://www.cmads.org. In this study, we used daily maximum temperature (T_{max}), daily minimum temperature (T_{min}), relative humidity (RH), solar radiation (Rs), and wind speed at 10m height (u_{10}) to estimate PET using PM. PET values derived from the CMADS datasets (PET_cma) were compared with PET_obs values and differences are referred to as under or overestimation (PET_cma < PET_obs or PET_cma > PET_obs).

Because the gridded CMADS data observations do not spatially correspond to the weather station observations, the CMADS grids were interpolated to the 836 stations using a polynomial interpolation method. Temperature differences caused by elevation are also considered in the interpolation: temperature was assumed to decrease by 0.65 °C for every 100 m increase in elevation.

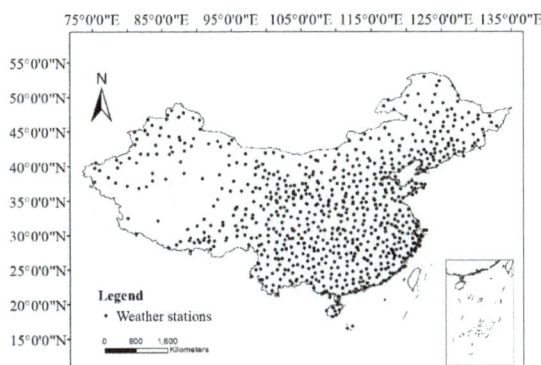

Figure 1. Spatial distribution of weather stations in China.

2.2. Penman–Monteith Equation for Potential Evapotranspiration

The Food and Agriculture Organization (FAO) Expert Consultation of Revision of FAO Methodologies of Crop Water Requirements standardized the types and characteristics of the vegetated surface based on the previous PET definition. The definition of vegetated surface was assumed to be a hypothetical reference crop with a crop height of 0.12 m, a fixed surface resistance of 70 s m^{-1} and albedo of 0.23 [50]. The daily potential evapotranspiration (PET, mm/day) estimated by the Penman–Monteith equation (PM) is:

$$\text{PET} = \frac{0.408\Delta(R_n - G) + \gamma\frac{900}{T_{mean}+273.3}u_2(e_s - e_a)}{\Delta + \gamma(1 + 0.34)u_2} \tag{1}$$

where: R_n is the net radiation at the crop surface (MJ m^{-2} day^{-1}); G is the soil heat flux density (MJ m^{-2} day^{-1}), which is assumed to be zero as the magnitude of G, in this case, is relatively small; T_{mean} is the mean daily air temperature (°C); u_2 is the wind speed at 2 m height (m s^{-1}); e_s is the saturation vapor pressure (kPa); e_a is the actual vapor pressure (kPa); $e_s - e_a$ is the vapor pressure deficit (kPa); Δ is the slope of the relationship between saturation vapor pressure and mean daily air temperature (kPa °C^{-1}); γ is the psychrometric constant which depends on the altitude of each location (kPa °C^{-1}).

Saturation vapor pressure is related to air temperature and can be calculated from air temperature. The relationship is expressed in Equation (2), in which: e_s is the mean of the saturation vapor pressure at T_{max} and T_{min}; e_a was calculated by multiplying the average values of the saturation vapor pressure at T_{max} and T_{min} by the mean daily relative humidity. The FAO recommendation is to calculate the actual vapor pressure by taking the average the product of vapor pressure at the higher temperature and daily low humidity and the product of vapor pressure at the lower temperature and the daily high humidity. However, only the mean relative humidity is available from the CMADS datasets, and in the case of missing maximum and minimum relative humidity, Equation (4) was used.

$$e_{(T)} = 0.6108exp\left(\frac{17.27T}{T + 237.3}\right) \tag{2}$$

$$e_s = \frac{e_{(T_{min})} + e_{(T_{max})}}{2} \tag{3}$$

$$e_a = \frac{RH_{mean}}{100}\left(\frac{e_{(T_{min})} + e_{(T_{max})}}{2}\right) \tag{4}$$

where: T_{max} and T_{min} are daily maximum and minimum temperatures; $e_{(T_{max})}$ and $e_{(T_{min})}$ are the saturation vapor pressures at daily minimum temperature and daily maximum temperature; and T is the temperature. Using Equation (2), the saturation vapor pressure at the daily maximum and minimum air temperatures can be calculated by:

$$e_{(T_{max})} = 0.618exp\left(\frac{17.27T_{max}}{T_{max} + 237.3}\right) \tag{5}$$

$$e_{(T_{min})} = 0.618exp\left(\frac{17.27T_{min}}{T_{min} + 237.3}\right) \tag{6}$$

The mean saturation vapor pressure is calculated as the mean of the saturation vapor pressures at the daily maximum and daily minimum air temperatures using Equation (3).

R_n is the net radiation which is expressed as the difference between the incoming net shortwave radiation (R_{ns}) and the outgoing net long wave radiation (R_{nt}):

$$R_n = R_{ns} - R_{nl} \tag{7}$$

R_{ns} is computed by:

$$R_{ns} = (1 - \alpha)R_s \tag{8}$$

where: α is the albedo, which is 0.23 for the hypothetic grass reference crop; and R_s is the solar radiation which is either computed the from the daily solar duration (n), using the Ångström–Prescott radiation equation (see Equation (9)) for the weather station data or is obtained directly from CMADS.

$$R_s = \left(a_s + b_s \frac{n}{N}\right) R_a \tag{9}$$

where: R_a is the extraterrestrial radiation which is calculated from the solar constant, the solar declination, and the time of the year as suggested by the FAO (the recommended values $a_s = 0.25$ and $b_s = 0.5$ are used in this study); n is the actual solar duration; N is the maximum possible solar duration which is related to the latitude and can be computed using the sunset hour angle in radians; and $\frac{n}{N}$ is the relative solar duration. The outgoing net long wave radiation (R_{nt}) is derived by the Stefan–Boltzmann law.

2.3. Evaluation Method

Several statistical measures are used to compare PET_cma with PET_obs: percentage bias (*PB*), the coefficient of determination (R^2), the normalized root mean square error (*NRMSE*), and the skill score (S_{score}). PB is a basic measure used to assess average annual PET and seasonal patterns of PET which provide an overview of the performance of the two models. For a more comprehensive analysis, R^2, NRMSE, and S_{score} are used to analyze the performance of daily, monthly, and annual PET_cma.

PB, which is the ratio between CMADS bias and observations, indicates the average magnitude of underestimation or overestimation of PET. Intuitively, PB is the average bias. It is given by:

$$PB = \frac{\sum_{i=1}^{n} (M_i - O_i)}{\sum_{i=1}^{n} O_i} \times 100\% \tag{10}$$

where the M_i are PET_cma and the O_i are PET_obs.

R^2 shows how well PET_cma approximates the real data points (PET_obs). It indicates the proportion of the variance in the dependent variable that is predictable from the independent variable. Ordinary least squares regression is used to fit the line to the data. The ideal fitted line is found when R^2 is very close to 1. Linear regression is suitable for a long time series. We have data for nine years, so R^2 is appropriate to use for daily and monthly PET_cma. R^2 is given by:

$$R^2 = 1 - \frac{SS_{res}}{SS_{tot}} \tag{11}$$

$$SS_{tot} = \sum_i (O_i - \overline{O})^2 \tag{12}$$

$$SS_{res} = \sum_i (O_i - M_i)^2 \tag{13}$$

NRMSE is a normalized version of root mean square deviation. It is a dimensionless indicator, which makes it suitable for a comparison between observations and simulations that have different scales. Lower absolute values of NRMSE represent less residual variance. The equation is:

$$NRMSE = \sqrt{\frac{\sum_i^n (O_i - M_i)^2}{n\overline{O}^2}} \tag{14}$$

S_{score} indicates the common area of the probability distribution function (PDF) of PET_cma and PET_obs. It is the cumulative minimum value of the two distributions at each binned value. The equation is:

$$S_{score} = \sum_1^n minimum \, (Z_M, Z_O) \tag{15}$$

where: n is the number of bins of the PDFs where the bin sizes are 0.01 mm, 0.1 mm, and 1 mm for daily, monthly, and annual PET; Z_M and Z_O are the frequency values in a given bin from PET_cma and PET_obs.

After analyzing the performance of PET_cma, control variables are used in analysis to identify the most important factors that influence PET_cma. The variables considered are input parameters for PM: T_{max}, T_{min}, R_h, R_s, and u_{10}. The control is PET_obs, which used all the input parameters with data from the weather stations. For comparison, we singly examine each input parameter as an independent variable using CMADS data, covering the same area and time period, for the variable instead of the weather station data while keeping the observed weather station data for all other input parameters.

3. Results

3.1. Spatial and Seasonal Patterns of Average Annual PET

Figure 2 shows the average annual values of PET_obs and PET_cma at all the stations across China and their histograms. There is large spatial variation in PET_obs. The northeast and midwest inland have smaller values of PET, mainly <950 mm. The south coastal area, southwest inland and some areas in the northwest have larger values of PET, >1150 mm. Some stations on the islands of the South China Sea, and in southern and northwestern China, are >1550 mm. PET_cma and PET_obs agree well spatially except for overestimation in midwest China. The multi-year average PET_obs is 1000 mm, and the multi-year average PET_cma is 1120 mm. PET_cma is an underestimate of the observed value of PET when PET_obs <1080 mm but it is an overestimate when PET_obs >1080 mm. The major overestimation is when PET_obs is in the interval 750 mm to 1000 mm. However, the maximum values of the multi-year average for PET_obs and PET_cma are close, approximately 1640 mm.

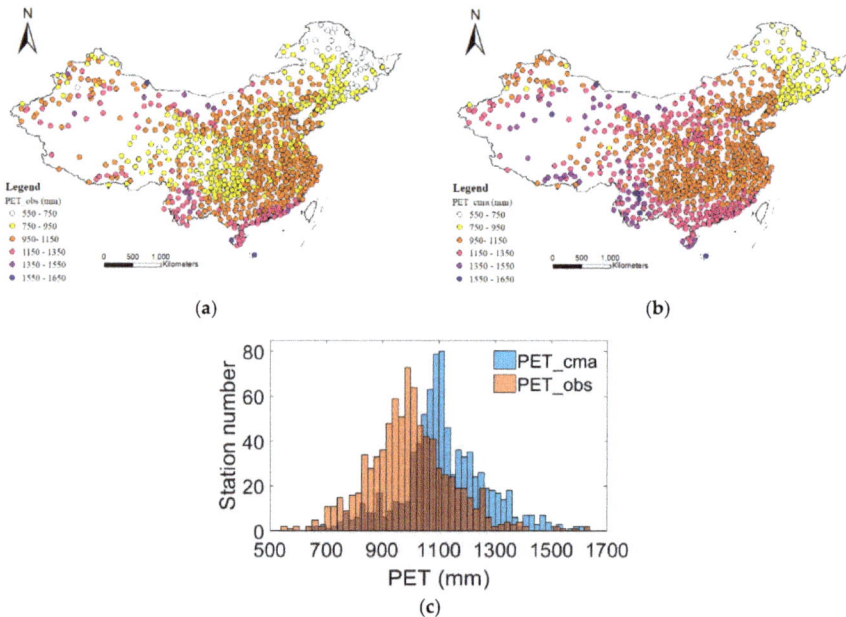

Figure 2. Average annual PET_obs (**a**) and PET_cma (**b**) across China, with their histograms (**c**).

Figure 3 shows the spatial distribution of mean seasonal PET values across China for observed (PET_obs) and reanalyzed (PET_cma) data for spring (March, April, May), summer (June, July, August), autumn (September, October, November), and winter (December, January, February). Most areas in China have a mean value of PET_obs <300 mm in the spring. Average PET_obs increased in the summer across China, almost everywhere >350 mm. Average PET_obs decreases to <320 mm in autumn except for some south coastal stations. Average PET_obs is lowest in winter at 110 mm and is approximately 100 mm for most stations.

CMADS data captures the general feature of the seasonal and spatial distributions of observed PET, as shown in the center of Figure 3, but compared to mean seasonal PET_obs, mean seasonal PET_cma is overestimated for all seasons. The bias is larger in spring and summer than that in autumn and winter. Most stations show mean seasonal PET_obs in the range 200–300 mm (spring) and 300–400 mm (summer), whereas most mean seasonal PET_cma values are in the range 300–350 mm (spring) and 400–450 mm (summer). The overestimates are seen when the PET_cam values are above the mode of PET_obs. The highest mean seasonal PET_cma value in summer is 755 mm which is very close to the highest mean seasonal PET_obs value. The closest PET_cma comes to PET_obs is in winter with a mean bias value of 5 mm.

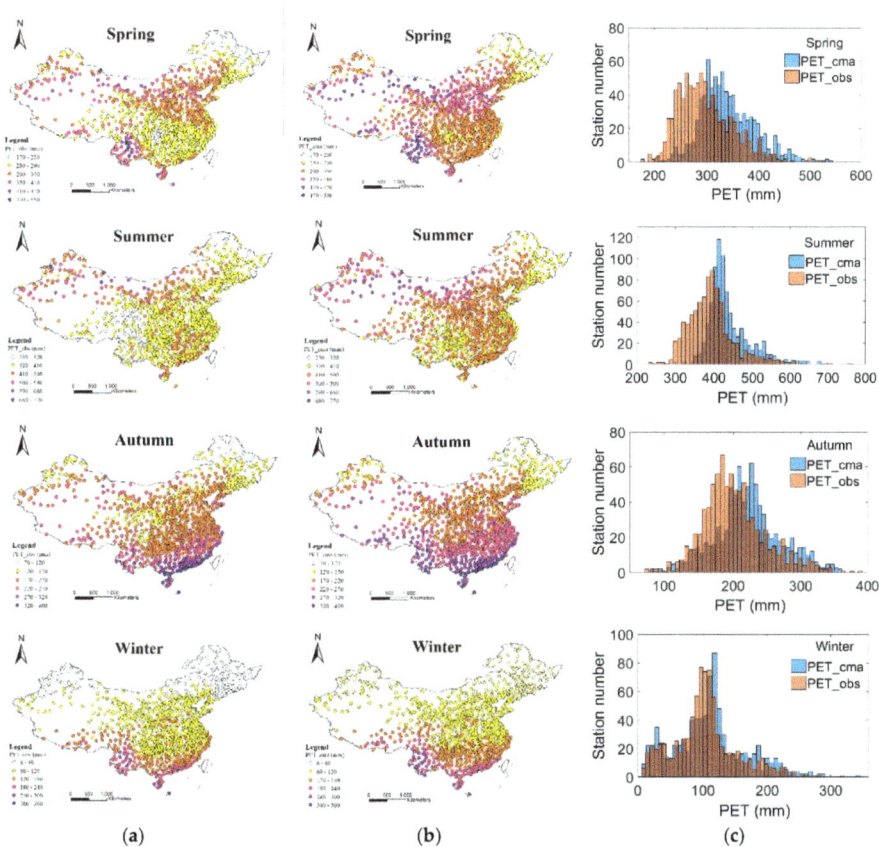

Figure 3. Mean seasonal potential evapotranspiration estimated from weather station data (PET_obs, (**a**)), CMADS datasets (PET_cma, (**b**)) with their histograms (**c**).

3.2. Evaluation of the Performance with Multiple Indicators

Figure 4 shows that the percentage bias for most stations is positive. This result indicates that average annual PET_cma is consistently overestimated; only two stations show an underestimate. The overall average percentage bias for the whole of China is 12.58%. Percentage bias varies spatially.

The range 5% to 15% is the most frequent and is seen mainly in the east and north of China. In the west of inland China, percentage bias is mainly in the range 15% to 30%. There are a few stations in the west at high elevation with high PET_obs where percentage bias >30%.

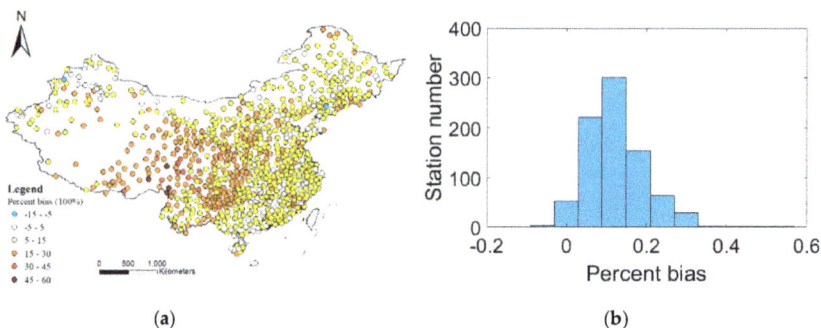

(a)   (b)

Figure 4. Spatial distribution of percentage bias, indicating the accuracy of average annual PET_cma values (**a**), and frequency distribution of percentage bias (**b**).

Figure 5 shows the spatial distribution of the percentage bias of mean seasonal PET_cma. There are different seasonal features. Percentage bias is least in winter at 4.3% and greatest in spring at 15.7% compared to the percentage bias of mean annual PET_obs; the spring distribution shows more stations in southeast China with higher percentage bias (i.e., PET_cma is greatly overestimated), mainly in the range 15% to 30%. In summer, there are more stations in the west of China with higher percentage bias but fewer in southeast China, which is reflected in the frequency distribution showing that there are more stations in the percentage bias ranges 0–5% and >30%. In autumn, there are more stations in the northeast of China with percentage bias in the range −2% to 5%; the few stations with negative percentage bias represent that PET_cma is underestimated in comparison to PET_obs. In winter, PET_cma is overestimated in the south and underestimated in the north. The number of stations with percentage bias in the range −5% to 5% is greater than the annual average.

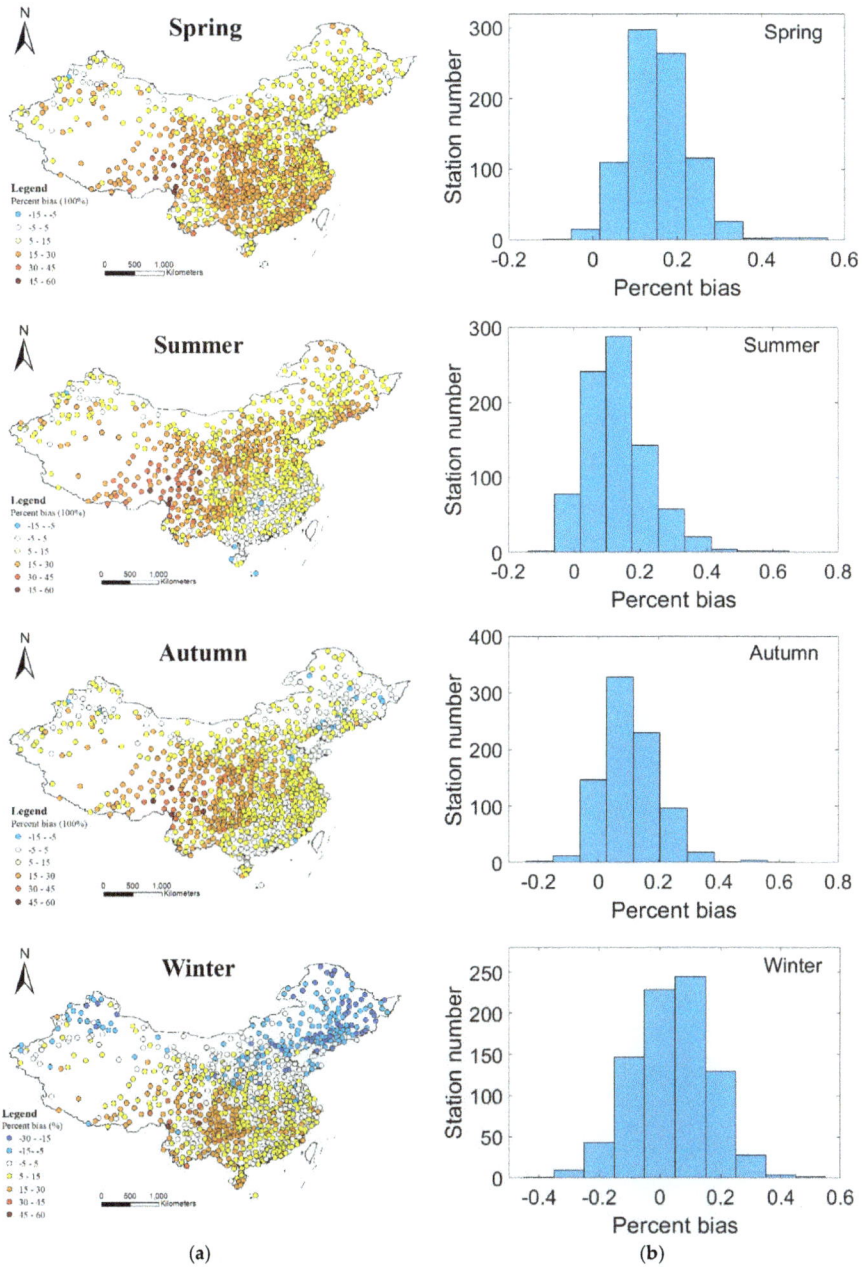

Figure 5. Spatial distribution of percentage bias showing the accuracy of mean seasonal potential evapotranspiration (PET) estimated using CMADS datasets, PET_cma (**a**) and the frequency distribution of percentage bias (**b**).

Percentage bias measures the trend of the average error distribution for a time series, but it cannot be used if there is a difference in time scales. The statistical measures $NRMSE$, R^2, and S_{score} are used to identify the variation in PET estimates with daily and monthly time scales. R^2 and S_{score} can be used to indicate differences at an annual scale, but the CMADS datasets cover only nine years, which is not long enough for adequate linear regression. Thus, annual behavior is only measured by NRMSE as reference. The three measures are used for all the stations. Figure 6 shows the cumulative distribution functions (CDF) of the measures for different time scales. The CDF for NMRSE shows that the estimate given by PET_cma is best at an annual time scale. It decreases as the time scale gets finer, but the difference is not very large. Almost 100% of the stations are <0.4 for every time scale. Up to 80% stations are <0.18, <0.23, and <0.27 for annual, monthly, and daily time scales, respectively. The R^2 values also show similar results for monthly and daily time scales with the monthly CDF better than the daily CDF. For 99% of the stations, monthly and daily R^2 values are >0.90 and >0.80. S_{score} shows that the difference between monthly and daily time scales is very small, but it has a broader range than the other two measures. The monthly and daily S_{score} values for most stations (99%) are >0.70 and >0.75, respectively.

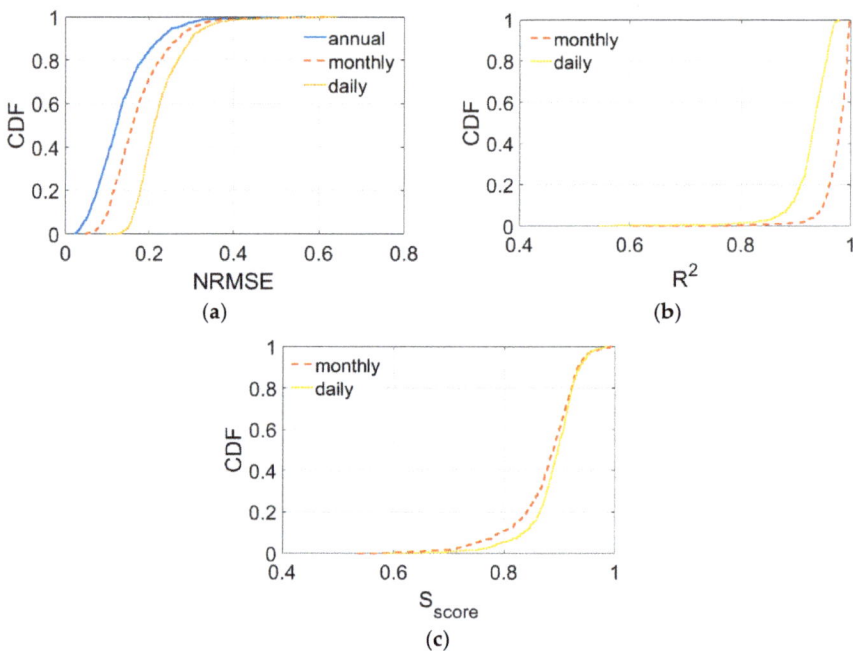

Figure 6. Cumulative distribution functions of the statistical measures *NRMSE* (**a**), R^2 (**b**), and S_{score} (**c**) indicating the accuracy of monthly and daily PET_cma estimates compared to PET_obs.

3.3. Effect of Different Variables on the Bias in Estimation of the PET

Figure 7 shows the contribution of each variable to the error in estimating mean annual PET_cma. Each is used in turn as an independent variable. PET_obs is the control variable. Percentage bias is the error measure. To estimate PET as a dependent variable, the observed data for one input variable of PM is replaced with reanalysis data from CMADS; all other input variables remain unchanged (i.e., they take the observed data used in the calculation of PET_obs). Percentage bias for most areas is in the range −5% to 5% when T_{max}, T_{min}, and R_h were the independent variables. This indicates that errors in T_{max}, T_{min}, and R_h from the CMADS data contributed little to the bias in PET_cma. Figure 7 shows that wind speed and solar radiation contribute to the error in PET_cma in different ways. When wind speed is the independent variable, PET is underestimated, and percentage bias is in the range −15% to −5%. Most underestimated values are found in eastern China. For solar radiation, R_s, PET is mainly overestimated with percentage bias in the range 5% to 30% over most of the area. In central and western China the overestimation is greater, with percentage bias predominantly in the range 15% to 30%.

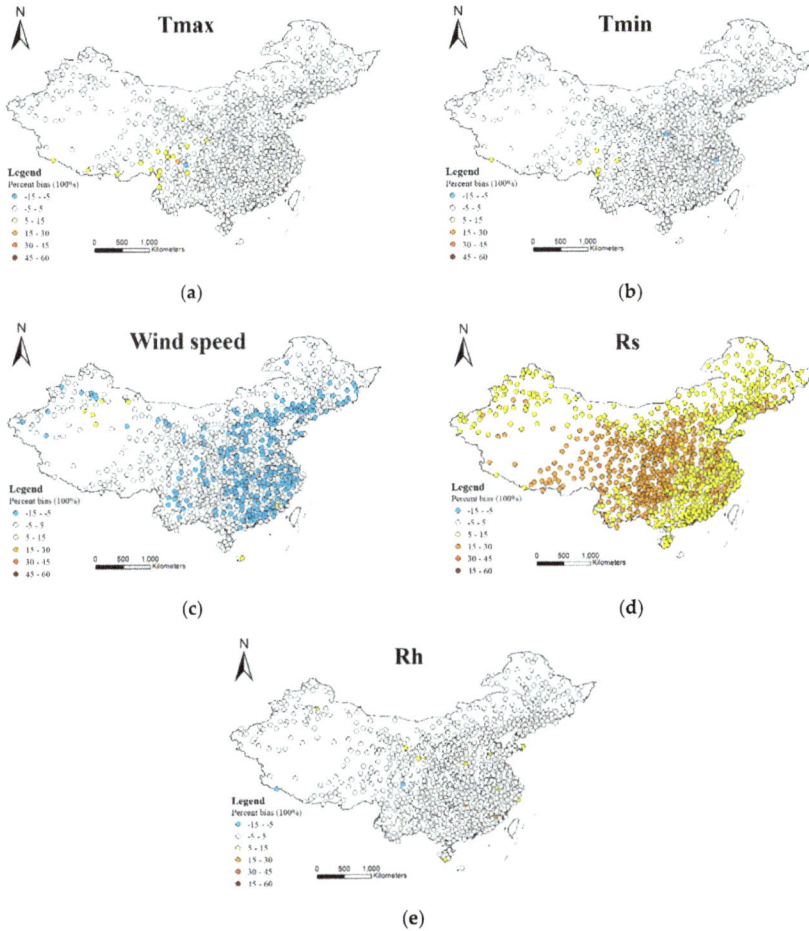

Figure 7. Spatial distribution of percentage bias showing the effects of maximum temperature (**a**), minimum temperature (**b**), wind speed (**c**), solar radiation (**d**), and relative humidity (**e**) on the bias of PET_cma.

Elevation is commonly an important influence on wind speed and solar radiation. For example, topography influences wind speed when wind speed increase as air moves around a hill or along a narrow valley. The pressure gradient, friction due to the earth's surface, and air density also influence wind speed. Figure 8 shows that when wind speed is the independent variable, the percentage bias is mainly <0. Bias <−10% is seen mainly at stations with elevation <2000 m when PET_obs is in the range 800–1300 mm. Solar radiation can be affected by atmospheric conditions, such as clouds and pollution, and topography can also cause substantial spatial variation in solar radiation [55]. PET was overestimated for most of the stations when R_s was the independent variable. When elevation increased, the lower boundary of percentage bias also increased. When elevation was >2000 m, the percentage bias of PET is >10%. Larger percentage bias, >20%, is found mainly for stations with PET_obs in the range 750–1250 mm. Windspeed, which causes underestimation of PET, and solar radiation, which causes overestimation of PET, offset each other, reducing the overestimation of PET_cma.

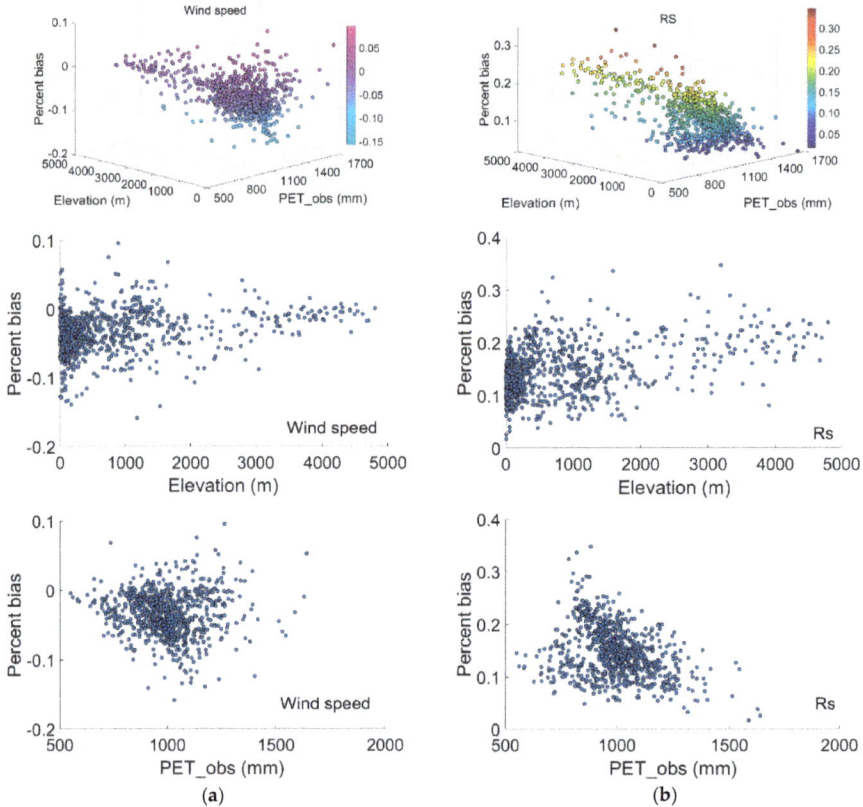

Figure 8. Relationship between elevation, PET_obs, and errors indicated by percentage bias for wind speed (**a**) and solar radiation (**b**).

4. Discussion

This study evaluated the use of CMADS datasets in estimating PET across China during the period 2008–2016. As mentioned previously, there are alternative methods and different datasets for estimating PET. Weiland et al. [35] compared six different methods using Climate Forecast System Reanalysis (CFSR) data and evaluated the results against global Climate Research Unit (CRU) data. They noted that PM has high data demands and is sensitive to inaccuracies in the input data, and so recommended a re-calibrated form of the Hargreaves equation which gave global reference PET values that were comparable to CRU-derived values for many climate conditions. Lang [22] compared eight PET models with PM for southwestern China and found that the Makkink and Hargreaves–Samani methods are good alternatives to PM. Droogers and Allen [50] compared global PET predicted by PM and Hargreaves and recommended that the Hargreaves method be used in regions where accurate weather data cannot be expected because the method requires fewer climate variables as input, which makes it less sensitive to errors in climate data. Thus, the Hargreaves method has advantages in an area for which data is scarce, such as Africa [37].

PET is affected by many climatic factors. Liu et al. [56] analyzed the sensitivity of PET to meteorological data in China for the period 1960–2007. They found that the particular factor which is most sensitive differs across the country but nation-wide the most sensitive factor was, on average, vapor pressure. They also found some correlation between factor sensitivity and elevation. Yao et al. [57] used meteorological reanalysis datasets from the Environmental and Ecological Science Data Center for West China and found that solar radiation was the largest contributor to change in PET, and that wind speed most affected inter-annual variation of PET in China. Xu et al. [58] found that as well as solar radiation, atmospheric dynamics also strongly influence PET. Vegetation degradation in many regions of China is highly correlated with thermodynamic and physical land surface changes, which intensify the uneven spatial distribution of PET in China [59]. Gao et al. [60] analyzed PET from 580 stations in China for 1956–2000 and obtained similar results to ours: less solar radiation and decreased wind speed are major causes of reduced PET in most areas; and solar radiation, wind speed, and relative humidity have a greater effect than temperature on PET.

The overestimation of PET using the CMADS datasets is mainly due to the effect of solar radiation. Different methods in obtaining solar radiation need to be addressed. Solar radiation was calculated by Ångström–Prescott radiation equation for the station data, and Solar radiation of CMADS were obtained from radiance data of the International Satellite Cloud Climatology Project (ISCCP) with combination of data retrieved from the FY-2E satellite using the discrete-ordinate radiative transfer (DISTORT) model. For the satellite derived solar radiation, visible-band observations obtained from the FY2C geostationary meteorological satellite are used to generate hourly ground-incident solar radiation data with spatial resolution of $0.1° \times 0.1°$. The discrete ordinate method [61] was used to calculate radiation transfer in the inversion algorithm for the ground-incident solar radiation output. A 5-layer planoparallel ideal atmospheric model nonuniform in the vertical direction was designed, which consists of five solar spectral intervals (0.2–0.4, 0.4–0.5, 0.5–0.6, 0.6–0.7, 0.7–4.0 μm) to calculate the scattering, absorption, and reflection of solar radiation [39]. Although satellite-derived radiation has better spatial resolution, a study in Nigeria found that estimated solar radiation from the radiation equation model has a better error range and fits the ground measured data better than the satellite-derived data [62]. The ground measured, model estimated and satellite-derived solar radiation data can complement each other.

PET is important in water resources management and hydrological modeling. Overestimation of PET can lead to overestimating the severity of drought. Simulated discharge in hydrological models can also be affected by overestimated PET. Parmele [63] used a Hiemstra watershed model and two versions of the Stanford model and found that a constant bias of 20% in PET input data has a cumulative effect and results in considerable error in the computed hydrograph peaks and recessions. Other studies have found that parameter calibration can reduce errors from input data to some degree [64]. Oudin et al. [65] found that systematic errors in PET predictions have a greater impact than random errors, but that such errors are reduced by soil moisture accounting (SMA) using the GR4J model and TOPMODEL.

5. Conclusions

The CMADS reanalysis dataset is a useful alternative to observed weather data, especially in remote areas where observations are not easy to make. Evaluating PET calculated from the reanalysis dataset by comparing it with China-wide observations is important for the applicability of the CMADS dataset. This study used observed data from 836 weather stations across China to evaluate PET estimated by PM using CMADS data (PET_cma) by spatiotemporal comparison with PET_obs.

For the average annual PET, PET_cma and PET_obs agree well in their spatial distribution for most of China. PET_cma is an overestimation, compared to PET_obs, in western inland China with percentage bias in the range 15% to 30%. Average annual PET_obs is 1000 mm while average annual

Water 2018, 10, 1126

PET_cma is 1120 mm. Mean seasonal PET is overestimated, in comparison to PET_obs, for all four seasons, although the spatial distribution of PET_cma captures the general seasonal features. Average percentage bias is least in winter at 4.3% and greatest in spring at 15.7%. Mean seasonal PET estimates differ from mean annual PET estimates. In spring there are more stations in southeastern China for which PET_cma greatly overestimates. The percentage bias for a number of stations is mainly in the range 15% to 25%. In winter, there is overestimation in the south and underestimation in the north. The statistical measures NRMSE, R^2, and S_{score} consistently show that the annual PET_cma values are better than those at shorter time scales when compared with PET_obs.

Wind speed and solar radiation are the major variables that contribute to the errors in PET_cma but they each influence estimated PET in a different way. Wind speed causes an underestimation of PET with a percentage bias in the range −15% to −5%, with the largest errors being found in eastern China. Solar radiation causes an overestimation in the range 15% to 30% in central and western China. A larger percentage bias due to wind speed is found mainly at elevations below 2000 m while the larger percentage bias due to solar radiation is spread evenly across elevations. Underestimation of PET due to wind speed and overestimation of PET due to solar radiation are offset, reducing the overestimation of PET.

Author Contributions: Y.T. conceived and designed the experiments and wrote the paper. K.Z. performed the experiments. X.G. performed data curation. Y.-P.X. and J.W. reviewed and edited the paper.

Funding: This work was financially supported by the National Key Research and Development Programs of China (Grant No. 2016YFA0601501), the National Natural Science Foundation of China (Grant No. 51709148; 41501029), and NUIST research startup fund (Grant No. 2016r12).

Acknowledgments: China Meteorological Administration and CMADS from www.cmads.org.

Conflicts of Interest: The authors declare no conflict of interest.

References

1. Xu, C. Modelling the effects of climate change on water resources in central Sweden. *Water Resour. Manag.* 2000, *14*, 177–189. [CrossRef]
2. Tsakiris, G.; Vangelis, H. Establishing a drought index incorporating evapotranspiration. *Eur. Water* 2005, *9*, 3–11.
3. Miao, Q.; Rosa, R.D.; Shi, H.; Paredes, P.; Zhu, L.; Dai, J.; Gonçalves, J.M.; Pereira, L.S. Modeling water use, transpiration and soil evaporation of spring wheat–maize and spring wheat–sunflower relay intercropping using the dual crop coefficient approach. *Agric. Water Manag.* 2016, *165*, 211–229. [CrossRef]
4. Sprenger, M.; Leistert, H.; Gimbel, K.; Weiler, M. Illuminating hydrological processes at the soil–vegetation–atmosphere interface with water stable isotopes. *Rev. Geophys.* 2016, *54*, 674–704. [CrossRef]
5. Li, Z.; Zhang, Y.; Wang, S.; Yuan, G.; Yang, Y.; Cao, M. Evapotranspiration of a tropical rain forest in Xishuangbanna, southwest China. *Hydrol. Process.* 2010, *24*, 2405–2416. [CrossRef]
6. Zhang, B.; Kang, S.; Li, F.; Zhang, L. Comparison of three evapotranspiration models to Bowen ratio-energy balance method for a vineyard in an arid desert region of northwest China. *Agric. For. Meteorol.* 2008, *148*, 1629–1640. [CrossRef]
7. Xu, C.; Chen, D. Comparison of seven models for estimation of evapotranspiration and groundwater recharge using lysimeter measurement data in Germany. *Hydrol. Process.* 2005, *19*, 3717–3734. [CrossRef]
8. Samain, B.; Pauwels, V.R.N. Impact of potential and (scintillometer-based) actual evapotranspiration estimates on the performance of a lumped rainfall–runoff model. *Hydrol. Earth Syst. Sci.* 2013, *17*, 4525–4540. [CrossRef]
9. Audrézet, M.P.; Robaszkiewicz, M.; Mercier, B.; Nousbaum, J.B.; Bail, J.P.; Hardy, E.; Volant, A.; Lozac'H, P.; Charles, J.F.; Gouéron, H. Uncertainty analysis of computational methods for deriving sensible heat flux values from scintillometer measurements. *Atmos. Meas. Tech.* 2009, *2*, 741–753.
10. Wang, K.; Dickinson, R.E. A review of global terrestrial evapotranspiration: Observation, modeling, climatology, and climatic variability. *Rev. Geophys.* 2012, *50*. [CrossRef]
11. Zhang, K.; Kimball, J.S.; Running, S.W. A review of remote sensing based actual evapotranspiration estimation: A review of remote sensing evapotranspiration. *WIREs Water* 2016, *3*, 834–853. [CrossRef]

12. Long, D.; Longuevergne, L.; Scanlon, B.R. Uncertainty in evapotranspiration from land surface modeling, remote sensing, and GRACE satellites. *Water Resour. Res.* **2014**, *50*, 1131–1151. [CrossRef]

13. Meng, X.; Ji, X.; Liu, Z.; Xiao, J.; Chen, X.; Wang, F. Research on improvement and application of snowmelt module in SWAT. *J. Nat. Resour.* **2014**, *29*, 528–539.

14. Zhao, L.; Xia, J.; Xu, C.; Wang, Z.; Leszek, S.; Long, C. Evapotranspiration estimation methods in hydrological models. *J. Geogr. Sci.* **2013**, *23*, 359–369. [CrossRef]

15. Lu, J.; Sun, G.; McNulty, S.G.; Amatya, D.M. A comparison of six potential evapotranspiration methods for regional use in the southeastern United States. *J. Am. Water Resour. Assoc.* **2005**, *41*, 621–633. [CrossRef]

16. Muniandy, J.M.; Yusop, Z.; Askari, M. Evaluation of reference evapotranspiration models and determination of crop coefficient for Momordica charantia and Capsicum annuum. *Agric. Water Manag.* **2016**, *169*, 77–89. [CrossRef]

17. Tabari, H.; Grismer, M.E.; Trajkovic, S. Comparative analysis of 31 reference evapotranspiration methods under humid conditions. *Irrig. Sci.* **2013**, *31*, 107–117. [CrossRef]

18. Paparrizos, S.; Maris, F.; Matzarakis, A. Sensitivity analysis and comparison of various potential evapotranspiration formulae for selected Greek areas with different climate conditions. *Theor. Appl. Climatol.* **2017**, *128*, 745–759. [CrossRef]

19. Arnold, J.G.; Srinivasan, R.; Muttiah, R.S.; Williams, J.R. Large area hydrologic modeling and assessment part I: Model development. *J. Am. Water Resour. Assoc.* **1998**, *34*, 73–89. [CrossRef]

20. Liang, X.; Xie, Z.; Huang, M. A new parameterization for surface and groundwater interactions and its impact on water budgets with the variable infiltration capacity (VIC) land surface model. *J. Geophys. Res. Atmos.* **2003**, *108*. [CrossRef]

21. Abbott, M.B.; Bathurst, J.C.; Cunge, J.A.; O'Connell, P.E.; Rasmussen, J. An introduction to the European hydrological system—Systeme Hydrologique Europeen, "SHE", 1: History and philosophy of a physically-based, distributed modelling system. *J. Hydrol.* **1986**, *87*, 45–59. [CrossRef]

22. Lang, D.; Zheng, J.; Shi, J.; Liao, F.; Ma, X.; Wang, W.; Chen, X.; Zhang, M. A comparative study of potential evapotranspiration estimation by eight methods with FAO Penman–Monteith method in southwestern China. *Water* **2017**, *9*, 734. [CrossRef]

23. Kite, G.W.; Droogers, P. Comparing evapotranspiration estimates from satellites, hydrological models and field data. *J. Hydrol.* **2000**, *229*, 3–18. [CrossRef]

24. Saha, S.; Moorthi, S.; Pan, H.-L.; Wu, X.; Wang, J.; Nadiga, S.; Tripp, P.; Kistler, R.; Woollen, J.; Behringer, D. The NCEP climate forecast system reanalysis. *Bull. Am. Meteorol. Soc.* **2010**, *91*, 1015–1058. [CrossRef]

25. Kanamitsu, M.; Ebisuzaki, W.; Woollen, J.; Yang, S.-K.; Hnilo, J.J.; Fiorino, M.; Potter, G.L. NCEP–DOE AMIP-II Reanalysis (R-2). *Bull. Am. Meteorol. Soc.* **2002**, *83*, 1631–1644. [CrossRef]

26. Kalnay, E.; Kanamitsu, M.; Kistler, R.; Collins, W.; Deaven, D.; Gandin, L.; Iredell, M.; Saha, S.; White, G.; Woollen, J.; et al. The NCEP/NCAR 40-year reanalysis project. *Bull. Am. Meteorol. Soc.* **1996**, *77*, 437–472. [CrossRef]

27. Bromwich, D.H.; Wang, S.-H. Evaluation of the NCEP–NCAR and ECMWF 15- and 40-yr reanalysis using rawinsonde data from two independent Arctic field experiments. *Mon. Weather Rev.* **2005**, *133*, 3562–3578. [CrossRef]

28. Uppala, S.M.; Kållberg, P.W.; Simmons, A.J.; Andrae, U.; Bechtold, V.D.C.; Fiorino, M.; Gibson, J.K.; Haseler, J.; Hernandez, A.; Kelly, G.A. The ERA-40 re-analysis. *Q. J. R. Meteorol. Soc.* **2005**, *131*, 2961–3012. [CrossRef]

29. Dee, D.P.; Uppala, S.M.; Simmons, A.J.; Berrisford, P.; Poli, P.; Kobayashi, S.; Andrae, U.; Balmaseda, M.A.; Balsamo, G.; Bauer, P. The ERA-Interim reanalysis: Configuration and performance of the data assimilation system. *Q. J. R. Meteorol. Soc.* **2011**, *137*, 553–597. [CrossRef]

30. Ebita, A.; Kobayashi, S.; Ota, Y.; Moriya, M.; Kumabe, R.; Onogi, K.; Harada, Y.; Yasui, S.; Miyaoka, K.; Takahashi, K.; et al. The Japanese 55-year reanalysis: An interim report. *Sola* **2011**, *7*, 149–152. [CrossRef]

31. Rienecker, M.M.; Suarez, M.J.; Gelaro, R.; Todling, R.; Bacmeister, J.; Liu, E.; Bosilovich, M.G.; Schubert, S.D.; Takacs, L.; Kim, G.-K.; et al. MERRA: NASA's Modern-Era Retrospective Analysis for Research and Applications. *J. Clim.* **2011**, *24*, 3624–3648. [CrossRef]

32. Bao, X.; Zhang, F. Evaluation of NCEP–CFSR, NCEP–NCAR, ERA-Interim, and ERA-40 reanalysis datasets against independent sounding observations over the Tibetan plateau. *J. Clim.* **2013**, *26*, 206–214. [CrossRef]

33. Lindsay, R.; Wensnahan, M.; Schweiger, A.; Zhang, J. Evaluation of seven different atmospheric reanalysis products in the Arctic. *J. Clim.* **2014**, *27*, 2588–2606. [CrossRef]

34. Ma, L.; Zhang, T.; Li, Q.; Frauenfeld, O.W.; Qin, D. Evaluation of ERA-40, NCEP-1, and NCEP-2 reanalysis air temperatures with ground-based measurements in China. *J. Geophys. Res. Atmos.* **2008**, *113*. [CrossRef]

35. Weiland, F.C.S.; Tisseuil, C.; Dürr, H.H.; Vrac, M.; Beek, L.P.H.V. Selecting the optimal method to calculate daily global reference potential evaporation from CFSR reanalysis data for application in a hydrological model study. *Hydrol. Earth Syst. Sci.* **2012**, *16*, 983–1000. [CrossRef]

36. Srivastava, P.K.; Han, D.; Ramirez, M.A.R.; Islam, T. Comparative assessment of evapotranspiration derived from NCEP and ECMWF global datasets through weather research and forecasting model. *Atmos. Sci. Lett.* **2013**, *14*, 118–125. [CrossRef]

37. Trambauer, P.; Dutra, E.; Maskey, S.; Werner, M.; Pappenberger, F.; Van Beek, L.P.H.; Uhlenbrook, S. Comparison of different evaporation estimates over the African continent. *Hydrol. Earth Syst. Sci.* **2014**, *18*, 193–212. [CrossRef]

38. Meng, X.; Wang, H.; Wu, Y.; Long, A.; Wang, J.; Shi, C.; Ji, X. Investigating spatiotemporal changes of the land-surface processes in Xinjiang using high-resolution CLM3.5 and CLDAS: Soil temperature. *Sci. Rep.* **2017**, *7*, 13286. [CrossRef] [PubMed]

39. Shi, C.; Xie, Z.; Hui, Q.; Liang, M.; Yang, X. China land soil moisture ENKF data assimilation based on satellite remote sensing data. *Sci. China Earth Sci.* **2011**, *54*, 1430–1440. [CrossRef]

40. Zhang, T. Multi-Source Data Fusion and Application Research Base on LAPS/STMAS. Master's Thesis, Nanjing University of Information Science & Technology, Nanjing, China, 2013. (In Chinese)

41. Meng, X.; Wang, H. Significance of the China Meteorological Assimilation Driving Datasets for the SWAT Model (CMADS) of East Asia. *Water* **2017**, *9*, 765. [CrossRef]

42. Meng, X.; Wang, H.; Lei, X.; Cai, S.; Wu, H.; Ji, X.; Wang, J. Hydrological modeling in the Manas river basin using soil and water assessment tool driven by CMADS. *Teh. Vjesn.* **2017**, *24*, 525–534.

43. Meng, X.; Yu, D.; Liu, Z. Energy balance-based swat model to simulate the mountain snowmelt and runoff—Taking the application in Juntanghu watershed (China) as an example. *J. Mt. Sci.* **2015**, *12*, 368–381. [CrossRef]

44. Meng, X.; Wang, H.; Cai, S.; Zhang, X.; Leng, G.; Lei, X.; Shi, C.; Liu, S.; Shang, Y. The China meteorological assimilation driving datasets for the SWAT model (CMADS) application in China: A case study in Heihe river basin. *Preprints* **2016**, *37*, 1–19.

45. Zhao, F.; Wu, Y. Parameter uncertainty analysis of the SWAT Model in a Mountain Loess transitional watershed on the Chinese Loess Plateau. *Water* **2018**, *10*, 690. [CrossRef]

46. Vu, T.T.; Li, L.; Jun, K.S. Evaluation of multi satellite precipitation products for streamflow simulations: A case study for the Han river basin in the Korean peninsula, east Asia. *Water* **2018**, *10*, 642. [CrossRef]

47. Liu, J.; Shanguan, D.; Liu, S.; Ding, Y. Evaluation and hydrological simulation of CMADS and CFSR reanalysis datasets in the Qinghai Tibet Plateau. *Water* **2018**, *10*, 513. [CrossRef]

48. Cao, Y.; Zhang, J.; Yang, M. Application of SWAT model with CMADS data to estimate hydrological elements and parameter uncertainty based on SUFI-2 Algorithm in the Lijiang river basin, China. *Water* **2018**, *10*, 742. [CrossRef]

49. Shao, G.; Guan, Y.; Zhang, D.; Yu, B.; Zhu, J. The impacts of climate variability and land use change on streamflow in the Hailiutu river basin. *Water* **2018**, *10*, 814. [CrossRef]

50. Allen, R.G.; Pereira, L.S.; Raes, D.; Smith, M. *Crop Evapotranspiration—Guidelines for Computing Crop Water Requirements*; FAO Irrigation and Drainage Paper 56; Food and Agriculture Organization of the United Nations (FAO): Rome, Italy, 1998.

51. Monteith, J.L. Evaporation and environment. *Symp. Soc. Exp. Biol.* **1965**, *19*, 205–234. [PubMed]

52. Gong, L.; Xu, C.; Chen, D.; Halldin, S.; Chen, Y. Sensitivity of the Penman–Monteith reference evapotranspiration to key climatic variables in the Changjiang (Yangtze River) basin. *J. Hydrol.* **2006**, *329*, 620–629. [CrossRef]

53. Yang, Y.; Cui, Y.; Luo, Y.; Lyu, X.; Traore, S.; Khan, S.; Wang, W. Short-term forecasting of daily reference evapotranspiration using the Penman-Monteith model and public weather forecasts. *Agric. Water Manag.* **2016**, *177*, 329–339. [CrossRef]

54. Almorox, J.; Senatore, A.; Quej, V.H.; Mendicino, G. Worldwide assessment of the Penman–Monteith temperature approach for the estimation of monthly reference evapotranspiration. *Theor. Appl. Climatol.* **2016**, *131*, 1–11. [CrossRef]

55. Kang, S.; Kim, S.; Lee, D. Spatial and temporal patterns of solar radiation based on topography a. *Can. J. For. Res.* **2002**, *32*, 487–497. [CrossRef]

56. Droogers, P.; Allen, R.G. Estimating reference evapotranspiration under inaccurate data conditions. *Irrig. Drain. Syst.* **2002**, *16*, 33–45. [CrossRef]

57. Liu, C.; Zhang, D.; Liu, X.; Zhao, C. Spatial and temporal change in the potential evapotranspiration sensitivity to meteorological factors in China (1960–2007). *J. Geogr. Sci.* **2012**, *22*, 3–14. [CrossRef]

58. Yao, Y.; Zhao, S.; Zhang, Y.; Jia, K.; Liu, M. Spatial and decadal variations in potential evapotranspiration of China based on reanalysis datasets during 1982–2010. *Atmosphere* **2014**, *5*, 737–754. [CrossRef]

59. Xu, X. Analyzing potential evapotranspiration and climate drivers in China. *Chin. J. Geophys.* **2011**, *54*, 125–134. [CrossRef]

60. Gao, G.; Chen, D.; Ren, G.; Chen, Y.; Liao, Y. Spatial and temporal variations and controlling factors of potential evapotranspiration in China: 1956–2000. *J. Geogr. Sci.* **2006**, *16*, 3–12. [CrossRef]

61. Stamnes, K.; Tsay, S.C.; Wiscombe, W.; Jayaweera, K. Numerically stable algorithm for discrete ordinate method radiative transfer in multiple scattering and emitting layered media. *Appl. Opt.* **1988**, *27*, 502–509. [CrossRef] [PubMed]

62. Olomiyesan, B.M.; Oyedum, O.D. Comparative study of ground measured, satellite-derived, and estimated global solar radiation data in Nigeria. *J. Sol. Energy* **2016**, *3*, 1–7. [CrossRef]

63. Parmele, L.H. Errors in output of hydrologic models due to errors in input potential evapotranspiration. *Water Resour. Res.* **1972**, *8*, 348–359. [CrossRef]

64. Andréassian, V.; Perrin, C.; Michel, C. Impact of imperfect potential evapotranspiration knowledge on the efficiency and parameters of watershed models. *J. Hydrol.* **2004**, *286*, 19–35. [CrossRef]

65. Oudin, L.; Perrin, C.; Mathevet, T.; Andréassian, V.; Michel, C. Impact of biased and randomly corrupted inputs on the efficiency and the parameters of watershed models. *J. Hydrol.* **2006**, *320*, 62–83. [CrossRef]

Article

An Integrated Methodology to Analyze the Total Nitrogen Accumulation in a Drinking Water Reservoir Based on the SWAT Model Driven by CMADS: A Case Study of the Biliuhe Reservoir in Northeast China

Guoshuai Qin [1], Jianwei Liu [1], Tianxiang Wang [1,2,*], Shiguo Xu [1] and Guangyu Su [1]

[1]   The Institution of Water and Environment Research, Dalian University of Technology, Dalian 116024, China; qgs1991@mail.dlut.edu.cn (G.Q.); jwliu@dlut.edu.cn (J.L.); sgxu@dlut.edu.cn (S.X.); sugy@mail.dlut.edu.cn (G.S.)

[2]   China Water Resources Pearl River Planning Surveying & Designing Co, Ltd., Guangzhou 510610, China

*   Correspondence: tianxiang@dlut.edu.cn; Tel.: +86-138-4285-7691

Received: 26 September 2018; Accepted: 26 October 2018; Published: 27 October 2018

Abstract: Human activities, especially dam construction, have changed the nutrient cycle process at the basin scale. Reservoirs often act as a sink in the basin and more nutrients are retained due to sedimentation, which induces the eutrophication of the surface water system. This paper proposes an integrated methodology to analyze the total nitrogen (TN) accumulation in a drinking water reservoir, based on the soil and water assessment tool (SWAT) model driven by the China Meteorological Assimilation Driving Datasets for the SWAT model (CMADS). The results show that the CMADS could be applied to drive the SWAT model in Northeast China. The dynamic process of TN accumulation indicates that the distribution of TN inputted into the reservoir fluctuated with the dry and wet seasons from 2009–2016, which was mainly governed by the amount of runoff. The annual average TN input and output fluxes of the Biliuhe reservoir were 274.41×10^4 kg and 217.14×10^4 kg, which meant that 19.76% of the TN input accumulated in the reservoir. Higher TN accumulation in the reservoir did not correspond to a higher TN load, due to the influence of flood discharge and the water supply. Interestingly, a higher TN accumulation efficiency was observed in normal hydrological years, because the water source reservoir always stores most of the water input for future multiple uses but rarely discharges surplus water. The non-point sources from fertilizer and atmospheric deposition and soils constituted the highest proportion of the TN input, accounting for 35.15%, 30.15%, and 27.72% of the average input. The DBWD (Dahuofang reservoir to Biliuhe reservoir water diversion) project diverted 32.03×10^4 kg year^{-1} TN to the Biliuhe reservoir in 2015–2016, accounting for 14.05% of the total annual input. The discharge output and the BDWD (Biliuhe reservoir to Dalian city water diversion) project output accounted for 48.75% and 47.74%, respectively. The effects of inter-basin water diversion projects should be of great concern in drinking water source water system management. There was a rising trend of TN level in the Biliuhe reservoir, which increases the eutrophication risk of the aquatic ecosystem. The TN accumulated in the sediment contributed to a large proportion of the TN accumulated in the reservoir. In addition to decreasing the non-point source nitrogen input from the upper basin, discharging anoxic waters and sediment with a high nitrogen concentration through the bottom hole of the dam could alleviate the nitrogen pollution in the Biliuhe reservoir.

Keywords: total nitrogen; accumulation; SWAT model; CMADS; Biliuhe reservoir

1. Introduction

On a global scale, human activities—such as agricultural fertilization, domestic and industrial sewage discharge, and fossil fuel combustion—have dramatically increased the nutrient loads transported by rivers, which has resulted in severe eutrophication problems in aquatic ecosystems [1,2]. Reservoirs, acting as the important engineering controls in rivers, have greatly changed the nutrient cycles in the basin. The building of dams impedes the transport of pollutants and nutrients from basins to oceans by rivers [3]. Due to the long water residence time (compared with rivers) and the enhanced particle settling velocity, pollutants will deposit in the reservoir along with other sediment [4]. Additionally, water source reservoirs always store most of the water input for future multiple uses, but rarely discharge surplus water, which leads to the accumulation of pollutants. After decades of operation, the contamination in the reservoir water and sediment become increasingly noticeable, until finally this limits the function of the water supply [5]. China is the country with the largest number of reservoirs, which has built 98,002 reservoirs with a total storage capacity of 932 billion m^3 to fulfill the growing demands for flood control, water resources, and power generation [6]. Due to rapid social and economic development in recent decades, the eutrophication problem of reservoirs is aggravated in China. The eutrophication status assessment of 943 reservoirs across the country demonstrates that 28.8% of the reservoirs are in a eutrophic state [7]. In addition, reservoirs in many other countries are also suffering the eutrophication and water quality deterioration problems [8–10].

Nutrient accumulation in reservoirs occurs when nutrients are retained due to the construction of a dam, which could be calculated by the nutrient input flux and output flux. A wide range of nitrogen retention rates have been reported in existing studies, which implies that the process of nutrient transportation varies in different basins [11,12]. Nutrient transportation in the basin is complicated, and sometimes the inter-basin water diversion project makes quantitative analysis more difficult. Physical-based models, such as the soil and water assessment tool (SWAT), agricultural non-point source pollution model (AGNPS), and Hydrological Simulation Program—Fortran (HSPF) have been developed to evaluate the nutrient loads since the 1970s [13–15]. The SWAT model has been widely used to estimate the non-point pollution yields, and the model performance has been confirmed in many typical basins [16]. However, the model requires high-resolution input data, especially meteorological data, which is the important drive factor of the nutrient cycle [17]. Sparse spatial data, measurement errors, and the sensitivity of the hydrologic parameters would limit the model accuracy [18]. In recent years, various atmospheric reanalysis datasets such as the JRA-55, the ERA-Interim, the CFSR, and the MERRA have been developed and used globally [19–22]. The China Meteorological Assimilation Driving Datasets for the SWAT model (CMADS) are the latest East Asia atmospheric reanalysis datasets developed by Dr. Xianyong Meng from the China Agricultural University (CAU), which have attracted widespread attention [23–25]. The CMADS series of datasets have been verified in different basins of China and Korea and have performed well in the Heihe basin, Manas River Basin, Qinghai–Tibet Plateau, Han River Basin and so on, however the application of CMADS mainly focuses on hydrological simulation and there are few studies about non-point source pollution simulation driven by the datasets, especially in the cold regions of Northeast China [26–33].

Excessive nitrogen is one of the main problems facing surface freshwater systems today [34]. There is widespread research on the monitoring, simulation and assessment of nitrogen in reservoirs [35,36]. However, the quantitative analysis of the dynamic accumulation process of nitrogen in the reservoirs is difficult because of the various nitrogen sources and complicated transportations. Dalian city is one of the most important cities in Northeast China, which is also a water-deficient city. Biliuhe reservoir, the drinking water reservoir of Dalian city, is facing severe nitrogen pollution and the concentration of TN has exceeded Grade V in China's water quality standard in recent years [37]. A systematic water transfer network has been built to ease the shortage of water resources in the Dalian area, which brings about new issues in environmental management between the different basins. In the Biliuhe reservoir basin, there are some studies about the runoff but few about nutrient transport process [38,39]. The existing research on nitrogen accumulation of the Biliuhe reservoir is based on an empirical model and the results are static and simple, so further analysis is necessary [40]. The effects of the inter-basin water diversion project on the reservoir total nitrogen (TN) accumulation are also rarely studied. The aims of this paper are (1) to propose an integrated methodology for analysis of the reservoir TN accumulation based on the SWAT model driven by CMADS, and (2) to analyze the dynamic process of the TN accumulation in the Biliuhe reservoir.

2. Materials and Methods

2.1. Study Area

The Biliuhe reservoir ($122°29'24.11''$ N; $39°49'12.52''$ E), located 175 km northeast of Dalian city, is a large canyon-shaped reservoir with multiple uses: water supply, flood control, power generation, and agricultural irrigation. The maximum storage capacity and the effective storage capacity of the reservoir are 9.34×10^8 m^3 and 6.44×10^8 m^3, respectively. The mean surface water area of the reservoir is 55.60 km^2 and the mean water depth is 12.84 m. With a designed annual water supply of 4.38×10^8 m^3, the Biliuhe reservoir accounts for 80% of the domestic and industrial water supply for Dalian city. The Yushi reservoir, located in the upper stream of the Biliuhe River, has a storage capacity of 0.88×10^8 m^3. There are two main inter-basin water diversion projects in the basin: the Biliuhe reservoir to Dalian city water diversion project (BDWD), and the Dahuofang reservoir to Biliuhe reservoir water diversion project (DBWD). The Biliuhe reservoir to Dalian city water diversion project began to divert water in 1984, and the designed water diversion of the project is 4.38×10^8 m^3 year^{-1}. The Dahuofang reservoir to Biliuhe reservoir inter-basin water diversion project was completed in October of 2014. The designed water diversion of the project is 3.29×10^8 m^3 year^{-1}. By the end of 2016, 3.2×10^8 m^3 water had been transferred into Biliuhe reservoir to alleviate the severe drought in the Dalian area.

The reservoir basin reaches an area of about 2085 km^2, with the three main tributary rivers being Biliuhe River, Geli River, and Bajia River (Figure 1). The reservoir catchment has a temperate monsoon climate characteristic, with an annual average temperature of 10.6 °C, hot summers and cold winters. The annual average precipitation is 743 mm and 75% of the precipitation is centralized in the flood period (June–September). The icebound season of the Biliuhe reservoir is long and lasts from November–March. The upper catchment of the reservoir is mainly covered by forest (72%) and agricultural land (19%), which can be seen from the Figure 2. There are about 256,000 residents in the basin and they mostly live along the bank of the river. Agriculture dominates the economic development in the basin.

2.2. SWAT Model and Data Source

The SWAT (soil and water assessment tool) model is a basin-scale, semi-distributed, and physically based model developed by the United States Department of Agriculture Research Service in the early 1990s [41]. It has been widely applied to predict the impacts of land management practices on water, sediment, and nutrient loss in large, complex basins. The model partitions the basin into multiple sub-basins, which are further divided into hydrological response units (HRUs) consisting of homogeneous land use, soil characteristic, and slope. The hydrological sub-model is based on the water-balance equation to simulate the processes of precipitation, infiltration, surface runoff, evapotranspiration, lateral flow, and percolation [42]. In this study, the Biliuhe reservoir upper basin was divided into 99 sub-basins and 805 HRUs. The simulation of nutrients in the catchment was fully considered in the nutrients module, including migration and transformation in soil, the transportation process from the upper catchment to downstream through the main river channel, and the surface runoff or interflow [19]. The SWAT model can also simulate the water and nutrient transfer between reservoirs, the reach, and sub-basins, or the water transfer between different basins.

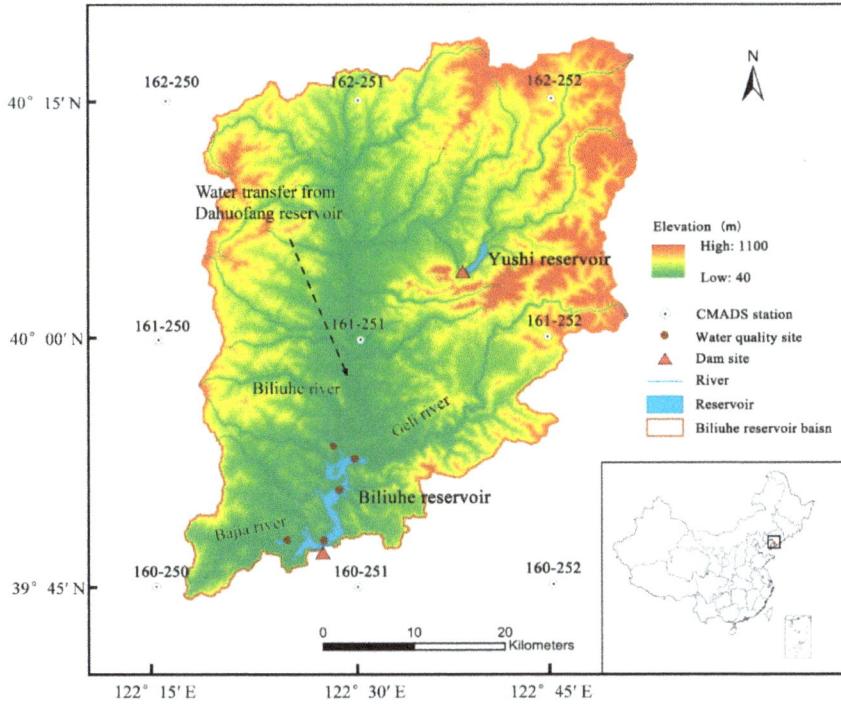

Figure 1. Geography of the Biliuhe reservoir basin.

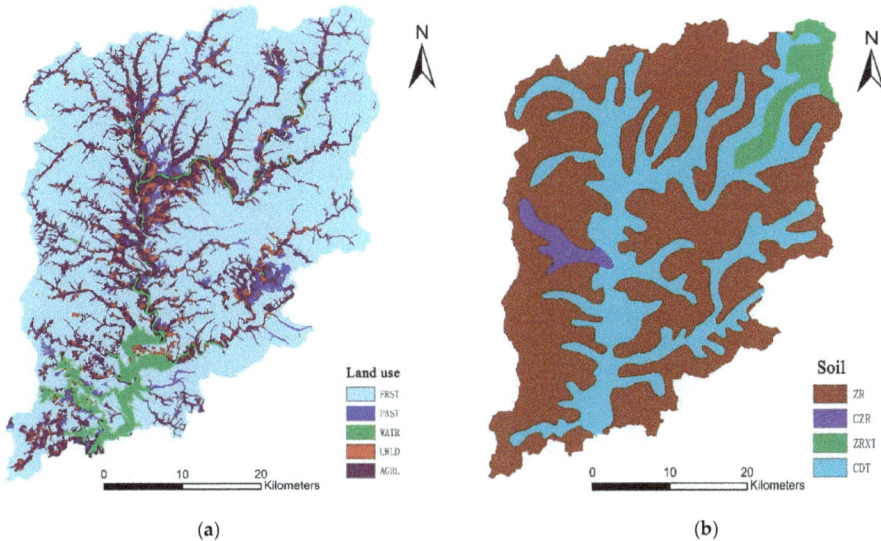

(a)   (b)

Figure 2. Land use types (**a**) and soil types (**b**) of the Biliuhe reservoir basin. Land use classes correspond to FRST (frost), PAST (pasture), WATR (water), URLD (low-density residential), and AGRL (agricultural land). Soil classes correspond to ZR (brown earth), CZR (young brown earth), ZRXT (meadow brown earth), and CDT (meadow soil).

The version of SWAT used in this study is the ArcSWAT 2012 (Texas A&M University, College Station, TX, USA), which is an ArcGIS-ArcView extension and graphical user input interface for the SWAT model. The spatial data used in the SWAT model includes digital elevation data (DEM), land cover data, and soil type data. The 90×90 m DEM data was obtained from the International Scientific & Technical Data Mirror Site, Computer Network Information Center, Chinese Academy of Sciences (http://www.gscloud.cn). The land cover data of the 2000s (1:100,000) and soil information map (1:1,000,000) were provided by the Data Center for Resources and Environmental Sciences, Chinese Academy of Sciences (RESDC) (http://www.resdc.cn). The land cover in the study area could be classified into five types: forest (72.30%), farmland (18.94%), grassland (2.40%), low-density residential (2.76%), and water body (3.60%). There are four soil types distributed within the basin: brown earth, young brown earth, meadow brown earth, and meadow soil (Figure 2). The main soil attributes were obtained from the China Soil Database or calculated by the soil–plant–atmosphere–water (SPAW) model. Some attributes that could not be obtained were taken from the soil database of the SWAT model. The soil particle size transformation from an international system to the US system was completed by the cubic spline interpolation method. The basic data of fertilization per hectare, manure of livestock and poultry breeding, and rural domestic sewage came from the report of the investigation of pollution sources and water quality in the Biliuhe reservoir (2012). After an investigation, there was found to be nine point-source pollution outlets within the catchment: Dalian Buyunshan hot spring bath center, Dalian Guanba Silk Food Co., Ltd., Dalian Jiantang hot spring bath center, Dalian Jiantang hot spring bath center east sewage, Dalian Xingzhi Canned Food Co., Ltd., Yingkou Epps Sewing Products Co., Ltd., Shizijie Town comprehensive sewage outlet, Wanfu Town east comprehensive sewage outlet, and Wanfu Town south comprehensive sewage outlet [40]. The regulation of the Yushi reservoir could be simulated directly by the reservoir module of the SWAT model. The model calibration, sensitivity analysis, and uncertainty analysis were achieved by the Sufi-2 (Sequential Uncertainty Fitting, version 2) algorithm, which was coupled with the SWAT-CUP. The model performance was evaluated by the coefficient of determination (R^2), Nash–Sutcliffe simulation efficiency (NSE), and percent bias (PBIAS), which are defined as [43]

$$R^2 = \sum_{i=1}^{n} (O_i - O_{avg})(S_i - S_{avg}) / \left\{ \left[\sum_{i=1}^{n} (O_i - O_{avg})^2 \right] \left[\sum_{i=1}^{n} (S_i - S_{avg})^2 \right] \right\}^{\frac{1}{2}} \tag{1}$$

$$NSE = 1 - \sum_{i=1}^{n} (O_i - S_i)^2 / \sum_{i=1}^{n} (O_i - O_{avg})^2 \tag{2}$$

$$PBIAS = \sum_{i=1}^{n} (O_i - S_i) / \sum_{i=1}^{n} (O_i) \times 100\% \tag{3}$$

where O_i and S_i are the observed and simulated data, respectively. O_{avg} and S_{avg} are the average values of the observed and simulated data, while n is the total number of data records. The discharge and TN data required for calibration and validation were derived from the Biliuhe reservoir management bureau, and the TN concentration was monitored by the national standard method.

The meteorological data used in the SWAT model was obtained from the China Meteorological Assimilation Driving Datasets for the SWAT model (CMADS V1.1), which can be downloaded from the website of Cold and Arid Regions Sciences Data Center (www.cmads.org). The CMADS integrated the air temperature, air pressure, humidity, and wind velocity data through the LAPS/STMAS system and other multiple techniques, such as data loop nesting, resampling, pattern estimation, and bilinear interpolation. Precipitation data were assimilated using the CMORPH's global precipitation product and the National Meteorological Information Center's data of China [23]. The CMADS V1.1 serials of datasets cover the entire East Asian region ($0°$–$65°$ N, $60°$–$160°$ E) and can provide high-resolution and high-quality meteorological data for the study area with sparse weather station coverage. In this study, the SWAT model made use of the meteorological data of nine CMADS stations in the study area, which included precipitation, relative humidity, solar radiation, temperature, and wind speed. The spatial resolution is 0.25 degrees, the time resolution is daily, and the time scale is 2008–2016. Detailed location information of the stations is shown in Figure 1.

2.3. Integrated Methodology of TN Accumulation

The TN load simulated by the SWAT model is the TN flux transported from the upper basin into the Biliuhe reservoir through the surface and underground runoff, which includes the point source input, non-point source input, and inter-basin water diversion input. Because the SWAT cannot properly conduct reservoir nitrogen simulations in general, it was mainly used to simulate the TN load of the Biliuhe reservoir upper watershed in this study, and the study reservoir was only taken as an outlet of the basin. To better study the nitrogen pollution of the Biliuhe reservoir, an integrated methodology based on the mass balance theory is proposed to calculate the TN accumulation in the Biliuhe reservoir. The TN accumulation of the Biliuhe reservoir is defined as the TN flux difference between input and output, which can be evaluated by accumulation quantity and accumulation efficiency. The reservoir nitrogen input includes the point source input, non-point source input, inter-basin water diversion project (DBWD) input, and aquaculture input due to the fish feed and manure, while the non-point source input includes soil source input, fertilizer application input, livestock and poultry breeding input, rural domestic sewage input, and atmospheric deposition input. The reservoir nitrogen output includes the water discharge output, water diversion project output, aquaculture output, and denitrification output. The TN accumulation model is described by Equations (4) to (6)

$$\Delta N = N_{in} - N_{out} \tag{4}$$

$$N_{in} = N_p + N_{np} + N_{div1} + N_{aqua1} \tag{5}$$

$$N_{out} = N_{dis} + N_{div2} + N_{aqua2} + N_{den} \tag{6}$$

where ΔN is the accumulation of TN, kg; N_{in} is the TN fluxes input the reservoir, kg; and N_{out} is the TN measured export from the reservoir, kg. In the composition of the TN input, N_p is the point source input, kg. The sewage discharge and nitrogen concentration of the nine outlets were measured in 2011, and used for the point source load simulation in this study. N_{np} is the non-point source input, which can be divided into soil source input (N_{soil}), fertilizer application input (N_{fer}), livestock and poultry breeding input (N_{liv}), rural domestic sewage input (N_{rur}), and atmospheric deposition input (N_{atm}). N_{div1} is the TN input of the inter-basin water diversion projects, kg. The mean discharge and nitrogen concentration in 2014–2016 were used to calculate the water diversion project load. N_{aqua1} is the aquaculture input. Because cage culture is prohibited in the Biliuhe reservoir and the fishery statistics data is unavailable, the aquaculture input and output are ignored in this study. All of the TN inputs except the direct rainfall N input were simulated by the SWAT model. The direct rainfall input was calculated by the precipitation, average TN concentration in the precipitation, and the surface water area. The precipitation data was derived from the CMADS. The TN concentration in the precipitation refers to the research of Yan and Shi [44]. The surface water area was obtained from the Biliuhe reservoir management bureau.

In the composition of the TN output, N_{dis} is the output of the water discharge to the downstream channel, kg. N_{aqua2} is the output of the aquaculture harvest, kg. N_{div2} is the output of the water diversion project, kg. N_{den} is the output of denitrification, kg. N_{aqua2} is the aquaculture output which was not considered in this study. The water discharge output and the inter-basin water diversion project output were calculated by the output water volume and the TN concentration in front of the dam, which were also obtained from the Biliuhe reservoir management bureau. The denitrification process mainly occurs in the sediment and varies greatly among different aquatic systems [45]. The denitrification output in this study can be calculated by the denitrification rate and reservoir surface area, described by Equation (7)

$$N_{den} = 0.014 \times V \times A \tag{7}$$

where V is the denitrification rate, referring to a lake in Canada with similar meteorological and hydrological characteristics, 15 μmol N m^{-2} h^{-1} [46], and A is the surface water area, km^2. The TN accumulation efficiency (R$_N$) can be defined by Equation (8)

$$R_N = (N_{in} - N_{out})/N_{in} \times 100\% \tag{8}$$

3. Results

3.1. SWAT Model Performance

The monthly observed data of discharge at the dam monitoring station and TN concentration at the entrance of reservoir were used for warm-up (2008), calibration (2009–2012), and validation (2013–2016). The hydrology and nitrogen parameters used for model calibration were shown in the Table S1 of the Supplementary Materials. For assessment of the SWAT model results, Wang and Melesse proposed that NSE >0.75 can be considered as "good", while values between 0.36 to 0.75 can be considered as "satisfactory" [47]. On the other hand, Moriasi et al. suggested a monthly model simulation can be judged as satisfactory if NSE values > 0.5 for runoff and 0.35 for nutrients, and if PBIAS was ±15% for discharge, and ±30% for N [43]. The evaluation results of the simulation discharge and TN are shown in Table 1. For the discharge, the monthly R^2 values were 0.96 and 0.90, the NSE values were 0.96 and 0.89, and PBIAS values were 8.68% and −11.53% in the calibration and validation periods, indicating a good performance of the SWAT model.

Table 1. Evaluation results of the discharge and TN simulation.

Parameter	Index	Calibration	Validation
	R^2	0.96	0.90
Discharge	NSE	0.96	0.89
	PBIAS	8.68%	−11.53%
	R^2	0.87	0.71
TN	NSE	0.85	0.53
	PBIAS	−13.49%	−21.71%

For the simulation of TN, the R^2 values were 0.87 and 0.71, the NSE values were 0.85 and 0.53, and the PBIAS values were −13.49% and −21.71% in the calibration and validation periods, respectively. The accuracy of the TN simulation was lower than that of the stream flow, but can still be considered as satisfactory according to the standard suggested by Moriasi. The observed and simulated monthly stream flow and TN are shown in Figure 3. The accuracy of the stream simulation results in the validation periods was worse than those in the calibration periods, which can be attributed to the drought that occurred in 2014–2015. Increased irrigation and domestic water consumption during droughts lead to a severe water intake from the river, and the observed discharge at the dam monitoring sites will be lower than the simulation value. The simulation of TN had similar characteristics. In general, the meteorological data of CMADS has a good applicability in the study area, and the CMADS-driven SWAT model can be used for runoff and TN simulation in the Biliuhe reservoir basin.

The sensitivity analysis result showed that the maximum canopy storage (CANMX), baseflow alpha factor for bank storage (ALPHA_BNK), Manning's n value for the main channel (CH_N2), SCS runoff curve number (CN2), and effective hydraulic conductivity in the main channel alluvium (CH_K2) were the most sensitive parameters for stream flow, while the nitrogen percolation coefficient (NPERCO), saturated hydraulic conductivity (SOL_K), moist bulk density (SOL_BD), Manning's n value for the main channel (CH_N2), and organic nitrogen enrichment ratio (ERORGN) were the most sensitive parameters for the TN load. The uncertainty analysis results showed that 88% (P-factor) of the discharge observations and 75% (P-factor) of the TN observations fell within the 95% confidence level uncertainty range (95PPU) in the calibration period, and the R-factors were 0.59 and 0.96, respectively. During the validation period, 79% of the discharge observations, and 71% of the TN observations fell within the 95% confidence interval (95PPU), and the R-factors were 0.58 and 0.90. The uncertainty of the SWAT model in the validation period was higher than that in the calibration period.

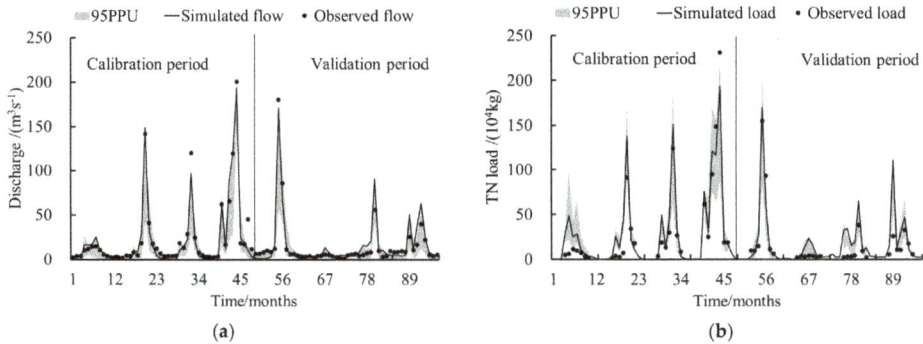

Figure 3. Observed and simulated (**a**) flow and (**b**) TN loads.

3.2. Temporal Characteristic of TN Fluxes of the Biliuhe Reservoir

The stream flow and TN fluxes of the Biliuhe reservoir are shown in Figure 4. The distribution of the TN input was very uneven between different years, and the TN input during the flood period was significantly higher than in the non-flood period. The TN input was consistent with the runoff. A simple Pearson's correlation analysis was performed to assess possible relationships between the TN input and the runoff, and the precipitation and the runoff. The results showed a strong and significant positive correlation between the TN input and the discharge ($r = 0.918$, $p < 0.01$), which meant that the nitrogen input was mainly governed by the runoff (Figure 5). This result was in a good agreement with other studies [48–50]. The correlation between the precipitation and the runoff was also high ($r = 0.872$, $p < 0.01$), but lower than that between the TN input and the runoff. This can be attributed to the severe drought and the inter-basin water diversion project. Due to the flushing of surface runoff, a large amount of nitrogen—along with the flood—enters the reservoir, resulting in the higher input during flood periods (August 2010, August 2011, August 2012, and July 2013). In contrast, the TN input in dry periods (July 2014–November 2015) was relatively lower due to the decrease in precipitation and runoff. For the Biliuhe reservoir, the TN output is greatly influenced by artificial regulation. Therefore, the TN output does not follow the variation of runoff. During wet years, surplus water quantity increases due to the flood input, leading to an increasing TN output. Additionally, affected by the disturbance of the density flow, the nitrogen stored in sediments will release into the water, then be exported from the reservoir with surplus water. In addition to the three years of 2011–2013, the TN output was very stable most of the time. Because in dry years there is no chance for the reservoir to discharge surplus water, water withdrawal becomes the main pathway for TN output, the TN output in dry years is stable and relatively lower than in wet years.

The seasonal TN input and output in the Biliuhe reservoir were assessed to analyze the temporal characteristics of TN fluxes during the year. As can be seen from Figure 6, the TN input changed from month to month and was high in wet seasons and fall and low in dry seasons, consistent with the precipitation and runoff, as well as with the intensive agriculture activities in the wet seasons. The wet seasons (April–October) accounted for 95% of the annual TN input, indicating an extremely uneven distribution of TN input during the year. There was an abrupt increase of TN input at the beginning of the wet season (April), meaning that TN stored in dry seasons had a considerable release during the dry-wet alternation process. The TN output in dry seasons was mainly due to the water supply output and changed slightly, while it increased a lot in wet seasons due to the flood discharge. In general, the

TN input was higher than the output during the wet seasons except for October, which meant that the Biliuhe reservoir acted as a sink during most of the wet season. The monthly input of TN was always lower than the output from September to the next March, suggesting that the Biliuhe reservoir acted as a source during dry seasons. The TN accumulation and the composition of TN input and output were analyzed in the discussion.

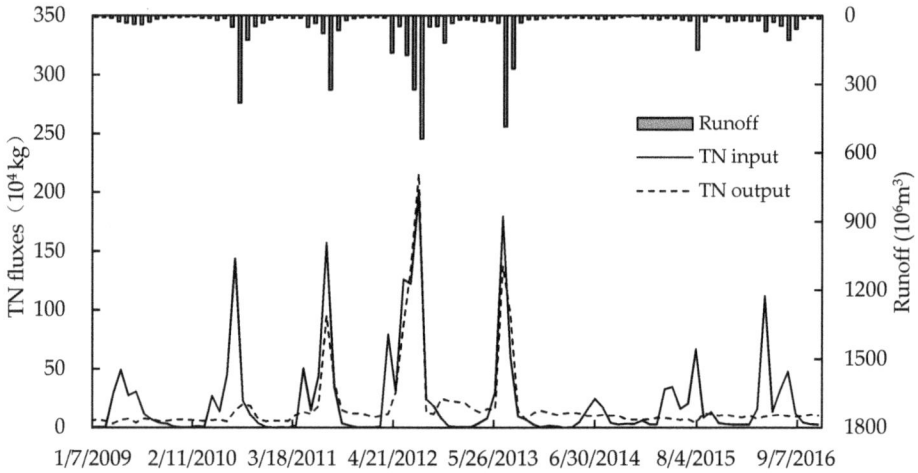

Figure 4. The runoff and TN fluxes of the Biliuhe reservoir.

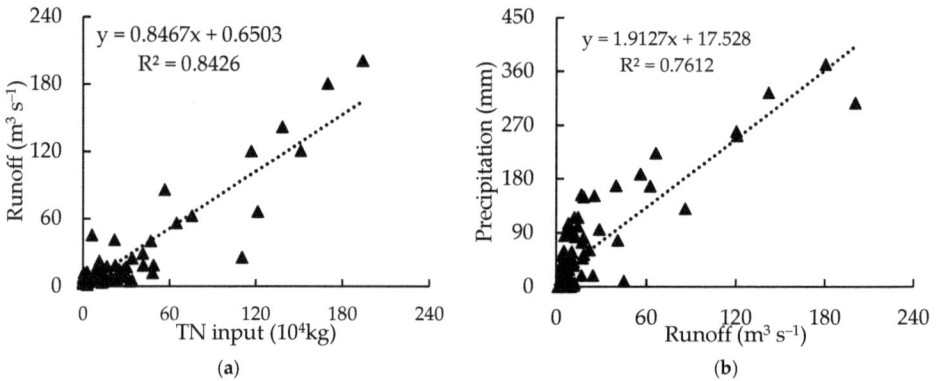

(a)

(b)

Figure 5. Relationships between runoff and TN input (**a**) and precipitation and runoff (**b**) from 2009–2016.

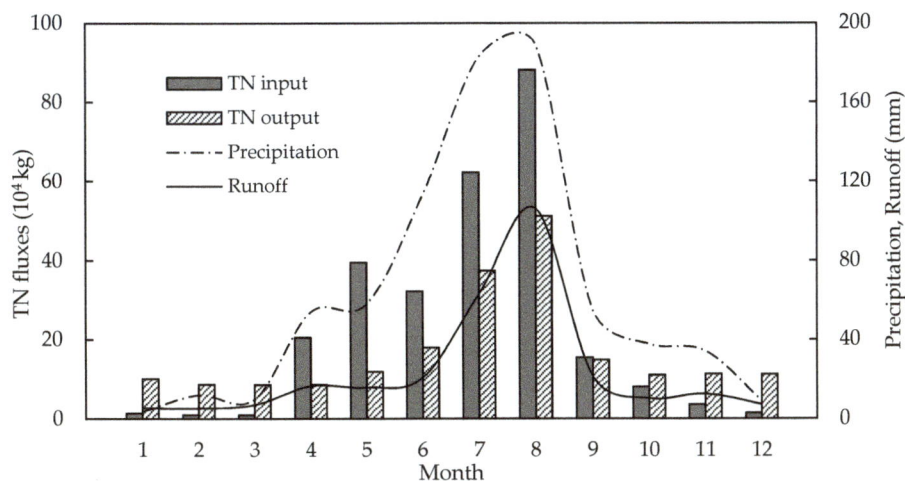

Figure 6. Seasonal variations of precipitation, runoff, and TN fluxes.

4. Discussion

4.1. TN Accumulation in the Biliuhe Reservoir

The dynamic TN accumulation process of the Biliuhe reservoir were analyzed based on the TN input and output fluxes. The annual TN accumulation of the Biliuhe reservoir are presented in Table 2. In the study period, the mean annual TN input flux was 274.41×10^4 kg, which was higher than the output flux of 217.14×10^4 kg. The average annual accumulation of TN was 57.27×10^4 kg, indicating that 19.76% of the TN input was retained by the Biliuhe reservoir, which was higher than the TN retained by the Three Gorges reservoir (China) and Wivenhoe reservoir (Australia) [51,52], but lower than the TN retained by Lake Shelbyville (USA) [53]. The maximum and minimum TN input fluxes were 60.98×10^4 kg in 2012 and 78.02×10^4 kg in 2014, while the maximum and minimum output fluxes were 582.88×10^4 kg in 2012 and 71.75×10^4 kg in 2009, indicating that TN accumulation in the study reservoir varied in different hydrological years. It can also be seen from Table 2 that the Biliuhe reservoir was not always a sink of TN—sometimes it could transform into a source. The accumulation of TN in reservoirs is influenced by the balance of input and output. The input TN is mainly driven by runoff, while the TN export from the reservoir is driven by the domestic water supply and flood discharge. Therefore, higher nitrogen input did not correspond to higher nitrogen accumulation in the study period. Especially in 2013, the accumulation of TN was negative due to the higher flood discharge output. Interestingly, higher TN accumulation efficiency was observed in normal hydrological years, because the Biliuhe reservoir always stores most of the water input for future multiple uses but rarely discharges surplus water. During extreme drought years, the reservoir may also act as a source, with the TN input flux decreasing and the water consumption increasing, as shown in 2014. Reservoirs act as a sink or source of nutrients due to the basin characteristics and reservoir regulation.

For the Biliuhe river basin, the dam decreases the ecological water volume of the downstream area and obstructs the dispersal and migration of nutrients, which has resulted in the degradation of the downstream ecosystem and extinction of species. In the study reservoir, there was a rising trend of TN accumulation. The TN retained by the reservoir partly stores in the water and partly deposits with sediment due to adsorption. The water TN concentration of the Biliuhe reservoir was 2.16 mg/L in 2009, rising to 2.92 mg/L in 2016, which meant an increase of the eutrophication risk of the water body. Despite the increase of the TN concentration in the water, the TN stored in the water decreased during the study period due to the decrease in water volume (from 562 million m^3 in 2009 to 183 million m^3 in 2016), which demonstrated that the TN that accumulated in the sediment contributed to a large proportion of the TN accumulation in the reservoir. Recent research has also shown that TN in the sediment of the Biliuhe reservoir is at a relatively high level, and endogenous nitrogen released from the sediment could contribute to the water nitrogen pollution [54]. Water supply and flood discharge are two major ways to export the TN from the reservoir. Water withdrawal for urban use is mainly from the upper water column in front of the dam, which has good water quality. Therefore, the nitrogen in the bottom water and sediment could not be discharged from the reservoir. For the above problems, in addition to decreasing the non-point source nitrogen input from upper basin, discharging anoxic waters and sediment with a high nitrogen concentration through the bottom hole of the dam could alleviate the nitrogen pollution in the Biliuhe reservoir.

Table 2. Annual TN accumulation of the Biliuhe reservoir (2009–2016).

Year	$N_i/(10^4 \text{ kg})$	$N_o/(10^4 \text{ kg})$	$\Delta N/(10^4 \text{ kg})$	$R_N/(\%)$
2009	165.59	71.75	93.84	56.67
2010	270.25	108.16	162.09	59.98
2011	308.44	245.12	63.32	20.53
2012	609.98	582.88	27.10	4.44
2013	301.95	384.96	−83.01	−27.49
2014	78.02	125.84	−47.82	−61.29
2015	211.07	100.20	110.87	52.53
2016	249.99	118.19	131.80	52.72
Mean value	274.41	217.14	57.27	19.76

4.2. Composition of TN Input and Output

The composition of TN input during 2009–2016 is shown in Figures 7a and 8a. It can be seen from these figures that the point source input accounted for a small proportion of TN input, which is only 0.16% of the average input. This means that the point-source nitrogen input is not serious and the non-point source input dominates the TN input of the Biliuhe reservoir. This result is consistent with the actual conditions wherein most of the untreated sewage outlets in the basin have been closed in recent years, but non-point sources like agricultural fertilization and livestock and poultry breeding are difficult to treat. The non-point sources from fertilizer and atmospheric deposition and soils constituted the highest proportion of the TN input, accounting for 35.15, 30.15, and 27.72% of the average input. According to the study result, reduction of the non-point source TN input during the flood period should be an effective measure to control the TN pollution of the study reservoir. In addition to the TN load from the reservoir control basin, the TN input of the inter-basin water diversion project was 32.03×10^4 kg year^{-1} during 2015–2016, accounting for 14.05% of the annual total input. According to the design water diversion capacity of 3.00×10^8 m^3 year^{-1} in the future, 63.90×10^4 kg year^{-1} TN will input into the reservoir via the DBWD project, which was approximately equal to the TN input in extreme drought years. The impact of the inter-basin water diversion project on TN accumulation should be of great concern in the future.

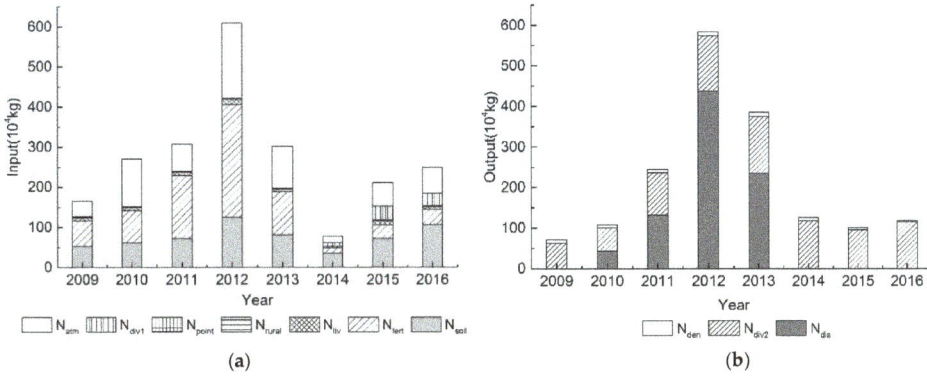

Figure 7. The composition of annual TN input (**a**) and output (**b**) fluxes of the Biliuhe reservoir from 2009–2016.

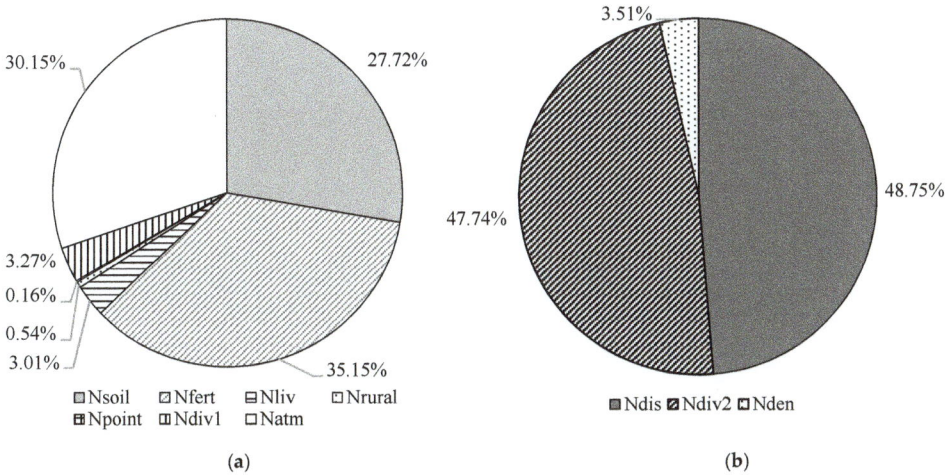

Figure 8. The average composition ratio of the TN input (**a**) and output (**b**) fluxes of the Biliuhe reservoir from 2009–2016.

The composition of the TN output during 2009–2016 is shown in Figures 7b and 8b. It can be seen from these figures that the average discharge output accounted for 48.75% of the average TN output, and mainly occurred in wet years with the flood discharge. The reservoir rarely discharges surplus water in normal or dry years due to the high water demand of Dalian city. The average TN output of the BDWD project was 103.66×10^4 kg year^{-1}, which contributed to 47.74% of the annual TN output. In the dry or normal years like 2014–2016, the TN output by the BDWD project accounted for more than 90% of the total TN output, indicating that a large proportion of the nitrogen was transported to other basins through the inter-basin water diversion project. The inter-basin water diversion project links different basins together and changes the original nutrient transportation within or between river basins, which creates new challenges for comprehensive environmental management across multiple basins [55]. The denitrification output accounted for 3.51% of the total nitrogen output, which was much lower than the inter-basin water diversion project and downstream water discharge output.

4.3. Analysis of the Proposed Methodology

This study explores an integrated methodology that is based on the SWAT model to quantify the nitrogen accumulation in the reservoir. It is convenient for the proposed methodology to analyze the TN accumulation process in the study reservoir. The CMADS has also been proven to be capable of driving the SWAT model in the study area. The methodology combines the mechanism model and the statistical model to obtain more reasonable results, which show that the Biliuhe reservoir serves as a sink of TN most of the time, and the TN accumulation process is significantly influenced by the reservoir operation mode. In this study, the uncertainty of the methodology may come from the meteorological data of CMADS, parameters of SWAT, and sparse distribution of water quality monitoring sites due to the complex mechanism of TN transformation and transportation. More work will be done to reduce the uncertainty and improve methodology accuracy. In this study, the reservoir was taken as a small point and the internal N transportation process of the reservoir was not taken into consideration. In the next step, we will improve the simulation by coupling the SWAT with the 2-D or 3-D hydrological and water quality model for further research.

5. Conclusions

The integrated methodology based on the SWAT model driven by CMADS was established in the Biliuhe river basin to analyze the TN accumulation of the Biliuhe reservoir. The calibrated model generally achieved a satisfactory performance, indicating that the CMADS can be successfully applied to drive the SWAT model in Northeast China. The model performance in the validation periods was worse than in calibration periods, which can be attributed to the drought in 2014–2015. The maximum canopy storage (CANMX) was the most sensitive parameter for runoff, and the nitrogen percolation coefficient (NPERCO) was the most sensitive parameter for TN. The uncertainty analysis results showed that 88% (P-factor) of discharge observations and 75% (P-factor) of TN observations fell within the 95% confidence level uncertainty range (95PPU) in the calibration period, and within 79% of the stream flow and 71% of the TN in the validation period. The uncertainty of the SWAT model in the validation period was higher than that in the calibration period.

The distributions of annual and seasonal TN input were very uneven and consistent with the runoff, indicating that TN input was mainly governed by the runoff. The TN output was greatly influenced by artificial regulation and did not follow the variation of runoff. The TN input was higher than the output during the wet seasons, and was always lower than the output during the dry seasons, meaning that the reservoir acted as a sink of nitrogen most of time in the wet seasons and as a source in the dry seasons. The mean annual TN accumulation of the Biliuhe reservoir was 57.27×10^4 kg during the study period, which meant that 19.76% of the TN input was retained in the Biliuhe reservoir. The TN accumulation varied in different hydrological years and higher TN accumulation in the reservoir did not correspond to a higher TN load due to the influence of artificial regulation. Higher TN accumulation efficiency is often observed in normal hydrological years because the water source reservoir stores most of the water input for future multiple uses, but rarely discharges surplus water. In addition, the non-point sources from fertilizer, atmospheric deposition, and soils constituted the highest proportion of the TN input, accounting for 35.15%, 30.15%, and 27.72% of the average input, respectively. The inter-basin water diversion project of DBWD diverted 32.03×10^4 kg year^{-1} TN to the Biliuhe reservoir in 2015–2016, accounting for 14.5% of the total annual input. According to the design water diversion capacity of 3.00×10^8 m^3 year^{-1} in the future, 63.90×10^4 kg year^{-1} TN will input into the reservoir via the DBWD project in the future. The discharge output and the inter-basin water diversion project output of the BDWD accounted for 48.75% and 47.74%, respectively. The inter-basin water diversion projects have noticeably influenced the TN accumulation in drought years, which should be of great concern in drinking water source water system management.

There was a rising trend of TN level in the Biliuhe reservoir, which increased the eutrophication risk of the aquatic ecosystem. However, the TN accumulated in the sediment contributed to a large proportion of the total TN accumulated in the reservoir. Decreasing the non-point source nitrogen input and discharging of anoxic waters and sediment with a high concentration of TN through the bottom of the dam should be effective measures to reduce the TN concentration in the Biliuhe reservoir. The integrated methodology proposed in this work provided a convenient way to quantify the TN accumulation in reservoirs, and the results could contribute to reservoir water quality improvement under the influence of inter-basin water diversion projects.

Supplementary Materials: The following are available online at http://www.mdpi.com/2073-4441/10/11/1535/s1, Figure S1: SWAT parameters calibrated for the monthly streamflow and the TN load in the Biliuhe reservoir basin.

Author Contributions: Conceptualization, G.Q. and T.W.; Data curation, T.W.; Formal analysis, G.Q.; Funding acquisition, J.L., S.X. and T.W.; Investigation, S.X. and G.S.; Methodology, G.Q. and T.W.; Project administration, T.W.; Resources, S.X.; Software, G.Q.; Supervision, J.L. and T.W.; Validation, J.L. and T.W.; Visualization, G.Q.; Writing—original draft, G.Q.; Writing—review & editing, G.S.

Funding: This research was supported by the National Key Research and Development Program of China (2016YFC0400903), the Natural Sciences Foundation of China (51679026;51809032;51879031), and the Fundamental Research Funds for the Central Universities (DUT17JC17).

Acknowledgments: The authors would like to thank the anonymous reviewers for their review and constructive comments related to this manuscript.

Conflicts of Interest: The authors declare no conflict of interest.

References

1. Saunders, D.L.; Kalff, J. Nitrogen retention in wetlands, lakes and rivers. *Hydrobiologia* **2001**, *443*, 205–212. [CrossRef]
2. Thieu, V.; Billen, G.; Garnier, J. Nutrient transfer in three contrasting NW European watersheds: The Seine, Somme, and Scheldt Rivers. A comparative application of the Seneque/River strahler model. *Water Res.* **2009**, *43*, 1740–1754. [CrossRef] [PubMed]
3. Zhang, Q.; Hirsch, R.M.; Ball, P.W. Long-term changes in sediment and nutrient delivery from Conowingo Dam to Chesapeake Bay: Effects of reservoir sedimentation. *Environ. Sci. Technol.* **2016**, *50*, 1877–1886. [CrossRef] [PubMed]
4. Chen, D.; Hu, M.; Dahlgren, R.A. A dynamic watershed model for determining the effects of transient storage on nitrogen export to rivers. *Water Resour. Res.* **2014**, *50*, 7714–7730. [CrossRef]
5. Ma, W.; Huang, T.; Li, X.; Zhou, Z.; Li, Y.; Zeng, K. The Effects of storm runoff on water quality and the coping strategy of a deep canyon-shaped source water reservoir in China. *Int. J. Environ. Res. Public Health* **2015**, *12*, 7839–7855. [CrossRef] [PubMed]
6. Ministry of Water Resources, China; National Bureau of Statistics, China. Bulletin of First National Census for Water, 2013. Available online: http://www.mwr.gov.cn/sj/tjgb/dycqgslpcgb/201701/t20170122_790650.html (accessed on 21 March 2013).
7. Ministry of Water Resources, China. Bulletin of National Water Resources, 2016. Available online: http://www.mwr.gov.cn/sj/tjgb/szygb/201707/t20170711_955305.html (accessed on 31 December 2016).
8. Zaragüeta, M.; Acebes, P. Controlling eutrophication in a Mediterranean Shallow Reservoir by phosphorus loading reduction: The need for an integrated management approach. *Environ. Manag.* **2017**, *59*, 635–651. [CrossRef] [PubMed]
9. Clark, V.E.; Odhiambo, K.B.; Ricker, C.M. Comparative analysis of metal concentrations and sediment accumulation rates in two Virginian reservoirs, USA: Lakes Moomaw and Pelham. *Water Air Soil Pollut.* **2014**, *225*, 1860. [CrossRef]
10. Lopez-Doval, J.C.; Montagner, C.C.; de Alburquerque, A.F.; Moschini-Carlos, V.; Umbuzeiro, G.; Pompeo, M. Nutrients, emerging pollutants and pesticides in a tropical urban reservoir: Spatial distributions and risk assessment. *Sci. Total Environ.* **2017**, *575*, 1307–1324. [CrossRef] [PubMed]
11. Zhang, E.R.; Zhang, J. Analysis of the Three-Gorge Reservoir impacts on the retention of N and P in the Yangtze River. *J. Lake Sci.* **2003**, *15*, 41–48. (In Chinese)
12. Cunha, D.F.; Calijuri, M.D.; Dodds, W.K. Trends in nutrient and sediment retention in Great Plains reservoirs (USA). *Environ. Monit. Assess.* **2014**, *186*, 1143–1155. [CrossRef] [PubMed]

13. Ullrich, A.; Volk, M. Application of the Soil and Water Assessment Tool (SWAT) to predict the impact of alternative management practices on water quality and quantity. *Agric. Water Manag.* **2009**, *96*, 1207–1217. [CrossRef]

14. Cho, J.; Park, S.; Im, S. Evaluation of Agricultural Nonpoint Source (AGNPS) model for small watersheds in Korea applying irregular cell delineation. *Agric. Water Manag.* **2008**, *95*, 400–408. [CrossRef]

15. Kourgialas, N.N.; Karatzas, G.P.; Nikolaidis, N.P. An integrated framework for the hydrologic simulation of a complex geomorphological river basin. *J. Hydrol.* **2010**, *381*, 308–321. [CrossRef]

16. Bosch, N.S. The influence of impoundments on riverine nutrient transport: An evaluation using the Soil and Water Assessment Tool. *J. Hydrol.* **2008**, *355*, 131–147. [CrossRef]

17. Xu, F.; Dong, G.; Wang, Q.; Liu, L.; Yu, W.; Men, C.; Liu, R. Impacts of DEM uncertainties on critical source areas identification for non-point source pollution control based on SWAT model. *J. Hydrol.* **2016**, *540*, 355–367. [CrossRef]

18. Abbaspour, K.C.; Rouholahnejad, E.; Vaghefi, S.; Srinivasan, R.; Yang, H.; Klove, B. A continental-scale hydrology and water quality model for Europe: Calibration and uncertainty of a high-resolution large-scale SWAT model. *J. Hydrol.* **2015**, *524*, 733–752. [CrossRef]

19. Ebita, A.; Kobayashi, S.; Ota, Y.; Moriya, M.; Kumabe, R.; Onogi, K.; Harada, Y.; Yasui, S.; Miyaoka, K.; Takahashi, K.; et al. The Japanese 55-Year Reanalysis "JRA-55": An interim report. *SOLA* **2011**, *7*, 149–152. [CrossRef]

20. Dee, D.P.; Uppala, S.M.; Simmons, A.J.; Berrisford, P.; Poli, P.; Kobayashi, S.; Andrae, U.; Balmaseda, M.A.; Balsamo, G.; Bauer, P.; et al. The ERA—Interim reanalysis: Configuration and performance of the data assimilation system. *Q. J. R. Meteor. Soc.* **2011**, *137*, 553–597. [CrossRef]

21. Saha, S.; Moorthi, S.; Pan, H.L.; Wu, X.; Wang, J.; Nadiga, S.; Tripp, P.; Kistler, R.; Woollen, J.; Behringer, D.; et al. The NCEP climate forecast system reanalysis. *Bull. Am. Meteorol. Soc.* **2010**, *91*, 1015–1057. [CrossRef]

22. Rienecker, M.M.; Suarez, M.J.; Gelaro, R.; Todling, R.; Bacmeister, J.; Liu, E.; Bosilovich, M.G.; Schubert, S.D.; Takacs, L.; Kim, G.; et al. MERRA: NASA's Modern-Era retrospective analysis for research and applications. *J. Clim.* **2011**, *24*, 3624–3648. [CrossRef]

23. Meng, X.; Wang, H. Significance of the China meteorological assimilation driving datasets for the SWAT Model (CMADS) of East Asia. *Water* **2017**, *9*, 765. [CrossRef]

24. Zhou, S.; Wang, Y.; Chang, J.; Guo, A.; Li, Z. Investigating the dynamic influence of hydrological model parameters on runoff simulation using Sequential Uncertainty Fitting 2-Based Multilevel-Factorial-Analysis Method. *Water* **2018**, *10*, 1177. [CrossRef]

25. Tian, Y.; Zhang, K.; Xu, Y.-P.; Gao, X.; Wang, J. Evaluation of Potential Evapo-transpiration Based on CMADS Reanalysis Dataset over China. *Water* **2018**, *10*, 1126. [CrossRef]

26. Meng, X.; Wang, H.; Cai, S.; Zhang, X.; Leng, G.; Lei, X.; Shi, C.; Liu, S.; Shang, Y. The China meteorological assimilation driving datasets for the SWAT Model (CMADS) application in China: A case study in Heihe River Basin. *Preprints* **2016**. [CrossRef]

27. Meng, X.; Wang, H.; Lei, X.; Cai, S.; Wu, H. Hydrological modeling in the Manas River Basin using soil and water assessment tool driven by CMADS. *Teh. Vjesn.* **2017**, *24*, 525–534. [CrossRef]

28. Liu, J.; Shanguan, D.; Liu, S.; Ding, Y. Evaluation and hydrological simulation of CMADS and CFSR reanalysis datasets in the Qinghai-Tibet Plateau. *Water* **2018**, *10*, 513. [CrossRef]

29. Zhao, F.; Wu, Y.; Qiu, L.; Sun, Y.; Sun, L.; Li, Q.; Niu, J.; Wang, G. Parameter uncertainty analysis of the SWAT model in a mountain-loess transitional watershed on the Chinese Loess Plateau. *Water* **2018**, *10*, 690. [CrossRef]

30. Vu, T.T.; Li, L.; Jun, K.S. Evaluation of multi-satellite precipitation products for streamflow simulations: A case study for the Han River Basin in the Korean Peninsula, East Asia. *Water* **2018**, *10*, 642. [CrossRef]

31. Cao, Y.; Zhang, J.; Yang, M.; Lei, X.; Guo, B.; Yang, L.; Zeng, Z.; Qu, J. Application of SWAT model with CMADS data to estimate hydrological elements and parameter uncertainty based on SUFI-2 algorithm in the Lijiang River Basin, China. *Water* **2018**, *10*, 742. [CrossRef]

32. Shao, G.; Guan, Y.; Zhang, D.; Yu, B.; Zhu, J. The impacts of climate variability and land use change on streamflow in the Hailiutu River Basin. *Water* **2018**, *10*, 814. [CrossRef]

33. Gao, X.; Zhu, Q.; Yang, Z.; Wang, H. Evaluation and hydrological application of CMADS against TRMM 3B42V7, PERSIANN-CDR, NCEP-CFSR, and Gauge-Based Datasets in Xiang River Basin of China. *Water* **2018**, *10*, 1225. [CrossRef]

34. Harrison, J.A.; Maranger, R.J.; Alexander, R.B.; Giblin, A.E.; Jacinthe, P.A.; Mayorga, E.; Seitzinger, S.P.; Sobota, D.J.; Wollheim, W.M. The regional and global significance of nitrogen removal in lakes and reservoirs. *Biogeochemistry* **2009**, *93*, 143–157. [CrossRef]

35. Komai, Y.; Umemoto, S.; Takeda, Y.; Inoue, T.; Imai, A. Budgets of major ionic species and nutrients on a dam reservoir in forested watershed. *Water Sci. Technol.* **2007**, *56*, 287–293. [CrossRef] [PubMed]

36. Han, H.; Lu, X.; Burger, D.F.; Joshi, U.M.; Zhang, L. Nitrogen dynamics at the sediment-water interface in a tropical reservoir. *Ecol. Eng.* **2014**, *73*, 146–153. [CrossRef]

37. Xu, S.; Wang, T.; Hu, S. Dynamic assessment of water quality based on a variable fuzzy pattern recognition model. *Int. J. Environ. Res. Public Health* **2015**, *12*, 2230–2248. [CrossRef] [PubMed]

38. Zhang, C.; Shoemaker, C.A.; Woodbury, J.D.; Cao, M.; Zhu, X. Impact of human activities on stream flow in the Biliu River Basin, China. *Hydrol. Process.* **2013**, *27*, 2509–2523. [CrossRef]

39. Zhu, X.; Zhang, C.; Qi, W.; Cai, W.; Zhao, X.; Wang, X. Multiple climate change scenarios and runoff response in Biliu River. *Water* **2018**, *10*, 126. [CrossRef]

40. Wang, T.; Xu, S.; Liu, J. Analysis of accumulation formation of sediment contamination in reservoirs after decades of running: A case study of nitrogen accumulation in Biliuhe Reservoir. *Environ. Sci. Pollut. R.* **2018**, *25*, 9165–9175. [CrossRef] [PubMed]

41. Arnold, J.G.; Fohrer, N. SWAT2000: Current capabilities and research opportunities in applied basin modelling. *Hydrol. Process.* **2005**, *19*, 563–572. [CrossRef]

42. Fan, M.; Shibata, H. Spatial and temporal analysis of hydrological provision ecosystem services for watershed conservation planning of water resources. *Water Resour. Manag.* **2014**, *28*, 3619–3636. [CrossRef]

43. Moriasi, D.N.; Gitau, M.W.; Pai, N.; Daggupati, P. Hydrologic and water quality models: Performance measures and evaluation criteria. *Trans. ASABE* **2015**, *58*, 1763–1785. [CrossRef]

44. Seitzinger, S.; Harrison, J.A.; Böhlke, J.K.; Bouwman, A.F.; Lowrance, R.; Peterson, B.; Tobias, C.; Van Drecht, G. Denitrification across landscapes and waterscapes: A synthesis. *Ecol. Appl.* **2006**, *16*, 2064–2090. [CrossRef]

45. Yan, W.J.; Shi, K. Program on the nitrogen concentrations in rain water in Dalian city. *Ecol. Environ.* **2013**, *22*, 517–522. (In Chinese)

46. Saunders, D.L.; Kalff, J. Denitrification rates in the sediments of Lake Memphremagog, Canada-USA. *Water Res.* **2001**, *35*, 1897–1904. [CrossRef]

47. Wang, X.; Melesse, A.M. Effects of STATSGO and SSURGO as inputs on SWAT model's snowmelt simulation. *J. Am. Water Resour. Assoc.* **2006**, *42*, 1217–1236. [CrossRef]

48. Helmreich, B.; Hilliges, R.; Schriewer, A.; Horn, H. Runoff pollutants of a highly trafficked urban road—Correlation analysis and seasonal influences. *Chemosphere* **2010**, *80*, 991–997. [CrossRef] [PubMed]

49. Liu, M.; Chen, X.; Yao, H.; Chen, Y. A coupled modeling approach to evaluate nitrogen retention within the Shanmei Reservoir basin, China. *Estuar. Coast. Shelf Sci.* **2015**, *166*, 189–198. [CrossRef]

50. Molina-Navarro, E.; Trolle, D.; Martinez-Perez, S.; Sastre-Merlin, A.; Jeppesen, E. Hydrological and water quality impact assessment of a Mediterranean limno-reservoir under climate change and land use management scenarios. *J. Hydrol.* **2014**, *509*, 354–366. [CrossRef]

51. Ran, X.; Bouwman, L.; Yu, Z.; Beusen, A.; Chen, H.; Yao, Q. Nitrogen transport, transformation, and retention in the Three Gorges Reservoir: A mass balance approach. *Limnol. Oceanogr.* **2017**, *62*, 2323–2337. [CrossRef]

52. Burford, M.A.; Green, S.A.; Cook, A.J.; Johnson, S.A.; Kerr, J.G.; O'Brien, K.R. Sources and fate of nutrients in a subtropical reservoir. *Aquat. Sci.* **2012**, *74*, 179–190. [CrossRef]

53. David, M.B.; Wall, L.G.; Royer, T.V.; Tank, J.L. Denitrification and the nitrogen budget of a reservoir in an agricultural landscape. *Ecol. Appl.* **2006**, *16*, 2177–2190. [CrossRef]

54. Wang, T.; Xu, S.; Liu, J. Dynamic assessment of comprehensive water quality considering the sediment release. *Water* **2017**, *9*, 275. [CrossRef]

55. Wang, Y.; Zhang, W.; Zhao, Y.; Peng, H.; Shi, Y. Modelling water quality and quantity with the influence of inter-basin water diversion projects and cascade reservoirs in the Middle-lower Hanjiang River. *J. Hydrol.* **2016**, *541*, 1348–1362. [CrossRef]

water

MDPI

Article

Application of SWAT Model with CMADS Data to Estimate Hydrological Elements and Parameter Uncertainty Based on SUFI-2 Algorithm in the Lijiang River Basin, China

Yang Cao [1,2], Jing Zhang [1,*], Mingxiang Yang [3], Xiaohui Lei [3], Binbin Guo [1], Liu Yang [2], Zhiqiang Zeng [3] and Jiashen Qu [4]

1   Key Laboratory of 3D Information Acquisition and Application of Ministry of Education, Capital Normal University, Beijing 100048, China; m171892@hiroshima-u.ac.jp (Y.C.); guobinbin@126.com (B.G.)
2   Graduate School of Integrate Arts and Sciences, Hiroshima University, Hiroshima 7398521, Japan; g170292@hiroshima-u.ac.jp
3   China Institute of Water Resource and Hydropower Research, Beijing 100038, China; yangmx@iwhr.com (M.Y.); lxh@iwhr.com (X.L.); zengzhiqiang@hust.edu.cn (Z.Z.)
4   Graduate School of Education, Hiroshima University, Hiroshima 7398521, Japan; jackykutsu@yahoo.com
*   Correspondence: maggie2008zj@yahoo.com; Tel.: +010-68903139

Received: 29 March 2018; Accepted: 4 June 2018; Published: 7 June 2018

Abstract: The China Meteorological Assimilation Driving Datasets for the Soil and Water Assessment Tool model (CMADS) have been widely applied in recent years because of their accuracy. An evaluation of the accuracy and efficiency of the Soil and Water Assessment Tool (SWAT) model and CMADS for simulating hydrological processes in the fan-shaped Lijiang River Basin, China, was carried out. The Sequential Uncertainty Fitting (SUFI-2) algorithm was used for parameter sensitivity and uncertainty analysis at the daily scale. The pair-wise correlation between parameters and the uncertainties associated with equifinality in model parameter estimation were investigated. The results showed that the SWAT model performed well in predicting daily streamflow for the calibration period (2009–2010). The correlation coefficient (R^2) was 0.92, and the Nash-Sutcliffe model efficiency coefficient (NSE) was 0.89. For the validation period (2011–2018), $R^2 = 0.89$, NSE = 0.88, and reasonable values for the P-factor, R-factor, and percent bias (PBIAS) were obtained. In addition, the spatial and temporal variation of evapotranspiration (ET), surface runoff, and groundwater discharge were analyzed. The results clearly showed that spatial variation in surface runoff and groundwater discharge are strongly related to precipitation, while ET is largely controlled by land use types. The contributions to the water budget by surface runoff, groundwater discharge, and lateral flow were very different in flood years and dry years.

Keywords: SWAT model; CMADS; Lijiang River; runoff; uncertainty analysis; hydrological elements

1. Introduction

The water cycle is one of the most important of the earth's cycles, and it plays a crucial role in biosphere changes. Water balance elements in a basin are affected by natural and human factors, such as the types of land use, soil properties [1], geological conditions, glacier [2] and human economic activity [3,4]. It is necessary to study the contribution to the water budget by different hydrological elements in a basin for the purpose of land use management, water resources management, and hydrological process analysis. Because the contribution to the water budget by different hydrological elements is hard to measure in the field, it is more practical to estimate the water cycle components of a watershed using a hydrological model [5].

The Soil and Water Assessment Tool (SWAT) model is an important tool in the development of water management strategies [6]. At the beginning of SWAT model establishment, it is difficult to calculate the water cycle components, especially groundwater [7]. Sophocleous et al. [8] simulated combined surface-water, ground-water, and stream-aquifer interactions using a comprehensive SWATMOD basin model, which was based on the Modular Three-Dimensional Finite-Difference Ground-Water Flow Model (MODFLOW). Because the SWAT model was established using the characteristics of a North American river basin, the accuracy of the model can be compromised in other areas. For example, the SWAT99.2 version could not satisfactorily calculate the runoff in low mountain regions of Germany. To address these shortcomings, Eckhardt et al. [9,10] developed the SWAT Giessen (SWAT-G) version for simulating the runoff in catchments with predominantly steep slopes, shallow soils, and consolidated rock aquifers. In addition, Easton et al. [11] established a Soil and Water Assessment Tool-Variable Source Area (SWAT-VSA) model for predicting runoff by modifying the curve number and available water content in variable source areas.

Although the SWAT hydrological model has been widely used for nutrient transport and hydrological modeling, the model is difficult to apply in areas where meteorological data are scarce, such as glacial and deserts areas [12]. Therefore, meteorological data are urgently needed for runoff simulation and prediction in non-data basins [13]. The CMADS was developed by Dr. Xianyong Meng from the China Institute of Water Resources and Hydropower Research (IWHR). The data range is from 2008 to 2016. It covers the entire East Asian region [14]. Some studies considered that CMADS+SWAT have better results for runoff simulation [15,16]. Meng et al. [17] evaluated the water cycle in an area without meteorological data using the CMADS meteorological data. They obtained satisfactory results through parameter calibration in areas with a high glacial recharge rate. Meng et al. also used three different datasets to simulate runoff in the Heihe Basin, and the results showed that the simulation accuracy of the CMADS was higher than other datasets [18]. The uncertainty analysis based on CMADS data has also been investigated [19]. In recent years, SWAT has been successfully applied in the study of hydrological elements in various watersheds. For example, the SWAT model was applied to study changes in the water budget caused by climate change [20–23]. The SWAT model was used to study hydrological elements in ice- and snow-covered mountainous area [24–26]. The SWAT model has also been used to study the main hydrological elements in agricultural areas [27–29].

Although the CMADS data have been applied worldwide since its release in 2016, the application of CMADS in abundant rainfall areas in southern China is lacking [30]. Further investigations of the applicability of the CMADS in the SWAT model are needed to better understand and evaluate the accuracy and efficacy of the dataset. The Lijiang River is an important water system in the Pearl River Basin, and the CMADS data have not been verified in this basin. To address this knowledge gap, the present study applied the SWAT model to explore the applicability of the CMADS in this basin. The Sequential Uncertainty Fitting (SUFI-2) algorithm was used for parameter sensitivity and uncertainty analysis at a daily scale. Pair-wise correlation between parameters and the uncertainties associated with equifinality in model parameter estimation was also investigated. The simulation results were used to investigate the water budget and its elements in the basin. The study also investigated the spatial variation and temporal variation of the water budget elements. In addition, the correlation between hydrological elements and precipitation were investigated.

2. Materials and Methods

2.1. Study Area

The Lijiang River Basin (23°23'–25°59' N, 110°18'–111°18' E) is located in the northeast of the Guangxi Province, within the upper reaches of the Guijiang River in the Pearl River system (Figure 1). From north to south, the basin runs through Xingan County, Lingchuan County, Guilin City, Yangshuo County, and Pingle County. The total area of the basin is about 6050 km^2, and the climate is characterized by high temperatures and rainfall in summer, cold and drier conditions in winter.

The annual average precipitation is about 1800 mm, and the annual average temperature is about 18 °C [31,32]. The terrain is high in the north and low in the south, and the water system in the river basin is fan-shaped. Floods may easily arise at the confluence of the river systems during heavy rainfall. Carbonate rocks in the basin are widely distributed, forming a typical Karst topography that accentuates droughts and floods in the basin [33]. As a result of the floods and droughts, the study of hydrological processes in this basin has become especially important.

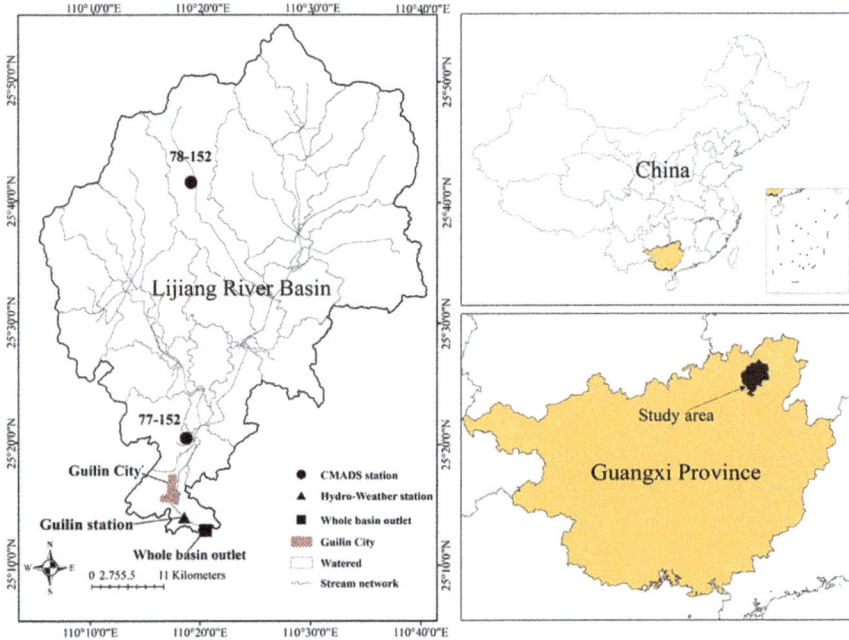

Figure 1. The location of the study area in China.

Guilin City is located in the lower reaches of the fan-shaped watershed, and in 2016 had a population of about 5.34 million and an urban area of 27,800 km^2. In recent years, heavy rainfall in the Lijiang River Basin has led to flood disasters in Guilin, resulting in huge losses of life and property. The section of the Lijiang River that flows though Guilin City was selected as the study area. The study area covers 2531 km^2, and accounts for about 42% of the total basin area.

2.2. SWAT Model Input

The Guilin Hydrological Station was used as the whole outlet. The SWAT 2012 version was used to divide the basin into 33 sub-basins and 355 hydrological response units. The basic data needed for the model included topography, soil, land use, and meteorological data. The data are presented in Table 1:

(i) The digital elevation model used is the first version of the Advanced Spaceborne Thermal Emission and Reflection Radiometer (ASTER) Global Digital Elevation Model (GDEM) (grid cell: 30 m × 30 m). The outliers have been processed, and the original Digital Elevation Model (DEM) has been spliced, cropped, and projected using ArcMAP (ESRI, Redlands, CA, America) software. Sub-watershed divisions, river formation, and slope reclassification were all generated from the pre-treated DEM.

(ii)   The soil data were taken from the 1:1 million soil dataset created by the Second National Land Survey Nanjing Soil Institute and were supplied by the Cold and Arid Regions Sciences Data Center at Lanzhou.

(iii) The land use data were derived from Landsat-8 remote sensing data (multi-spectral band resolution of 30 m) after supervised classification and post-processing steps. The remote sensing data were provided by the Geospatial Data Cloud site, the Computer Network Information Center, the Chinese Academy of Sciences.

(iv)   The meteorological data are taken from the CMADS version 1.1 (http://www.cmads.org). This dataset includes precipitation, temperature, relative humidity, solar radiation, wind speed, location, and the elevation of each site. The data of temperature, relative humidity and wind speed were generated using the information from 2421 national automatic stations and 39,439 regional automatic stations. Precipitation was achieved through the integration of multiple satellite data and precipitation from ground automatic stations. The production of radiation data was based on the Discrete Ordinates Radiative Transfer (DISORT) radiative transfer model and the acquisition of products from the FY2E satellite primary product for inversion of solar shortwave radiation. Two CMADS weather stations are used in the study area.

(v)   The hydrological data were provided by the Guangxi Water Conservancy, and comprise measured daily and monthly data from 2008 to 2016 at the Guilin Hydrological Station.

Table 1. Data description for the study area.

Data Type	Source	Spatial Resolution
DEM	ASTER GDEM https://earthexplorer.usgs.gov/	30 m
Land use	Landsat-8 https://earthexplorer.usgs.gov/	30 m
Soil	HWSD http://westdc.westgis.ac.cn/data/	30 m
Weather	CMADS version 1.1 http://www.cmads.org/	28 km

The SWAT database was constructed using CMADS meteorological data, DEM, and land use and soil data (Figure 2). Daily and monthly scale simulations of the hydrological processes in the Lijiang River Basin were conducted using the measured data from the Guilin Hydrological Station. The model calibration included a 1 year warm-up period (2008), and then the calibration was performed for a period of 2 years (2009–2010), followed by a validation period of 6 years (2011–2016). The simulation results were evaluated using the NSE, R^2, and PBIAS.

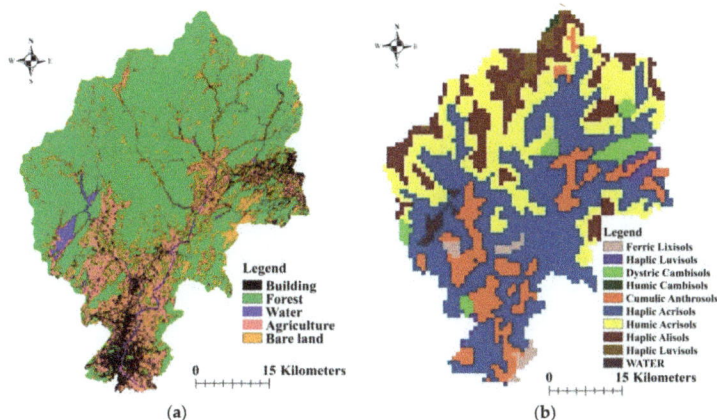

Figure 2. (**a**) Land use data; (**b**) Soil data.

3. Results and Analysis

3.1. Model Calibration and Validation

The Computer Program for Calibration of Soil and Water Assessment Tool Models (SWAT-CUP) software SUFI-2 algorithm was used to calibrate and validate the model. SWAT-CUP is a program that does automatic calibration and uncertainty analysis, and was developed by EWAGE research institute for the SWAT model [34]. The SUFI-2 algorithm uses an inversion modeling method that defines a large range of parameters and then performs multiple iterations. By comparing the results of each iteration, the most suitable parameter range of the model is determined, and uncertainty analysis is conducted by evaluating the range results for each parameter [35]. SUFI-2 is an iterative procedure that accounts for parameter uncertainty from all kinds of sources (e.g., weather, model parameters, and model structure). It provides a comprehensive optimization and uncertainty analysis through the global search method [36]. The calibration and validation of the modeled simulation results are needed for the satisfactory assessment of watershed characteristics.

There are many parameters in SWAT-CUP that affect the simulation of the hydrological cycle. Choosing appropriate parameters can play a crucial role in determining the effectiveness of the calibration. After comparing the efficiency of each parameter, we chose 8 parameters for the monthly simulation, and 13 parameters for the daily simulation (Tables 2 and 3).

Table 2. Ranking of the most sensitive parameters and their monthly simulation variation ranges.

Parameter Name	Description	Min	Max	Value Adopted	Calibration		
					t-Stat	*p*-Value	Rank
R__OV_N	Manning's "n" value for overland flow	10.00	20.00	17.25	−2.68	0.02	1
V__ALPHA_BF	Baseflow alpha factor (days)	0.00	0.50	0.41	2.48	0.03	2
R__CN2	SCS runoff curve number for moisture condition II	0.00	0.60	0.41	1.57	0.14	3
V__CH_K2	Effective hydraulic conductivity in main channel alluvium	100.00	150.00	131.25	−1.25	0.24	4
V__GWQMN	Treshold depth of water in the shallow aquifer required for return flow to occur (mm)	0.00	3.00	2.63	−0.45	0.66	5
R__ESCO	Soil evaporation compensation factor	0.00	0.80	0.30	0.41	0.69	6
R__SOL_AWC(1)	Available water capacity of the soil layer	0.00	0.60	0.11	−0.31	0.76	7
V__GW_DELAY	Groundwater delay (days)	0.00	170.00	46.75	0.04	0.97	8

Table 3. Ranking of the most sensitive parameters and their daily simulation variation ranges.

Parameter Name	Description	Min	Max	Value Adopted	Calibration		
					t-Stat	*p*-Value	Rank
R__CN2	SCS runoff curve number for moisture condition II	−0.30	0.01	−0.17	8.77	0.00	1
R__HRU_SLP	Average slope steepness	−0.98	0.10	−0.35	6.08	0.00	2
R__SOL_K(1)	Saturated hydraulic conductivity	0.00	5.00	0.03	−4.99	0.00	3

Table 3. *Cont.*

Parameter Name	Description	Min	Max	Value Adopted	Calibration		
					t-Stat	*p*-Value	Rank
V_RCHRG_DP	Deep aquifer percolation fraction	0.10	0.40	0.15	−3.65	0.00	4
V_GW_DELAY	Groundwater delay (days)	0.00	2.00	0.17	−3.14	0.00	5
V_OV_N	Manning's "n" value for overland flow	3.00	6.00	5.47	−2.55	0.01	6
V_ALPHA_BF	Baseflow alpha factor (days)	0.10	0.20	0.16	2.50	0.01	7
V_GWQMN	Treshold depth of water in the shallow aquifer required for return flow to occur (mm)	10.00	500.00	46.75	−2.40	0.02	8
R_SOL_Z(1)	Depth from soil surface to bottom of layer	−0.25	0.25	−0.11	1.52	0.13	9
V_CH_K2	Effective hydraulic conductivity in main channel alluvium	0.00	220.00	212.30	1.14	0.25	10
V_REVAPMN	Threshold depth of water in the shallow aquifer for "revap" to occur (mm)	0.00	500.00	367.5	−0.65	0.51	11
R_SOL_AWC(1)	Available water capacity of the soil layer	0.20	0.40	0.21	0.58	0.55	12
R_ESCO	Soil evaporation compensation factor	0.00	0.10	0.04	0.46	0.64	13

The results for the evaluation index after the model runs are shown in Table 4. Previous research results show that if R^2 and NSE are close to 1, then the simulated value of the model is close to the true value. PBIAS is also used as a model evaluation criteria, and an absolute value of less than 10 is usually considered a good result [37]. In the monthly simulation, R^2 and NSE were both 0.96 during the calibration period, while R^2 was 0.96 and NSE was 0.95 during the validation period. PBIAS was less than 10 during the calibration and the validation period. Figure 3 shows the results calculated for the monthly simulated and observed values. The figure shows that the trend and values of the simulated results are very close to the measured results. The figure also shows that runoff characteristics (large summer runoff and a small winter runoff) of the Lijiang River Basin are similar to those found in most regions with a monsoon climate. It is noteworthy that there was a large peak in the Lijiang River runoff in November 2015. This was the result of a rare winter storm in Guilin. Figure 4 shows the calculated results of the daily simulated and observed values. In the daily simulation, R^2 was 0.92 for the calibration period and 0.89 for the validation period. NSE values for the calibration period were 0.89 and 0.88, respectively, and the corresponding PBIAS values were 20.70 and 14.40, respectively. Thus, good daily simulation results were obtained. It can be concluded that the SWAT model, driven by CMADS meteorological data, provided good results for the Lijiang River Basin in Guangxi Province, and the data set and model can be used to further study hydrological processes in this basin.

Table 4. Performance statistics for the flow simulations.

Object	Calibration (2009–2010)	Validation (2011–2016)
P-factor (Monthly)	0.79	0.63
R-factor (Monthly)	0.33	0.37
R^2 (Monthly)	0.96	0.96
NSE (Monthly)	0.96	0.95
PBIAS (Monthly)	7.70	7.80
RSR (Monthly)	0.20	0.22
P-factor (Daily)	0.70	0.77
R-factor (Daily)	0.30	0.43
R^2 (Daily)	0.92	0.89
NSE (Daily)	0.89	0.88
PBIAS (Daily)	20.70	14.40
RSR (Daily)	0.33	0.35

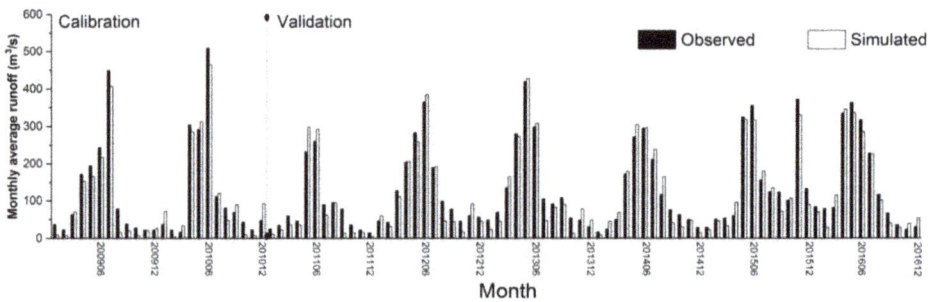

Figure 3. Comparison of monthly runoff using Soil and Water Assessment Tool (SWAT).

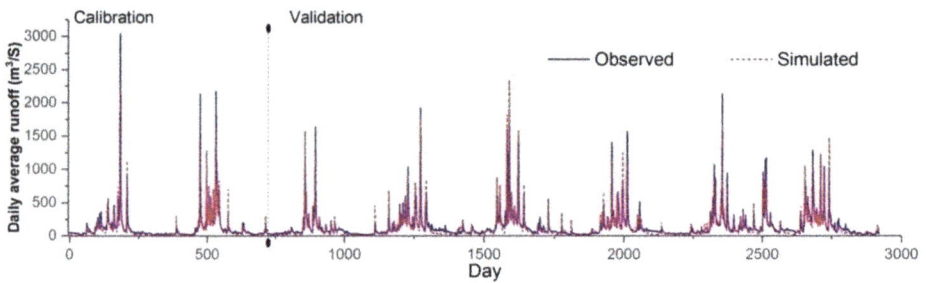

Figure 4. Comparison of daily runoff using SWAT.

3.2. Uncertainty Analysis

The SUFI-2 algorithm was iterated three times, with each calibration and validation iteration running 100 times. The R-factor and P-factor are important indicators for evaluating the uncertainty of simulation results. It is generally assumed that the closer the R-factor is to 0 and the closer the P-factor is 1, the closer the simulation results are to the measured data, and the lower the uncertainty in the model results [38]. Table 3 shows that in the runoff simulation of the Guilin sites, both the R-factor and the P-factor had reached optimal values in the calibration and validation periods, thus indicating that the uncertainty in the simulation results was small.

In addition, the correlation between parameters is an indicator of their redundancy. Figure 5 shows the relationship between the parameters and also the relationship between the parameters and the objective function using the NSE. The correlation between most of the parameters was very small, indicating that the redundancy was small in the parameterization for the Lijiang River Basin. The relationship between the NSE and the parameters in Figure 5, shows that the NSE was always above 0.7, and usually higher than 0.8. In the areas with a high NSE, there are many parameters exhibiting the equifinality phenomenon. These characteristics also indicate that most of the parameters from the model simulation results have a low uncertainty. It should be noted that the degree of aggregation of NSE decreases as the value of the curve number for moisture condition II (CN2) decreases, indicating that CN2 has a greater influence on the uncertainty of the simulation results. Parameter CN2 is associated with soil permeability, land use, and initial soil water condition, and indicates the potential for surface runoff from precipitation in a river basin. To take into account the important impact of CN2 on the hydrological elements of surface runoff, we chose relatively stable values in the −0.3–0.01 range for the final values of CN2.

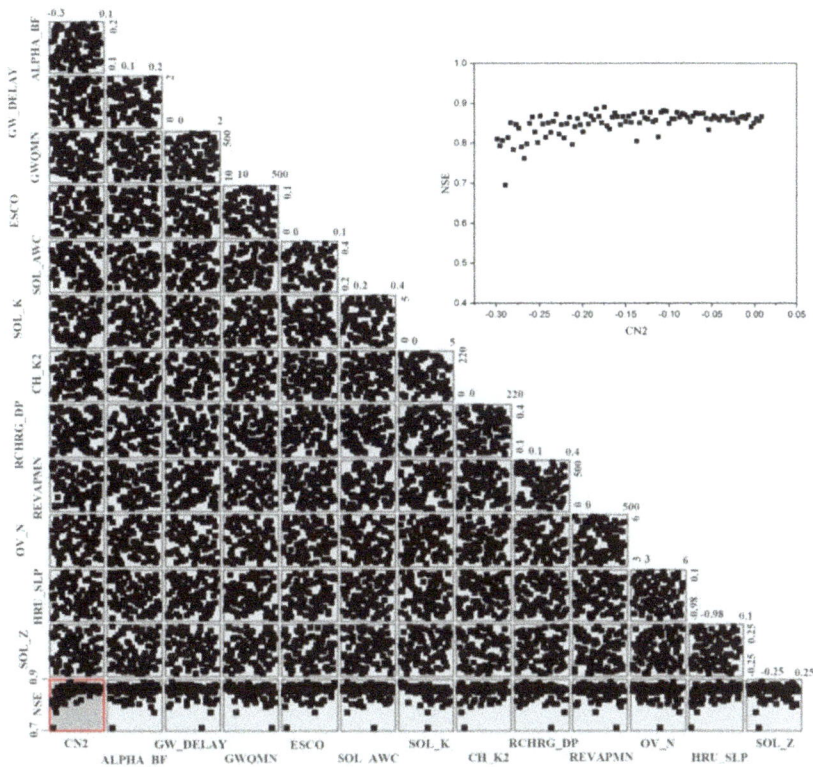

Figure 5. Pair-wise correlations between parameters, and correlations between parameters and Nash-Sutcliffe model efficiency coefficient (NSE).

To show the results of the operation of the SUFI-2 algorithm, we used kernel smoothing to represent the distribution of NSE. Figure 6 shows that the NSE values for each simulation are larger than the SUFI-2 algorithm's default value of 0.5. Most of the values were concentrated between 0.82–0.87. These distributions show that the SUFI-2 algorithm performed well, and that the uncertainty in the model results was low.

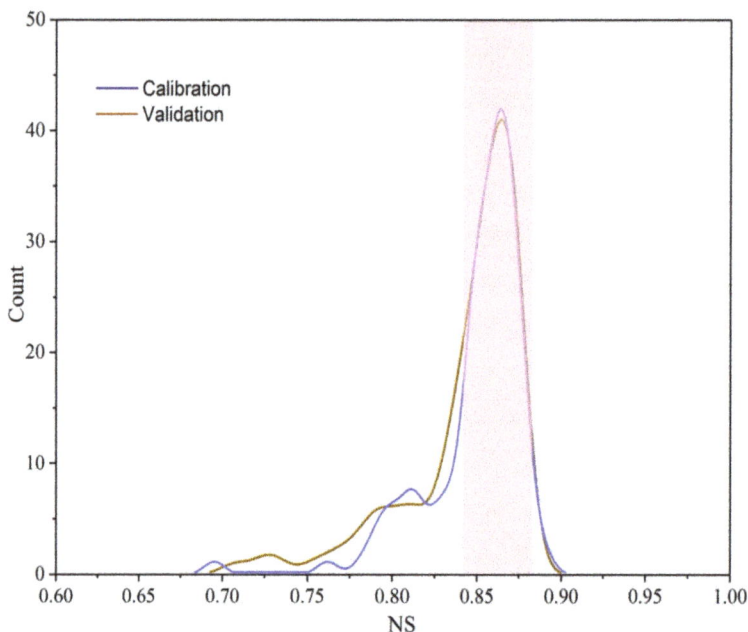

Figure 6. Kernel smoothing fit of the distribution of NSE.

3.3. Water Balance Components

The SWAT model often overestimates or underestimates some elements of hydrological budgets. Calibration ensures that the simulated values are closer to the observed values, and also ensures that the hydrological elements are in a reasonable range. Table 5 shows the average annual contributions to the water budget for the main hydrological elements. From 2009–2016, the average annual precipitation was up to 2150.20 mm. The average annual values of surface runoff, evapotranspiration (ET), lateral flow, and shallow groundwater in the Lijiang River Basin were 518.36 mm, 750.60 mm, 129.21 mm, and 555.34 mm, respectively. Figure 7 shows the average annual values of the hydrological elements as a relative percentage of precipitation for uncalibrated and calibrated periods. The figure shows that the percentage of deep aquifer recharge, deep aquifer flow, shallow aquifer flow, and lateral flow increased. Actual ET and surface runoff decreased. In the calibration period, ET caused major water losses, and the average proportion of ET to precipitation was 34.9% per year. The low latitude and high temperature of the basin location contributed to the high ET, and the wide distribution of agriculture further increased the ET. The average annual contribution of lateral flow as a relative percentage of precipitation was 6.0%. Shallow groundwater flow to streamflow accounted for 25.8% of precipitation. Deep aquifer recharge accounted for 8.8% of the total precipitation.

It was possible to view the variation in the model's output across the basin. Figure 8 shows the spatial distribution of precipitation, actual ET, surface runoff contribution to streamflow, and groundwater contribution to streamflow during the study period. When the spatial distribution of ET (Figure 8) and land use (Figure 2) are compared, it can be seen that the lake area and the agricultural area have high ET values. The distribution of surface runoff contribution to streamflow, and of groundwater contribution to streamflow, is related to precipitation. There is more precipitation in the northern part of the basin and less precipitation in the south. The contribution of surface runoff and groundwater to streamflow in the sub-basins upstream is also consistent with the spatial distribution of precipitation in the basin that was studied.

Table 5. Average annual contribution by the hydrological elements to the water budget.

Hydrological Elements	Calibration
Precipitation	2150.20 mm
Surface runoff	518.36 mm
Lateral flow	129.21 mm
Shallow groundwater contribute to streamflow	555.34 mm
Deep groundwater contribute to streamflow	188.62 mm
Total aquifer recharge	746.08 mm
Deep groundwater recharge	189.88 mm
Water yield	1391.51 mm
Evapotranspiration	750.6 mm

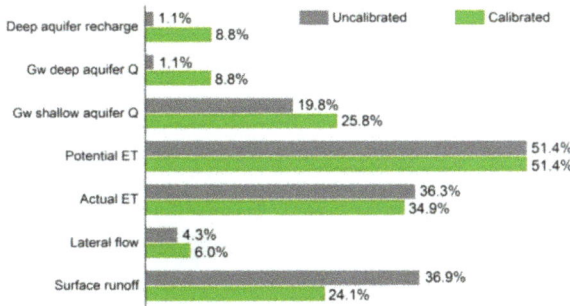

Figure 7. Average annual values of hydrological elements as a percentage of precipitation for pre- and post-calibration periods.

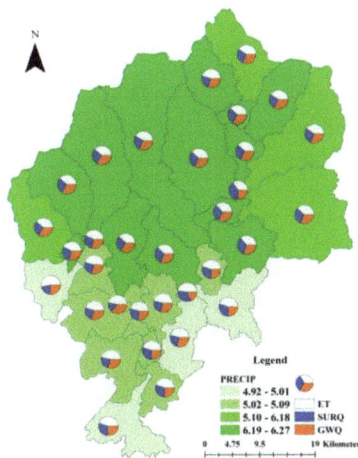

Figure 8. Spatial distribution of actual evapotranspiration (ET), surface runoff, groundwater discharge, and precipitation for the study period.

Figure 9 shows the variation in the contribution to the water budget by the main hydrological elements in the Lijiang River Basin from 2009–2016. The contribution to the water budget by groundwater discharge, lateral flow, and surface runoff decreased significantly in 2011, which was a dry year. The contributions by groundwater, lateral flow, and surface runoff reached their highest values in 2015, which was a flood year. The annual changes in the contributions by the different components of the water budget are consistent with the annual changes in precipitation. The change rates for surface runoff, lateral flow, and groundwater discharge decreased the most in 2011, and increased the most in 2012. The contribution by surface runoff decreased by about 38.4% in 2010–2011 and increased by 80.0% in 2011–2012. The contribution by lateral flow decreased by about 28.8% in 2010–2011 and increased by 56.4% in 2011–2012. The contribution by groundwater decreased by about 39.4% in 2010–2011 and increased by 81.9% in 2011–2012. It is worth noting that contributions to the water budget may be carried over from year to year. For example, precipitation in 2012 was more than in 2013, but surface runoff in 2012 was less than in 2013. The water balance in 2012 may have been affected by the drought in the previous year. However, there was no significant change in ET, which remained stable during the calibration and validation periods. The average annual change rate for ET was only about 1.6%.

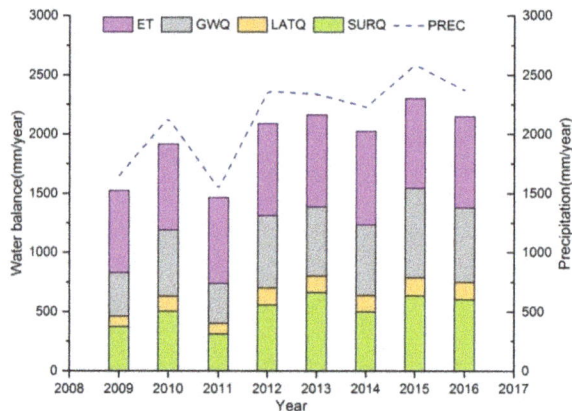

Figure 9. Annual change in contribution by the hydrological elements to the water budget.

Figure 10 shows the variation in each element of the water balance at a monthly scale. ET was an important loss in the basin water balance, and its change shows clear seasonal variations. There was little ET in winter and high ET in summer. The seasonal variation in surface runoff, lateral flow, and groundwater discharge was consistent with the change in precipitation. The proportion of surface flow to precipitation was 0.1–40.4% at the monthly scale, and there was a noticeable difference between the winter and summer percentages. The proportion of lateral flow to precipitation was 4.2–21.7%. Lateral flow was a large proportion of precipitation in summer and a small proportion in winter. The groundwater discharge was also consistent with changes in the precipitation. The proportion of groundwater discharge to precipitation varies widely between summer and winter.

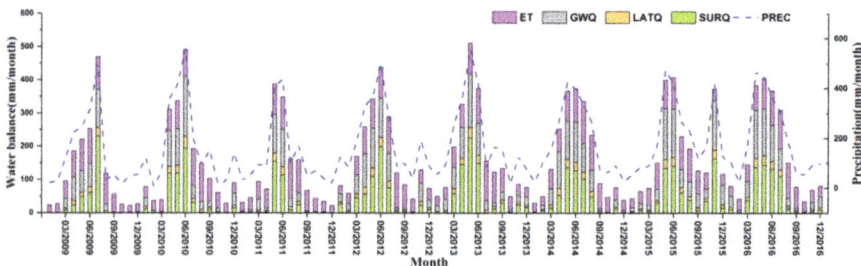

Figure 10. Monthly change in the contribution by the hydrological elements to the water budget.

The relationship between each water balance element and precipitation is shown in Figure 11. The correlation between the contribution of each hydrological element and precipitation was analyzed separately. In addition, the P-value is less than 0.01 in the significance test of precipitation and these hydrological elements. Precipitation has an important influence on surface runoff. The surface flow and precipitation in the Lijiang River Basin maintained a curvilinear relationship with a R^2 of 0.903. This relationship between surface runoff and precipitation implies that high precipitation is likely to rapidly increase surface runoff and cause floods in the Lijiang River Basin. Lateral runoff and precipitation maintained a linear correlation, with a R^2 of 0.971. The relationship between groundwater discharge and precipitation was also analyzed, and a R^2 of 0.926 was obtained. The relationship between ET and precipitation in summer and winter was also analyzed. The correlation between ET and precipitation in summer was 0.153. The correlation between ET and precipitation in winter was 0.094. The results show that there was no significant relationship between ET and precipitation in the Lijiang River Basin.

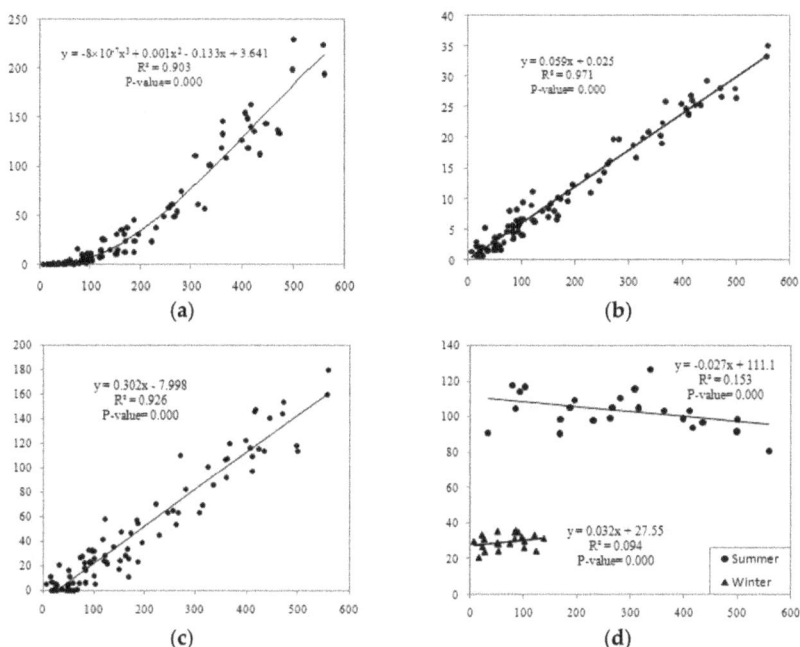

Figure 11. Regression analysis results for the hydrological elements and precipitation. (**a**) Surface flow and precipitation; (**b**) Lateral runoff and precipitation; (**c**) Groundwater discharge and precipitation; (**d**) ET and precipitation.

4. Conclusions

The present study used CMADS data and the SWAT model to successfully generate daily and monthly scale runoff simulations for the Lijiang River Basin. The analysis of pair-wise correlations between the parameters shows that the redundancy was small in the parameterization. Both the R-factor and P-factor reached ideal values in the calibration and validation periods, which indicated that there was low uncertainty in the simulation results and model parameters.

Using the model's output, the average annual contribution to the water budget by hydrological elements was analyzed. From 2009–2016, the average annual value of surface runoff, ET, lateral flow, and shallow groundwater in the Lijiang River Basin were 518.36 mm, 750.60 mm, 129.21 mm,

555.34 mm, respectively. The spatial distribution of surface runoff and groundwater discharge was related to precipitation. The highest ET values were obtained in the west of the basin, where agriculture is prevalent. The water budget of groundwater discharge, lateral flow, and surface runoff reached the highest values in the flood year, and reached the lowest values in the driest year. The high correlation between these elements and precipitation was reflected in the regression analysis. ET remained stable during the calibration and validation period. The results for the hydrological elements could provide valuable reference information for water resources management in the Lijiang River Basin.

Author Contributions: The modeling and writing of this work by Y.C.; J.Z. designed the experiments; Observed data was provided by M.Y. and Z.Z.; The improvement of manuscript writing by J.Z., B.G., L.Y., J.Q. and X.L.

Acknowledgments: This work is supported by National Key R&D Program of China (2017YFC0406004), NSFC (41271004) and the project of Ecological Dispatching Technology of Flood Prevention in Guilin City and Lijiang River Replenishment Reservoir (GXZC2016-G3-2344-JHZJ; AB16380313).

Conflicts of Interest: The authors declare no conflicts of interest.

References

1. Meng, X.; Wang, H.; Wu, Y.; Long, A.; Wang, J.; Shi, C.; Ji, X. Investigating spatiotemporal changes of the land-surface processes in Xinjiang using high-resolution CLM3.5 and CLDAS: Soil temperature. *Sci. Rep.* **2017**, *7*, 13286. [CrossRef] [PubMed]
2. Meng, X.; Sun, Z.; Zhao, H.; Ji, X.; Wang, H.; Xue, L.; Wu, H.; Zhu, Y. Spring flood forecasting based on the WRF-TSRM mode. *Teh. Vjesn.* **2018**, *25*, 141–151. [CrossRef]
3. Allen, M.R.; Ingram, W.J. Constraints on future changes in climate and the hydrologic cycle. *Nature* **2002**, *419*, 224–232. [CrossRef] [PubMed]
4. Sandra, G.; Sathian, K.K. Assessment of water balance of a watershed using swat model for water resources management. *Int. J. Eng. Sci. Res. Technol.* **2016**, *5*, 177–184. [CrossRef]
5. Arnold, J.G.; Allen, P.M. Estimating hydrologic budgets for three Illinois watersheds. *J. Hydrol.* **1996**, *176*, 57–77. [CrossRef]
6. Zhang, J.; Li, Q.; Guo, B.; Gong, H. The comparative study of multi-site uncertainty evaluation method based on SWAT model. *Hydrol. Process.* **2015**, *29*, 2994–3009. [CrossRef]
7. Arnold, J.G.; Srinivasan, R.; Muttiah, R.S.; Williams, J.R. Large area hydrologic modeling and assessment. Part I: Model development. *J. Am. Water Resour. Assoc.* **1998**, *34*, 73–89. [CrossRef]
8. Sophocleous, M.A.; Koelliker, J.K.; Govindaraju, R.S. Integrated numerical modeling for basin-wide water management: The case of the Rattlesnake Creek basin in south-central Kansans. *J. Hydrol.* **1999**, *214*, 179–196. [CrossRef]
9. Eckhardt, K.; Haverkamp, S.; Fohrer, N.; Frede, H.G. SWAT-G, a version of SWAT99.2 modified for application to low mountain range catchments. *Phys. Chem. Earth* **2002**, *27*, 641–644. [CrossRef]
10. Eckhardt, K.; Nicola, F.; Frede, H.G. Automatic model calibration. *Hydrol. Process.* **2005**, *19*, 651–658. [CrossRef]
11. Easton, Z.M.; Fuka, D.R.; Walter, M.T.; Cowan, D.M.; Schneiderman, E.M.; Steenhuis, T.S. Re-conceptualizing the Soil and Water Assessment Tool (SWAT) model to predict runoff from variable source areas. *J. Hydrol.* **2008**, *348*, 279–291. [CrossRef]
12. Meng, X.; Long, A.; Wu, Y.; Yin, G.; Wang, H.; Ji, X. Simulation and spatiotemporal pattern of air temperature and precipitation in Eastern Central Asia using RegCM. *Sci. Rep.* **2018**, *8*, 3639. [CrossRef] [PubMed]
13. Meng, X.; Yu, D.; Liu, Z. Energy balance-based SWAT model to simulate the mountain snowmelt and runoff—Taking the application in Juntanghu watershed (China) as an example. *J. Mt. Sci.* **2015**, *12*, 368–381. [CrossRef]
14. Meng, X.; Wang, H. Significance of the China Meteorological Assimilation Driving Datasets for the SWAT Model (CMADS) of East Asia. *Water* **2017**, *9*, 765–769. [CrossRef]
15. Liu, J.; Shangguan, D.; Liu, S.; Ding, Y. Evaluation and hydrological simulation of CMADS and CFSR reanalysis datasets in the Qinghai-Tibet Plateau. *Water* **2018**, *10*, 513. [CrossRef]

16. Liu, J.; Liu, S.; Shangguan, D.; Xu, J. Applicability evaluation of precipitation datasets from CMADS, ITPCAS and TRMM 3B42 in Yurungkax River Basin. *J. North China Univ. Water Res. Electr. Power* **2017**, *38*, 28–37. (In Chinese) [CrossRef]

17. Meng, X.; Wang, H.; Lei, X.; Cai, S.; Wu, H.; Ji, X.; Wang, J. Hydrological modeling in the Manas River Basin using soil and water assessment tool driven by CMADS. *Teh. Vjesn.* **2017**, *24*, 525–534. [CrossRef]

18. Meng, X.; Wang, H.; Cai, S.; Zhang, X.; Leng, G.; Lei, X.; Shi, C.; Liu, S.; Shang, Y. The China Meteorological Assimilation Driving Datasets for the SWAT Model (CMADS) application in China: A case study in Heihe River Basin. Available online: https://www.preprints.org/manuscript/201612.0091/v2 (accessed on 18 December 2016).

19. Zhao, F.; Wu, Y.; Qiu, L.; Sun, Y.; Sun, L.; Li, Q.; Niu, J.; Wang, G. Parameter uncertainty analysis of the SWAT model in a mountain-loess transitional watershed on the Chinese Loess Plateau. *Water* **2018**, *10*, 690. [CrossRef]

20. Leta, O.T.; EI-Kadi, A.I.; Dulai, H. Implications of climate change on water budgets and reservoir water harvesting of Nuuanu Area Watersheds, Oahu, Hawaii. *J. Water Resour. Plan. Manag.* **2017**, *143*. [CrossRef]

21. Cuceloglu, G.; Abbaspour, K.C.; Ozturk, I. Assessing the water-resources potential of Istanbul by using a Soil and Water Assessment Tool (SWAT) hydrological model. *Water* **2017**, *9*, 814–833. [CrossRef]

22. Zhou, G.-Y.; Wei, X.-H.; Wu, Y.-P.; Liu, S.-G.; Huang, Y.-H.; Yan, J.-H.; Zhang, D.-Q.; Zhang, Q.-M.; Liu, J.-X.; Meng, Z. Quantifying the hydrological responses to climate change using an intact forested small watershed in southern China. *Glob. Chang. Biol.* **2011**, *17*, 3736–3746. [CrossRef]

23. Wu, Y.-P.; Liu, S.-G.; Yan, W.-D.; Xia, J.-Z.; Xiang, W.-H.; Wang, K.-L.; Luo, Q.; Fu, W.; Yuan, W.-P. Climate change and consequences on the water cycle in the humid Xiangjiang River Basin, China. *Stoch. Environ. Res. Risk Assess.* **2016**, *30*, 225–235. [CrossRef]

24. Dhami, B.; Himanshu, S.K.; Pandey, A.; Gautam, A.K. Evaluation of the SWAT model for water balance study of a mountainous snowfed river basin of Nepal. *Environ. Earth Sci.* **2018**, *77*, 1–20. [CrossRef]

25. Troin, M.; Caya, D. Evaluating the SWAT's snow hydrology over a Northern Quebec watershed. *Hydrol. Process.* **2014**, *28*, 1858–1873. [CrossRef]

26. Zhang, Y.-Q.; Luo, Y.; Sun, L.; Liu, S.-Y.; Chen, X.; Wang, X.-L. Using glacier area ratio to quantify effects of melt water on runoff. *J. Hydrol.* **2016**, *538*, 269–277. [CrossRef]

27. Golmohammadi, G.; Rudra, R.; Prasher, S. Impact of tile drainage on water budget and spatial distribution ofsediment generating areas in an agricultural watershed. *Agric. Water Manag.* **2017**, *184*, 124–134. [CrossRef]

28. Ale, S.; Bowling, L.C.; Brouder, S.M.; Frankenberger, J.R.; Youssef, M.A. Simulated effect of drainage water management operational strategy onhydrology and crop yield for Drummer soil in the Midwestern United States. *Agric. Water Manag.* **2009**, *96*, 653–665. [CrossRef]

29. Singh, R.; Helmers, M.J.; Crumpton, W.G.; Lemke, D.W. Predicting effects of drainage water management in Iowa's subsurface drained landscapes. *Agric. Water Manag.* **2007**, *92*, 162–170. [CrossRef]

30. Vu, T.T.; Li, L.; Jun, K.S. Evaluation of multi-satellite precipitation products for streamflow simulations: A case study for the Han River Basin in the Korean Peninsula, East Asia. *Water* **2018**, *10*, 642. [CrossRef]

31. Gao, M.-H.; Wu, Z.-Q.; Huang, L.-L.; Ding, Y.; Zhu, Z.-J. Length–weight relationships of 13 fish species from the Lijiang River, China. *Tech. Contrib.* **2018**, *34*, 180–182. [CrossRef]

32. Li, J.; Zhang, Y.; Qin, Q.-M.; Yan, Y.-G. Investigating the impact of human activity on land use/cover change in China's Lijiang River Basin from the perspective of flow and type of population. *Sustainability* **2017**, *9*, 383. [CrossRef]

33. Liu, G.; Jin, Q.-W.; Li, J.-Y.; Li, L.; He, C.-X.; Huang, Y.-Q.; Yao, Y.-F. Policy factors impact analysis based on remote sensing data and the CLUE-S model in the Lijiang River Basin, China. *CATENA* **2017**, *158*, 286–297. [CrossRef]

34. Schuol, J.; Abbaspour, K.C. Using monthly weather statistics to generate daily data in a SWAT model application to West Africa. *Ecol. Model.* **2007**, *201*, 301–311. [CrossRef]

35. Abbaspour, K.C.; Genuchten, M.T.; Schulin, R.; Schlappi, E. A sequential uncertainty domain inverse procedure for estimating subsurface flow and transport parameters. *Water Resour. Res.* **1997**, *33*, 1879–1892. [CrossRef]

36. Abbaspour, K.C.; Johnson, C.A.; Van Genuchten, M.T. Estimating uncertain flow and transport parameters using a sequential uncertainty fitting procedure. *Vadose Zone J.* **2004**, *3*, 1340–1352. [CrossRef]

37. Moriasi, D.; Arnold, J.G.; Vanliew, M.W.; Bingner, R.L.; Harmel, R.D.; Veith, T.L. Model evaluation guidelines for systematic quantification of accuracy in watershed simulations. *Trans. ASABE* **2007**, *50*, 885–900. [CrossRef]

38. Bekele, E.G.; Nicklow, J.W. Multi-objective automatic calibration of SWAT using NSGA-II. *J. Hydrol.* **2007**, *341*, 165–176. [CrossRef]

water

Article

Applicability Assessment and Uncertainty Analysis of Multi-Precipitation Datasets for the Simulation of Hydrologic Models

Binbin Guo [1,2], Jing Zhang [1,*], Tingbao Xu [3], Barry Croke [3,4], Anthony Jakeman [3], Yongyu Song [1], Qin Yang [1,2], Xiaohui Lei [5] and Weihong Liao [5]

[1]   Beijing Key Laboratory of Resource Environment and Geographic Information System, Capital Normal University, Beijing 100048, China; guobinbin@126.com (B.G.); songdy1006@gmail.com (Y.S.); yqinss@163.com (Q.Y.)
[2]   College of City and Tourism, Hengyang Normal University, Hengyang 421008, China
[3]   Fenner School of Environment and Society, The Australian National University, Canberra, ACT 2601, Australia; tingbao.xu@anu.edu.au (T.X.); barry.croke@anu.edu.au (B.C.); Tony.Jakeman@anu.edu.au (A.J.)
[4]   Mathematical Sciences Institute, The Australian National University, Canberra, ACT 2601, Australia
[5]   State Key Laboratory of Simulation and Regulation of Water Cycle in River Basin, China Institute of Water Resources and Hydropower Research, Beijing 100038, China; Lxh@iwhr.com (X.L.); liaowh@iwhr.com (W.L.)
*   Correspondence: 5607@cnu.edu.cn; Tel.: +86-10-6890-3139

Received: 10 October 2018; Accepted: 5 November 2018; Published: 9 November 2018

Abstract: Hydrologic models are essential tools for understanding hydrologic processes, such as precipitation, which is a fundamental component of the water cycle. For an improved understanding and the evaluation of different precipitation datasets, especially their applicability for hydrologic modelling, three kinds of precipitation products, CMADS, TMPA-3B42V7 and gauge-interpolated datasets, are compared. Two hydrologic models (IHACRES and Sacramento) are applied to study the accuracy of the three types of precipitation products on the daily streamflow of the Lijiang River, which is located in southern China. The models are calibrated separately with different precipitation products, with the results showing that the CMADS product performs best based on the Nash–Sutcliffe efficiency, including a much better accuracy and better skill in capturing the streamflow peaks than the other precipitation products. The TMPA-3B42V7 product shows a small improvement on the gauge-interpolated product. Compared to TMPA-3B42V7, CMADS shows better agreement with the ground-observation data through a pixel-to-point comparison. The comparison of the two hydrologic models shows that both the IHACRES and Sacramento models perform well. The IHACRES model however displays less uncertainty and a higher applicability than the Sacramento model in the Lijiang River basin.

Keywords: precipitation; TMPA-3B42V7; CMADS; hydrologic model; uncertainty

1. Introduction

Hydrologic models are essential tools for understanding processes of the hydrologic cycle and provide useful information for sustainable water-resource management [1]. Precipitation is the main driving factor of hydrologic processes. Accurate estimation of precipitation is crucial for reliable hydrologic predictions [2]. Traditionally, precipitation data from a ground observational network have been used as the source of areal precipitation estimates used in watershed modelling. However, ground-based precipitation observation networks are sparsely distributed and may be unable to represent the spatial variability of the precipitation completely. Moreover, precipitation measurements are frequently missing because of malfunctioning of devices [3]. Remote sensing [4] and modelling [5] of precipitation have become viable approaches to address these problems effectively and are often used as input data to hydrologic models.

With regard to the development of remote-sensing technology, satellite-derived precipitation data are an attractive alternative in data-sparse regions because of the relatively high resolution and complete spatial coverage. A number of such remotely sensed precipitation products are currently available. These include, for example, the Climate Prediction Center morphing method (CMORPH, [6]), the Global Satellite Mapping of Precipitation (GSMaP) project [7], the Tropical Rainfall Measuring Mission (TRMM) Multi-satellite Precipitation Analysis (TMPA) [8] and the Global Precipitation Measurement (GPM) products [9]. Among them, the TMPA products developed by the National Aeronautics and Space Administration (NASA) Goddard Space Flight Center (GSFC), with a spatial resolution of 0.25° × 0.25° for multiple timescales (3 hourly, daily and monthly), has received much more attention [10]; the latest research product of TMPA for post-real-time research (3B42) is version 7. Most of the applications using the TMPA-3B42V7 product indicate an excellent potential to supply reasonably high spatial and temporal resolution data for hydrometeorological applications [10–13]. However, remotely sensed precipitation data suffer from uncertainty in their retrieval algorithms and observation errors [14] due to the inference of rainfall based on observations of the conditions at the top of clouds.

While precipitation modelling is fairly accurate for coarse-scale (global-scale), organized, synoptic systems, the modelling accuracy decreases rapidly for more localized events as spatial and temporal features cannot be explicitly resolved by global models [2]. Reanalysis datasets, which are produced by assimilating multi-source data into a climate model, are a viable option of deriving reliable precipitation estimates [5]. Commonly used reanalysis datasets include the Climate Forecast System Reanalysis (CFSR) [15] from National Center for Environmental Prediction (NCEP), the European Centre for Medium-Range Weather Forecasts (ECMWF) Reanalysis from September 1957 to August 2002 (ERA-40 [16]) and the ECMWF Reanalysis-Interim (ERA-Interim [17]) products. While these reanalysis datasets provide important basic data for global researchers for the analysis of climate–water cycles, the spatial resolution of global reanalysis datasets is often too coarse to be used reliably in local-scale studies. Hydrologic modelling forced by reanalysis datasets has been conducted by, for example, Andreadis et al. (2017) [5], who reproduced flooding over large scales by using the Twentieth Century Reanalysis (20CRv2, [18]) dataset and downscaling techniques. Fuka et al. (2014) found the CFSR precipitation product provides a relatively reliable precipitation input for the hydrologic modelling of large-area basins.

Given the strongly underconstrained nature of precipitation inversion, data assimilation based on the large number of stations on regional scales has the potential to resolve fine-scale structures and microphysical processes with more details. The China Meteorological Assimilation Driving Datasets for the Soil and Water Assessment Tool (SWAT) model (CMADS,) developed by Dr. Xianyong Meng from the China Agricultural University (CAU), has received worldwide attention [19,20]. CMADS incorporate Space and Time Mesoscale Analysis System (STMAS) assimilation techniques [21,22] and multiple other techniques, such as loop nesting of data, projection of resampling models and bilinear interpolation. The precipitation data of the CMADS product is generated by the assimilation of multi-satellite data and precipitation from ground stations. Using CMORPH satellite products as the background field, the CMADS product assimilates hourly precipitation products of nearly 40,000 regional automatic stations and 2421 national automatic stations in China. Relative studies found the CMADS product significantly reduces the uncertainties of precipitation input for the hydrologic modelling [19]. CMADS has been verified in several basins in China and Korea [20,23–29]. However, reanalysis datasets are limited by the quality of precipitation observations and the uncertainty from the assimilation model.

A number of spatial-interpolation methods [30] are commonly used for estimating precipitation based on ground-observation data, even in data-sparse regions [31,32]. Conventional interpolation methods, such as the Thiessen polygon [33,34] and inverse-distance weighting [35], are widely used for precipitation interpolation [32]. Ordinary kriging [36] is a geostatistical technique requiring prior calibration of a semivariogram for its parameters (range, nugget and sill). These methods are suitable for application over relatively flat areas. These methods assume that other potential drivers (particularly topography) of the spatial variation in rainfall is captured by the gauge data and information on other drivers is not needed. In data-sparse regions this is not correct and methods that explicitly include the topography are preferred. Hutchinson found that the interpolating accuracy of a precipitation surface would be enhanced significantly with an appropriate digital elevation model (DEM) [37–39]. The advantages of the ANUSPLIN package (Hutchinson and Xu, 2013) over kriging are its simplicity and there is no requirement of a separate calibration of the spatial-covariance structure. The ANUSPLIN interpolation technique has been applied in a number of studies, proving to be one of the best techniques for interpolating point precipitation data [40–42]. However, if the meteorological stations are very sparse, obtaining an accurate distribution of precipitation values through interpolation is impossible. The low density of precipitation stations is a major uncertainty source, which potentially impacts the result. Moreover, interpolation of precipitation data is unable to capture some extreme weather conditions. All interpolation techniques have difficulty in simulating sharply varying climate transitions.

Spatially distributed precipitation datasets incorporate uncertainties or errors resulting from the interpolation and retrieval algorithms, the quality of precipitation observations and the uncertainty from the assimilation model. As different precipitation datasets are limited by quantitative inaccuracies, they exhibit significant bias [43]. Smith and Kummerow [44] analyzed the water budgets of precipitation datasets from in situ, reanalysis and satellite data over the upper Colorado River basin and found the reanalysis datasets tend to overestimate in situ data, while satellite-derived precipitation data underestimate in situ data. Pfeifroth et al. [45] evaluated satellite-based and reanalysis precipitation data in the tropical Pacific and found reanalysis products overestimate small and medium precipitation amounts but underestimate high amounts. Some studies have found that runoff-generation is highly sensitive to the spatial and temporal variability of precipitation data, as a result this is found to be the main source of uncertainty in rainfall–runoff modelling [46]. Therefore, assessing the accuracy of different precipitation products and their applicability and uncertainty for hydrologic models is of great importance; the uncertainties associated with hydrologic models also play a role in the performance of hydrologic simulations [47].

Our main objective here is to assess and evaluate three general precipitation datasets in terms of their accuracy and efficacy, including the CMADS, TMPA-3B42V7 and gauge-interpolated product. The assessment is based on the simulation results from two well-known hydrologic models (IHACRES and Sacramento models). In addition, the precipitation detection capability of TRMA-3B42V7 and CMADS datasets is also evaluated through their pixel-to-point comparison to the ground-based data. The applicability of these two models is assessed. Moreover, the parameter uncertainty of each hydrologic model is also explored as this is another source of uncertainty in modelling streamflow. This research will provide more insight into precipitation analyses and hydrologic modelling.

2. Material and Methods

2.1. Study Area

The Lijiang River basin ($25°12'$–$25°55'$ N, $110°5'$–$110°40'$ E) is located in the northeastern Guangxi Zhuang Autonomous Region of China and belongs to the upper reaches of the Guijiang River in the Pearl River system, with an area of 2591 km^2. The basin is an important headwater for the downstream Guilin City and has a sparse and unevenly distributed meteorological observation network.

The climate in this area is mainly sub-tropical monsoonal, with the wet season from March to August and the dry season from September to February. From 1961 to 2016, the average daily temperature was 19.10 ± 0.06 °C and the yearly average precipitation was 1900 ± 50 mm, varying between 1253.6 and 3011 mm. With elevation ranging from 32 m to 2037 m (Figure 1), the terrain is high in the north and low in the south, with highly complex topography consisting of steep mountains and floodplains.

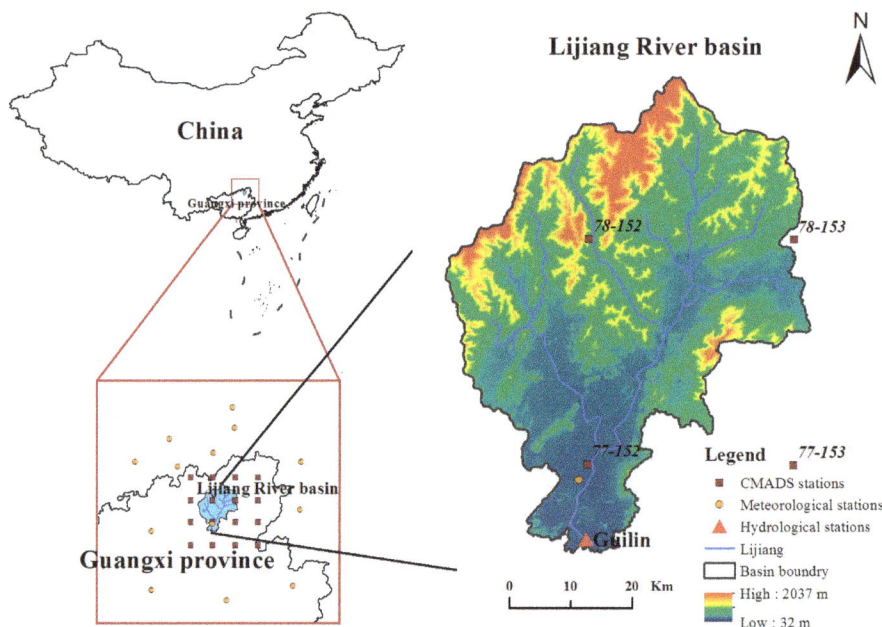

Figure 1. Location of Lijiang River basin, China and meteorological stations for the ANUSPLIN interpolation technique.

The basin is one of the most famous karst areas in the world but suffers from the fragile ecology of the karst geomorphology. The rapid development of tourism and urbanization has promoted the economy, while also causing serious environmental issues. Therefore, the development of hydrologic models in the basin is important for aiding understanding of the hydrologic processes and the formulation of scientific strategies for the management of its water resources.

2.2. Dataset Acquisition

Assessment and analysis have been conducted using three kinds of precipitation data, including in situ measurements, remote-sensing products and reanalysis data. The in situ measurements of daily precipitation from 13 meteorological stations of the national weather station network are interpolated with the ANUSPLIN technique. The mean density of meteorological stations (number of gauges per 10^4 km^2) is 0.91 (calculated using the kernel-density estimation for a search radius of 100 km; see Supplementary Materials for more details). The remote-sensing products are using the TRMM–TMPA product (also denoted TMPA-3B42V7) available from the National Aeronautics and Space Administration (NASA) official website (https://pmm.nasa.gov/trmm). The reanalysis data originate from the CMADS V1.0 product available from World Data System for Cold and Arid Regions (CARD) official website (http://westdc.westgis.ac.cn). The precipitation and maximum temperature values from 13 national meteorological stations are available from the Meteorological Data network (http://data.cma.cn/).

The ANUSPLIN method interpolates meteorological data from the station to the grid scale at the surface. Precipitation and maximum temperature data from meteorological stations, as well as precipitation from CMADS grid points, are interpolated to the surface grid by use of the ANUSPLIN package version 4.4, which interpolates precipitation and temperature as a function of latitude, longitude and elevation, while accounting for the effect of topography [37–39,48]. This method interpolates precipitation and the maximum temperature to grids of size of 0.01° × 0.01° (≈1 km × 1 km), before combination with a digital elevation model (DEM) of the same resolution. The DEM for integration of the final climate grid products is derived from Global Multi-resolution Terrain Elevation Data 2010 (GMTED2010) 7.5-arc-second dataset using the Australian National University Digital Elevation Model version 5.3 (ANUDEM5.3) [49–51]. This the latest and possibly best global-terrain product to date, since it uses ground surface elevation rather than a canopy top surface as found in some satellite terrain products. Due to the smoother nature of climate surfaces relative to the underlying topography, a rebuilt 1-km resolution DEM has been used to generate the resulting climate surfaces, yielding climate grids with more realistic spatial-distribution patterns. Cross-validation statistics were calculated to evaluate the overall predictive error of the ANUSPLIN precipitation data from meteorological stations, with results demonstrating that the distributed precipitation interpolated to the surface grid by the ANUSPLIN method provides a reliable precipitation distribution for input into hydrologic models (see Supplementary Materials for more details).

The latest research product of TMPA for post-real-time research version 7 (TMPA-3B42V7), which has a spatial resolution of 0.25°, is used here. With a spatial resolution of the CMADS1.0 product of 1/3°, this study includes 16 CMADS grid points within and around the basin (Figure 1).

The daily discharge data of the Guilin hydrologic station from 2008 to 2016 (9 years) is provided by the Hydrological Bureau of the Guangxi Zhuang Autonomous Region. Time series of streamflow, maximum temperature and rainfall were used as input for hydrologic models. The spatial distributions of annual precipitation derived from CMADS, TRMM and ground-observation and maximum temperature are shown in Figure 2 (the annual average precipitation and average daily maximum temperature were calculated based on grid data from 2008 to 2016). The spatial distribution of observed precipitation and maximum temperature is interpolated from meteorological stations shown in Figure 1 using the ANUSPLIN interpolation technique. The annual precipitation estimation for CMADS, TRMM and ground-observation have similar spatial distribution patterns, with a decreasing pattern from south to north and from west to east. The average daily maximum temperature is higher in the south than in the north.

2.3. Rainfall–Runoff Models

We used two well-known conceptual rainfall–runoff models with different complexities ranging from 6–13 parameters (Table 1) to assess if the performance of the model improved with greater complexity, as well as their applicability in streamflow prediction in the Lijiang River basin.

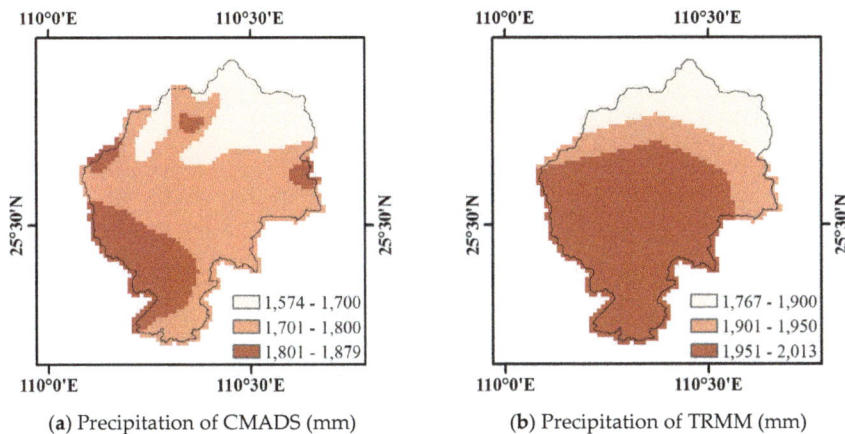

(**a**) Precipitation of CMADS (mm)   (**b**) Precipitation of TRMM (mm)

Figure 2. *Cont.*

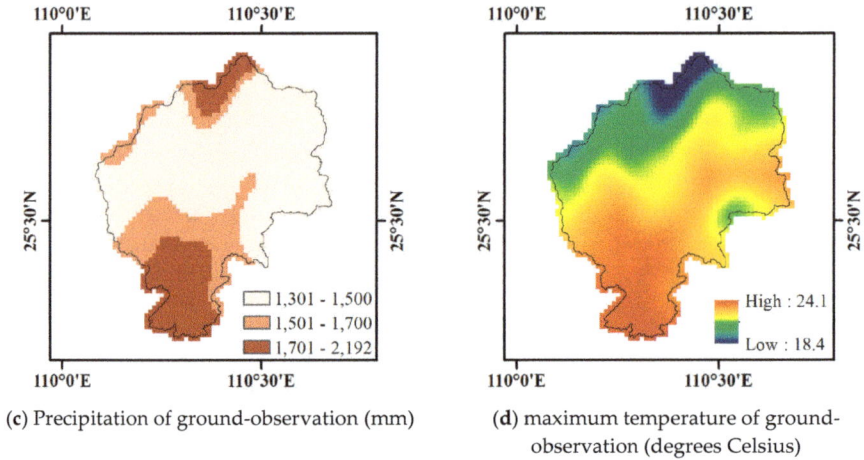

(**c**) Precipitation of ground-observation (mm)

(**d**) maximum temperature of ground-observation (degrees Celsius)

Figure 2. Spatial distribution of annual precipitation and maximum temperature estimation.

The IHACRES model (which has been used in various studies [52–54]) catchment moisture deficit (CMD) version [55] is used here as it has more consistent physical meaning for the parameters. There are two stores in the IHACRES model: the nonlinear store for the generation of effective rainfall, which uses a nonlinear function to deal with the raw rainfall, as well as using accounting equations to calculate the CMD output, and the linear store which converts the effective rainfall into quick and slow flow using unit hydrographs.

The Sacramento model, which has been used in many studies [56,57], has five runoff components: a direct runoff from an impervious area, surface runoff, interflow, supplementary base flow and primary base flow, with the 13-parameter version of the Sacramento model used here [58]. Briefly, excess rainfall becomes runoff through a unit hygrograph, with the rest of the rainfall filling various depths of interconnected soil-moisture stores. Loss through evapotranspiration occurs at the soil stores, with the remaining water becoming interflow and groundwater; the summation of the surface and lateral flow forms the streamflow.

Table 1 gives a description of the parameters of the two rainfall–runoff models. We used the Hydrological Model Assessment and Development (Hydromad) [59] modelling package to help us construct the hydrologic models. Hydromad is an open-source software package in R and is available at http://hydromad.catchment.org. It provides a modelling framework for environmental hydrology and supports simulation, estimation, assessment and visualization of flow response to time series of rainfall and other drivers.

2.4. Model Performance Evaluation Criteria

The Nash–Sutcliffe efficiency (*NSE*) performance measure [60] is a form of the mean squared error widely used in hydrology as a criterion for assessing hydrologic-model performance. The *NSE* objective function focuses on fitting high flowrates [61]. Here, it is also computed in terms of square-root-transformed and logarithmic-transformed flows (denoted NSE_{sq} and NSE_{log} hereafter), which makes it possible to assess the model efficiency for low flowrates [62]. The use of these three criteria (*NSE*, NSE_{sq} and NSE_{log}) gives a more general overview of the model efficiency and are defined as

$$NSE = 1 - \frac{\sum_{i=1}^{n}(obs_i - sim_i)^2}{\sum_{i=1}^{n}\left(obs_i - \overline{obs}\right)^2},$$

(1)

$$NSE_{sq} = 1 - \frac{\sum_{i=1}^{n}\left(\sqrt{obs_i} - \sqrt{sim_i}\right)^2}{\sum_{i=1}^{n}\left(\sqrt{obs_i} - \overline{\sqrt{obs}}\right)^2}, \tag{2}$$

$$NSE_{log} = 1 - \frac{\sum_{i=1}^{n}\left(\log(obs_i + \varepsilon) - \log(sim_i + \varepsilon)\right)^2}{\sum_{i=1}^{n}\left(\log(obs_i + \varepsilon) - \overline{\log(obs + \varepsilon)}\right)^2}, \tag{3}$$

respectively. The relative bias (*rel.bias* [mm]) is also used as a model-performance criterion, with the optimal value of zero and is defined as

$$rel.bias = \frac{\sum_{i=1}^{n}\left(sim_i - obs_i\right)}{\sum_{i=1}^{n} obs_i}, \tag{4}$$

where i is the index for individual days in the period, n the total number of days, *sim* denotes the simulated runoff, *obs* the observed runoff and $\overline{obs}$ the mean observed runoff averaged over the period.

Table 1. Parameters for each model.

Parameter Name	Unit	Range	Description
IHACRES-CMD			
f	-	0.01–3	CMD stress threshold as a proportion of d
e	-	0.01–1.5	Temperature to potential evapotranspiration (PET) conversion factor
d	mm	50–550	CMD threshold for producing flow
tau_s	day	30–600	Time constant for slow flow store
tau_q	day	1–10	Time constant for quick flow store
v_s	-	0.1–1	Fractional volume for slow flow
Sacramento			
UZTWM	mm	1–150	Upper zone tension water maximum capacity
UZFWM	mm	1–150	Upper zone free water maximum capacity
UZK	1/day	0.1–0.5	Upper zone free water lateral depletion rate
PCTIM	-	0.000001–0.1	Fraction of the impervious area
ADIMP	-	0–0.4	Fraction of the additional impervious area
ZPERC	-	1–250	Maximum percolation rate coefficient
REXP	-	0–5	Exponent of the percolation equation
LZTWM	mm	1–500	Lower zone tension water maximum capacity
LZFSM	mm	1–1000	Lower zone supplementary free water maximum capacity
LZFPM	mm	1–1000	Lower zone primary free water maximum capacity
LZSK	1/day	0.01–0.25	Lower zone supplementary free water depletion rate
LZPK	1/day	0.0001–0.25	Lower zone primary free water depletion rate
PFREE	-	0–0.6	Fraction percolating from upper to lower zone free water storage

The optimization evolutionary technique used to calibrate parameter values is the Shuffled Complex Evolution (SCE) [63] algorithm, which is a popular method for parameter calibration and has proven to be both effective and relatively efficient, providing a similar performance to other evolutionary optimization algorithms [64]. The value of the objective functions for the calibration of parameters can be used as model-performance statistics.

2.5. Performance of Precipitation Detection

The expression for these statistical measures are based on a contingency table (Table 2).

Table 2. Contingency table for the ground observations and the Satellite/reanalysis estimate with a threshold of 1.0 mm.

Satellite/Reanalysis Estimate	Ground Observation	
	Observation ≥ 1.0 mm	Observation < 1.0 mm
Estimate ≥ 1.0 mm	H	F
Estimate < 1.0 mm	M	Z

Where H, F, M and Z represent the number of hits (true positives), the number of misses (false positives), the number of false alarms (false negatives) and true negatives respectively, based on a threshold of 1.0 mm. The precipitation detection capability of TRMA-3B42V7 and CMADS datasets is evaluated through their pixel-to-point comparison with the ground-based data. We use six statistical measures, including the Proportion Correct (PC), Probability of Detection (POD), Frequency Bias Index (FBI), False Alarm Ratio (FAR), Critical Success Index (CSI) and Heidke skill score (HSS) to estimate their precipitation detection capability. Those statistical measures are defined in Table 3 [14,43].

Table 3. Statistical measures for validation of precipitation detection capability.

Metric	Formula	Range	Optimal Value
Proportion Correct	$PC = \frac{H+Z}{N}$	0–1	1
Probability of Detection	$POD = \frac{H}{H+M}$	0–1	1
Frequency Bias Index	$FBI = \frac{H+F}{H+M}$	0–+∞	1
False Alarm Ratio	$FAR = \frac{F}{F+H}$	0–1	0
Critical Success Index	$CSI = \frac{H}{M+H+F}$	0–1	1
Heidke Skill Score	$HSS = \frac{2*(Z*H-F*M)}{(Z+F)*(F+H)+(M+H)*(Z+M)}$	$-∞–1$	1

Note: $N = H + F + M + Z$.

The PC, POD, FBI and FAR were used to measure the misdetection and false alarms from satellite/reanalysis data. CSI and HSS are more comprehensive contingency metric were used to evaluate the strength of the correlation between the ground observations and the satellite/reanalysis estimate. CSI combines the advantages of both POD and FAR. HSS can safely be compared on different datasets and also measures the overall detection skill accounting for matches due to random chance.

2.6. Generalized Likelihood Uncertainty Estimation Method

Precipitation input and the uncertainties in the parameter values of a hydrologic model are the two major factors affecting the performance of hydrologic and water-resource modelling in a basin. These are assessed here using the generalized likelihood uncertainty estimation (GLUE) method [65,66], which is a stochastic method for quantifying the uncertainty of model predictions. It can be summarized in the following steps:

(1) A large number of models are run with randomly chosen parameter sets selected from a probability distribution; here, 100,000 group parameters are chosen obeying a uniform distribution.
(2) Definition of the "likelihood" function (here, the performance measures NSE and NSElog) and calculation of likelihood values corresponding to each parameter set.
(3) Definition of a cut-off threshold value for the likelihood function to distinguish between the "behavioral" parameter sets and the "non-behavioral" parameter sets.
(4) Rescaling of the cumulative likelihood values of all behavioral models to unity.
(5) Calculation of the percentiles of the cumulative distribution of the likelihood measure. The GLUE method integrates the outputs of all behavioral models in an ensemble prediction. For each timestep of the simulation, the output prediction is obtained as the median of the distribution of all ensemble members, with its uncertainty bounds estimated as the 5% and 95% percentiles of the distribution.

3. Results

3.1. Evaluation of Model Performance

By setting one year as the warm-up period, with 2008–2012 as the calibration period and *NSE* as the objective function, the shuffled complex evolution algorithm is used to calibrate the parameter values of the two hydrologic models. Using the resulting calibrated parameters, the overall performance (*NSE*) of observed and simulated values for the precipitation datasets and models for the period 2008–2016 is shown in Table 4. Tables 5 and 6 depict the calibrated optimal parameters sets and daily *NSE* for each precipitation dataset applied respectively to the IHACRES model and Sacramento model (2008–2012).

Table 4. Overall performance (daily *NSE* (monthly *NSE*)) of precipitation datasets for models using *NSE* as the objective function for calibration.

	IHACRES	Sacramento
Gauged	0.57 (0.83)	0.52 (0.80)
TRMM	0.56 (0.89)	0.56 (0.87)
CMADS	0.69 (0.93)	0.70 (0.92)

Table 5. Calibrated optimal parameters sets and daily *NSE* for each precipitation dataset applied to the IHACRES model (2008 to 2012).

Datasets	*f*	*e*	*d*	*tau_q*	*tau_s*	*v_s*	NSE
Gauged	1.132	0.05149	80.55	2.420	30.00	0.10	0.61
TRMM	1.060	0.06742	147.45	5.061	30.00	0.10	0.52
CMDAS	3.000	0.08322	50.00	3.055	30.00	0.10	0.69

Table 6. Calibrated optimal parameter sets and daily *NSE* values for each precipitation dataset applied to the Sacramento model (2008–2012).

Datasets	*uztwm*	*uzfwm*	*uzk*	*pctim*	*adimp*	*zperc*	*rexp*
Gauged	1.000	93.5	0.322	0.0499	0.0656	149.7	3.420
TRMM	1.000	140.1	0.102	1.01×10^{-6}	1.76×10^{-8}	140.8	1.205
CMDAS	1.002	150.0	0.158	0.0509	9.48×10^{-8}	159.4	4.844

Datasets	*lztwm*	*lzfsm*	*lzfpm*	*lzsk*	*lzpk*	*pfree*	NSE
Gauged	1.000	998.9	944.7	0.250	0.250	0.0100	0.57
TRMM	1.320	1000.0	119.1	0.152	0.212	0.2156	0.51
CMDAS	1.963	1000.0	1.00	0.227	0.228	0.0842	0.68

Selecting the years 2008–2012 as the calibration period and the years 2012–2016 as the validation period, the daily and monthly performance of the IHACRES model for different precipitation products are shown in Table 7, with daily and monthly observed and modelled flow and rainfall shown in Figure 3. The additive merit of NSE_{sq} and NSE_{log} is also calculated. Since NSE_{sq} and NSE_{log} shift the focus from high flows to progressively lower flows, using NSE_{sq} and NSE_{log} help us judge the performance of the models in simulating over a broader range of flows. The performance of the IHACRES model shows that the CMADS dataset has the best performance among all three precipitation products during the calibration period. The TMPA-3B42V7 and gauge-interpolated product have a similar performance but perform slightly worse than the CMADS dataset. During the validation period, the *NSE*, NSE_{sq} and NSE_{log} values when using the CMADS dataset show a better performance than other precipitation datasets, which indicates the CMADS dataset performs better than other precipitation datasets in simulating both high flow and low flows. Overall, all the three precipitation datasets perform well, with the CMADS dataset performing slightly better.

Table 7. Model performance of the IHACRES model (calibrated using *NSE*) for the calibration period and validation periods.

	Datasets	*Daily rel.bias*	*DailyNSE*	*DailyNSE$_{sq}$*	*Daily NSE$_{log}$*	*Monthly NSE*
Calibration period	Gauged	−0.18	0.61	0.63	0.51	0.86
Validation period	Gauged	−0.13	0.52	0.54	0.45	0.81
Calibration period	TMPA-3B42V7	−0.12	0.52	0.62	0.56	0.89
Validation period	TMPA-3B42V7	−0.15	0.61	0.59	0.49	0.89
Calibration period	CMADS	−0.21	0.69	0.57	0.32	0.93
Validation period	CMADS	−0.07	0.70	0.63	0.49	0.93

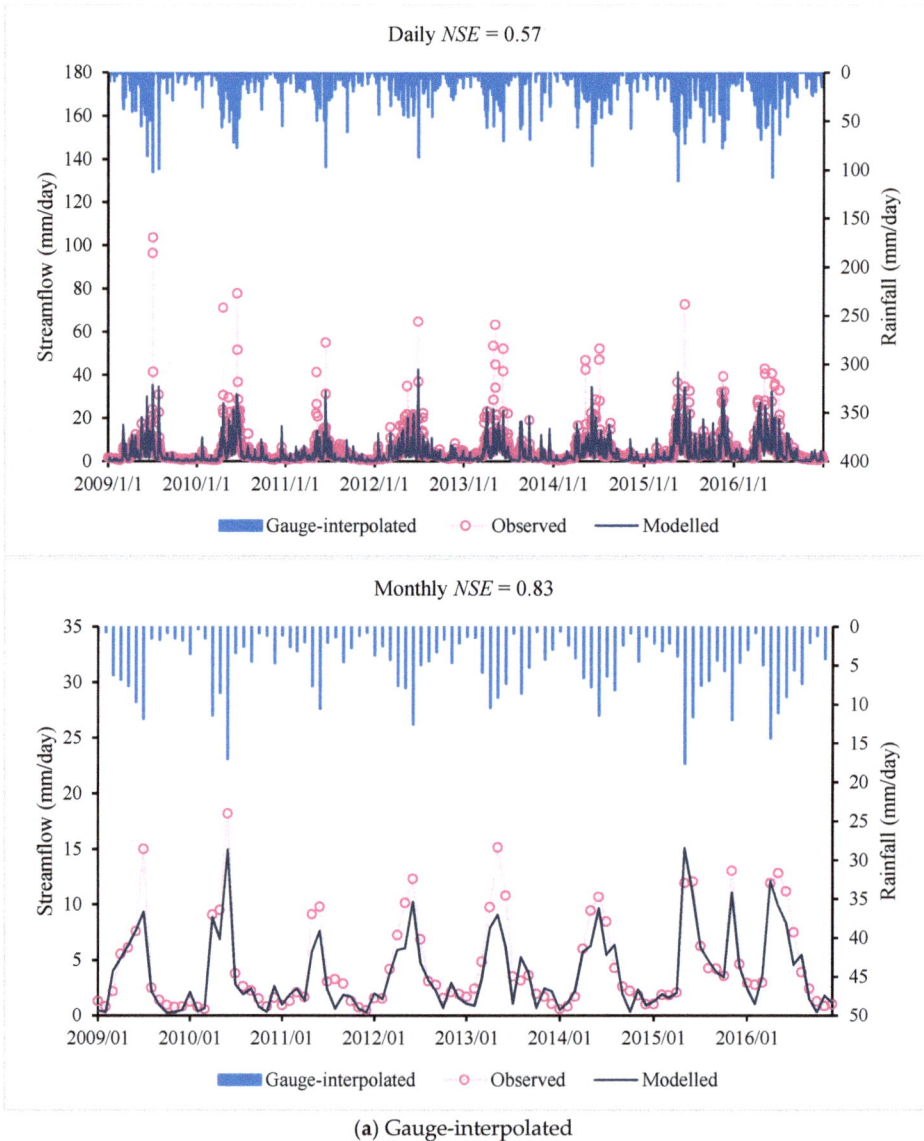

(**a**) Gauge-interpolated

Figure 3. *Cont.*

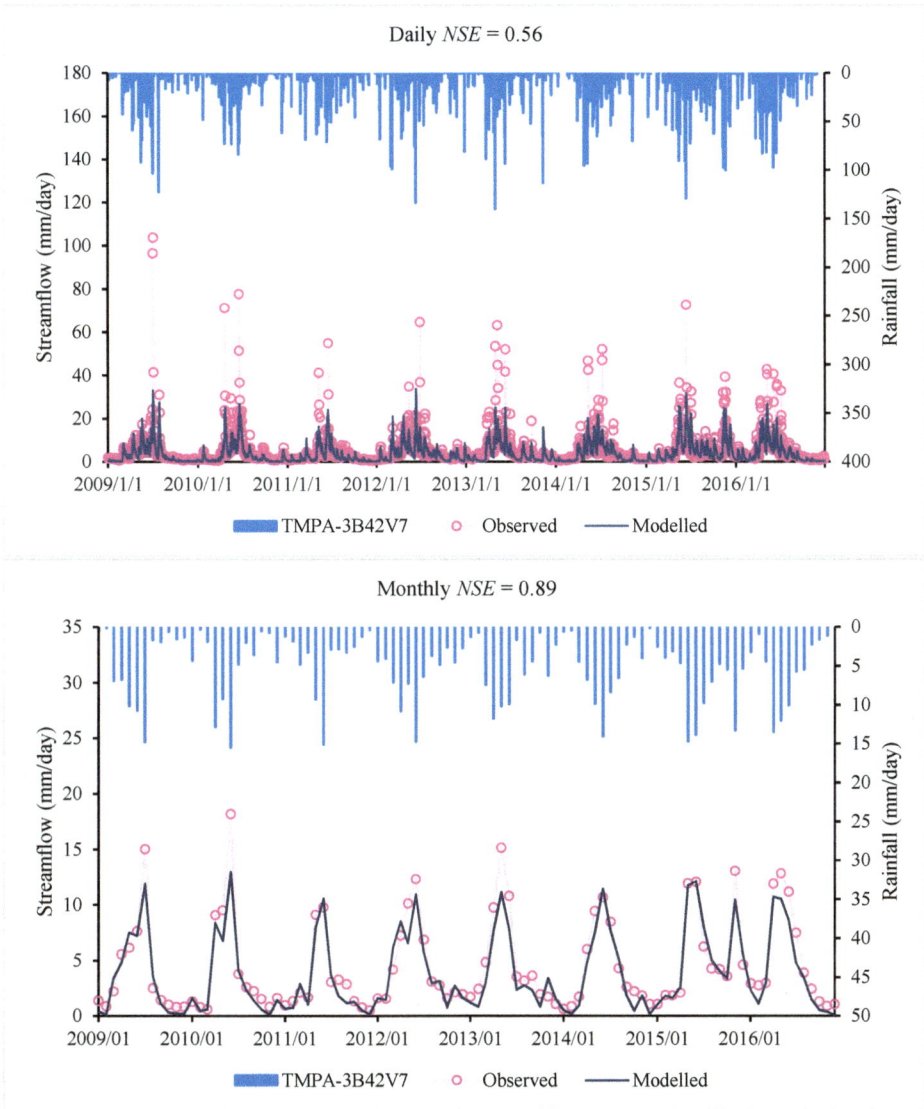

(**b**) TMPA-3B42V7

Figure 3. *Cont.*

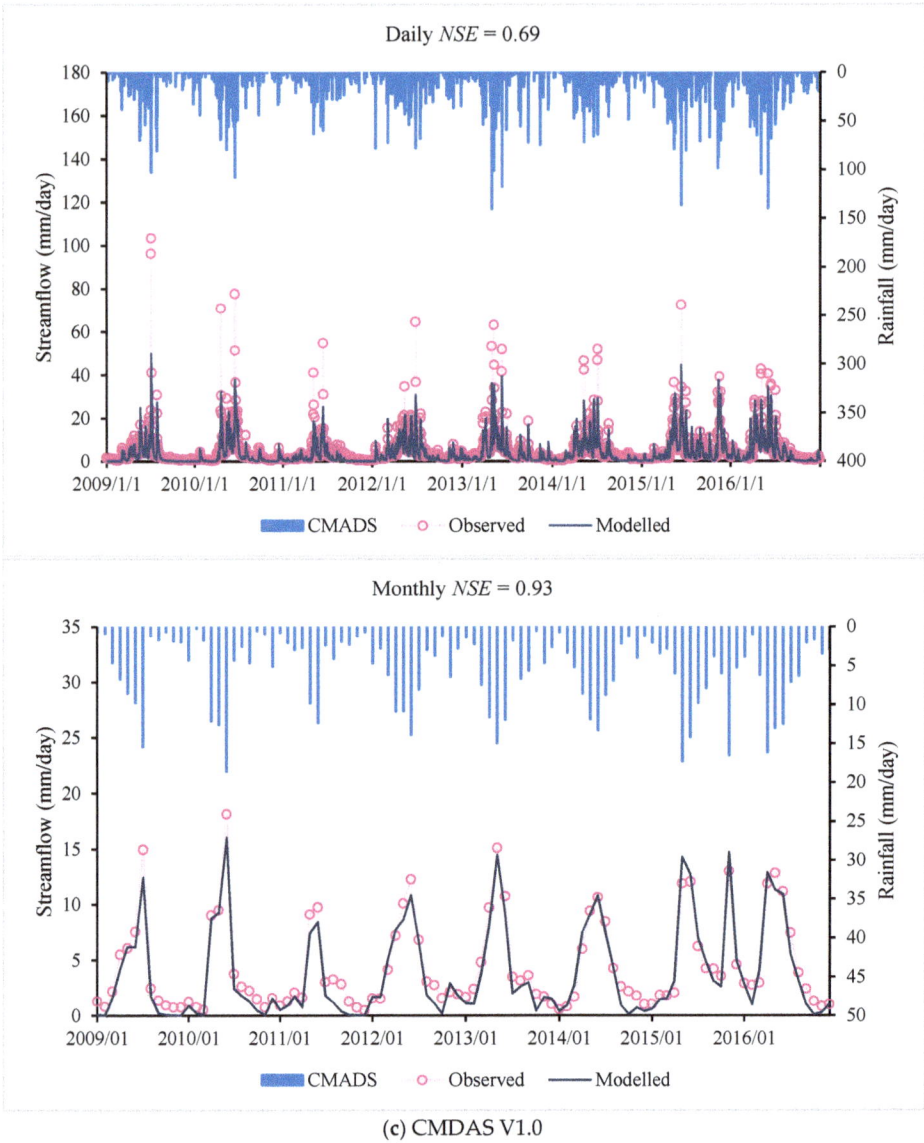

(c) CMDAS V1.0

Figure 3. Observed and IHACRES-model-simulated daily and monthly runoffs for (**a**) Gauge-interpolated, (**b**) TMPA-3B42V7 and (**c**) CMDAS rainfall datasets (for the IHACRES model calibrated using *NSE* as the objective function).

The daily and monthly model performance of the Sacramento model for different precipitation products are shown in Table 8, with time series plotted in Figure 4. Again, the performance of the Sacramento model show that the precipitation product of the CMADS dataset performs best for both the calibration and validation periods, followed by the TMPA-3B42V7 datasets, with the performance of the gauge-interpolated product slightly worse.

Table 8. Model performance of the Sacramento model (calibrated using *NSE*) for the calibration and verification periods.

	Datasets	Daily rel.bias	DailyNSE	DailyNSE$_{sq}$	Daily NSE$_{log}$	Monthly NSE
Calibration period	Gauged	−0.12	0.57	0.56	0.41	0.84
Validation period	Gauged	−0.11	0.47	0.41	0.29	0.77
Calibration period	TMPA-3B42V7	0.02	0.51	0.56	0.47	0.86
Validation period	TMPA-3B42V7	−0.06	0.61	0.54	0.42	0.89
Calibration period	CMADS	−0.05	0.68	0.63	0.46	0.93
Validation period	CMADS	0.01	0.71	0.57	0.40	0.91

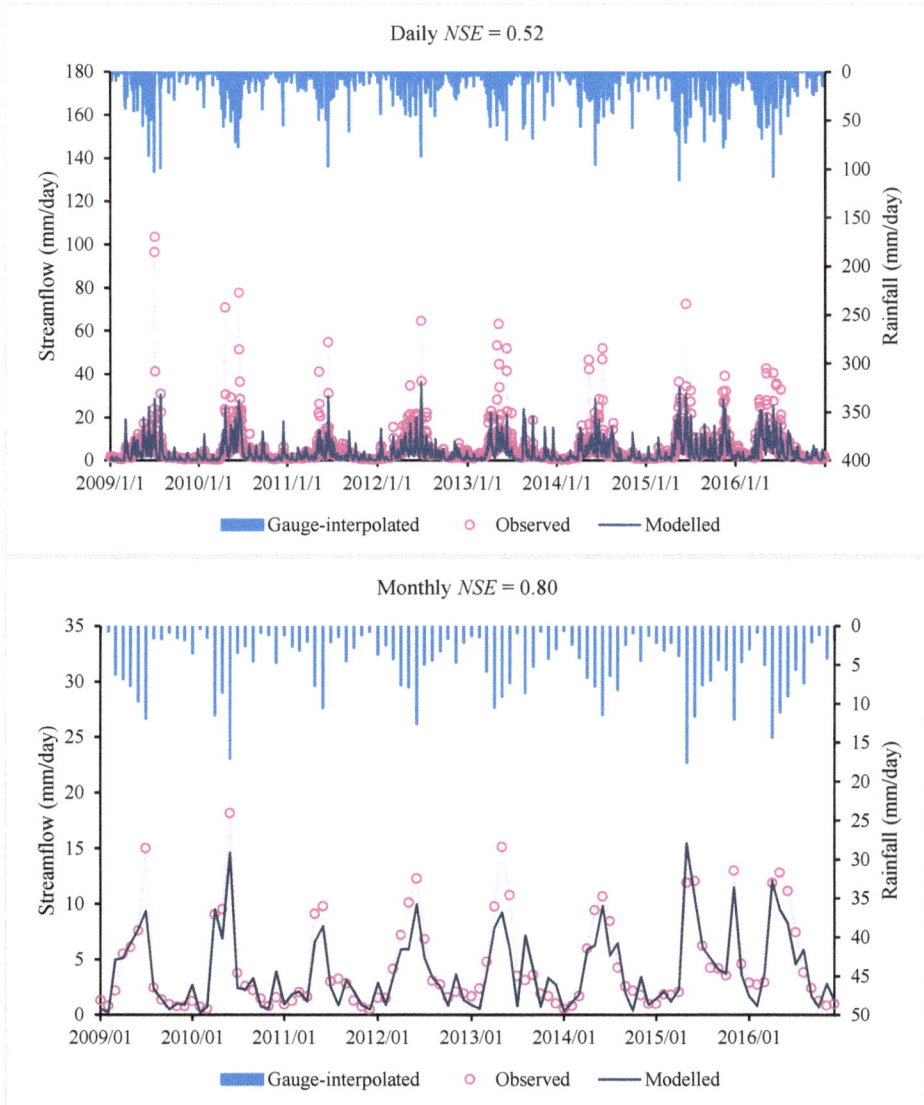

(**a**) Gauge-interpolated

Figure 4. *Cont.*

95

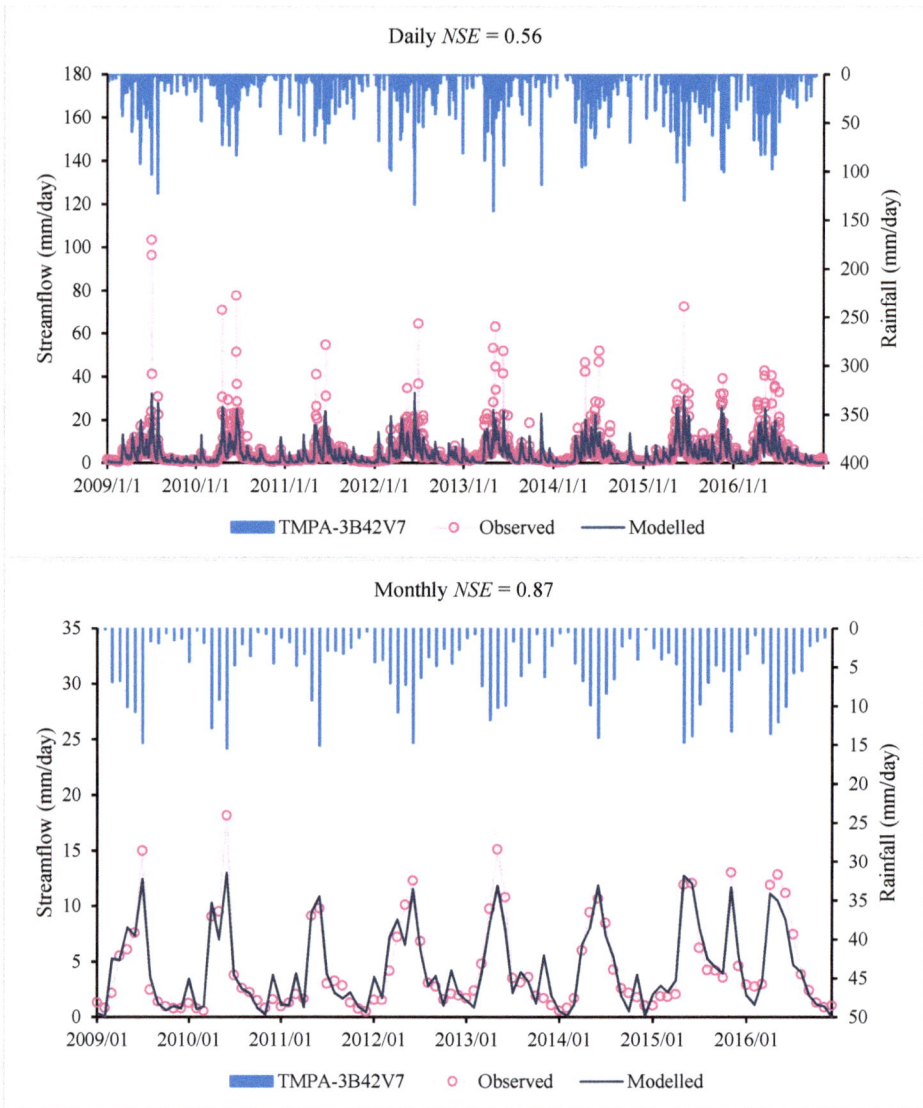

(**b**) TMPA-3B42V7

Figure 4. *Cont.*

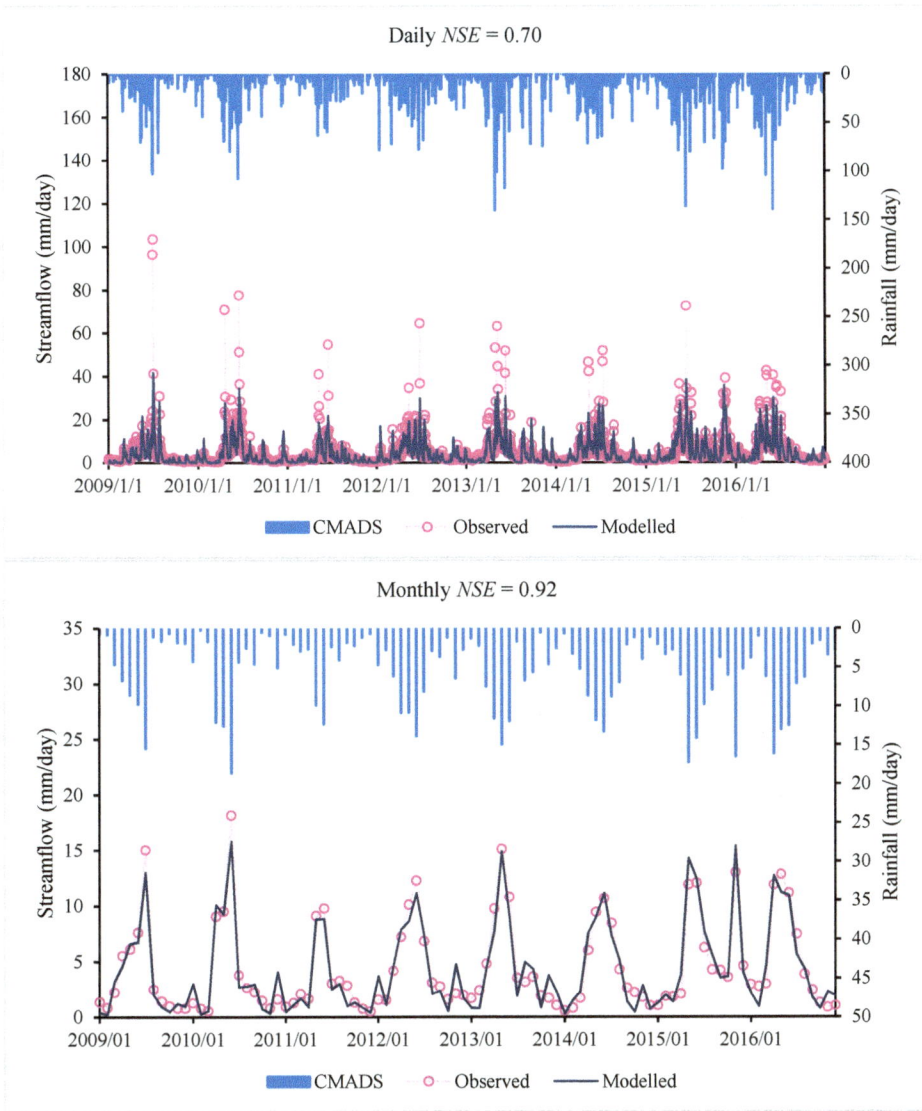

Daily *NSE* = 0.70

Monthly *NSE* = 0.92

(c) CMDAS V1.0

Figure 4. *Cont.*

Figure 4. Observed and Sacramento-model-simulated daily and monthly runoffs for (**a**) Gauge-interpolated, (**b**) TMPA-3B42V7 and (**c**) CMDAS rainfall datasets (for the Sacramento model calibrated using *NSE* as the objective function).

These results show that, among the three precipitation datasets considered here for the Lijiang River basin, the CMADS precipitation datasets have a higher accuracy and better applicability in calibrating and validating the rainfall–runoff models. The reason that the gauge-interpolated rainfall always provides the worst result in flow simulations is due to the meteorological stations available in the Lijiang river basin being too sparsely distributed to permit reliable interpolation.

3.2. Precipitation Detection

We examined the performance of TRMA-3B42V7 and CMADS using the six statistical measures (PC, POD, FBI, FAR, CSI and HSS) through the pixel-to-point comparison with the ground-based data. Following the studies of Dai [67] and Vu et al. [24], a minimum precipitation threshold of 1.0 mm per day was used for the precipitation and non-precipitation event for ground observation and satellite/reanalysis estimate. The contingency statistics of TRMA-3B42V7 and CMADS were evaluated each year through the precipitation datasets from 2008 to 2016. Figure 5 shows the contingency statistics calculated for the TRMA-3B42V7 and CMADS.

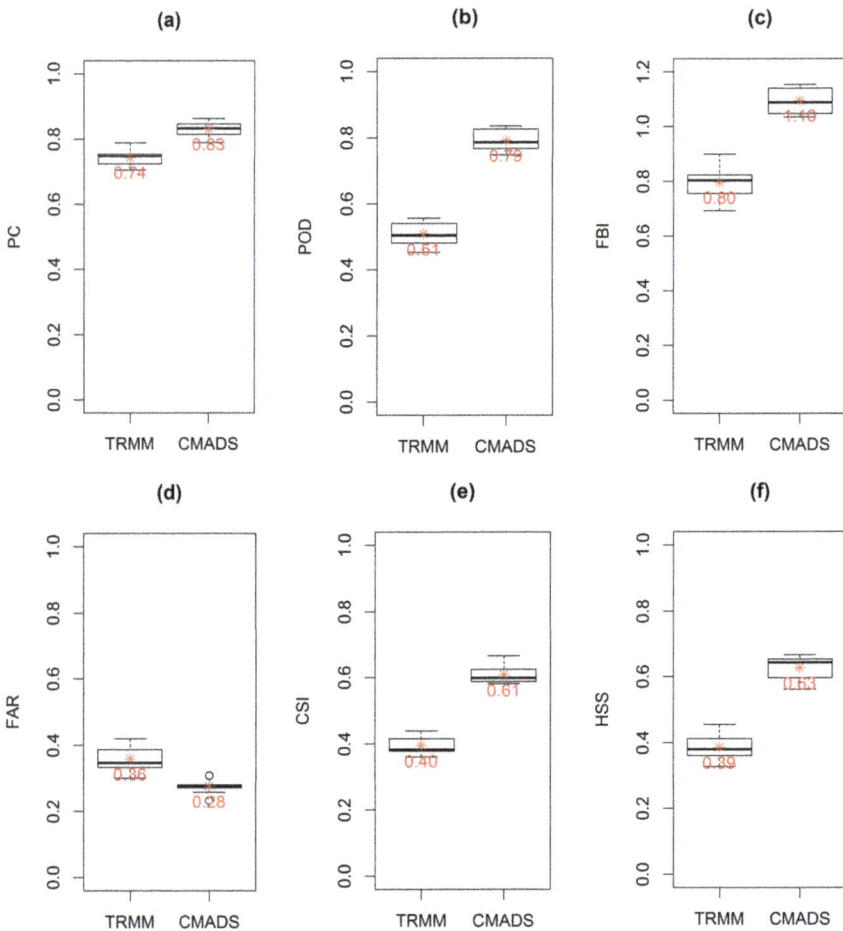

Figure 5. The box plots for the contingency statistics of (**a**) Proportion Correct (PC), (**b**) Probability of Detection (POD), (**c**) Frequency Bias Index (FBI), (**d**) False Alarm Ratio (FAR), (**e**) Critical Success Index (CSI), (**f**) Heidke Skill Score (HSS). The labelled asterisk dot represent the mean value and the middle line in the box represent the median value. Each box ranges from the lower (25th) to upper quartile (75th).

For the statistics PC, POD and CSI, the CMADS scheme (with an average of 0.83, 0.79 and 0.61) shows higher values than TRMM (with an average of 0.74, 0.51 and 0.40). For the statistic FBI, CMADS gives a mean of 1.1 (ranging from 1.03 to 1.16) and is closer to the perfect score than the FBI of TRMM with a mean of 0.8 (ranging from 0.69 to 0.9). With respect to the FAR statistic, CMADS has a smaller FAR with an average of 0.28 (ranging from 0.23 to 0.31) than TRMM with an average of 0.36 (ranging from 0.30 to 0.42). Finally, the HSS statistic for CMADS has a larger value with a mean of 0.63 (ranging 0.56 to 0.67) than TRMM with a mean of 0.39 (ranging from 0.33 to 0.45). These results indicate that the CMADS scheme shows better performance than TRMA-3B42V7 for all the six contingency statistics. Overall, compared to TRMM data, CMADS show better agreement with the ground observation data in Lijiang river basin.

3.3. Uncertainty Analysis

All precipitation products are limited by quantitative inaccuracies and they can exhibit significant bias and errors in spatial and temporal variability. As the runoff-generation is highly sensitive to the spatial and temporal variability of precipitation data, the spatial and temporal variability of precipitation is one of the main source of uncertainty in rainfall–runoff modelling.

Parameter uncertainty is another source of uncertainty in rainfall-runoff modelling. The parameter uncertainty of each hydrologic model has been explored in this study. A GLUE uncertainty analysis is applied to assess parameter uncertainty of hydrologic models here.

In the first case, for the IHACRES model, 100,000 samples are chosen from a uniform distribution for each parameter and the performance measures NSE and NSE_{log} are used as the "likelihood" functions. Using a threshold value of $NSE > 0.67$ (or $NSE_{log} > 0.78$) for the CMADS product, the GLUE algorithm finds 2000 (1416) behavioral solutions in 100,000 simulations with the IHACRES model. Using a threshold value of $NSE > 0.56$ (or $NSE_{log} > 0.69$) for the TMPA-3B42V7 product, the GLUE algorithm finds 3180 (2819) behavioral solutions in 100,000 simulations with the IHACRES model.

The red (blue) dots and lines in Figure 6 represent the distribution and boundary of behavioral parameters when using NSE_{log} (NSE) as the likelihood function, with the calibrated parameter set for each rainfall dataset indicated. From the distribution of behavioral parameter sets, we see the parameters d, f and tau_s have behavioral values distributed across the full parameter range, indicating these have the greatest uncertainty. In comparison, the distribution of v_s for behavioral parameter sets is constricted to smaller values ($<\sim0.5$), particularly when using considering NSE. The values of tau_q are more constrained when considering NSE, due to the focus NSE gives to high flows compared to $NSElog$. Generally, there is little interaction between most of the parameters. The main exception is the e and f parameters. The value of the e parameter is constrained to $<\sim0.1$ providing $f > \sim1$, increasing rapidly for smaller values of f. This indicates a highly non-linear interaction between these parameters. It should also be noted that the optimal value of the e parameter is considerably smaller than that found in Australia (0.166) found by Chapman (2001) [68], due to the influence of other factors (e.g., atmospheric transmissivity).

A similar analysis is applied for the Sacramento model, where 100,000 samples are chosen obeying a uniform distribution, and NSE and NSE_{log} are again defined as the likelihood functions. Using a threshold value of $NSE > 0.45$ (or $NSE_{log} > 0.53$), the GLUE algorithm finds 929 (359) behavioral solutions in 100,000 simulations for the CMADS product with the Sacramento model. Using a threshold value of $NSE > 0.32$ (or $NSE_{log} > 0.43$), the GLUE algorithm finds 1294 (140) behavioral solutions in 100,000 simulations for the TMPA-3B42V7 product with the Sacramento model. The pairwise correlation of behavioral parameters for the Sacramento model is shown in Figure 7.

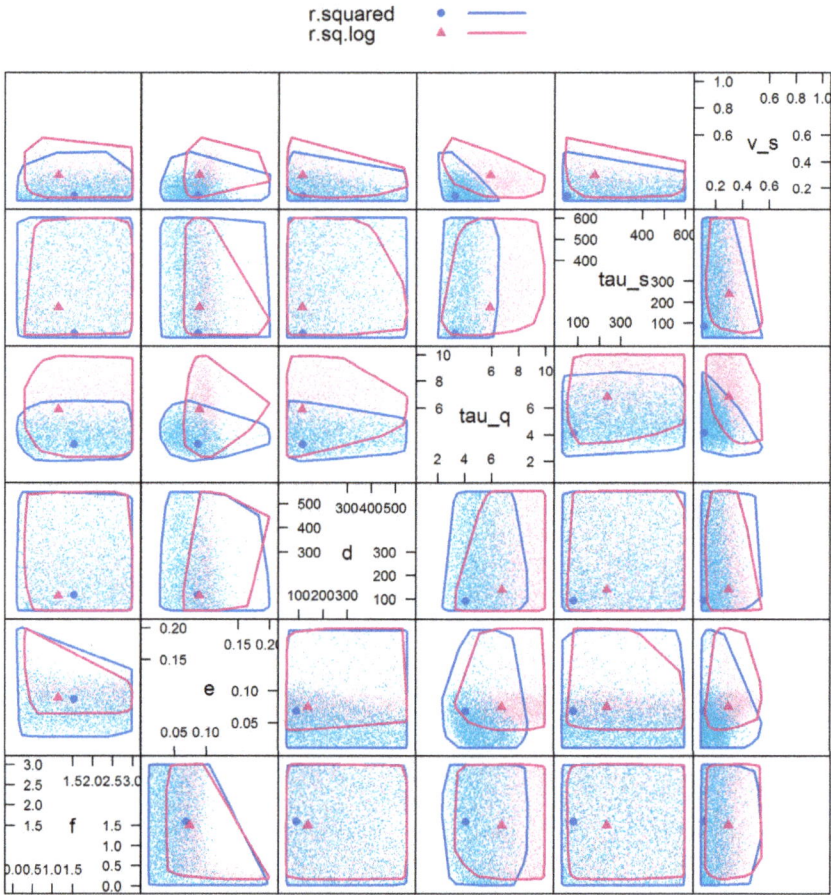

Figure 6. Two-dimensional projections of pairwise correlation of behavioral parameters for the IHACRES model using the CMADS (above diagonal) and TMPA-3B42V7 (below diagonal) precipitation datasets. The heavy dots represent the location of the best objective function value obtained from the GLUE sample.

The meaning of the red (blue) dot and line in Figure 7 is similar to that in Figure 6. From the distribution of behavioral parameters, we can see the parameters *uztwm*, *adimp* and *lztwm* show less uncertainties overall. When *NSE* is selected as the likelihood function, the distribution of parameters *lzwm*, *uztwm* and *adimp* are relatively low, while the distributions of parameters *lzpk* and *lzsk* are relatively high; when NSE_{log} is selected as the likelihood function, the distribution of the parameters *uztwm*, *lztwm* and *adimp* is relatively low and the distribution of the parameter *pfree* is relatively high, which indicates that these parameters may be more sensitive and less uncertain.

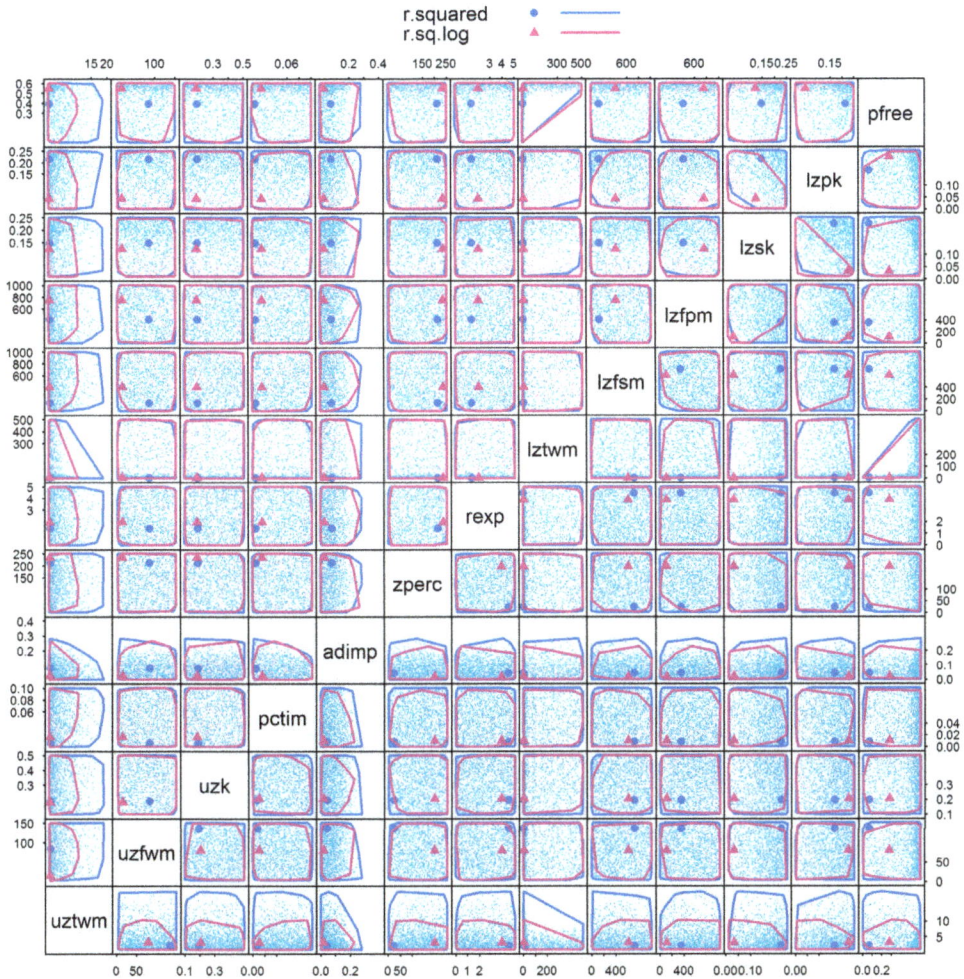

Figure 7. Two-dimensional projections of the pairwise correlation of behavioral parameters for the Sacramento model using CMADS (above diagonal) and TMPA-3B42V7 (below diagonal) precipitation datasets. The heavy dots represent the location of the best objective function value obtained from the GLUE sample.

In the second case, model performance is considered satisfactory when *NSE* is greater than 0.5 [69,70], with likelihood-function values > 0.5 defined as behavioral parameter sets. The behavioral parameter space may be used as further criteria for the evaluation of different precipitation products. The behavioral parameter space describes the number (and percentage) of behavioral solutions in 100,000 simulations, with 100,000 parameter sets generated by the same Monte Carlo random sampling method, with the same criteria of acceptability employed (i.e., the same threshold value and the objective function) for the different precipitation datasets. The statistics of the number and percentage of behavioral parameter sets for different precipitation schemes driving the IHACRES model are shown in Table 9.

Table 9. Number (percentage) of behavioral parameter sets for the IHACRES model.

	NSE	NSE_{log}
Gauged	576 (0.58%)	9144 (9.14%)
TMPA-3B42V7	1056 (1.06%)	9929 (9.93%)
CMADS	5186 (5.19%)	12,095 (12.10%)

Using a threshold value of $NSE > 0.5$ (or $NSE_{log} > 0.5$), the GLUE algorithm finds 5186 (12,095) behavioral solutions in 100,000 simulations for the CMADS precipitation product with the IHACRES model, while the GLUE algorithm achieves 1056 (9929) and 576 (9144) behavioral solutions in 100,000 simulations for the TMPA-3B42V7 and gauge-interpolated products with the IHACRES model. Therefore, the behavioral parameter space of the IHACRES model driven by the CMADS precipitation is larger than the behavioral parameter space driven by the other two precipitation inputs. The CMADS product gives a better performance than the TMPA-3B42V7 and gauge-interpolated products, because CMADS assimilated datasets are based on the large number of stations (nearly 40,000 regional automatic stations and 2421 national automatic stations in China), which gives it any priority in reflecting the actual processes of areal precipitation.

A similar analysis is applied for the Sacramento model as well. The statistics of the number and percentage of behavioral parameter sets for different precipitation schemes driving the Sacramento model are shown in Table 10. For the Sacramento model, the behavioral parameter space is very sparse when using the GLUE method for the Lijiang River basin. Using a threshold value of $NSE > 0.5$ (or $NSE_{log} > 0.5$), the GLUE algorithm finds 32 (60) behavioral solutions in 100,000 simulations for the CMADS precipitation dataset within the Sacramento model, while the GLUE algorithm achieves 0 (11) and 0 (5) behavioral solutions in 100,000 simulations for the TMPA-3B42V7 and gauge-interpolated products with the Sacramento model. Similar to their performance with the IHACRES model, the behavioral parameter space of the Sacramento model driven by the CMADS precipitation dataset is larger than the behavioral parameter space driven by the other two precipitation inputs. The CMADS product shows a better performance than the TMPA-3B42V7 and gauge-interpolated products, which, as mentioned before, is probably because the CMADS assimilated datasets are based on the strongly underconstrained large number of stations.

Table 10. Number (percentage) of behavioral parameter sets for the Sacramento model.

	NSE	NSE_{log}
Gauged	0(0.00%)	5(0.01%)
TMPA-3B42V7	0(0.00%)	11(0.01%)
CMADS	32(0.03%)	60(0.06%)

4. Discussion

Our work presents a comparative analysis for different precipitation datasets and their applicability for hydrologic modelling, including gauge-interpolated datasets, TMPA-3B42V7 and CMADS precipitation products. Two hydrologic models; IHACRES and Sacramento, are evaluated in the Lijiang River basin, as well as the accuracy of different precipitation datasets for hydrologic modelling.

The results show that the IHACRES and Sacramento models demonstrate a good and similar performance in the Lijiang River basin. Driven by the CMADS precipitation, the *NSE* values of the IHACRES (Sacramento) model are 0.69 (0.68) and 0.70 (0.71) for the calibration and validation periods, respectively. Figure 6 shows there are three sensitive parameters (*f*, *e* and *v_s*) for the IHACRES model. Figure 7 shows there are three sensitive parameters (*uztwm*, *adimp* and *lztwm*) for the Sacramento model. The number of effective parameters is similar in both Sacramento model (with more parameters) and IHACRES model (with less parameters), consistent with sensitivity results in Shin et al. [64].

The uncertainty analysis carried out for the IHACRES and Sacramento models show that the uncertainty in the model predictions is greater for the Sacramento model. To sum up, IHACRES and Sacramento perform similarly in terms of simulation performance and number of effective parameters, the latter model having far more insensitive parameters. What's more, the IHACRES model has a reduced uncertainty compared with the Sacramento model. Based on these analyses, the authors conclude that IHACRES generally outperforms Sacramento in Lijiang river basin. It confirms previous findings (e.g., Orth et al. [71]) that more parameters may lead to over-fitting without an improved performance of the hydrologic model.

Of the three precipitation datasets (gauge-interpolated product, TMPA-3B42V7 and CMADS products), the CMADS product gives the best performance in simulating the rainfall–runoff process in the Lijiang River basin. The overall performance (based on DailyNSE values) of the CMADS product is 0.69 and 0.70 for the IHACRES and Sacramento models, respectively, with the overall performance of the TMPA-3B42V7 (0.56 and 0.56) and gauge-interpolated (0.57 and 0.52) products correspondingly much lower. From the analysis of Figures 3 and 4, the hydrologic model driven by the CMADS product shows a superior skill in capturing the flow peaks because the CMADS reanalysis data are based on a large number of stations. However, these datasets overestimate the simulated flood peak and underestimate low flowrates, which is probably because the models are calibrated using the performance measure NSE, being an objective function that puts more emphasis on high flowrates. The model calibrated using Nash–Sutcliffe efficiency on transformed streamflow NSE_{log}, which gives more emphasis to low flowrates, is presented in the Supplementary Materials, as well as the model performance calibrated using NSE_{log} as an objective function. Similar conclusions can be reached with the models calibrated using the performance measure NSE_{log} as the objective function. The CMADS precipitation datasets perform best in all three precipitation datasets, followed by the TMPA-3B42V7 precipitation and then the gauge-interpolated product. Comparing the model performance using NSE_{log} and NSE as the objective functions for calibration, the model calibrated using the performance measure NSE performed better in simulating peak flows, while underestimating low flowrates. In contrast, the model calibrated using the performance measure NSE_{log} performs well in simulating low flowrates but underestimates the flood peak in the simulations.

The GLUE pairwise correlation of behavioral parameters (Figures 5 and 6) give us an intuitional view of parameter uncertainties. From the distribution of behavioral parameters for the IHACRES model (Figure 5), we see the parameters d, tau_s and tau_q have the greatest uncertainties overall. Further analysis reveals that different precipitation products reshape the distribution of tau_q greatly. The parameter tau_q, which represents the time constant for quick flow store, is very sensitive to the precipitation input. For the Sacramento model (Figure 6), the parameters $uztwm$, $adimp$ and $lztwm$ show less uncertainties overall than other parameters.

The superiority of the CMADS product can also be found in the number of GLUE behavioral parameters (or their occupation percentage among all the uniformly distributed parameter sets), as well as their GLUE relative measurements coverage. From Tables 7 and 8, we find the CMADS driven hydrologic models are responsible for more behavioral parameters than the hydrologic models driven by the other two precipitation datasets.

Although the two hydrologic models introduced here are widely used, the precipitation input data are basin-averaged precipitation. The comparison and applicability of different precipitation datasets maybe be affected by this "average" precipitation, since the spatial distribution and variability of different precipitation datasets may be weakened by the effect of spatial averaging. The improved accuracy of the CMADS precipitation dataset may be more obvious with the simulation of a distributed or semi-distributed model as the rainfall input. Moreover, different uncertainty-analysis methods may affect the efficiency of the uncertainty analysis, which thus requires further research.

5. Conclusions

Precipitation is a fundamental component of the global water cycle. Precipitation datasets range from conventional ground-based datasets to remote-sensing products and reanalysis datasets, such as the gauge-interpolated product, the TMPA-3B42V7 precipitation products and the CMADS datasets invoked here. The two hydrologic models (IHACRES and Sacramento) are introduced to evaluate their applicability in the basin, the impact of model complexity and the applicability of different precipitation datasets on the hydrologic modelling.

CMADS gives best results when used in IHACRES and Sacramento to simulate flow in the Lijiang River basin. The CMADS precipitation datasets (DailyNSE = 0.69 for the IHACRES model; DailyNSE = 0.70 for the Sacramento model) give improved applicability and accuracy compared with the gauge-interpolated datasets (DailyNSE = 0.57 for the IHACRES model; DailyNSE = 0.52 for the Sacramento model) and TMPA-3B42V7 datasets (DailyNSE = 0.56 for the IHACRES model; DailyNSE = 0.56 for the Sacramento model) in the Lijiang River basin. From the analysis of Figures 3 and 4, we conclude that the CMADS precipitation-driven hydrologic models give better skill in capturing the streamflow peaks. Interpolation of gauge data performed worst, reflecting the impact of low gauge density

The precipitation detection ability of TRMA-3B42V7 and CMADS is also evaluated using six statistical measures (*PC*, *POD*, *FBI*, *FAR*, *CSI* and *HSS*) through a pixel-to-point comparison to the ground-based data. CMADS (with an average of 0.83, 0.79, 1.1, 0.28, 0.61 and 0.63) shows better performance and is closer to the perfect score than TRMM (with an average of 0.74, 0.51, 0.80, 0.36, 0.4 and 0.39).

Based on the analysis of Table 7, for the IHACRES model and using *NSE* as the likelihood function, the number and percentage of behavioral parameters for the CMADS, TMPA-3B42V7 and gauge-interpolated product are 5186 (5.19%), 1056 (1.06%) and 576 (0.58%). Using NSE_{log} as the likelihood function, the number and percentage of behavioral parameters for the corresponding precipitation datasets are 12,095 (12.1%), 9929 (9.93%) and 9144 (9.14%). Similar phenomena can be found with the same analysis of the Sacramento model in Table 8 but the behavioral parameter sets for that model are very sparse. We conclude that the CMADS precipitation-driven hydrologic models are more accurate, as they are responsible for more behavioral parameters than the hydrologic models driven by the other two precipitation datasets. The TMPA-3B42V7 datasets show slightly better performance than the gauge-interpolated product in this case study, indicating that global datasets are particularly useful in poorly gauged areas.

The performance of the IHACRES model (DailyNSE = 0.69 driven by the CMADS product) and Sacramento model (DailyNSE = 0.70 driven by the CMADS product) give respectable results. While both models work well, IHACRES gives lower predictive uncertainty compared to Sacramento, which implies that the general applicability of the IHACRES model is preferable to the Sacramento model in the Lijiang River basin.

Supplementary Materials: The following are available online at http://www.mdpi.com/2073-4441/10/11/1611/s1, Figure S1: Density map of precipitation stations over the region (units: gauges per km^2)), Table S1: Mean absolute error (MAE), root mean square error (RMSE) and relative error of cross-validation for daily interpolated precipitation grids, Table S2: Overall performance (daily NSE_{log} (monthly NSE)) of precipitation datasets for models using NSE_{log} as the objective function for calibration, Table S3 Model performance of IHACRES model (calibrated using NSE_{log}) for the calibration period and validation periods, Table S4 Model performance of Sacramento model (calibrated using NSE_{log}) for the calibration period and validation periods, Figure S2: Observed and IHACRES-model-simulated daily and monthly runoffs for (a) Gauge-interpolated, (b) TMPA-3B42V7 and (c) CMDAS rainfall datasets (for the IHACRES model calibrated using NSE_{log} as the objective function), Figure S3: Observed and Sacramento-model-simulated daily and monthly runoffs for (a) Gauge-interpolated, (b) TMPA-3B42V7 and (c) CMDAS rainfall datasets (for the Sacramento model calibrated using NSE_{log} as the objective function)).

Author Contributions: B.G. carried out the experiments, B.G. and J.Z. wrote the draft manuscript. B.G., T.X., B.C. and A.J. performed data analysis and designed the experiments. Y.S., Q.Y., X.L. and W.L. collected the data for the experiments. All authors contributed to the discussions and reviewed the manuscript.

Funding: This work is supported by National Key R&D Program of China (2017YFC0406004) and NSFC (41271004).

Conflicts of Interest: The authors declare no conflicts of interests.

References

1. Feng, D.; Zheng, Y.; Mao, Y.; Zhang, A.; Wu, B.; Li, J.; Tian, Y.; Wu, X. An integrated hydrological modeling approach for detection and attribution of climatic and human impacts on coastal water resources. *J. Hydrol.* **2018**, *557*, 305–320. [CrossRef]

2. Michaelides, S.; Levizzani, V.; Anagnostou, E.; Bauer, P.; Kasparis, T.; Lane, J.E. Precipitation: Measurement, remote sensing, climatology and modeling. *Atmos. Res.* **2009**, *94*, 512–533. [CrossRef]

3. Steiner, M.; Smith, J.A.; Burges, S.J.; Alonso, C.V.; Darden, R.W. Effect of bias adjustment and rain gauge data quality control on radar rainfall estimation. *Water Resour. Res.* **1999**, *35*, 2487–2503. [CrossRef]

4. Himanshu, S.K.; Pandey, A.; Patil, A. Hydrologic Evaluation of the TMPA-3B42V7 Precipitation Data Set over an Agricultural Watershed Using the SWAT Model. *J. Hydrol. Eng.* **2018**, *23*, 5018003. [CrossRef]

5. Andreadis, K.M.; Schumann, G.J.P.; Stampoulis, D.; Bates, P.D.; Brakenridge, G.R.; Kettner, A.J. Can Atmospheric Reanalysis Data Sets Be Used to Reproduce Flooding Over Large Scales? *Geophys. Res. Lett.* **2017**, *44*, 10369–10377. [CrossRef]

6. Joyce, R.J.; Janowiak, J.E.; Arkin, P.A.; Xie, P. CMORPH: A Method that Produces Global Precipitation Estimates from Passive Microwave and Infrared Data at High Spatial and Temporal Resolution. *J. Hydrometeorol.* **2004**, *5*, 487–503. [CrossRef]

7. Kubota, T.; Shige, S.; Hashizume, H.; Aonashi, K.; Takahashi, N.; Seto, S.; Hirose, M.; Takayabu, Y.N.; Ushio, T.; Nakagawa, K.; et al. Global Precipitation Map Using Satellite-Borne Microwave Radiometers by the GSMaP Project: Production and Validation. *IEEE Trans. Geosci. Remote Sens.* **2007**, *45*, 2259–2275. [CrossRef]

8. Huffman, G.J.; Bolvin, D.T.; Nelkin, E.J.; Wolff, D.B.; Adler, R.F.; Gu, G.; Hong, Y.; Bowman, K.P.; Stocker, E.F. The TRMM Multisatellite Precipitation Analysis (TMPA): Quasi-Global, Multiyear, Combined-Sensor Precipitation Estimates at Fine Scales. *J. Hydrometeorol.* **2007**, *8*, 38–55. [CrossRef]

9. Kidd, C.; Huffman, G. Global precipitation measurement. *Meteorol. Appl.* **2011**, *18*, 334–353. [CrossRef]

10. Himanshu, S.K.; Pandey, A.; Yadav, B. Assessing the applicability of TMPA-3B42V7 precipitation dataset in wavelet-support vector machine approach for suspended sediment load prediction. *J. Hydrol.* **2017**, *550*, 103–117. [CrossRef]

11. Erazo, B.; Bourrel, L.; Frappart, F.; Chimborazo, O.; Labat, D.; Dominguez-Granda, L.; Matamoros, D.; Mejia, R. Validation of Satellite Estimates (Tropical Rainfall Measuring Mission, TRMM) for Rainfall Variability over the Pacific Slope and Coast of Ecuador. *Water* **2018**, *10*, 213. [CrossRef]

12. Jiang, S.; Liu, S.; Ren, L.; Yong, B.; Zhang, L.; Wang, M.; Lu, Y.; He, Y. Hydrologic Evaluation of Six High Resolution Satellite Precipitation Products in Capturing Extreme Precipitation and Streamflow over a Medium-Sized Basin in China. *Water* **2018**, *10*, 25. [CrossRef]

13. Yang, Y.; Tang, G.; Lei, X.; Hong, Y.; Yang, N. Can Satellite Precipitation Products Estimate Probable Maximum Precipitation: A Comparative Investigation with Gauge Data in the Dadu River Basin. *Remote Sens.* **2018**, *10*, 41. [CrossRef]

14. Tang, G.; Behrangi, A.; Long, D.; Li, C.; Hong, Y. Accounting for spatiotemporal errors of gauges: A critical step to evaluate gridded precipitation products. *J. Hydrol.* **2018**, *559*, 294–306. [CrossRef]

15. Saha, S.; Moorthi, S.; Pan, H.; Wu, X.; Wang, J.; Nadiga, S.; Tripp, P.; Kistler, R.; Woollen, J.; Behringer, D.; et al. The NCEP Climate Forecast System Reanalysis. *Bull. Am. Meteorol. Soc.* **2010**, *91*, 1015–1058. [CrossRef]

16. Uppala, S.M.; KÅllberg, P.W.; Simmons, A.J.; Andrae, U.; Bechtold, V.D.C.; Fiorino, M.; Gibson, J.K.; Haseler, J.; Hernandez, A.; Kelly, G.A.; et al. The ERA-40 re-analysis. *Q. J. R. Meteorol. Soc.* **2005**, *131*, 2961–3012. [CrossRef]

17. Dee, D.P.; Uppala, S.M.; Simmons, A.J.; Berrisford, P.; Poli, P.; Kobayashi, S.; Andrae, U.; Balmaseda, M.A.; Balsamo, G.; Bauer, P.; et al. The ERA-Interim reanalysis: Configuration and performance of the data assimilation system. *Q. J. R. Meteorol. Soc.* **2011**, *137*, 553–597. [CrossRef]

18. Compo, G.P.; Whitaker, J.S.; Sardeshmukh, P.D.; Matsui, N.; Allan, R.J.; Yin, X.; Gleason, B.E.; Vose, R.S.; Rutledge, G.; Bessemoulin, P.; et al. The Twentieth Century Reanalysis Project. *Q. J. R. Meteorol. Soc.* **2011**, *137*, 1–28. [CrossRef]

19. Meng, X.; Wang, H. Significance of the China Meteorological Assimilation Driving Datasets for the SWAT Model (CMADS) of East Asia. *Water* **2017**, *9*, 765. [CrossRef]

20. Zhou, S.; Wang, Y.; Chang, J.; Guo, A.; Li, Z. Investigating the Dynamic Influence of Hydrological Model Parameters on Runoff Simulation Using Sequential Uncertainty Fitting-2-Based Multilevel-Factorial-Analysis Method. *Water* **2018**, *10*, 1177. [CrossRef]

21. Meng, X.; Wang, H.; Wu, Y.; Long, A.; Wang, J.; Shi, C.; Ji, X. Investigating spatiotemporal changes of the land-surface processes in Xinjiang using high-resolution CLM3.5 and CLDAS: Soil temperature. *Sci. Rep.* **2017**, *7*, 13286. [CrossRef] [PubMed]

22. Meng, X.; Wang, H.; Shi, C.; Wu, Y.; Ji, X. Establishment and Evaluation of the China Meteorological Assimilation Driving Datasets for the SWAT Model (CMADS). *Water* **2018**, *10*, 1555. [CrossRef]

23. Zhao, F.; Wu, Y.; Qiu, L.; Sun, Y.; Sun, L.; Li, Q.; Niu, J.; Wang, G. Parameter Uncertainty Analysis of the SWAT Model in a Mountain-Loess Transitional Watershed on the Chinese Loess Plateau. *Water* **2018**, *10*, 690. [CrossRef]

24. Vu, T.; Li, L.; Jun, K. Evaluation of Multi-Satellite Precipitation Products for Streamflow Simulations: A Case Study for the Han River Basin in the Korean Peninsula, East Asia. *Water* **2018**, *10*, 642. [CrossRef]

25. Cao, Y.; Zhang, J.; Yang, M.; Lei, X.; Guo, B.; Yang, L.; Zeng, Z.; Qu, J. Application of SWAT Model with CMADS Data to Estimate Hydrological Elements and Parameter Uncertainty Based on SUFI-2 Algorithm in the Lijiang River Basin, China. *Water* **2018**, *10*, 742. [CrossRef]

26. Shao, G.; Guan, Y.; Zhang, D.; Yu, B.; Zhu, J. The Impacts of Climate Variability and Land Use Change on Streamflow in the Hailiutu River Basin. *Water* **2018**, *10*, 814. [CrossRef]

27. Gao, X.; Zhu, Q.; Yang, Z.; Wang, H. Evaluation and Hydrological Application of CMADS against TRMM 3B42V7, PERSIANN-CDR, NCEP-CFSR, and Gauge-Based Datasets in Xiang River Basin of China. *Water* **2018**, *10*, 1225. [CrossRef]

28. Tian, Y.; Zhang, K.; Xu, Y.; Gao, X.; Wang, J. Evaluation of Potential Evapotranspiration Based on CMADS Reanalysis Dataset over China. *Water* **2018**, *10*, 1126. [CrossRef]

29. Meng, X.; Wang, H.; Lei, X.; Cai, S.; Wu, H.; Ji, X.; Wang, J. Hydrological modeling in the Manas River Basin using soil and water assessment tool driven by CMADS. *Teh. Vjesn.* **2017**, *24*, 525–534.

30. Li, J.; Heap, A.D. Spatial interpolation methods applied in the environmental sciences: A review. *Environ. Model. Softw.* **2014**, *53*, 173–189. [CrossRef]

31. Benoit, L.; Mariethoz, G. Generating synthetic rainfall with geostatistical simulations. *Wiley Interdiscip. Rev. Water* **2017**, *4*, e1199. [CrossRef]

32. Ly, S.; Charles, C.; Degré, A. Different methods for spatial interpolation of rainfall data for operational hydrology and hydrological modeling at watershed scale: A review. *Biotechnol. Agron. Soc. Environ.* **2013**, *17*, 392–406.

33. Vente, C. *Handbook of Applied Hydrology: A Compendium of Water-Resources Technology*; McGraw-Hill: New York, NY, USA, 1964.

34. Thiessen, A.H. Precipitation averages for large areas. *Mon. Weather Rev.* **1911**, *39*, 1082–1089. [CrossRef]

35. Philip, G.M.; Watson, D.F. A precise method for determining contoured surfaces. *APPEA J.* **1982**, *22*, 205–212. [CrossRef]

36. Cressie, N.A. *Statistics for Spatial Data*; Wiley Online Library: New York, NY, USA, 1993.

37. Hutchinson, M.F. Interpolation of rainfall data with thin plate smoothing splines. Part II: Analysis of topographic dependence. *J. Geogr. Inf. Decis. Anal.* **1998**, *2*, 152–167.

38. Hutchinson, M.F. Interpolation of rainfall data with thin plate smoothing splines. Part I: Two dimensional smoothing of data with short range correlation. *J. Geogr. Inf. Decis. Anal.* **1998**, *2*, 139–151.

39. Hutchinson, M.F. Interpolating mean rainfall using thin plate smoothing splines. *Int. J. Geogr. Inf. Syst.* **1995**, *9*, 385–403. [CrossRef]

40. Arowolo, A.O.; Bhowmik, A.K.; Qi, W.; Deng, X. Comparison of spatial interpolation techniques to generate high-resolution climate surfaces for Nigeria. *Int. J. Climatol.* **2017**, *371*, 179–192. [CrossRef]

41. Taesombat, W.; Sriwongsitanon, N. Areal rainfall estimation using spatial interpolation techniques. *Sci. Asia* **2009**, *35*, 268–275. [CrossRef]

42. Guo, B.; Xu, T.; Zhang, J.; Barry, C.; Jakeman, A.; Seo, L.; Lei, X.; Liao, W. A comparative analysis of precipitation estimation methods for streamflow prediction. In Proceedings of the 22nd International Congress on Modelling and Simulation (MODSIM2017), Hobart, Tasmania, Australia, 3–8 December 2017; Modelling and Simulation Society of Australia and New Zealand Inc.: Canberra, Australia, 2017; pp. 43–49.

43. Sahlu, D.; Moges, S.A.; Nikolopoulos, E.I.; Anagnostou, E.N.; Hailu, D. Evaluation of High-Resolution Multisatellite and Reanalysis Rainfall Products over East Africa. *Adv. Meteorol.* **2017**, *2017*, 4957960. [CrossRef]

44. Smith, R.A.; Kummerow, C.D. A Comparison of in Situ, Reanalysis, and Satellite Water Budgets over the Upper Colorado River Basin. *J. Hydrometeorol.* **2013**, *14*, 888–905. [CrossRef]

45. Pfeifroth, U.; Mueller, R.; Ahrens, B. Evaluation of Satellite-Based and Reanalysis Precipitation Data in the Tropical Pacific. *J. Appl. Meteorol. Clim.* **2013**, *52*, 634–644. [CrossRef]

46. Smith, M.B.; Koren, V.I.; Zhang, Z.; Reed, S.M.; Pan, J.; Moreda, F. Runoff response to spatial variability in precipitation: An analysis of observed data. *J. Hydrol.* **2004**, *298*, 267–286. [CrossRef]

47. Yang, J.; Jakeman, A.; Fang, G.; Chen, X. Uncertainty analysis of a semi-distributed hydrologic model based on a Gaussian Process emulator. *Environ. Model. Softw.* **2018**, *101*, 289–300. [CrossRef]

48. Hutchinson, M.F.; Xu, T. *Anusplin Version 4.4 User Guide*; Fenner School of Environment and Society, The Australian National University: Canberra, Australia, 2013.

49. Hutchinson, M.F. *Anudem Version 5.3 User Guide*; Fenner School of Environment and Society, The Australian National University: Canberra, Australia, 2011.

50. Hutchinson, M.F. A new procedure for gridding elevation and stream line data with automatic removal of spurious pits. *J. Hydrol.* **1989**, *106*, 211–232. [CrossRef]

51. Zheng, X.; Xiong, H.; Yue, L.; Gong, J. An improved ANUDEM method combining topographic correction and DEM interpolation. *Geocarto Int.* **2016**, *31*, 492–505. [CrossRef]

52. Post, D.A.; Chiew, F.H.S.; Teng, J.; Viney, N.R.; Ling, F.L.N.; Harrington, G.; Crosbie, R.S.; Graham, B.; Marvanek, S.; McLoughlin, R. A robust methodology for conducting large-scale assessments of current and future water availability and use: A case study in Tasmania, Australia. *J. Hydrol.* **2012**, *412–413*, 233–245. [CrossRef]

53. Kim, K.B.; Kwon, H.; Han, D. Exploration of warm-up period in conceptual hydrological modelling. *J. Hydrol.* **2018**, *556*, 194–210. [CrossRef]

54. Kan, G.; Zhang, M.; Liang, K.; Wang, H.; Jiang, Y.; Li, J.; Ding, L.; He, X.; Hong, Y.; Zuo, D.; et al. Improving water quantity simulation & forecasting to solve the energy-water-food nexus issue by using heterogeneous computing accelerated global optimization method. *Appl. Energy* **2018**, *210*, 420–433.

55. Croke, B.; Jakeman, A.J. A catchment moisture deficit module for the IHACRES rainfall-runoff model. *Environ. Model. Softw.* **2004**, *19*, 1–5. [CrossRef]

56. Petheram, C.; Rustomji, P.; Chiew, F.H.S.; Vleeshouwer, J. Rainfall–runoff modelling in northern Australia: A guide to modelling strategies in the tropics. *J. Hydrol.* **2012**, *462–463*, 28–41. [CrossRef]

57. Huang, C.; Newman, A.J.; Clark, M.P.; Wood, A.W.; Zheng, X. Evaluation of snow data assimilation using the ensemble Kalman filter for seasonal streamflow prediction in the western United States. *Hydrol. Earth Syst. Sci.* **2017**, *21*, 635–650. [CrossRef]

58. Burnash, R.J.; Ferral, R.L.; McGuire, R.A. *A Generalized Streamflow Simulation System, Conceptual Modeling for Digital Computers*; U. S. Dept. of Commerce, National Weather Service: Sacramento, CA, USA, 1973.

59. Andrews, F.T.; Croke, B.F.W.; Jakeman, A.J. An open software environment for hydrological model assessment and development. *Environ. Model. Softw.* **2011**, *26*, 1171–1185. [CrossRef]

60. Nash, J.E.; Sutcliffe, J.V. River flow forecasting through conceptual models part I—A discussion of principles. *J. Hydrol.* **1970**, *10*, 282–290. [CrossRef]

61. Krause, P.; Boyle, D.P.; Se, F.B. Comparison of different efficiency criteria for hydrological model assessment. *Adv. Geosci.* **2005**, *5*, 89–97. [CrossRef]

62. Pushpalatha, R.; Perrin, C.; Moine, N.L.; Andréassian, V. A review of efficiency criteria suitable for evaluating low-flow simulations. *J. Hydrol.* **2012**, *420–421*, 171–182. [CrossRef]

63. Duan, Q.; Sorooshian, S.; Gupta, V. Effective and efficient global optimization for conceptual rainfall-runoff models. *Water Resour. Res.* **1992**, *28*, 1015–1031. [CrossRef]

64. Shin, M.; Guillaume, J.H.A.; Croke, B.F.W.; Jakeman, A.J. Addressing ten questions about conceptual rainfall–runoff models with global sensitivity analyses in R. *J. Hydrol.* **2013**, *503*, 135–152. [CrossRef]

65. Beven, K.; Binley, A. The future of distributed models: Model calibration and uncertainty prediction. *Hydrol. Process.* **1992**, *6*, 279–298. [CrossRef]

66. Hornberger, G.M.; Spear, R.C. An Approach to the Preliminary Analysis of Environmental Systems. *J. Environ. Manag.* **1981**, *12*, 7–18.

67. Dai, A. Precipitation Characteristics in Eighteen Coupled Climate Models. *J. Clim.* **2006**, *19*, 4605–4630. [CrossRef]

68. Chapman, T.G. Estimation of daily potential evaporation for input to rainfall-runoff models. In *MODSIM2001: Integrating Models for Natural Resources Management across Disciplines*; Modelling and Simulation Society of Australia and New Zealand Inc.: Canberra, Australia, 2001; pp. 293–298.

69. Safeeq, M.; Mauger, G.S.; Grant, G.E.; Arismendi, I.; Hamlet, A.F.; Lee, S. Comparing Large-Scale Hydrological Model Predictions with Observed Streamflow in the Pacific Northwest: Effects of Climate and Groundwater. *J. Hydrometeorol.* **2014**, *15*, 2501–2521. [CrossRef]

70. Moriasi, D.N.; Arnold, J.G.; Van Liew, M.W.; Bingner, R.L.; Harmel, R.D.; Veith, T.L. Model Evaluation Guidelines for Systematic Quantification of Accuracy in Watershed Simulations. *Trans. ASABE* **2007**, *50*, 885–900. [CrossRef]

71. Orth, R.; Staudinger, M.; Seneviratne, S.I.; Seibert, J.; Zappa, M. Does model performance improve with complexity? A case study with three hydrological models. *J. Hydrol.* **2015**, *523*, 147–159. [CrossRef]

water

MDPI

Article

Parameter Uncertainty Analysis of the SWAT Model in a Mountain-Loess Transitional Watershed on the Chinese Loess Plateau

Fubo Zhao [1], Yiping Wu [1,*], Linjing Qiu [1], Yuzhu Sun [1], Liqun Sun [2], Qinglan Li [2], Jun Niu [3] and Guoqing Wang [4]

[1] Department of Earth and Environmental Science, School of Human Settlements and Civil Engineering, Xi'an Jiaotong University, Xi'an 710049, China; zfubo789@163.com (F.Z.); qiulinjing@mail.xjtu.edu.cn (L.Q.); sunyuzhu12@xjtu.edu.cn (Y.S.)
[2] Shenzhen Institutes of Advanced Technology, Chinese Academy of Sciences, Shenzhen 518055, China; lq.sun@siat.ac.cn (L.S.); ql.li@siat.ac.cn (Q.L.)
[3] Center for Agricultural Water Research in China, China Agricultural University, Beijing 100083, China; niuj@cau.edu.cn
[4] Nanjing Hydraulic Research Institute, Nanjing 210029, China; guoqing_wang@163.com
* Correspondence: rocky.ypwu@gmail.com or yipingwu@xjtu.edu.cn

Received: 12 April 2018; Accepted: 22 May 2018; Published: 25 May 2018

Abstract: Hydrological models play an important role in water resource management, but they always suffer from various sources of uncertainties. Therefore, it is necessary to implement uncertainty analysis to gain more confidence in numerical modeling. The study employed three methods (i.e., Parameter Solution (ParaSol), Sequential Uncertainty Fitting (SUFI2), and Generalized Likelihood Uncertainty Estimation (GLUE)) to quantify the parameter sensitivity and uncertainty of the SWAT (Soil and Water Assessment Tool) model in a mountain-loess transitional watershed—Jingchuan River Basin (JCRB) on the Loess Plateau, China. The model was calibrated and validated using monthly observed streamflow at the Jingchuan gaging station and the modeling results showed that SWAT performed well in the study period in the JCRB. The parameter sensitivity results demonstrated that any of the three methods were capable for the parameter sensitivity analysis in this area. Among the parameters, CN2, SOL_K, and ALPHA_BF were more sensitive to the simulation of peak flow, average flow, and low flow, respectively, compared to others (e.g., ESCO, CH_K2, and SOL_AWC) in this basin. Although the ParaSol method was more efficient in capturing the most optimal parameter set, it showed limited ability in uncertainty analysis due to the narrower 95CI and poor P-factor and R-factor in this area. In contrast, the 95CIs in SUFI2 and GLUE were wider than ParaSol, indicating that these two methods can be promising in analyzing the model parameter uncertainty. However, for the model prediction uncertainty within the same parameter range, SUFI2 was proven to be slightly more superior to GLUE. Overall, through the comparisons of the proposed evaluation criteria for uncertainty analysis (e.g., P-factor, R-factor, NSE, and R^2) and the computational efficiencies, SUFI2 can be a potentially efficient tool for the parameter optimization and uncertainty analysis. This study provides an insight into selecting uncertainty analysis method in the modeling field, especially for the hydrological modeling community.

Keywords: GLUE; hydrological model; ParaSol; SUFI2; uncertainty analysis

---

1. Introduction

Watershed systems are complex due to multiple influencing factors (e.g., climate, land use, and other anthropogenic disturbances), and an accurate prediction of the hydrological processes is indispensable to watershed management [1,2]. Hydrological models have been developed and applied to mathematical representation of hydrological processes, because they can improve the understanding of the impact of natural and anthropogenic disturbances on hydrological features and forecast water resource changes, thus supporting decisions in water resource management [3–5]. However, the model simulation can be highly uncertain due to the defects of the model itself and the complexities of the watershed system, which is now a big concern in the hydrological modeling community [6–9]. Without a realistic assessment of model uncertainty, it is hard to gain confidence in modeling tasks, such as evaluating the responses of the water cycle to future shifts of climate and land use [10]. Therefore, the uncertainty analysis is quite necessary to improve the accuracy and credibility of hydrological simulation.

Uncertainties in hydrological modeling are associated with three possible sources: input data, such as the precipitation data, who can alter the hydrological modeling procedure and simulation results directly (e.g., surface runoff); model structure, which is mainly caused by the assumptions and simplification of the model; and model parameters [6,11–13]. Among these three sources, parameter uncertainty is the most common but relatively easy to control through appropriate calibrations [14]. In general, there exist numerous key parameters in a certain watershed, depicting watershed properties and hydrological processes. These parameters are usually difficult to measure directly, and they are generally derived from the empirical estimation and literature reference, which may introduce uncertainties into the modeling system [12,15,16]. In addition, parameters obtained from calibration are also affected by several factors such as correlations among parameters, sensitive or insensitive in parameters, spatial and temporal scales and statistical features of model residuals, and these may lead to so-called equifinality [17,18].

Numerous studies have focused on parameter uncertainty issues in the hydrological modeling [1,10,19–23]. Several techniques for addressing model uncertainty have been proposed over recent decades. Among those, Parameter Solution (ParaSol) [24], sequential uncertainty fitting (SUFI2) [25], and generalized likelihood uncertainty estimation (GLUE) [26] are three robust ones in the parameter sensitivity and uncertainty analysis in the hydrological simulation [12,14,27,28]. In recent years, there have been a number of studies involving uncertainty analysis using these three methods as well as comparisons of the capabilities for the methods in hydrological simulation and uncertainty analysis [12,14,27,29]. However, the key parameters' identification and the magnitude of their uncertainties vary with the study area/location; it is, therefore, necessary to implement the parameter sensitivity and uncertainty analysis before further hydrological analyses, especially in some distinctive watersheds. The present study aimed to apply these three methods to a distributed hydrological model—SWAT (Soil and Water Assessment Tool) [30]—a physically based distributed hydrological model, which has been increasingly applied to simulate most of the key hydrological processes and assess the water resource management at the watershed scale. We took a typical mountain-loess transitional watershed (Jingchuan River Basin, JCRB) on the Loess Plateau as a case study to: (1) examine the performance and feasibility of SWAT in simulating the streamflow in the JCRB; (2) implement the sensitivity and uncertainty of the parameters using ParaSol, SUFI2, and GLUE; (3) compare the capabilities of these three methods in the parameter uncertainty analysis.

2. Materials and Methods

2.1. Study Area

The JCRB (Figure 1), controlled by the Jingchuan gaging station, lies in the western part of the Jinghe River Basin (106°11′~107°21′ E, 35°15′~35°45′ N). The JCRB is a mountain-loess transitional zone (Figure 1) with a total area of 3164 km². In this basin, 39% is mountainous/rocky terrain, which is mainly located in the high-elevation (>2000 m) area, while 61% is loess area [31]. The region is controlled by the continental climate, which is hot and humid in summer and cold and dry in winter. The mean annual temperature and precipitation is 8.8 °C and 475 mm in the loess area, and 6.5 °C and 614 mm in the mountainous area, respectively [31]. Topographically, the elevation drops from the mountainous area to the loess area with a range of 2898 to 1022 m. The major land use and land cover (LULC) types of this region are forest, cropland, and grassland. The forests are mainly distributed in the mountainous area, whereas the grasslands and croplands are mainly in the loess area. The dominant soil type of the JCRB is Cambisols (Figure 1), which is mainly distributed in the loess area.

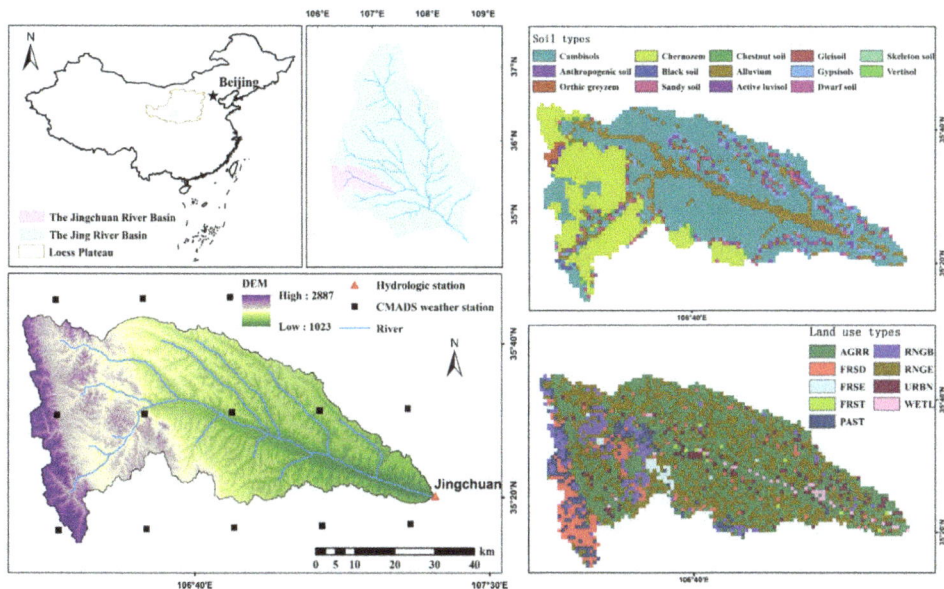

Figure 1. DEM, soil types, and land use types of the Jingchuan River Basin (JCRB).

2.2. Model Description

The SWAT model is a continuous, spatially distributed simulator developed to assist water resource managers in predicting impacts of land management practices on water, sediment, and agricultural chemical yields [30,32]. Fundamentally, the water cycle simulated by SWAT is based on the water balance, whose mathematical equation was reported by Neitsch [32]. The SWAT model is operated at the hydrologic response unit (HRU), which consists of same land use, management, and soil characteristics. The model has been successfully applied around the world for addressing numerous watershed issues under climate shifts and human activities [32,33]. Major outputs of SWAT include surface runoff, baseflow, lateral flow, evapotranspiration (ET), soil water, and water yield.

2.3. Model Input and Setup

The SWAT model requires several specific information such as Digital Elevation Map (DEM), weather, soil properties, and land use and cover types [34]. The DEM with a 90-m resolution was from Shuttle Radar Topography Mission (SRTM). The soil and LULC maps (1 km × 1 km) were from the Ecological and Environmental Science Data Center for West China (http://westdc.westgis.ac.cn). The daily meteorological data from 2008 to 2014 were from the China Meteorological Assimilation Driving Datasets for the SWAT model Version 1.1 (CMADS V1.1, http://www.cmads.org), which was developed by Dr. Xianyong Meng from the China Institute of Water Resources and Hydropower Research (IWHR) and has received worldwide attention [35]. The CMADS V1.1 provides daily precipitation, maximum/minimum temperature, relative humidity, wind speed, and solar radiation. In this study, to ensure that an equilibrium state is attained before the actual simulation (that is, the year of 2008), we took both the year 2006 and 2007 as the warm-up period using the actual weather information in this area. The Geographic Information System (GIS) interface was used to delineate the watershed, resulting in 30 sub-basins and 813 HRUs. The average monthly runoff data from 2008 to 2012 were obtained from the Yellow River Hydrology Year Book.

2.4. Methodology

All of the uncertainty analysis techniques (i.e., ParaSol, SUFI2, and GLUE) used in this study are embedded into a platform—SWAT-CUP [25]—an interface that allows the users to implement the uncertainty analysis for SWAT with multiple-methodological choices. A brief introduction of the three methods is provided in the following sections.

2.4.1. ParaSol

The ParaSol method combines the objective functions (OFs) with a global optimization criterion and implements the simulation and uncertainty analysis using the Shuffle Complex (SCE-UA) algorithm [36]. The SCE-UA is a global search algorithm for the minimization of a specific function [36]. It combines the direct search method of the simplex procedure with the concept of a controlled random search, a systematic evolution of points in the direction of global improvement, competitive evolution, and the concept of complex shuffling [25]. In the operation of SCE-UA, it firstly selects the initial 'population' by random sampling to optimize a certain parameter in feasible parameter space. After the optimization, the simulations are divided into behavioral and non-behavioral simulations according to the criterion value. The ParaSol is efficient in seeking the optimal parameters, because the algorithm samples over the entire parameter space with a focus on solutions near the optimum/optima [24]. The method has been widely applied in the uncertainty analysis in the hydrological simulation, especially for the SWAT model.

2.4.2. SUFI2

Based on a Bayesian framework, SUFI2 quantifies the uncertainties through the sequential and fitting processes. In SUFI2, the parameter uncertainty is calculated from all sources such as the indeterminacy of input variables (e.g., rainfall data, temperature and land use), model structure, and measured data (e.g., surface runoff) [12]. The P-factor, the percentage of observed data bracketed by 95% prediction uncertainty (95PPU), is used to quantify the degree of all uncertainties. The 95PPU is calculated at the 2.5% and 97.5% levels of the cumulative distribution of output variables through Latin hypercube sampling method [18]. For streamflow, a value of P-factor > 0.7 or 0.75 has been reported to be adequate, which illustrates most of the observed data within 95PPU band and the model have been well calibrated [18,23,37]. The R-factor is another index to quantify the strength of a calibration and uncertainty analysis and it reflects the average thickness of the 95PPU band divided by the standard deviation of the measured data. Theoretically, a P-factor of 1 and R-factor of 0 indicate that the simulation exactly corresponds to the measured data [18,25]. Further goodness of fit can be quantified by the R^2 and/or Nash-Sutcliffe model efficiency (NSE) between the observations and the best simulation. SUFI2 can currently handle six different objective functions (e.g., two types of root mean square error, Chi square, NSE, R^2, and bR^2) and the step-by-step operation of SUFI2 can be found in Abbaspour [25].

2.4.3. GLUE

The Generalized Likelihood Uncertainty Estimation (GLUE) method is an uncertainty analysis technique which was introduced by Beven and Binley [26] to allow for the possible non-uniqueness of parameter sets during the estimation in over-parameterized models. The method is used to derive the predictive probability of output variables based on the estimation of the weights or probabilities associated with different parameter sets [26]. In the GLUE operation, it assumes that in the case of the large over-parameterized models, there is no unique set of parameters. In addition, GLUE determines 'good' or 'not good' simulations by a combination of parameters, and the capability of the GLUE method in uncertainty analysis can also be evaluated by the P-factor and R-factor. GLUE can currently support a likelihood measure expressed as the NSE, and the method has also been increasingly applied in the parameter uncertainty analysis and hydrological simulation.

3. Results

3.1. Global Sensitivity

Based on the previous publications related to hydrological simulation using SWAT [38–40] as well as our own experience [41], we selected six key parameters (see Table 1) to implement sensitivity and uncertainty analysis by using the three methods, and the sensitivity ranks were shown in Figure 2. It is important to point out that we performed 1800 model runs in ParaSol and 2000 model runs in SUFI2 or GLUE. Obviously, the ranks of the six parameters yielded by the three methods showed that CN2 was the most sensitive parameter, followed by SOL_K, and the other four parameters showed relatively less sensitivity for streamflow. To accurately identify the parameter sensitivity towards the streamflow, we also tested the individual effect of the six parameters at three levels—the 25th (1st Quantile), the 50th (medium), and the 75th (3rd Quantile) percentiles of the parameter distributions, and their relationships were shown in Figure 3. For the peak flow, the parameter CN2 showed obviously positive relationship, especially in SUFI2, suggesting that CN2 played a key role in simulating the peak flow in this basin. For the average flow, SOL_K exhibited slightly positive relationship, while others showed no obvious relationships with the average flow. Significantly, the parameter ALPHA_BF negatively correlated with the low flow using the three methods. However, the obvious relationships were not found for other parameters. Additionally, the ranks of the sensitivity and relationships between the streamflow and each parameter yielded by ParaSol, SUFI2, and GLUE demonstrated that all the three methods can be used for parameter sensitivity analysis.

Table 1. Calibrated parameter values for monthly streamflow in the Jingchuan River Basin using Parasol, SUFI2, and GLUE.

Parameter	Description	Range	Calibrated Value		
			ParaSol	SUFI2	GLUE
r_CN2	SCS curve number for soil condition II	−50% to +10%	−46%	−38%	−29%
v_ALPHA_BF	Baseflow alpha factor (day)	0.01–0.1	0.04	0.05	0.04
v_ESCO	Soil evaporation percolation fraction	0.1–1.0	0.93	0.5	0.16
v_CH_K2	Effective hydraulic conductivity in main channel alluvium	8.0–18.0	8.0	8.4	8.06
r_SOL_AWC	Available water capacity of soil layer	−20% to +10%	−17%	9%	6%
r_SOL_K	Saturated hydraulic conductivity (mm/h)	−10% to +40%	−10%	−4%	−4%

Note: r means the relative change (%); v means replacing the existing parameter value with the given value.

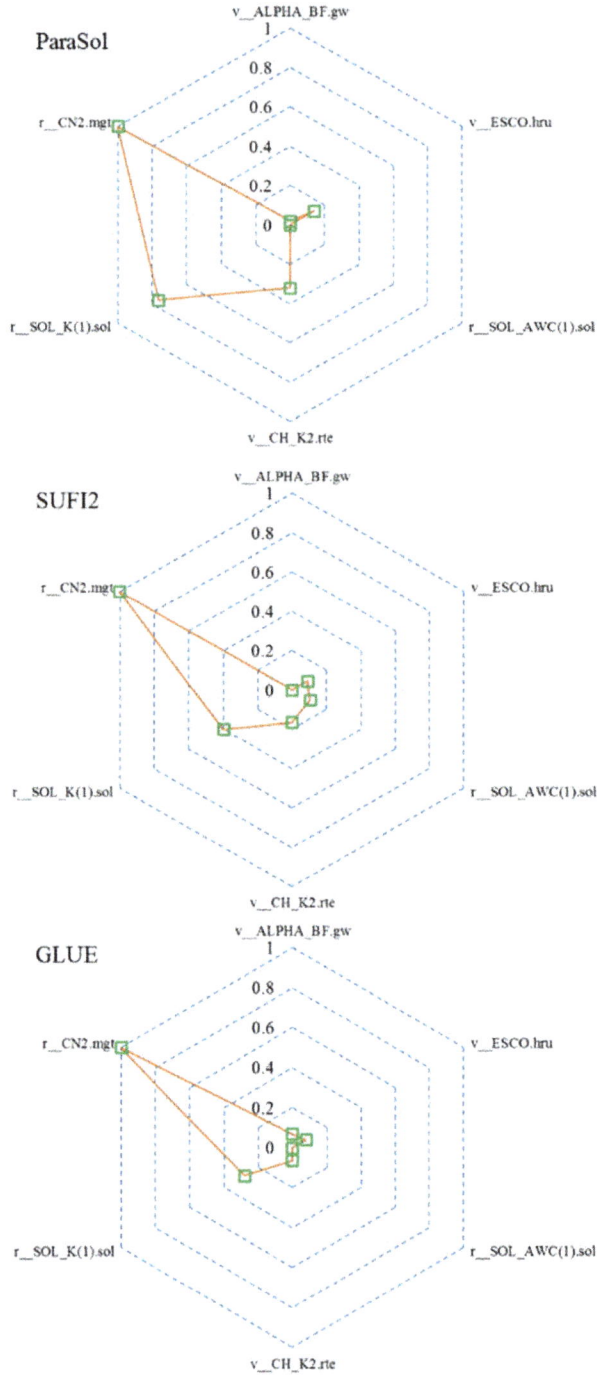

Figure 2. Graphical representation of the parameter sensitivity rank for streamflow yielded by ParaSol, SUFI2, and GLUE. The sensitivity result has been normalized, and a value close to one indicates the parameter is more sensitive to streamflow. The green box means the sensitivity magnitude of a parameter.

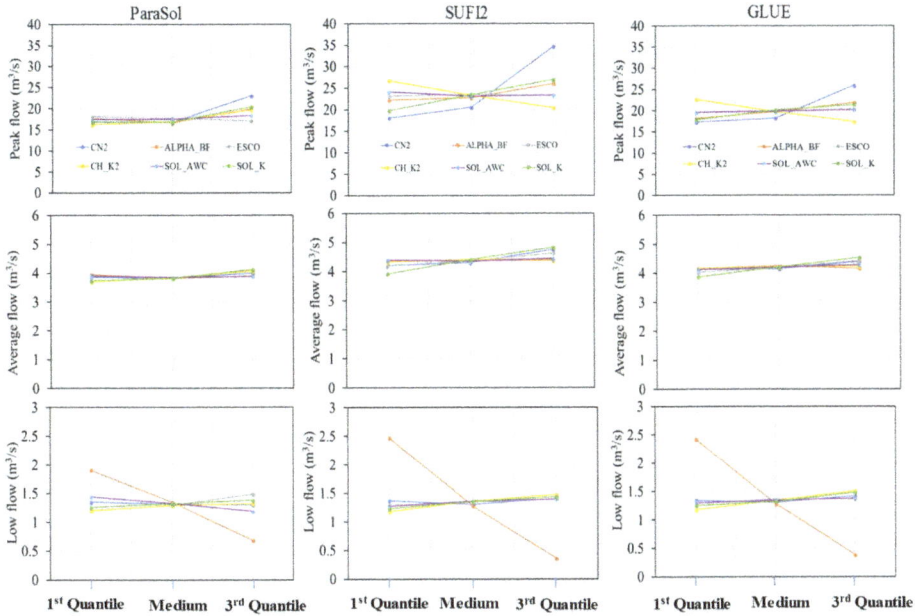

Figure 3. The effect of each parameter generated by the three methods on the peak flow, average flow, and low flow. 1st Quantile, medium, and 3rd Quantile are denoted by 25th, 50th, and 75th percentiles of the parameter distributions, respectively.

3.2. Model Calibration and Validation Results

We compared the capabilities of ParaSol, SUFI2, and GLUE in capturing the optimal parameter sets (in terms of the evaluation criteria) during both the calibration and validation periods in the JCRB. A three-year (2008–2010) record of monthly streamflow at the basin outlet was used for calibration and another two-year (2011–2012) dataset was used for validation. The three sets of the calibrated parameter values derived from the methods were listed in Table 1 and the graphical comparisons (scatterplots) between the observed streamflow and the best simulation were shown in Figure 4. It can be seen from Table 1 and Figure 4 that the calibrated parameter sets of the three methods were not completely in accordance with each other, implying that the three algorithms could recognize the different parameter sets that were able to produce similarly good performance. As can be seen from Table 2, in calibration, the RMSE and RSR yielded by ParaSol (1.24 m^3/s and 0.31) were less than those generated by SUFI2 and GLUE (1.3 m^3/s and 0.33 for SUFI2 and GLUE, respectively). Also, the NSE and R^2 in ParaSol (0.90 and 0.91) were higher than those yielded by SUFI2 (0.89 and 0.89) and GLUE (0.89 and 0.89), suggesting that ParaSol had its advantage on accurately seeking the optimized parameter set compared to SUFI2 and GLUE. In addition, based on the evaluation criteria (see Table 2) and according to Moriasi et al. (2007) [42], the overall model performance can be rated as "good" in both the calibration and validation periods.

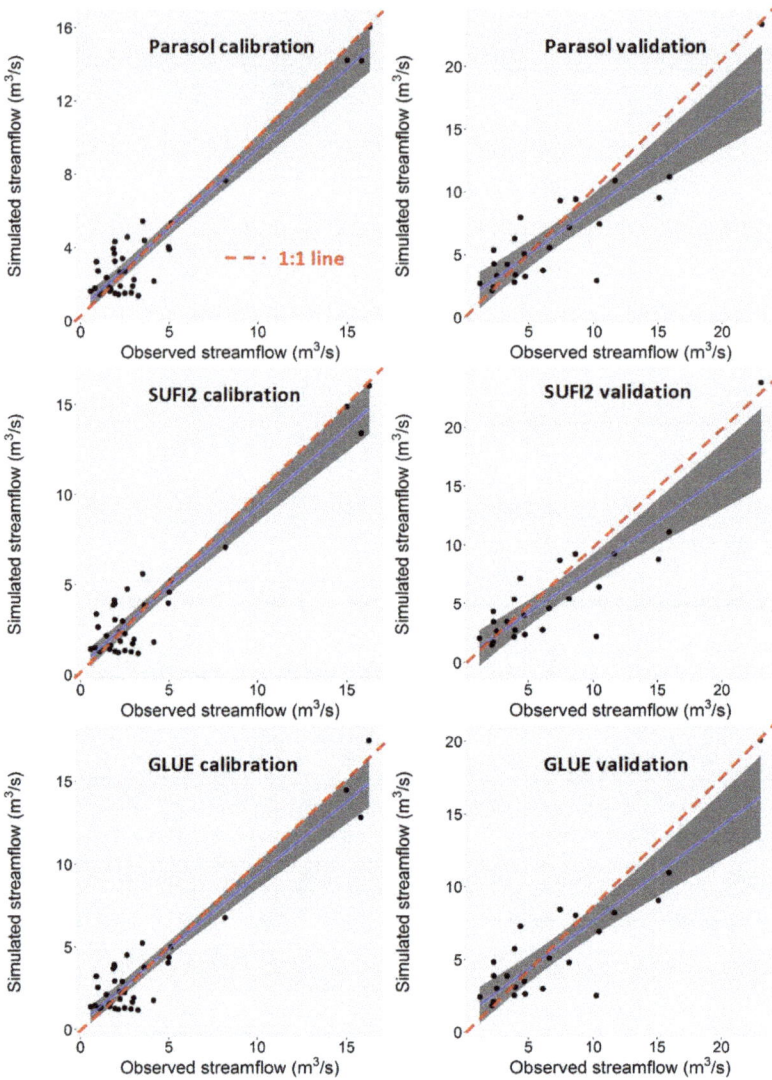

Figure 4. Scatterplots between the best simulation and the observation with the 95% confidence interval (the shaded area).

Table 2. Evaluation of model performance in streamflow simulation during the 3-year (2008–2010) calibration and 2-year (2011–2012) validation periods.

Method	Period	RMSE (m^3/s)	NSE	R^2	RSR	PB (%)
ParaSol	Calibration	1.2	0.90	0.91	0.31	6.6
	Validation	2.6	0.74	0.75	0.50	−6.8
SUFI2	Calibration	1.3	0.89	0.89	0.33	2.3
	Validation	2.9	0.69	0.75	0.54	−18.2
GLUE	Calibration	1.3	0.89	0.89	0.33	0.9
	Validation	2.9	0.68	0.76	0.55	−18.8

3.3. Uncertainty Analysis

It is important to point out that the uncertainty analysis was firstly implemented in the calibration period (2008–2010) and then enlarged to validation (2011–2012) due to the short time of the observed data.

3.3.1. Parasol

Implementation of ParaSol is relatively easy and the computation depends only on the convergence of the optimization process. The upper (97.5%) and lower (2.5%) bounds of the posterior parameter values, expressed as the 95% confidence interval (95CI), and the model prediction uncertainty were shown in Table 3 and Figure 5 (top panel), respectively. The 95CI widths for most parameters were narrower than the initial ranges, except for ESCO and SOL_AWC. Since the optimal values of the latter parameters remained either at the lower or at the upper bounds (see Table 3), indicating that ESCO and SOL_AWC were more uncertain. In general, a higher P-factor means more observations fall inside 95PPU. As can be seen from Figure 5, the uncertainty band was very narrow and the P-factor was only 0.39 and 0.46 in the calibration and validation periods, respectively. This demonstrated that ParaSol had the limited ability for conducting uncertainty analysis though the best simulation matched the observation very well with good NSE and R^2. Figure 6a, b showed the distribution of the model response as a function of the parameter values and the change of standard residuals following the simulated streamflow. It was significant that there existed an overestimation of prediction uncertainty in the wet month (high streamflow), suggesting that more attention should be paid to the wet season in the hydrological simulation. This phenomenon can also be seen from Figure 5 (top panel), where the width of 95PPU band was relatively larger in the high-rainfall seasons. Also, the variance of the residuals was not constant and changed with the streamflow, and this may illustrate that there existed heteroscedasticity in ParaSol. In addition, the correlation matrix showed relatively strong correlations (r ranging from −0.40 to 0.57) among the model parameters, especially the r_SOL_K and v_CH_K2 (r = 0.57), the v_CH_K2 and r_CN2 (r = 0.48).

3.3.2. SUFI2

The SUFI2 method is also convenient to use, though it is semi-distributed and needed for some knowledge of parameters' effects on model output. For the SUFI2 approach, we did one iteration with 2000 model runs using the same parameter ranges for the sake of comparison of the three methods. The 95CI of most parameters yielded by SUFI2 showed a narrower range, though the parameter ALPHA_BF was the same as the initial setting (Table 3), suggesting ALPHA_BF was more uncertain in SUFI2. For the model prediction, it can be seen from Figure 5 (medium panel) that the 95PPU bracketed 83% and 71% of the observations in the calibration and validation periods, respectively, illustrating that SUFI2 was more capable of capturing the observations in spite of a large R-factor. Further, the 95PPU was more suitable to bracket the observations of year 2010, while it slightly overestimated the runoff in winter seasons of year 2008 and 2009. For validation, it underestimated the streamflow from the autumn in 2011 to the summer in 2012, resulting in a relatively poor performance. Similarly, the sensitivity range (Figure 6c) also indicated that the effect of parameter on model outcome was relatively higher in wet seasons (i.e., months with high precipitation). This may be attributed to the uncertainty involved in computing baseflow recession in SWAT and the coarse observations [27]. The residuals were also not normally distributed with constant variance (Figure 6d), which may lead to biased parameter estimation due to the systematic error [43,44]. In addition, the correlation matrix (Table 4) showed very weak correlations among the parameters, and thus, the parameter correlations can be neglected in SUFI2.

Table 3. Uncertainty ranges of aggregate parameters from the three methods.

Parameter	Initial Range	95CI (Confidence Interval)		
		ParaSol	SUFI2	GLUE
r_CN2	−50% to +10%	(−48.0, 0.2)	(−48.5, 8.5)	(−48.5, 8.5)
v_ALPHA_BF	0.01–0.1	(0.01, 0.09)	(0.01, 0.1)	(0.01, 0.1)
v_ESCO	0.1–1.0	(0.14, 0.98)	(0.13, 0.98)	(0.12, 0.98)
v_CH_K2	8.0–18.0	(8.5, 16.7)	(8.3, 17.7)	(8.3, 17.7)
r_SOL_AWC	−20% to +10%	(−18.2, 8.5)	(−19.2, 9.2)	(−19.2, 9.2)
r_SOL_K	−10% to +40%	(−8.9, 29.5)	(−8.7, 38.7)	(−8.7, 38.7)

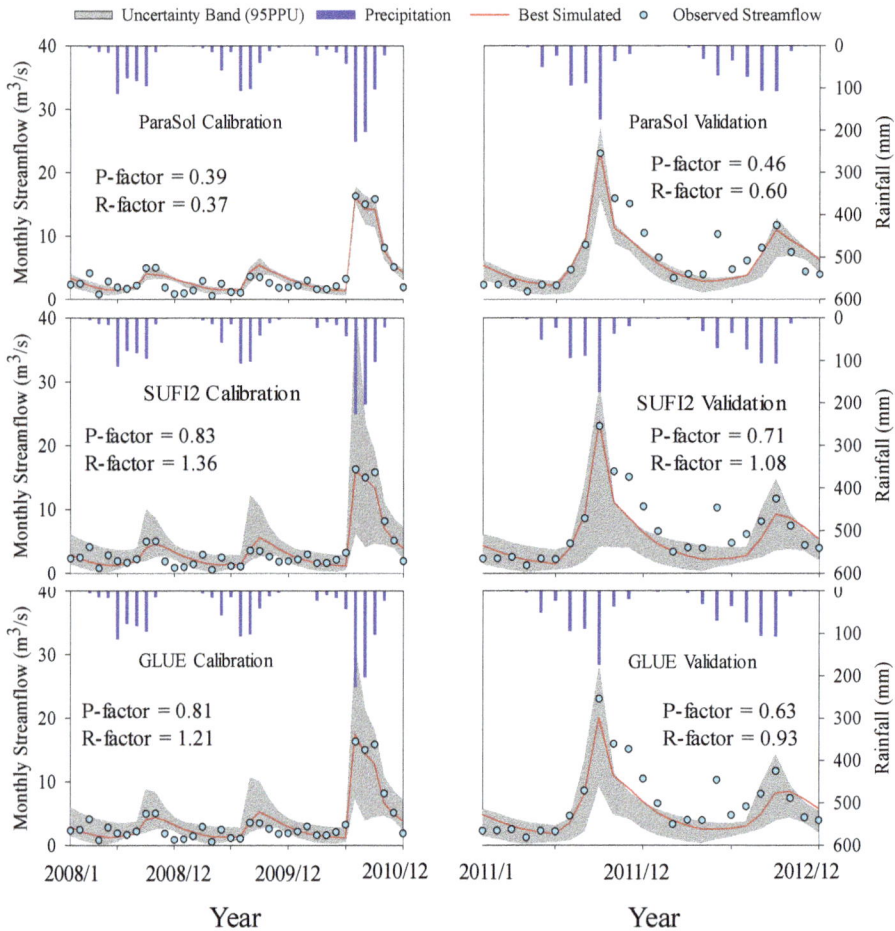

Figure 5. Comparison of best-simulated monthly streamflow with 95PPU against observed streamflow by ParaSol (**top**), SUFI2 (**medium**), and GLUE (**bottom**). P-factor indicates the percentage of observed data bracketed by 95% prediction uncertainty; R-factor reflects the average thickness of 95PPU band divided by the standard deviation of the measured data.

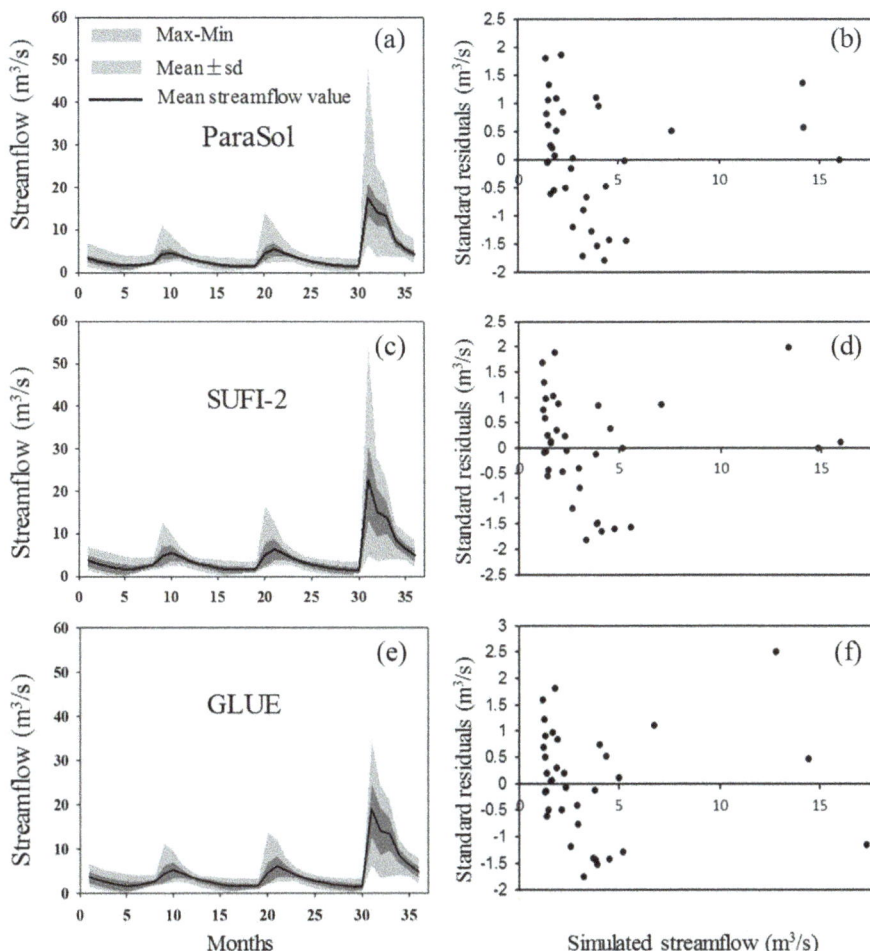

Figure 6. Sensitivity range of monthly streamflow (left column) based on the parameter distribution as generated by ParaSol (**a**), SUFI2 (**c**) and GLUE (**e**) during the 3-year calibration (2008–2010), sd refers to the standard deviation of the model response. The light grey shade by Mine-Max represents the minimum and maximum model response at each time step, whereas the dark grey shade by Mean ± sd refers to the mean model response plus/minus one standard deviation. The right column indicates the standard residuals versus simulated streamflow obtained from ParaSol (**b**), SUFI2 (**d**) and GLUE (**f**) during the calibration period.

3.3.3. GLUE

GLUE is convenient and easy to use and has been widely applied in hydrological field. We also did 2000 runs in GLUE implementation within the same parameter ranges to compare the capabilities of the three methods in parameter uncertainty analysis. The 95CI showed that the parameter uncertainty ranges generated by GLUE were similar with those yielded by SUFI2 but obviously larger than ParaSol, especially for CN2 and SOL_K (Table 3). It can be seen from Figure 5 (bottom panel) that 81% of the observations were bracketed by the 95PPU and the R-factor equaled 1.21 in calibration, which was similar to SUFI2, suggesting that GLUE was also able to capture the observations in calibration. In validation, 63% of the observations were bracketed by the 95PPU, which was slightly less than

SUFI2. Also, GLUE overestimated the streamflow in the winter seasons of year 2008 and 2009, while it somewhat underestimated the streamflow of year 2012. The parameter uncertainty was also found to be higher in the wet seasons during the calibration period (see Figure 6e). The change of residuals demonstrated that the parameter uncertainty estimation may be somewhat biased (Figure 6f). Similar to SUFI2, there was almost no correlations among the parameters yielded by GLUE, and thus the parameter correlations could also be neglected in GLUE.

Table 4. Correlation matrix of the streamflow parameters yielded by ParaSol, SUFI2, and GLUE.

Method	Parameter	r_CN2	v_ALPHA_BF	v_ESCO	v_CH_K2	r_SOL_AWC	r_SOL_K
ParaSol	r_CN2	1	0.11	−0.24	0.48	0.11	0.37
	v_ALPHA_BF		1	−0.27	0.27	0.25	0.21
	v_ESCO			1	−0.24	−0.40	−0.25
	v_CH_K2				1	0.20	0.57
	r_SOL_AWC					1	0.15
	r_SOL_K						1
SUFI2	r_CN2	1	−0.01	0.02	−0.02	−0.02	−0.02
	v_ALPHA_BF		1	0.01	0.02	−0.03	−0.02
	v_ESCO			1	0.03	−0.00	−0.00
	v_CH_K2				1	−0.02	−0.02
	r_SOL_AWC					1	0.03
	r_SOL_K						1
GLUE	r_CN2	1	−0.02	−0.03	−0.02	0.01	−0.02
	v_ALPHA_BF		1	0.01	0.02	−0.02	−0.01
	v_ESCO			1	0.01	−0.01	−0.00
	v_CH_K2				1	0.03	0.01
	r_SOL_AWC					1	−0.00
	r_SOL_K						1

4. Discussion

4.1. Model Parameterization and Performance

For a better streamflow simulation, the accurate identification of key parameters is important. In this study, we identified six key parameters related to streamflow simulation by using the three popular methods in the JCRB. As can be seen from Table 1, ParaSol provided the least CN2, SOL_AWC, and SOL_K values compared to those yielded by SUFI2 and GLUE. Most of the parameter values generated by SUFI2 were similar with those yielded by GLUE. The phenomenon may be attributed to the objective functions of the methods and the initial parameters' ranges. In our study, the objective function in ParaSol was limited to the sum of the squares of the residuals [25], indicating that ParaSol aimed to find the least bias when seeking the best parameter set. However, the NSE was the objective function in both SUFI2 and GLUE, which may handle a different pathway in finding the best fitting parameter set. In addition, the initial parameter ranges may also play an important role in seeking appropriate parameter sets because the initial ranges can decide both the parameters' combination and applicability. Therefore, the above conditions may result in different calibrated parameter values using the three methods.

The comparisons of the SWAT model performance generated by the three methods were listed in Table 2. In terms of the evaluation criteria, ParaSol provided slightly higher NSE and R^2 and achieved less predicting errors (see Table 2) in both the calibration and validation periods, showing its advantage in accurately capturing the optimal parameter set, which was also confirmed by others [1,14,27]. This is because ParaSol is based on the global optimization algorithms and thus samples over the entire parameter space with a focus on solutions near the optimum [45]. This algorithm is much more efficient in finding the maximum or minimum of the objective function than random or Latin hypercube sampling [27,46]. Therefore, ParaSol can be a reasonable choice in seeking the best parameter set in hydrological modeling.

4.2. Parameter Sensitivity and Uncertainty

For parameter sensitivity, the results showed that the parameter CN2 was the most sensitive (Figure 2) using any of the three methods and CN2 played a key role in peak flow simulation (Figure 3). CN2 is a function of soil's permeability, land use, and initial soil water condition, which suggests the potential of surface runoff from rainfall in a watershed [47]. CN2 had positive effects on the peak flow (Figure 3), and this may be because of concentrated distribution of rainfall in the wet season, causing infiltration excessed surface flow and a significant increase in peak flow. The parameter SOL_K, which represents soil hydraulic conductivity and closely relates to the movement of water in soil profiles, had a positive effect on average streamflow. In the study area, Cambisols is widely distributed (see Figure 1) and is mainly made up of sandy loam that is characterized with medium percolation capacity. Water in this soil can be easier to percolate to the shallow aquifer and further contributed to the baseflow and then the streamflow. ALPHA_BF is the baseflow recession factor, a high value of ALPHA_BF means quick recession of baseflow (i.e., the less water retention in the aquifer), and this was why this parameter played a key role in the relationship between the low flow and ALPHA_BF.

For parameter uncertainty, our study showed that GLUE and SUFI2 provided the wider 95CIs than ParaSol (see Table 3). Most of the uncertainty intervals derived by GLUE and SUFI2 contained the corresponding intervals from ParaSol. Based on SCE-UA, ParaSol was very efficient in seeking the most suitable parameter set near the maximum or minimum objective function value [27], which suggested that the parameter can be narrowed to a relatively small extent. The wider parameter ranges in SUFI2 and GLUE may be because they considered all sources of uncertainty and thus may lead to relatively larger ranges of parameter uncertainty [46]. As we know, the GLUE method considers the parameter correlation in uncertainty analysis, but Table 4 showed that there were almost no correlations among the parameters, which was the same as SUFI2. Therefore, through the comparisons of the parameter uncertainty ranges as well as the parameter correlations, both SUFI2 and GLUE showed advantages in providing similarly good parameter uncertainty ranges [1,12,14,27].

4.3. Model Prediction Uncertainty

For model prediction uncertainty analysis, we found that SUFI2 was a superior tool because of its relatively larger P-factor and reasonable R-factor. As seen from Figure 5, ParaSol did not derive reasonable prediction uncertainty and only 39% and 46% measurements were bracketed by the 95PPU in calibration and validation, respectively, in spite of the good R^2 and NSE. This was because ParaSol does not consider the error in the measured data, model structure, and measured response, leading to an underestimation of the prediction uncertainty [17,27]. As stated previously, the parameter uncertainty yielded by Parasol only accounted for a small part of the whole uncertainty; whereas, SUFI2 and GLUE took into account all sources of uncertainties, and the corresponding parameter ranges (95CI) were also larger than ParaSol, leading to the wider 95PPU bands. In addition, according to Abbaspour et al. [18], the 95PPU should bracket at least 80% of the observed data if the measurements are of high quality. In terms of our results, SUFI2 and GLUE bracketed above 80% of the observed streamflow in the calibration period, although we recognized that there still existed a certain uncertainty in SWAT (1.36 and 1.21 of R-factor in SUFI2 and GLUE, respectively), which may be because of the overestimation of the errors in the input, output, and model structure. It was also worth noting that the coverage (P-factor) of GLUE can be increased at the expense of increasing R-factor, and in SUFI2 this can be done by performing one more iteration. Compared to GLUE, the 95PPU in SUFI2 bracketed 83% and 71% measurements in calibration and validation, respectively, suggesting that SUFI2 was more capable of capturing the observations. The main reason could be all these sampled parameter sets were taken as behavioral samples and contributed to the 95PPU [12]. Additionally, based on the previous studies and our own experience [1,14,16,27], the SUFI2 method has a high efficiency in computation because of the advantages in taking into account the discrete parameter space of the Latin hypercube sampling [27]. In contrast, GLUE makes use of the Monte Carlo simulation for random sampling and needs a certain number of sampling runs to derive the most reasonable outputs, especially for the complex models [6,12,14]. Therefore, SUFI2 is more efficient for uncertainty analysis in the hydrological simulation when handling some high dimensional and complex hydrological models.

5. Conclusions

This study examined the capabilities of three uncertainty analysis methods through a distributed hydrological model—SWAT with a case study in the JCRB on the Chinese Loss Plateau. The modeling results showed that the SWAT model was acceptable in the streamflow simulation in the JCRB with NSE and R^2 being 0.90 and 0.91 for calibration, and 0.74 and 0.75 for validation, respectively. The sensitivity analysis of the selected six key parameters indicated that ParaSol, SUFI2, and GLUE could be used for parameter sensitivity analysis in the study area. The sensitivity results showed that CN2, SOL_K, and ALPHA_BF were more sensitive to the simulation of peak flow, average flow, and low flow, respectively, compared to others (e.g., ESCO, CH_K2, and SOL_AWC) in this area. Although ParaSol was more efficient in capturing the optimal parameter set, it did not derive the suitable parameter and prediction uncertainty ranges due to its relatively narrower 95CI and poor P-factor and R-factor. Compared to ParaSol, SUFI2 and GLUE were proven to be more capable in predicting the parameter uncertainty, and SUFI2 was superior to GLUE in terms of the P-factor and R-factor. In summary, through the comparisons of the evaluation criteria for uncertainty analysis (e.g., P-factor, R-factor, NSE, and R^2) and the computational efficiencies, the SUFI2 method performed better than the other two methods for the parameter uncertainty analysis of the SWAT model in the JCRB. The study provides an insight into the identifiability of more reliable methods for uncertainty analysis, especially in the hydrological modeling community.

Finally, although this study was informative by implementing the three popular parameter uncertainty analysis methods, the generality of such findings is to be evaluated with more applications in other areas. Moreover, in addition to the parameters' uncertainty, the uncertainty in model structure and input data should be examined for the complete and deep understanding of the modeling behavior.

Author Contributions: F.Z. and Y.W. designed and performed the study, interpreted the results, and wrote the paper. L.Q., Y.S., L.S., Q.L., J.N. and G.W. contributed to data collection, results presentation, and draft revision.

Funding: This study was funded by the National Thousand Youth Talent Program of China, the Hundred Youth Talent Program of Shaanxi Province, the Young Talent Support Plan of Xi'an Jiaotong University, National Natural Science Foundation of China (31741020), National Key Research and Development Project (2016YFA0601501), and Natural Science Foundation of Guangdong Province (2016A050503035).

Conflicts of Interest: The authors declare no conflict of interest.

References

1. Uniyal, B.; Jha, M.K.; Verma, A.K. Parameter identification and uncertainty analysis for simulating streamflow in a river basin of eastern India. *Hydrol. Process.* **2015**, *29*, 3744–3766. [CrossRef]

2. Gyamfi, C.; Ndambuki, J.; Salim, R. Hydrological responses to land use/cover changes in the Olifants Basin, South Africa. *Water* **2016**, *8*, 588. [CrossRef]

3. Viviroli, D.; Zappa, M.; Gurtz, J.; Weingartner, R. An introduction to the hydrological modelling system prevah and its pre- and post-processing-tools. *Environ. Model. Softw.* **2009**, *24*, 1209–1222. [CrossRef]

4. Wu, K.; Xu, Y.J. Evaluation of the applicability of the SWAT model for coastal watersheds in southeastern Louisiana. *J. Am. Water Resour. Assoc.* **2006**, *42*, 1247–1260. [CrossRef]

5. Zhu, X.; Zhang, C.; Qi, W.; Cai, W.; Zhao, X.; Wang, X. Multiple climate change scenarios and runoff response in Biliu River. *Water* **2018**, *10*, 126. [CrossRef]

6. Beven, K.; Freer, J. Equifinality, data assimilation, and uncertainty estimation in mechanistic modelling of complex environmental systems using the glue methodology. *J. Hydrol.* **2001**, *249*, 11–29. [CrossRef]

7. Zheng, Y.; Han, F. Markov Chain Monte Carlo (MCMC) uncertainty analysis for watershed water quality modeling and management. *Stoch. Environ. Res. Risk Assess.* **2015**, *30*, 293–308. [CrossRef]

8. Li, Z.; Xu, Z.; Shao, Q.; Yang, J. Parameter estimation and uncertainty analysis of SWAT model in upper reaches of the Heihe River Basin. *Hydrol. Process.* **2009**, *23*, 2744–2753. [CrossRef]

9. Song, X.; Zhang, J.; Zhan, C.; Xuan, Y.; Ye, M.; Xu, C. Global sensitivity analysis in hydrological modeling: Review of concepts, methods, theoretical framework, and applications. *J. Hydrol.* **2015**, *523*, 739–757. [CrossRef]

10. Li, Z.; Shao, Q.; Xu, Z.; Cai, X. Analysis of parameter uncertainty in semi-distributed hydrological models using bootstrap method: A case study of SWAT model applied to Yingluoxia watershed in Northwest China. *J. Hydrol.* **2010**, *385*, 76–83. [CrossRef]

11. Yen, H.; Wang, X.; Fontane, D.G.; Harmel, R.D.; Arabi, M. A framework for propagation of uncertainty contributed by parameterization, input data, model structure, and calibration/validation data in watershed modeling. *Environ. Model. Softw.* **2014**, *54*, 211–221. [CrossRef]

12. Xue, C.; Chen, B.; Asce, M.; Wu, H. Parameter uncertainty analysis of surface flow and sediment yield in the Huolin Basin, China. *J. Hydrol. Eng.* **2014**, *19*, 1224–1236. [CrossRef]

13. Refsgaard, J.C.; van der Sluijs, J.P.; Brown, J.; van der Keur, P. A framework for dealing with uncertainty due to model structure error. *Adv. Water Res.* **2006**, *29*, 1586–1597. [CrossRef]

14. Wu, H.; Chen, B. Evaluating uncertainty estimates in distributed hydrological modeling for the Wenjing River watershed in China by GLUE, SUFI-2, and ParaSol methods. *Ecol. Eng.* **2015**, *76*, 110–121. [CrossRef]

15. Nandakumar, N.; Mein, R.G. Uncertainty in rainfall-runoff model simulations and the implications for predicting the hydrologic effects of land-use change. *J. Hydrol.* **1997**, *193*, 211–232. [CrossRef]

16. Zhou, J.; Liu, Y.; Guo, H.; He, D. Combining the SWAT model with sequential uncertainty fitting algorithm for streamflow prediction and uncertainty analysis for the lake Dianchi Basin, China. *Hydrol. Process.* **2014**, *28*, 521–533. [CrossRef]

17. Zhang, J.; Li, Q.; Guo, B.; Gong, H. The comparative study of multi-site uncertainty evaluation method based on SWAT model. *Hydrol. Process.* **2015**, *29*, 2994–3009. [CrossRef]

18. Abbaspour, K.C.; Yang, J.; Maximov, I.; Siber, R.; Bogner, K.; Mieleitner, J.; Zobrist, J.; Srinivasan, R. Modelling hydrology and water quality in the pre-alpine/alpine Thur watershed using SWAT. *J. Hydrol.* **2007**, *333*, 413–430. [CrossRef]

19. Vilaysane, B.; Takara, K.; Luo, P.; Akkharath, I.; Duan, W. Hydrological stream flow modelling for calibration and uncertainty analysis using SWAT model in the Xedone River Basin, Lao PDR. *Procedia Environ. Sci.* **2015**, *28*, 380–390. [CrossRef]

20. Zhang, X.; Srinivasan, R.; Bosch, D. Calibration and uncertainty analysis of the swat model using Genetic Algorithms and Bayesian Model Averaging. *J. Hydrol.* **2009**, *374*, 307–317. [CrossRef]

21. Wu, Y.; Liu, S. Automating calibration, sensitivity and uncertainty analysis of complex models using the R package Flexible Modeling Environment (FME): SWAT as an example. *Environ. Model. Softw.* **2012**, *31*, 99–109. [CrossRef]

22. Wu, Y.; Liu, S.; Yan, W. A universal model-R coupler to facilitate the use of r functions for model calibration and analysis. *Environ. Model. Softw.* **2014**, *62*, 65–69. [CrossRef]

23. Abbaspour, K.C.; Rouholahnejad, E.; Vaghef, S.; Srinivasan, R.; Yang, H.; Kløve, B. A continental-scale hydrology and water quality model for Europe: Calibration and uncertainty of a high-resolution large-scale SWAT model. *J. Hydrol.* **2015**, *524*, 733–752. [CrossRef]

24. van Griensven, A.; Meixner, T. Methods to quantify and identify the sources of uncertainty for river basin water quality models. *Water Sci. Technol.* **2006**, *53*, 51–59. [CrossRef] [PubMed]

25. Abbaspour, K.C. *SWAT Calibration and Uncertainty Programs: A User Mannual*; Swiss Federal Institute of Aquatic Science and Technology (Eawag): Dübendorf, Switzerland, 2011.

26. Beven, K.; Binley, A. The future of distributed models-model calibration and uncertainty prediction. *Hydrol. Process.* **1992**, *6*, 279–298. [CrossRef]

27. Yang, J.; Reichert, P.; Abbaspour, K.C.; Xia, J.; Yang, H. Comparing uncertainty analysis techniques for a SWAT application to the Chaohe Basin in China. *J. Hydrol.* **2008**, *358*, 1–23. [CrossRef]

28. Khalid, K.; Ali, M.F.; Rahman, N.F.A.; Mispan, M.R.; Haron, S.H.; Othman, Z.; Bachok, M.F. Sensitivity analysis in watershed model using SUFI-2 algorithm. *Procedia Eng.* **2016**, *162*, 441–447. [CrossRef]

29. Kouchi, D.H.; Esmaili, K.; Faridhosseini, A.; Sanaeinejad, S.H.; Khalili, D.; Abbaspour, K.C. Sensitivity of calibrated parameters and water resource estimates on different objective functions and optimization algorithms. *Water* **2017**, *9*, 384. [CrossRef]

30. Arnold, J.G.; Srinivasn, R.; Muttiah, R.S.; Willians, J.R. Large area hydrologic modeling and assessment—Part I: Model development. *J. Am. Water Resour. Assoc.* **1998**, *34*, 73–89. [CrossRef]

31. Zhang, L.; Podlasly, C.; Feger, K.-H.; Wang, Y.; Schwärzel, K. Different land management measures and climate change impacts on the runoff—A simple empirical method derived in a mesoscale catchment on the loess plateau. *J. Arid Environ.* **2015**, *120*, 42–50. [CrossRef]

32. Neitsch, S.L.; Arnold, J.G.; Kiniry, J.R.; Willianms, J.R. *Soil and Water Assessment Tool Theoretical Documentation Version 2009*; Texas Water Resources Institute: Ollege Station, TX, USA, 2011.

33. Panagopoulos, Y.; Gassman, P.W.; Arritt, R.W.; Herzmann, D.E.; Campbell, T.D.; Valcu, A.; Jha, M.K.; Kling, C.L.; Srinivasan, R.; White, M.; et al. Impacts of climate change on hydrology, water quality and crop productivity in the Ohio-Tennessee River Basin. *Int. J. Agric. Biol. Eng.* **2015**, *8*, 36–53.

34. Arnold, J.G.; Muttiah, R.S.; Srinivasan, R.; Allen, P.M. Regional estimation of base flow and groundwater recharge in the upper Mississippi River Basin. *J. Hydrol.* **2000**, *227*, 21–40. [CrossRef]

35. Meng, X.; Wang, H. Significance of the china meteorological assimilation driving datasets for the SWAT model (CMADS) of East Asia. *Water* **2017**, *9*, 765. [CrossRef]

36. Duan, Q.; Sorooshian, S.; Gupta, V. Effective and efficient global optimation for conceptual rainfall-runoff models. *Water Resour. Manag.* **1992**, *28*, 1015–1031. [CrossRef]

37. Ashraf Vaghefi, S.; Abbaspour, K.; Faramarzi, M.; Srinivasan, R.; Arnold, J. Modeling crop water productivity using a coupled SWAT–MODSIM model. *Water* **2017**, *9*, 157. [CrossRef]

38. Li, Z.; Liu, W.-Z.; Zhang, X.-C.; Zheng, F.-L. Impacts of land use change and climate variability on hydrology in an agricultural catchment on the Loess Plateau of China. *J. Hydrol.* **2009**, *377*, 35–42. [CrossRef]

39. Wang, H.; Sun, F.; Xia, J.; Liu, W. Impact of lucc on streamflow based on the SWAT model over the Wei River Basin on the Loess Plateau in China. *Hydrol. Earth Syst. Sci.* **2017**, *21*, 1929–1945. [CrossRef]

40. Zuo, D.; Xu, Z.; Peng, D.; Song, J.; Cheng, L.; Wei, S.; Abbaspour, K.C.; Yang, H. Simulating spatiotemporal variability of blue and green water resources availability with uncertainty analysis. *Hydrol. Process.* **2015**, *29*, 1942–1955. [CrossRef]

41. Zhao, F.; Wu, Y.; Qiu, L.; Bellie, S.; Zhang, F.; Sun, Y.; Sun, L.; Li, Q.; Alexey, V. Spatiotemporal features of the hydro-biogeochemical cycles in a typical loess gully watershed. *Ecol. Indic.* **2018**, *91*, 542–554. [CrossRef]

42. Moriasi, D.N.; Arnold, J.G.; Van Liew, M.W.; Bingner, R.L.; Harmel, R.D.; Veith, T.L. Model evaluation guidelines for systematic quantification of accuracy in watershed simulations. *Trans. ASABE* **2007**, *50*, 885–900. [CrossRef]

43. Datta, A.R.; Bolisetti, T. Second-order autoregressive model-based likelihood function for calibration and uncertainty analysis of SWAT model. *J. Hydrol. Eng.* **2015**, *20*, 04014045. [CrossRef]

44. Neupane, R.P.; Kumar, S. Estimating the effects of potential climate and land use changes on hydrologic processes of a large agriculture dominated watershed. *J. Hydrol.* **2015**, *529*, 418–429. [CrossRef]

45. Van Griensven, A.; Meixner, T.; Grunwald, S.; Bishop, T.; Di luzio, M.; Srinivasan, R. A global sensitivity analysis tool for the paramters of multi-variable catchment models. *J. Hydrol.* **2006**, *324*, 10–23. [CrossRef]

46. Abbaspour, K.; Vaghefi, S.; Srinivasan, R. A guideline for successful calibration and uncertainty analysis for soil and water assessment: A review of papers from the 2016 International SWAT Conference. *Water* **2017**, *10*, 6. [CrossRef]

47. Liu, Y.R.; Li, Y.P.; Huang, G.H.; Zhang, J.L.; Fan, Y.R. A bayesian-based multilevel factorial analysis method for analyzing parameter uncertainty of hydrological model. *J. Hydrol.* **2017**, *553*, 750–762. [CrossRef]

water

MDPI

Article

Investigating the Dynamic Influence of Hydrological Model Parameters on Runoff Simulation Using Sequential Uncertainty Fitting-2-Based Multilevel-Factorial-Analysis Method

Shuai Zhou, Yimin Wang *, Jianxia Chang, Aijun Guo and Ziyan Li

State Key Laboratory of Eco-hydraulics in Northwest Arid Region of China, Xi'an University of Technology, Xi'an 710048, China; zhoushuai0113@163.com (S.Z.); chxiang@xaut.edu.cn (J.C.); aijunguo619@gmail.com (A.G.); liziyan94@163.com (Z.L.)
* Correspondence: wangyimin@xaut.edu.cn; Tel.: +86-136-7927-9030

Received: 26 June 2018; Accepted: 23 July 2018; Published: 3 September 2018

Abstract: Hydrological model parameters are generally considered to be simplified representations that characterize hydrologic processes. Therefore, their influence on runoff simulations varies with climate and catchment conditions. To investigate the influence, a three-step framework is proposed, i.e., a Latin hypercube sampling (LHS-OAT) method multivariate regression model is used to conduct parametric sensitivity analysis; then, the multilevel-factorial-analysis method is used to quantitatively evaluate the individual and interactive effects of parameters on the hydrologic model output. Finally, analysis of the reasons for dynamic parameter changes is performed. Results suggest that the difference in parameter sensitivity for different periods is significant. The soil bulk density (SOL_BD) is significant at all times, and the parameter Soil Convention Service (SCS) runoff curve number (CN2) is the strongest during the flood period, and the other parameters are weaker in different periods. The interaction effects of CN2 and SOL_BD, as well as effective hydraulic channel conditions (CH_K2) and SOL_BD, are obvious, indicating that soil bulk density can impact the amount of loss generated by surface runoff and river recharge to groundwater. These findings help produce the best parameter inputs and improve the applicability of the model.

Keywords: CMADS dataset; parameter sensitivity; SUFI-2; Yellow River

1. Introduction

Hydrological models play a crucial role in simulating the hydrological process of river basins. These hydrological models are generally composed of several parameters, whose values cannot be directly determined by field observations but which can be calibrated through input/output records, which inevitably contain the basin error response [1]. Among many hydrological models, the Soil and Water Assessment Tool (SWAT) model has been widely used in many countries for its ability to completely reflect the influence of spatiotemporal heterogeneity, such as topography, soil, and land use, on the water cycle of the river basin [2–4]. However, the uncertainty of the SWAT model parameter is difficult to evaluate, because the model parameters are numerous and difficult to obtain. It also brings difficulties to decision-makers when the hydrological process is accurately described, as well as the regional relationship between the model parameters and the watershed characteristics. Thus, more effort is required to quantify the uncertainty in the hydrological simulation.

To date, a variety of optimization algorithms have been developed for calibration and uncertainty analysis, and good results have been achieved [5–8]. Kouchi et al. [9] use three different optimization algorithms (sequential uncertainty fitting 2 (SUFI-2), particle swarm optimization (PSO), and generalized likelihood uncertainty estimation (GLUE)), as well as eight evaluation indexes in a

SWAT model for emphasizing that the combination of target functions of each optimization algorithm may lead to different optimal parameter sets, and have the same performance at the same time. Trudel et al. [10] use seven different objective functions to study hydrological model calibration and the structural uncertainty in low-flow simulations under climate change conditions. Muleta et al. [11] use comparative analysis of parameter sensitivity in the high- and low-flow period, and found that there was a significant difference in the parameter sensitivity between the high- and the low-flow periods, and the same parameters were different in different periods. However, the algorithm limitations, such as with GLUE and SUFI-2, lie in the quantitative analysis of the impact of their parameters on system performance. In fact, model parameters describing different hydrological processes have different individual effects on the model output. Factorial analysis can help study the individual and interaction effects of the parameters [12]. The factorial-analysis-of-variance method is used to diagnose the curve relationship between the parameters and the response [13–15]. Nevertheless, no previous study has been conducted to investigate the dynamic influence of hydrological model parameters on runoff simulation using the SUFI-2-based multilevel-factorial-analysis method.

As a case study, this study regarded the source region of the Yellow River, which is known as the "water tower" of the Yellow River basin and contributes 35% of total annual runoff from about 16.2% of the basin area [16]. More importantly, the Yellow River plays a key role in the water supply for 107 million people and for about 13% of the agricultural production of the country's total cultivated area.

The objective of this study was to develop a SUFI-2-based multilevel-factorial-analysis method to address the dynamic influence of hydrological model parameters on runoff simulation. The main steps of the study included (i) the China Meteorological Assimilation Driving Datasets (CMADS), which were used to drive the SWAT model; (ii) a multivariate regression model to conduct parametric sensitivity analysis, which was based on the results of the Latin hypercube sampling (LHS) method; (iii) the confidence interval of each sensitive parameter, found using the SUFI-2 algorithm; and (iv) using the SUFI-2-based multilevel-factorial-analysis method, we quantitatively evaluated the individual and interactive effects of parameters on the hydrologic model output. The results of the study are helpful for improving the simulation and prediction ability of the hydrologic model for water resources.

2. Study Area

The source region of the Yellow River (Figure 1) is located in the northeast Qinghai–Tibet Plateau between longitudes $95°50'$ E and $103°30'$ E and latitudes $30°30'$ E and $35°0'$ E, covering 12.19×104 km^2 and occupying 16.2% of the entire Yellow River basin (75.24×104 km^2) [17,18]. The average temperature is about 5 °C, and the temperature here varies greatly between day and night. The average annual precipitation varies between 320 and 750 mm. Precipitation in June to September accounts for 80% of the total year. Alpine vegetation and alpine meadows are the major vegetation types, accounting for the total area of 70% in 2010. The major soil type in the watershed is loam, and most of the soil has poor water retention and low fertility.

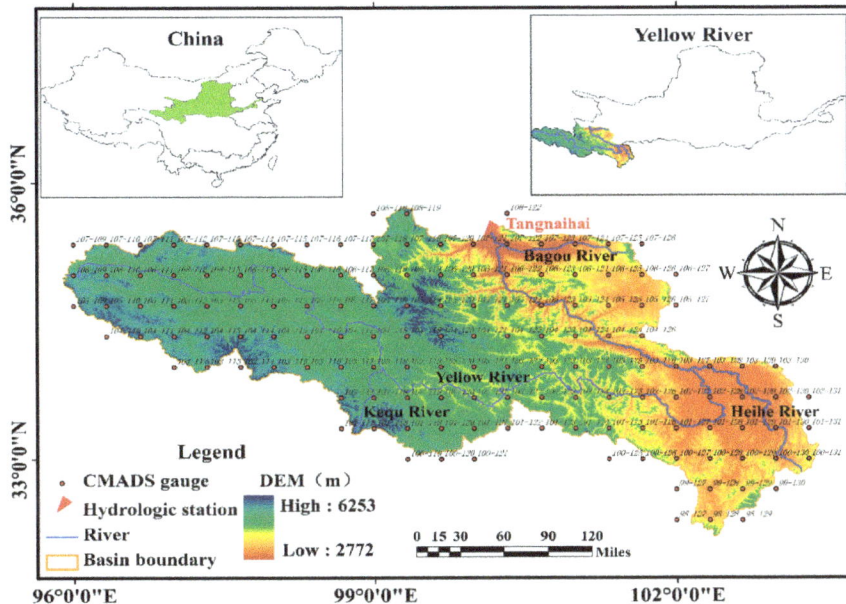

Figure 1. Locations of the Yellow River source region.

3. Methodology

This paper uses the CMADS data set to drive the SWAT model. The constructed SWAT model employs parameters to depict the characteristics of the hydrological process. To identify the parameter sensitivity to the runoff simulation of the SWAT model, we applied the LHS-based multivariate regression model. Then, the SUFI-2 algorithm was used to explore the confidence interval of these identified sensitive parameters. We employed the multilevel-factorial-analysis method to quantitatively evaluate the individual and interactive effects of the parameters on the hydrological model output, i.e., the simulated discharge. Finally, the physical mechanism of the parameters involved in the hydrological process was traced.

3.1. Construction of the Soil and Water Assessment Tool Model

The inputs required for the SWAT model include a DEM (digital elevation model), land use, and soil data sets. The DEM is the Shuttle Radar Topographic Mission (SRTM) (90) DEM, which comes from the geospatial data cloud (http://www.gscloud.cn). The soil data were obtained from the China Soil Data Set (v1.0), based on the World Soil Database (HSDW). The land use data (LCC2010) of the study region were derived from the dry area scientific data center in the cold region. The China Meteorological Assimilation Driving Datasets for the SWAT model Version 1.1 (CMADS V1.1, http://www.cmads.org), which was developed by Dr. Xianyong Meng from the China Agricultural University (CAU), has received worldwide attention [19]. This data set is widely used by countries throughout the world [20–26]. Meng et al. [27] chose the Manas River Basin (MRB) in China as a research area, in order to verify the adaptability of the China Meteorological Assimilation Driving Datasets for the Soil and Water Assessment Tool model (CMADS); the results showed that the SWAT model could reproduce the runoff process of two stations (Kenswat and Hongshanzui) in the research area well using data from CMADS. Zhang et al. [28] used CMADS to drive the SWAT model for a runoff simulation in the Hunhe River Basin, and the results showed that both Nash-Sutcliffe efficiency coefficient (NSE) values and R^2 were found to be greater than 0.74 in calibration, and are greater than 0.58 in validation. Related results indicate that CMADS performs particularly well in runoff simulation.

3.2. Parameter Sensitivity Analysis

There are many parameters in the SWAT model. Different parameters have different sensitivities to the model simulation [29–31]. The model can eliminate the parameters that have little influence on the model results from sensitivity analysis, and then reduce the influence of the uncertain transmissions. This study used NSEs as an objective function, defined as

$$\text{NSE} = 1 - \frac{\sum\limits_{i=1}^{n} (Q_{obs,i} - Q_{sim,i})^2}{\sum\limits_{i=1}^{n} (Q_{obs,i} - \overline{Q}_{obs})^2} \tag{1}$$

where $Q_{sim,i}$ is the ith simulated discharge, $Q_{obs,i}$ is the ith observed discharge, $\overline{Q}_{obs}$ is the mean of the observed data, and n is the simulation period.

According to the physical meaning of each parameter, we calibrated the SWAT model using the same initial parameter ranges, according to the calibration protocol presented by Abbaspour [32]. The sensitivity analysis of the parameter values is generated by LHS sampling, and the value of the target function by the multiple regression model. The calculation expression of the parameter sensitivity is written as

$$g = \alpha + \sum\limits_{i=1}^{m} \beta_i b_i \tag{2}$$

where g is the objective function value, α is the regression constant, β is the coefficient of parameters, b_i is the parameter value, and the m is the number of parameters. The t-test method is used to determine the sensitivity of each parameter.

The following method is used to deduce the confidence interval of each parameter. First, the sensitivity matrix calculation formula for the objective function is

$$J_{ij} = \Delta g_i / \Delta b_j \quad i = 1, \cdots, C_2^n, j = 1, \cdots, m \tag{3}$$

where C_2^n indicates the number of rows in the sensitivity matrix, j represents the number of parameters, i means the group number, Δb_j is the parameter of the j rate, and Δg_i represents the parameter sensitivity. The Hessian matrix calculation formula for the objective function is

$$H = J^T J \tag{4}$$

where H indicates the haessen matrix, J represents the matrix of the number of parameter columns.

According to Kramer's theorem, the covariance matrix C for estimating the lower limit of the parameter is calculated as

$$C = s_g^2 \left(J^T J\right)^{-1} \tag{5}$$

where s_g^2 is the deviation of the result of the n simulation of the target function.

The standard variance of parameter b_i and its 95% confidence interval (CI) are calculated by the diagonal elements in C, as follows:

$$s_j = \sqrt{C_{jj}} \tag{6}$$

$$b_{j,lower} = b_j^* - t_{v,0.025} \cdot S_j \tag{7}$$

$$b_{j,upper} = b_j^* + t_{v,0.025} \cdot S_j \tag{8}$$

where b_j^* is the optimal solution of the parameter b, v is the degree of freedom $(n - m)$, $b_{j,lower}$ is the lower limit of the confidence interval, and $b_{j,upper}$ is the upper limit of confidence interval.

3.3. Parameter Uncertainty Evaluation Index

In order to judge the influence of parameter uncertainty on runoff simulation under different levels, this paper uses the two indexes of variation rate (VR) and the relative length of the confidence interval (RL) as the model uncertainty evaluation index. These can be expressed as

$$VR = \frac{Abs|Q_{sim} - Q_{obs}|}{Q_{obs}} \tag{9}$$

$$RL = \frac{Q_{upper} - Q_{lower}}{Q_{obs}} \tag{10}$$

where VR is the variation rate, RL is the relative length of confidence interval, Q_{upper} is the upper limit of the runoff simulation under the 95% confidence interval, and Q_{lower} is the lower limit of the runoff simulation under the 95% confidence interval.

3.4. Multilevel Factorial Analysis

The factorial analysis is a multivariable reasoning method. It performs excellently in testing the effects of individual variables and their interactions on the dependent variable [33–35]. Factors A and B have m and n levels, respectively. Thus, a full-factor factorial design contains all possible factor combinations. The factor model for this factorial experiment can be expressed as

$$Y_{ijk} = \mu + \tau_i + \beta_j + (\tau\beta)_{ij} + \varepsilon_{ijk} \begin{cases} i = 1,2\ldots a \\ j = 1,2\ldots b \\ k = 1,2\ldots n \end{cases} \tag{11}$$

where μ is the total average effect, ε_{ijk} is a random error effect, τ_i is the effect of factor A at the ith level, β_j is the effect of factor B at the jth level, and $(\tau\beta)_{ij}$ is the interaction effect when A is at the ith level and B is at the jth level. There is a total of abn experiments, where n is the number of repeated experiments. In order to test the influence of the parameter main effect and the interaction effect on the runoff simulation, the F-statistic can be used as follows:

$$F_A = \frac{MS_A}{MS_E} = \frac{SS_A/a - 1}{SS_E/ab(n-1)} \tag{12}$$

$$F_B = \frac{MS_B}{MS_E} = \frac{SS_B/b - 1}{SS_E/ab(n-1)} \tag{13}$$

$$F_{AB} = \frac{MS_{AB}}{MS_E} = \frac{SS_{AB}/(a-1)(b-1)}{SS_E/ab(n-1)} \tag{14}$$

where MS_A, MS_B, MS_{AB}, and MS_E are the mean squares for factors A, B, their interaction with each other, and the error component, respectively. The SS_A, SS_B, SS_{AB}, and SS_E are the sum of squares for factors A and B, their interaction, and the error component, respectively. Each mean square deviation is the squared sum of the corresponding effects, divided by its degree of freedom. SS_T is the sum of the total effect square. This can be calculated by

$$SS_A = \frac{1}{bn}\sum_{i=1}^{a} y_{i..}^2 - \frac{y_{...}^2}{abn} \tag{15}$$

$$SS_B = \frac{1}{an}\sum_{j=1}^{b} y_{.j.}^2 - \frac{y_{...}^2}{abn} \tag{16}$$

$$SS_{AB} = \frac{1}{n} \sum_{i=1}^{a} \sum_{j=1}^{b} y_{ij.}^2 - \frac{y_{...}^2}{abn} - SS_A - SS_B \qquad (17)$$

$$SS_T = \sum_{i=1}^{a} \sum_{j=1}^{b} \sum_{k=1}^{n} y_{ijk}^2 - \frac{y_{...}^2}{abn} \qquad (18)$$

$$SS_E = SS_T - SS_{AB} - SS_A - SS_B \qquad (19)$$

where $y_{i..}$, $y_{.i.}$, and $y_{ij.}$ represent the ith level of the factor A, the jth level of the factor B, and the ijth interaction between factors A and B, respectively.

In particular, there are k factors in the 3^k factorial design. Each factor has three levels, so there is a total of $3k$ factor level combinations and $3k$ degrees of freedom. There is a total of k main effects, each of which is two degrees of freedom. There is an interaction effect of k factors, and the degree of freedom is $2k − 1$. If n repeated tests are performed for each factor level combination, the total degree of freedom is $n3k − 1$, and the degree of freedom of the error is $3k(n − 1)$. The sum of squares for the main effects and interaction effects is usually obtained by the factorial analysis method.

4. Results and Discussion

4.1. Parameter Sensitivity Analysis, Calibration, and Verification of Model

Parameter sensitivity analysis is an indispensable part of the evaluation model, and it is helpful in developing a deep understanding of the model characteristics. Hence, this paper takes NSE as the objective function, and uses LHS-based multiple regression models to analyze the parameter sensitivity. The t-test method is used to determine the parameter sensitivity—the higher the absolute value of t, the stronger the sensitivity of the parameters. The related CI of the parameters was determined by the SUFI-2 algorithm. Parameters' descriptions and their CIs are listed in Table 1. For the most part, specific values are given for the parameters. However, some parameters, such as the runoff curve number, can take on a variety of values. During calibrations, these parameters may be changed by increasing or reducing them by a certain percentage, until the calibration objective function is met. Thus, values of these parameters have been reported as a percentage change from a specified value. The Latin Hypercube sampling (LH-OAT) method was performed for 10 parameters, in order to screen out the most sensitive parameters of the model, and the results are illustrated in Figure 2.

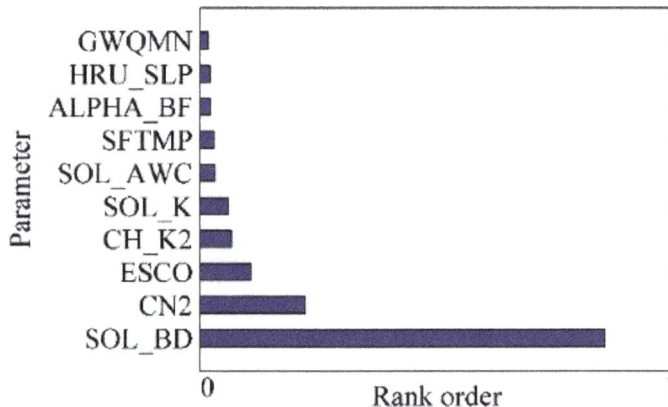

Figure 2. Graphical representation of the sensitivity ranking of the parameters (longer bars indicate greater parameter sensitivities).

As we can see from Figure 2 and Table 1, the most sensitive parameters for runoff simulation are followed by SOL_BD, CN2, ESCO, CH_K2, SOL_K, SOL_AWC, SFTMP, ALPHA_BF, HRU_SLP, and GWQMN. In order to design the experimental scheme, the four most sensitive parameters are selected: The soil bulk density (SOL_BD) was the most sensitive parameter, followed by the Soil Convention Service (SCS) runoff curve number for moisture conditions II (CN2), the soil evaporation compensation coefficient (ESCO), and effective hydraulic channel conductivity (CH_K2). A detailed interpretation of these parameters will be provided in the following section.

Table 1. Final value range of sensitivity parameters.

Parameter	Description	CI		Calibrated Value	*t* Value
		Min	Max		
GWQMN	Threshold depth of water in the shallow required for return flow to occur (mm)	700.61	780.61	742.93	1.17
HRU_SLP	Average slope steepness (m/m)	0.24	0.26	0.25	1.5
ALPHA_BF	Baseflow regression constant (days)	0.00	0.20	0.02	−1.52
SFTMP	Snow temperature (°C)	3.43	3.63	3.45	−2
SOL_AWC	Effective water capacity of soil layer (mmH$_2$O/mm soil)	0.15	0.17	0.16	2.19
SOL_K	Soil hydraulic conductivity (mm·hr^{-1})	−0.32	−0.29	−0.32	−4.04
CH_K2	Effective hydraulic conductivity of channel (mm/h)	110.97	125.97	121.19	4.59
ESCO	Soil evaporation compensation coefficient (mm/h)	0.73	0.75	0.73	−7.36
CN2	Initial SCS runoff curve number to moisture conditions II	0.26	0.28	0.26	−15.28
SOL_BD	Soil bulk density (g/cm^3)	0.45	0.47	0.46	−58.26

SWAT parameter ranges (and allowable percentage changes): $0 \leq$ GWQMN ≤ 5000; $0 \leq$ HRU_SLP ≤ 0.6; $0 \leq$ ALPHA_BF ≤ 1; $-5 \leq$ SFTMP ≤ 5; $0 \leq$ SOL_AWC ≤ 1 ($-50\% \leq$ SOL_AWC $\leq 50\%$); $0 \leq$ SOL_K ≤ 2000 (-80% $\leq$ SOL_K $\leq 80\%$); $0.01 \leq$ CH_K2 ≤ 150; $0 \leq$ ESCO ≤ 1; $35 \leq$ CN2 ≤ 98 ($-50\% \leq$ CN2 $\leq 50\%$); $0.9 \leq$ SOL_BD ≤ 2.5 ($-50\% \leq$ SOL_BD $\leq 50\%$); reference [36].

On the basis of CMADS (2008–2015), this paper selected years 2008–2009, 2010–2013, and 2014–2015 as warm-up, calibration, and validation periods, respectively. The NSE, R^2, and the absolute value of relative error (|Re|) indicators were used to evaluate the model for calibration and validation periods. The results show that the NSE values were 0.73 and 0.81, respectively, R^2 was 0.82 and 0.87, respectively, and |Re| was less than 10% for both periods. The results indicate a good performance of SWAT in describing the runoff simulation, based on the CMADS data in the source region of the Yellow River. Che [37] investigated the source area of the Yellow River, and used the SWAT model to simulate the daily runoff. The results of the study showed that both NSE values and R^2 were less than 0.74 in calibration (validation), which means that CMADS data is superior to other data in watershed runoff simulation.

Figure 3 shows the simulated daily runoff from 2010–2015 (i.e., the calibration and validation periods) for the Tangnaihai hydrological station. Specifically, this figure indicates that the 95% CI width of the daily runoff simulation varies with the flow amount. To further examine the above results, three different frequencies are set, in order to identify the influence of parameter uncertainty on flow of different levels (i.e., more than 75%, between 25% and 75%, and less than 75%).

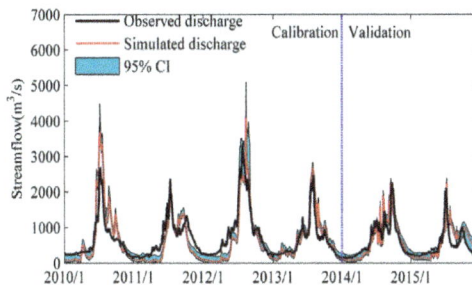

Figure 3. The 95% confidence interval (CI) for daily discharge.

Figure 4 shows that when the flow at the high level, the variation rate and the relative length of confidence interval are small; on the contrary, as the flow reaches the lower level, the variation rate and the relative length of confidence interval are larger. Given this, we can preliminarily infer that the physical mechanism of runoff generation dynamically changes during different periods.

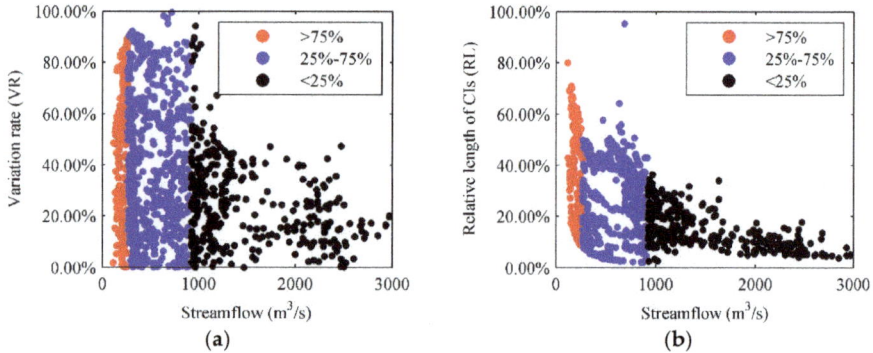

Figure 4. (a) The influence of parameter uncertainty on variable rates under different levels; (b) the influence of parameter uncertainty on the relative length of the confidence interval under different levels.

4.2. Multilevel Factorial Analysis and Dynamic Changes in Parameter Sensitivity

As mentioned, the physical mechanism of runoff generation varies at different periods. In the SWAT model, parameters reflect the characteristics of the hydrological processes. Thus, based on the above identified sensitive parameters, we used the analysis of variance method (ANOVA) to quantify the response of individual and interactive parameters in the runoff simulation.

Four parameters, CN2, ESCO, CH_K2, and SOL_BD, were selected as factorial experimental factors (denoted by A~D, respectively). Subsequently, we designed the 3^4 factorial design scheme shown in Table 2.

Table 2. The 3^4 factorial design scheme.

Parameter	Level		
	Low	Medium	High
CN2	0.257	0.267	0.277
ESCO	0.726	0.736	0.746
CH_K2	110.97	120.96	130.97
SOL_BD	0.431	0.452	0.471

Figure 5 shows the parameter contributions for the hydrological responses. The results show that parameter D is significant at different times, especially in the non-flood period (November to March), when its contribution reached 0.98. The contribution of parameter A in the pre-flood period (April to May) increased, but the effect was relatively small, and D was still dominant. The contributions of A, B, and C during the flood period (June to September) gradually increased, whereas D gradually weakened. The linear individual effects of A, B, and D, as well as the AC and CD interaction effects, are thus significant for modelling runoff simulation. During the post-flood period (October), the contributions of A, B, and C weaken, while parameter D increases.

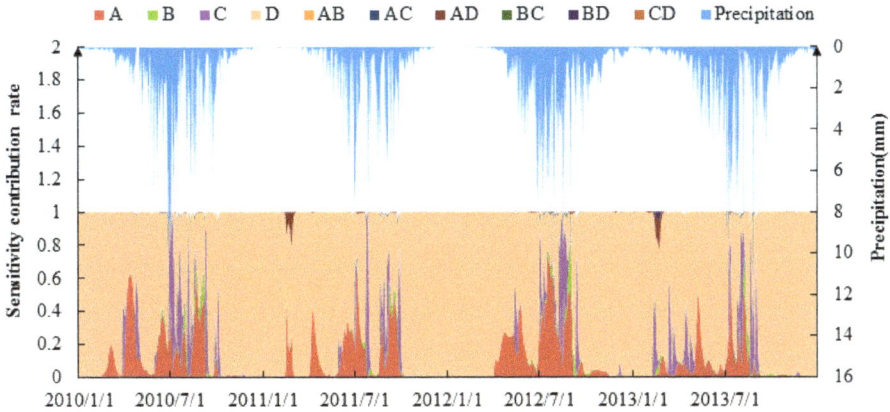

Figure 5. Dynamic characteristics of the parameter sensitivity contribution.

4.3. The Individual and Interactive Effects of Parameters on the Hydrologic Model Output in Different Periods

From the previous section, we find that the parameter sensitivity in different years shows similar characteristics in the non-flood, pre-flood, flood, and post-flood periods. To quantitatively evaluate the dynamic effect of the parameters on the runoff simulation, we selected 2012 as a typical year, and three consecutive days of non-flood, pre-flood, flood, and post-flood periods in a typical year. The objective is to explore the dynamic parameter individual and interaction effects on the runoff simulation. For simplicity, this article only presents the first-day results of the four periods in a typical year, and the other results are attached to Supplementary Materials.

4.3.1. The Statistically Significant Individual and Interaction Effects on Runoff Simulation in Non-Flood Period

Table 3 presents the results of variance analysis for the simulation of daily runoff in the non-flood period. In this paper, a p-value of less than 0.05 indicates that parameters A, B, and D have significant effects on the model output in the non-flood period. In particular, the influence of D is more significant. Moreover, the interaction effects of AB, AD, and BD have statistical significance.

Table 3. Results of ANOVA for the runoff simulation in the non-flood period.

Model Term	Sum of Squares	F Value	p-Value	Significance
A	28.47	16,998.37	0.00	**
B	0.76	456.16	0.00	*
C	0.00	0.00	1.00	
D	12,890.57	7,695,860.81	0.00	***
AB	0.12	34.59	0.00	*
AC	0.00	0.00	1.00	
AD	2.73	814.23	0.00	*
BC	0.00	0.00	1.00	
BD	0.12	34.83	0.00	*
CD	0.00	0.00	1.00	
Error	0.04			
Total	12,922.80			

For the non-flood period, the main effects of these parameters on the runoff simulation are shown in Figure 6. From this, we see that changes in the levels of *A*, *B*, and *C* have a weak effect on the runoff simulation, while changes in the level of parameter *D* have significant negative effects on the runoff simulation.

In essence, during the non-flood period, *D* is an important parameter affecting the model simulation, which is closely related to the high elevation and cold temperatures of the source region of the Yellow River, characterized by low rainfall and low temperature. Moreover, lower flow values relate to lower compensation for glaciers and snowmelt in soil water, resulting in decreased streamflow [38,39].

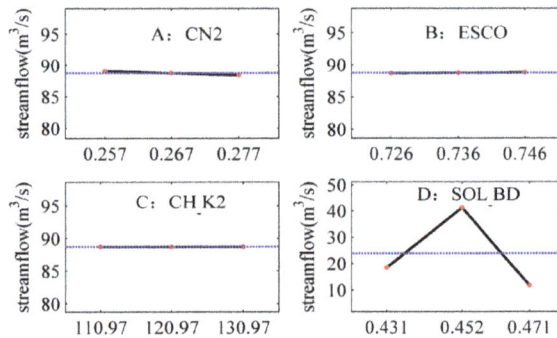

Figure 6. Individual effects of parameters in the non-flood period.

Figure 7 presents the parameter interaction effects in the non-flood period of the runoff simulation. Results show that the *AD*, *BD*, and *CD* interaction effects are significant in this period, while the others have less influence on the runoff simulation. The main reason for this is that the soil bulk density (i.e., parameter *D*) is the parameter that has the greatest impact on the runoff simulation.

Note that for the *AD* interaction plot in Figure 7 (bottom left), the red, blue, and black lines represent parameter *D* at low, medium, and high levels, respectively. This plot discloses that the changes differ across the three levels of parameter *D*, depending on the level of parameter *A*. When *D* is at the high level, *A* has obvious negative effects. However, when *D* is at the middle level, the negative effect of *A* is weakened. This result further indicates that *A* has a significant impact on the runoff simulation.

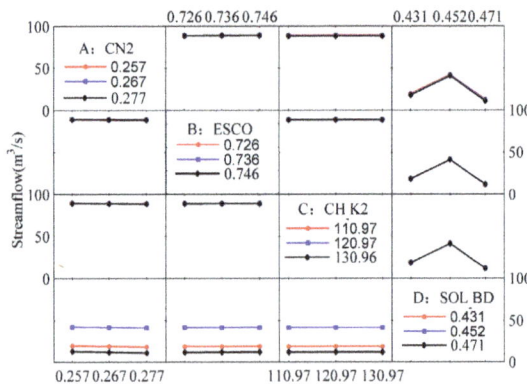

Figure 7. Interaction effects of parameters in the non-flood period.

4.3.2. The Statistically Significant Individual and Interaction Effects on the Runoff Simulation in the Pre-Flood Period

The parameters and their interactions have statistically significant effects on the runoff simulation (Table 4). The linear individual effects of *A*, *B*, and *D* on the runoff simulation are significant in the pre-flood period, while other parameter effects (individual and interaction) have little impact on the response.

Table 4. Results of ANOVA for the pre-flood period runoff simulation.

Model Term	Sum of Squares	F Value	p-Value	Significance
A	145.81	591,251.63	0.00	***
B	2.36	9586.61	0.00	**
C	0.00	5.42	0.01	*
D	420.67	1,705,827.37	0.00	***
AB	0.00	7.40	0.00	*
AC	0.00	0.94	0.45	
AD	0.03	66.12	0.00	*
BC	0.00	1.39	0.25	
BD	0.00	1.96	0.12	
CD	0.00	0.46	0.76	
Error	0.01			
Total	568.88			

Figure 8 presents the individual effect plot for the four parameters in the pre-flood period. Results show that *A* has a positive effect on the runoff simulation, while *D* has a negative effect. Nevertheless, parameters *B* and *C* have a slight effect on the flow.

The flow value reaches 112.12 m^3/s when *A* is at the high level. In contrast, the simulated flow value is only 96.20 m^3/s. Furthermore, compared with other levels of runoff simulation, the three levels of *D* lead to smaller runoff simulation values. These results are attributed to a higher compensation for snow melt, and thus increased flow [40]. However, the slope of *D* is irregular, implying an obvious nonlinear effect on the runoff simulation, owing to the complex topography and soil characteristics of the basin.

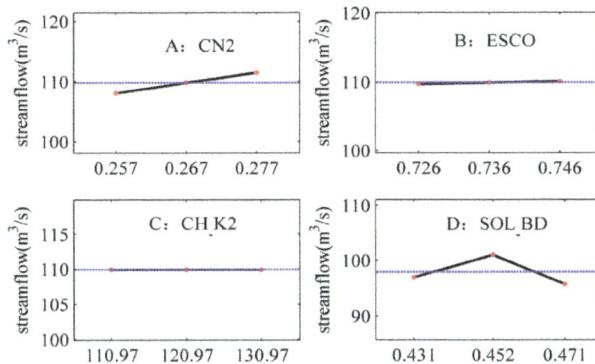

Figure 8. Individual effects of parameters in the pre-flood period.

Figure 9 depicts the parameter interaction effects in the pre-flood period. If the slope of one curve differs from another, there are interactive effects between the parameters. The parallel lines in the interaction plots of parameters *A*, *B*, *C*, and *D* indicate tiny parameter interactions. Specifically, the interaction of *A* and *B*, as well as *B* and *C*, are not apparent in Figure 9.

In addition, the results indicate that parameter *A* has the largest nonlinear effect on the runoff simulation, while the other nonlinear effects are weak. The identical results of these plots can be used to confirm the findings in the above numerical value model.

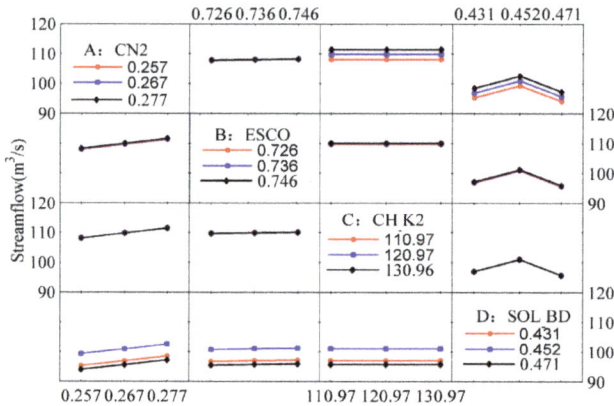

Figure 9. Interaction effects of parameters in the pre-flood period.

4.3.3. The Statistically Significant Individual and Interaction Effects on the Runoff Simulation in the Flood Period

The results of ANOVA for the runoff simulation during the flood period are shown in Table 5. Results indicate that parameters *A*, *B*, *C*, and *D* have significant linear main effects on the runoff simulation. Among them, the significance of *A* is the strongest, followed by *D*, *B*, and *C*. Meanwhile, parameters *AC*, *AD*, and *CD* interaction effects are of important significance.

Table 5. Results of ANOVA for the flood period runoff simulation.

Model Term	Sum of Squares	*F* Value	*p*-Value	Significance
A	31,490.77	3587.56	0.00	***
B	3800.32	432.95	0.00	**
C	745.95	84.98	0.00	*
D	15,926.84	1814.45	0.00	***
AB	1.53	0.09	0.99	
AC	287.46	16.37	0.00	*
AD	177.23	10.10	0.00	*
BC	37.23	2.12	0.09	
BD	26.79	1.53	0.21	
CD	263.16	14.99	0.00	*
Error	210.67			
Total	52,967.95			

The influence of the main parameter effects on the runoff simulation during the flood period is shown in Figure 10. It shows the influence of the parameters on the runoff simulation and compares the relative size of the influence. Parameter *A* has the largest positive individual effect when *A* is at low, middle, and high levels, when the simulated flow value is 2375.24, 2398.45, and 2420.22 m^3/s, respectively. Parameter *A* is a comprehensive reaction of the underlying surface characteristics, which directly determines the size of the flow. However, parameter *D* has the greatest nonlinear effect.

Therefore, the smaller the soil bulk density, the weaker the ability to resist precipitation and thus increased flow. However, with the increase of *D*, the ability to resist precipitation is enhanced, which aids infiltration; however, the flow is then increased, due to a sharp decrease in soil porosity [41,42].

Figure 11 plots three level interaction effects of the flood period for the four parameters. Results disclose that *AC*, *AD*, and *CD* have noticeable interaction effects, while the others have almost no interaction effect.

Take the plot on the bottom left, with the interaction of parameters *A* and *D* as an example. The plot depicts the changes in *D* at its low, middle, and high levels, depending on the level of *A*, as well as the interaction between the low level and the middle level curve of *D*. This reveals the interaction of *A* and *D* has a significant influence on the runoff simulation. Therefore, the parameter interaction must be emphasized at calibration.

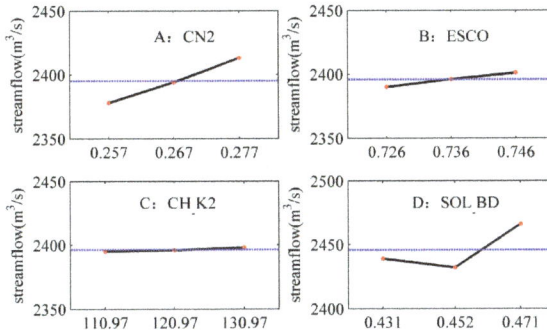

Figure 10. Individual effects of parameters in the flood period.

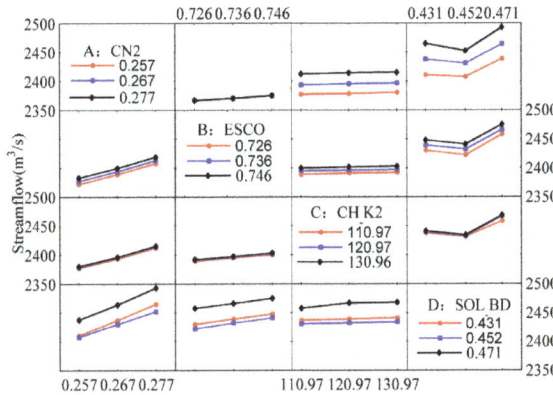

Figure 11. Interaction effects of parameters in the flood period.

4.3.4. The Statistically Significant Individual and Interaction Effects on the Runoff Simulation in the Post-Flood Period

Table 6 shows the ANOVA results for the runoff simulation in the post-flood period. Results reveal that the selected parameters, *A*, *B*, and *D*, have a significant ($p < 0.05$) effect on the runoff simulation; among them, the significance of *D* is the strongest. Similarly, the interaction effects of *A* and *C* have an important impact on the runoff simulation (Table 6).

The individual parameter effects for the runoff simulation in the post-flood period are shown in Figure 12. The result shows that the level change of *D* has the greatest negative effect on the runoff simulation. In essence, the above results are attributable to the continuous precipitation decrease, and the river runoff mainly depends on the recharge of the interflow, e.g., the greater the *A* value, the less recharge in the soil. Moreover, a decrease in soil evaporation is caused by lower temperature.

The nonlinear variation of the *D* curve is mainly due to the complex physical characteristics of soil. When *D* is at the low level, the soil viscosity is larger, and the interflow has less recharge for the runoff. As *D* reaches the high level, less precipitation and lower porosity combine to reduce runoff in the post-flood period [43].

Table 6. Results of ANOVA for the post-flood period runoff simulation.

Model Term	Sum of Squares	F Value	p-Value	Significance
A	1267.75	258,334.54	0.00	***
B	662.22	134,942.54	0.00	**
C	0.03	5.48	0.01	*
D	66,212.33	13,492,324.58	0.00	***
AB	0.05	4.88	0.00	*
AC	0.00	0.35	0.84	
AD	27.18	2769.41	0.00	*
BC	0.00	0.50	0.73	
BD	4.98	507.67	0.00	*
CD	0.00	0.50	0.73	
Error	0.12			
Total	68,174.67			

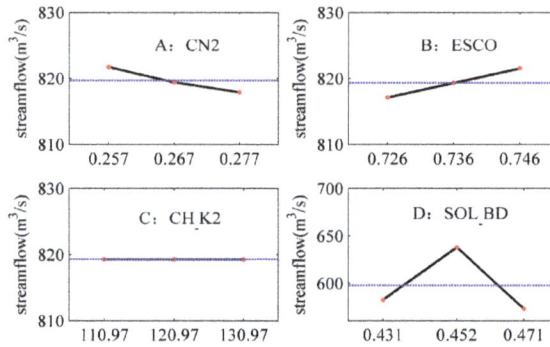

Figure 12. Individual effects of parameters in the post-flood period.

The interaction effects of parameters in the post-flood period are shown in Figure 13. The results disclose that the interaction effects of *D* and other parameters are negative for the runoff simulation, particularly the interaction effects of *A* and *C*.

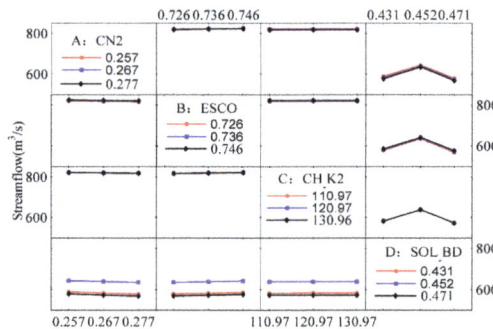

Figure 13. Interaction effects of parameters in the post-flood period.

4.3.5. Contributions of Parameter Individual and Interaction Effects for the Runoff Simulation in Four Periods

Figure 14 accurately quantifies the contribution of individual and interaction parameter effects to runoff in the four periods. The results show that the parameter contribution to the runoff simulation is significant for the different periods.

In detail, the parameter D have the greatest impact on the runoff simulation in the non-flood and post-flood period, contributing 0.99 and 0.97, respectively. The contribution of D to the runoff simulation is reduced while others increase in the pre-flood period—especially A, which contributes 0.26 (i.e., 26%). However, the contribution of A to the runoff simulation is the most significant in the flood period (0.60), implying that the infiltration-excess runoff production caused by concentrated precipitation has a great influence on this parameter, while soil porosity and soil moisture have great influence on D. The influence of temperature on B is also significant.

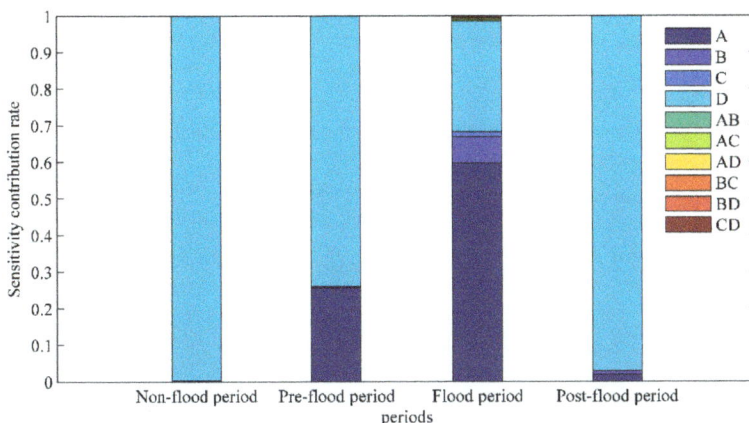

Figure 14. Contribution of individual and interaction parameters to the runoff simulation in different periods.

5. Conclusions

In this study, using CMADS to drive the SWAT model, we developed a SUFI-2-based multilevel-factorial-analysis method in order to disclose the effect of parameters in the SWAT model on runoff simulations. Here, the SUFI-2 was used to explore the CI of identified sensitive parameters.

Subsequently, we applied the multilevel-factorial-analysis method to explore the individual and interactive effects of parameters on the runoff simulation in different periods (i.e., non-flood, pre-flood, flood, and post-flood). The developed method was exemplarily applied to the source region of the Yellow River, due to its key role in water resource supply. Important conclusions drawn from this study are as follows:

(1) The influence of parameters CN2, ESCO, CH_K2, and SOL_BD (i.e., A, B, C, and D, respectively) on the runoff simulation is significant in different periods (Figure 2, Tables 3–6). In general, the linear individual effects of factors A, B, and D, as well as the AD interaction effects, are thus significant, while the others have little influence on the response.

(2) The contributions of different parameters to the runoff simulation are different in different periods (Figure 14). The effect of soil bulk density (D) on the runoff simulation is significant in four periods, contributing 0.99, 0.73, 0.30, and 0.97, respectively. The effect of the initial SCS runoff curve number (A) on the runoff simulation is significant in the non-flood and flood periods, contributing 0.26 and 0.60, respectively.

(3) The interaction effects of parameters on runoff simulation are significant in the flood period. Take parameters *A* and *D* as an example: The changes differ across the three levels of parameter *D*, depending on the level of parameter *A*. The slope curve is distinctly different between parameters *A* and *D*. This reveals the interaction of *A* and *D* has a significant influence on the runoff simulation. Therefore, the parameter interaction must be emphasized in flood periods.

(4) In essence, the soil bulk density moisture content and infiltration-excess runoff production are important water inputs for the hydrological system in the source region of the Yellow River. It is further explained that soil bulk density will affect the loss of surface runoff and river recharge groundwater.

Supplementary Materials: The following are available online at http://www.mdpi.com/2073-4441/10/9/1177/s1. Figure S1: Individual effect of non-flood period parameters for the second days, Figure S2: Individual effect of pre-flood period parameters for the second days, Figure S3: Individual effect of flood period parameters for the second days, Figure S4: Individual effect of post-flood period parameters for the second days, Figure S5: Individual effect of non-flood period parameters for the third days, Figure S6: Individual effect of pre-flood period parameters for the third days, Figure S7: Individual effect of flood period parameters for the third days, Figure S8: Individual effect of post-flood period parameters for the third days, Figure S9: Interaction effect of non-flood period parameters for the second days, Figure S10: Interaction effect of pre-flood period parameters for the second days, Figure S11: Interaction effect of flood period parameters for the second days, Figure S12: Interaction effect of post-flood period parameters for the second days, Figure S13: Interaction effect of non-flood period parameters for the third days, Figure S14: Interaction effect of pre-flood period parameters for the third days, Figure S15: Interaction effect of flood period parameters for the third days, Figure S16: Interaction effect of post-flood period parameters for the third days, Figure S17: Dynamic change characteristics of second day parameters on the time scale, Figure S18: Dynamic change characteristics of third day parameters on the time scale, Table S1: ANOVA results for the second day of the non-flood period runoff simulation, Table S2: ANOVA results for the second day of the pre-flood period runoff simulation, Table S3: ANOVA results for the second day of the flood period runoff simulation, Table S4: ANOVA results for the second day of the post-flood period runoff simulation, Table S5: ANOVA results for the third day of the non-flood period runoff simulation, Table S6: ANOVA results for the third day of the pre-flood period runoff simulation, Table S7: ANOVA results for the third day of the flood period runoff simulation, Table S8: ANOVA results for the third day of the post-flood period runoff simulation.

Author Contributions: S.Z. designed the framework and analyzed the data of this study; Y.W. and J.C. provided significant suggestions on the methodology and structure of the manuscript; A.G. and Z.L. collected the data; S.Z. wrote the paper.

Funding: This work was supported by the National Natural Science Foundation of China (grant number: 51679187, 51647112 and 51679189), the National Key R&D Program of China (2017YFC0405900, 2017YFC0404404 and 2017YFC0404406).

Conflicts of Interest: The authors declare no conflict of interest.

References

1. Joseph, J.F.; Guillaume, J.H. Using a parallelized MCMC algorithm in R to identify appropriate likelihood functions for SWAT. *Environ. Model. Softw.* **2013**, *46*, 292–298. [CrossRef]
2. Gashaw, T.; Tulu, T.; Argaw, M.; Worqlul, A.W. Modeling the hydrological impacts of land use/land cover changes in the Andassa watershed, Blue Nile Basin, Ethiopia. *Sci. Total Environ.* **2018**, *619*, 1394–1408. [CrossRef] [PubMed]
3. Her, Y.; Chaubey, I.; Frankenberger, J.; Jeong, J. Implications of spatial and temporal variations in effects of conservation practices on water management strategies. *Water Manag.* **2017**, *180*, 252–266. [CrossRef]
4. Vilaysane, B.; Takara, K.; Luo, P.; Akkharath, I.; Duan, W. Hydrological Stream Flow Modelling for Calibration and Uncertainty Analysis Using SWAT Model in the Xedone River Basin, Lao PDR. *Procedia Environ. Sci.* **2015**, *28*, 380–390. [CrossRef]
5. Li, Z.L.; Shao, Q.X.; Xu, Z.X.; Cai, X.T. Analysis of parameter uncertainty in semi-distributed hydrological models using bootstrap method: A case study of SWAT model applied to Yingluoxia watershed in northwest China. *J. Hydrol.* **2010**, *385*, 76–83. [CrossRef]
6. Ficklin, D.L.; Barnhart, B.L. SWAT hydrologic model parameter uncertainty and its implications for hydroclimatic projections in snowmelt-dependent watersheds. *J. Hydrol.* **2014**, *519*, 2081–2090. [CrossRef]

7. Wu, Y.; Liu, S. Automating calibration, sensitivity and uncertainty analysis of complex models using the R package Flexible Modeling Environment (FME): SWAT as an example. *Environ. Model. Softw.* **2012**, *317*, 99–109. [CrossRef]
8. Zhang, D.; Chen, X.; Yao, H.; James, A. Moving SWAT model calibration and uncertainty analysis to an enterprise Hadoop-based cloud. *Environ. Model. Softw.* **2016**, *84*, 140–148. [CrossRef]
9. Kouchi, D.H.; Esmaili, K.; Faridhosseini, A.; Sanaeinejad, S.H.; Khalili, D.; Abbaspour, K.C. Sensitivity of Calibrated Parameters and Water Resource Estimates on Different Objective Functions and Optimization Algorithms. *Water* **2017**, *9*, 384. [CrossRef]
10. Trudel, M.; Doucet-Généreux, P.L.; Leconte, R. Assessing River Low-Flow Uncertainties Related to Hydrological Model Calibration and Structure under Climate Change Condition. *Climate* **2017**, *5*, 19. [CrossRef]
11. Muleta, M.K. Improving Model Performance Using Season-Based Evaluatio. *J. Hydrol. Eng.* **2012**, *17*, 191–200. [CrossRef]
12. Şahan, T.; Öztürk, D. Investigation of Pb (II) adsorption onto pumice samples: Application of optimization method based on fractional factorial design and response surface methodology. *Clean Technol. Environ. Policy* **2014**, *16*, 819–831. [CrossRef]
13. Thiele, J.E.; Haaf, J.M.; Rouder, J.N. Is there variation across individuals in processing? Bayesian analysis for systems factorial technology. *J. Math. Psychol.* **2017**, *81*, 40–54. [CrossRef]
14. Saleh, T.A.; Tuzen, M.; Sarı, A. Polyamide magnetic palygorskite for the simultaneous removal of Hg (II) and methyl mercury; with factorial design analysis. *J. Environ. Manag.* **2018**, *211*, 323–333. [CrossRef] [PubMed]
15. Tang, X.; Dai, Y.; Sun, P.; Meng, S. Interaction-based feature selection using Factorial Design. *Neurocomputing* **2018**, *281*, 47–54. [CrossRef]
16. Meng, F.; Su, F.; Yang, D.; Tong, D.; Hao, Z. Impacts of recent climate change on the hydrology in the source region of the Yellow River basin. *J. Hydrol.* **2018**, *558*, 301–313. [CrossRef]
17. Wang, T.; Yang, H.; Yang, D.; Qin, Y.; Wang, Y. Quantifying the streamflow response to frozen ground degradation in the source region of the Yellow River within the Budyko framework. *J. Hydrol.* **2018**, *558*, 301–313. [CrossRef]
18. Lan, Y.; Zhao, G.; Zhang, Y.; Wen, J.; Hu, X. Response of runoff in the source region of the Yellow River to climate warming. *Quat. Int.* **2010**, *226*, 60–65. [CrossRef]
19. Meng, X.; Wang, H. Significance of the China Meteorological Assimilation Driving Datasets for the SWAT Model (CMADS) of East Asia. *Water* **2017**, *9*, 765. [CrossRef]
20. Meng, X.; Wang, H.; Cai, S.; Zhang, X.; Leng, G.; Lei, X.; Shi, C.; Liu, S.; Shang, Y. The China Meteorological Assimilation Driving Datasets for the SWAT Model (CMADS) Application in China: A Case Study in Heihe River Basin. *Preprints* **2016**. [CrossRef]
21. Meng, X.; Wang, H.; Wu, Y.; Long, A.; Wang, J.; Shi, C.; Ji, X. Investigating spatiotemporal changes of the land-surface processes in Xinjiang using high-resolution CLM3.5 and CLDAS: Soil temperature. *Sci. Rep.* **2017**, *7*, 13286. [CrossRef] [PubMed]
22. Zhao, F.; Wu, Y.; Qiu, L.; Sun, Y.; Sun, L.; Li, Q.; Niu, J.; Wang, G. Parameter Uncertainty Analysis of the SWAT Model in a MountainLoess Transitional Watershed on the Chinese Loess Plateau. *Water* **2018**, *10*, 690. [CrossRef]
23. Vu, T.T.; Li, L.; Jun, K.S. Evaluation of MultiSatellite Precipitation Products for Streamflow Simulations: A Case Study for the Han River Basin in the Korean Peninsula, East Asia. *Water* **2018**, *10*, 642. [CrossRef]
24. Liu, J.; Shanguan, D.; Liu, S.; Ding, Y. Evaluation and Hydrological Simulation of CMADS and CFSR Reanalysis Datasets in the QinghaiTibet Plateau. *Water* **2018**, *10*, 513. [CrossRef]
25. Cao, Y.; Zhang, J.; Yang, M.; Lei, X.; Guo, B.; Yang, L.; Zeng, Z.; Qu, J. Application of SWAT Model with CMADS Data to Estimate Hydrological Elements and Parameter Uncertainty Based on SUFI-2 Algorithm in the Lijiang River Basin, China. *Water* **2018**, *10*, 742. [CrossRef]
26. Shao, G.; Guan, Y.; Zhang, D.; Yu, B.; Zhu, J. The Impacts of Climate Variability and Land Use Change on Streamflow in the Hailiutu River Basin. *Water* **2018**, *10*, 814. [CrossRef]
27. Meng, X.Y.; Wang, H.; Lei, X.H.; Cai, S.Y.; Wu, H.J. Hydrological modeling in the Manas River Basin using Soil and Water Assessment Tool driven by CMADS. *Tehnički Vjesnik* **2017**, *24*, 525–534.
28. Zhang, L.M.; Wang, H.; Meng, X.Y. Application of SWAT Model Driven by CMADS in Hunhe River Basin in Liaoning Province. *J. N. China. Univ. Water Resour. Electr. Power* **2017**, *385*, 1–9.

29. Zuo, D.P.; Xu, Z.X. Distributed hydrological simulation using swat and sufi-2 in the wei river basin. *J. Beijing Norm. Univ.* **2012**, *48*, 490–496.

30. Zhang, Y.Q.; Chen, C.C.; Yang, X.H.; Yin, Y.X.; Du, J.K. Application of SWAT Model Based SUFI-2Algorithm to Runoff Simulation in Xiushui Basin. *Water Resour. Power* **2013**, *31*, 24–28.

31. Wu, H.; Chen, B. Evaluating uncertainty estimates in distributed hydrological modeling for the Wenjing River watershed in China by GLUE, SUFI-2, and ParaSol methods. *Ecol. Eng.* **2015**, *76*, 110–121. [CrossRef]

32. Abbaspour, K.C.; Rouholahnejad, E.; Vaghefi, S.; Srinivasan, R.; Kløve, B. A continental-scale hydrology and water quality model for Europe: Calibration and uncertainty of a high-resolution large-scale SWAT model. *J. Hydrol.* **2015**, *524*, 733–752. [CrossRef]

33. Yang, J.; Reichert, P.; Abbaspour, K.C.; Xia, J.; Yang, H. Comparing uncertainty analysis techniques for a SWAT application to the Chaohe Basin in China. *J. Hydrol.* **2008**, *358*, 1–23. [CrossRef]

34. Zhou, Y.; Huang, G.H. Factorial two-stage stochastic programming for water resources management. *Stoch. Environ. Res. Risk Assess.* **2011**, *25*, 67–78. [CrossRef]

35. Martens, H.; Måge, I.; Tøndel, K.; Isaeva, J.; Hoy, M.; Saebo, S. Multi-level binary replacement (MBR) design for computer experiments in high-dimensional nonlinear systems. *J. Chemometr.* **2010**, *24*, 748–756. [CrossRef]

36. Gitau, M.W.; Chaubey, I. Regionalization of SWAT model parameters for use in ungauged watersheds. *Water* **2010**, *2*, 849–871. [CrossRef]

37. Che, Q. Distributed Hydrological Simulation Using SWAT in Yellow River Source Region. Master's Thesis, Lanzhou University, Lanzhou, China, 2006.

38. Jia, D.Y.; Wen, J.; Ma, Y.M.; Liu, R.; Wang, X.; Zhou, J.; Chen, J.L. Impacts of vegetation on water and heat exchanges in the source region of Yellow River. *Plateau Meteorol.* **2017**, *36*, 424–435.

39. Zeng, Y.N.; Feng, Z.D.; Cao, G.C.; Xue, L. The Soil Organic Carbon Storage and Its Spatial Distribution of Alpine Grassland in the Source Region of the Yellow River. *Acta Geogr. Sin.* **2004**, *59*, 497–504.

40. Chen, X.; Song, Q.F.; Gao, M.; Sun, Y.M. Vegetation-soil-hydrology interaction and expression of parameter variations in ecohydrological models. *J. Beijing Norm. Univ.* **2016**, *52*, 362–368.

41. Yi, X.S.; Li, G.S.; Yin, Y.Y.; Wang, B.L. Preliminary Study for the Influences of Grassland Degradation on Soil Water Retention in the Source Region of the Yellow River. *J. Nat. Res.* **2012**, *27*, 1708–1719.

42. Liu, X.M. Effect of Soil Sample Initial State on Hydrological Process on Red Soil Slope. Master's Thesis, Hunan University, Changsha, China, 2013.

43. Hu, H.H. The Research on the Hydrological Cycle Based on the Synergistic Effect of Vegetation and Frozen Soil in the Source Region of Yangtze and Yellow River. Ph.D. Thesis, Lanzhou University, Lanzhou, China, 2011.

water

MDPI

Article

Evaluation and Hydrological Application of CMADS against TRMM 3B42V7, PERSIANN-CDR, NCEP-CFSR, and Gauge-Based Datasets in Xiang River Basin of China

Xichao Gao [1,2], Qian Zhu [3], Zhiyong Yang [1,2,*] and Hao Wang [1,2]

[1]   China Institute of Water Resources and Hydropower Research, Beijing 100038, China;
       pandagxc@zju.edu.cn (X.G.); wanghao@iwhr.com (H.W.)
[2]   State Key Laboratory of Simulation and Regulation of Water Cycle in River Basin, Beijing 100038, China
[3]   School of Civil Engineering, Southeast University, Nanjing 211189, China; zhuqian@seu.edu.cn
*   Correspondence: yangzy@iwhr.com; Tel.: +86-010-6878-1178

Received: 14 June 2018; Accepted: 8 September 2018; Published: 11 September 2018

Abstract: Satellite-based and reanalysis precipitation products provide a practical way to overcome the shortage of gauge precipitation data because of their high spatial and temporal resolution. This study compared two reanalysis precipitation datasets (the China Meteorological Assimilation Driving Datasets for the Soil and Water Assessment Tool (SWAT) model (CMADS), the National Centers for Environment Prediction Climate Forecast System Reanalysis (NCEP-CFSR)) and two satellite-based datasets (the Tropical Rainfall Measuring Mission 3B42 Version 7 (3B42V7) and the Precipitation Estimation from Remotely Sensed Information using Artificial Neural Networks–Climate Data Record (PERSIANN-CDR)) with observed precipitation in the Xiang River basin in China at two spatial (grids and the whole basin) and two temporal (daily and monthly) scales. These datasets were then used as inputs to a SWAT model to evaluate their usefulness in hydrological prediction. Bayesian model averaging was used to discriminate dataset performance. The results show that: (1) for daily timesteps, correlations between reanalysis datasets and gauge observations are >0.55, better than satellite-based datasets; The bias values of satellite-based datasets are <10% at most evaluated grid locations and for the whole baseline. PERSIANN-CDR cannot detect the spatial distribution of rainfall events; the probability of detection (POD) of PERSIANN-CDR at most evaluated grids is <0.50; (2) CMADS and 3B42V7 are better than PERSIANN-CDR and NCEP-CFSR in most situations in terms of correlation with gauge observations; satellite-based datasets are better than reanalysis datasets in terms of bias; and (3) CMADS and 3B42V7 simulate streamflow well for both daily (The Nash-Sutcliffe coefficient (NS) > 0.70) and monthly (NS > 0.80) timesteps; NCEP-CFSR is worst because it substantially overestimates streamflow; PERSIANN-CDR is not good because of its low NS (0.40) during the validation period.

Keywords: reanalysis products; satellite-based products; hydrological model; bayesian model averaging; Xiang River basin

1. Introduction

Precipitation is one of the primary drivers of the hydrological cycle and, thus of great importance in hydrological simulation [1], which is a major water resources management tool for forecasting floods and droughts. The accuracy of hydrological simulation depends on the spatial and temporal resolution of precipitation data [2]. Precipitation is more difficult than other atmospheric variables, such as temperature and relative humidity, to measure accurately because of its great spatial and temporal variability. Precipitation data are usually observed and collected using rainfall gauges and meteorological radar networks, but these measurement devices are usually geographically sparse and inadequate to fully capture the spatial and temporal variability of precipitation [3,4]. This situation is serious in China because of the country's complex topography and relatively unevenly distributed economic resources [5]. Satellite-based and reanalysis precipitation datasets have been effective in complementing traditionally obtained precipitation data as remote sensing and computing technologies have developed [6–8].

Satellite-based precipitation measurement technology uses visible data, infrared imaging, and passive microwave detection to gather precipitation data [9,10]. However, satellite-based datasets inevitably contain errors due to the measurement technology [11], the sampling method [12], and the retrieval algorithms [13]. Reanalysis datasets are created from a combination of observed data and model forecasts [14]. The accuracy of reanalysis datasets is determined by the observed forcing data, the data assimilation method, and the prediction model(s) used [15]. Many studies have shown that the accuracy of reanalysis datasets is highly related to both the observing system and the assimilated data [16,17]. Hodges et al. [18] showed that newer reanalysis datasets, including the European Centre for Medium-Range Weather Forecasts (ECMWF) Interim Re-Analysis (ERA-Interim, https://www.ecmwf.int/en/forecasts/datasets/archive-datasets/reanalysis-datasets/era-interim), the National Aeronautics and Space Administration Modern Era Retrospective-Analysis for Research and Applications (NASA's MERRA, https://climatedataguide.ucar.edu/climate-data/nasa-merra), and the NCEP-CFSR (http://globalweather.tamu.edu) perform better than older datasets (such as the 25-year Japanese Reanalysis (JRA-25, http://jra.kishou.go.jp/JRA-25/index_en.html)) in identifying recurrent extratropical cyclones because of the improvements in models, observations, and data assimilation in numerical weather prediction model (NWP) systems. Ebisuzaki and Zhang [19] compared NCEP-CFSR to a set of operational analyses for 2007 and found that NCEP-CFSR captured daily variability in precipitation better than the older reanalyses. The performance of NCEP-CFSR was attributed to major improvements in modeling, observation, and the method of data assimilation. Dee et al. [20] found that observed data have a significant effect on the initialization of an NWP model and thus on the quality of reanalysis data. They also found that successive generations of atmospheric reanalysis data have improved in quality as a result of better models, better input data, and better assimilation methods. Smith et al. [21] showed that even when the model and data assimilation method do not change, observational data density, type, and quality change over time. These observational changes can introduce spurious errors into reanalysis data. Model bias can also act on the data to introduce errors, as can the method of observations. There are many widely used satellite-based and reanalysis datasets, such as PERSIANN-CDR, 3B42V7, and NCEP-CFSR, available on the internet. Detailed information about these datasets, such as resolution, coverage, and data sources, is shown in Table 1. However, because of the errors inherent in satellite-based datasets and the high dependency of reanalysis datasets on the observation system, these datasets may not be suitable for hydrological applications in East Asia [22]. The CMADS were developed by Dr. Xianyong Meng from the China Agricultural University (CAU) and has received worldwide attention [5,23–30]. It using STMAS assimilation techniques as well as big data projection and processing methods to compensate for the fact that few specialized meteorological products were developed for East Asia [22].

In this study, four precipitation products that include two precipitation reanalysis datasets (NCEP-CFSR and CMADS) and two satellite-based precipitation datasets (3B42V7 and PERSIANN-CDR) were analyzed and evaluated in a hydrological application for the Xiang River basin, a humid watershed in central China. These datasets all have high spatial and temporal resolution. 3B42V7 is the latest release of the post-real time product (ftp://disc2.nascom.nasa.gov/ftp/data/s4pa/TRMM_L3/). Precipitation estimates from 3B42V7 have been evaluated in many studies [31–34]. PERSIANN-CDR is a new retrospective multi satellite-based precipitation dataset for long hydrological and climate studies [35], which is available online (ftp://data.ncdc.noaa.gov/cdr/persiann/files/). The dataset is produced by the PERSIANN algorithm [36] using gridded satellite (GridSat-B1) infrared data. Studies of precipitation estimates from PERSIANN-CDR are relatively few, and most of them compare PERSIANN-CDR predictions with ground-based precipitation observations [37–39]. only a small number are related to hydrological applications of PERSIANN-CDR [40]. The daily NCEP-CFSR data, which are in a format that the SWAT model can use, are available online (http://globalweather.tamu.edu). NCEP-CFSR datasets are widely used in many studies [41–43]. CMADS, which we use in this study, is a new reanalysis product which can be downloaded from the internet (www.cmads.org). There are very few studies that use CMADS because of its novelty, and they investigate northern arid areas such as the Juntanghu watershed [26,27,44], the Manas River basin [28], and the Qinghai-Tibet Plateau [30]. However, the error characteristics of precipitation products vary with climatic regions, seasons, surface conditions, storm regimes, and altitudes [45] which necessitates the analysis and evaluation of hydrological applications of precipitation dataset products in different regions. To the best of our knowledge, this is the first study of a hydrological application of CMADS in Central China.

This paper is organized as follows. Section 2 describes the materials and methods used in the study. Section 3 presents a detailed evaluation of the results given by the precipitation products, and a further discussion is given in Section 4. Lastly, Section 5 provides a short conclusion based on the results of our study.

2. Materials and Methods

2.1. Study Area

The Xiang River basin was selected as our study area. Xiang River is one of the largest tributaries of Yangtze River, flowing northward towards Dongting Lake, the second largest freshwater lake in China. The Xiang River Basin is located in Hunan Province, between between 24.5–28.25° N and 110.5–114.25° E. The outlet of the Xiang River Basin is Xiangtan station and the area of the basin is 82,375 km^2. The basin is dominated by subtropic monsoon climate, with a meaning annual precipitation of 1400 to 1700 mm and an average annual temperature of 17 °C. Most of the rainfall occurs between April and June. The basin suffers from frequent floods and droughts due to the uneven seasonal distribution of rainfall. The primary terrain of the Xiang River Basin is plain while the elevation of this area ranges from 1 m to more than 2000 m (based on China National Height Datum). The overview of Xiang River basin is shown in Figure 1. For more information about the study area, readers are referred to Zhu et al. [40].

2.2. Meteorological Data

In this section, the data used in the study, including gauge observations, that are required for the SWAT model, as well as the four precipitation dataset products, are briefly described. The spatial distribution of the CMADS, the locations of the precipitation and discharge gauges used in the study are shown in Figure 1. The spatial distribution of other datasets in the studied basin refers to Zhu et al. [40].

2.2.1. Satellite-Based and Reanalysis Precipitation Estimates

PERSIANN is a satellite-based precipitation retrieval algorithm based on infrared brightness temperature imagery generated by geostationary satellites [36]. The PERSIANN-CDR dataset is generated by the PERSIANN algorithm using gridded satellite (GridSat-BI) infrared data. NCEP Stage IV radar data is used to train the Artificial Neural Networks model and create nonlinear regression parameters. The model prediction (precipitation estimates) is then calibrated using the monthly Global Precipitation Climatology Project (GPCP) version 2.2 product that contains precipitation gauge data generated by the GPCP mission in order to increase the reliability of the PERSIANN-CDR data [35].

The Tropical Rainfall Measuring Mission (TRMM) is a joint mission between the U.S. National Aeronautics and Space Administration (NASA) and the Japan Aerospace Exploration Agency (JAXA) to study rainfall for weather and climate research. To increase the accuracy of the precipitation estimates, 3B42V7 integrates microwave and infrared measurements and incorporates the new Global Precipitation Climatology Center monthly precipitation data [40,46]. The TRMM satellite stopped collecting data on 15 April 2015 (https://trmm.gsfc.nasa.gov/). The Global Precipitation Measurement Mission Integrated MultisatellitE Retrievals for Global Precipitation Measurement (GPM IMERG), which has more accurate spatiotemporal resolution (half-hourly and 0.1) is a successor to TRMM [47].

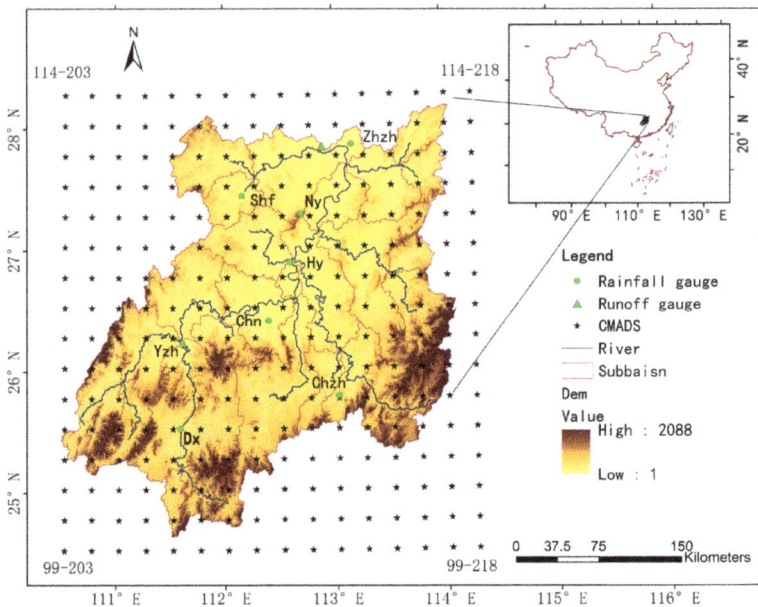

Figure 1. Spatial distribution of CMADS, precipitation gauge stations, and runoff stations in the Xiang River basin with elevations and subbasin divisions (Zhzh represents Zhuzhou site, Shf represents Shuangfeng site, Ny represents Nanyue site, Hy represents Hengyang site, Chn represents Changning site, Yzh represents Yongzhou site, Chzh represents Chenzhou site, and Dx represents Daoxian site, the site after is denoted by the above abbreviation).

CFSR is a global coupled atmosphere–land–ocean–sea-ice assimilation system developed at NCEP [48]. Its spatial resolution is approximately 38 km. CFSR includes the coupling of atmosphere and ocean during the generation of the 6-h guess field, an interactive sea-ice model, and assimilation of satellite radiance data by grid point statistical interpolation over the entire period (https://rda.ucar.edu/#!pub/cfsr.html). All available conventional and satellite observations were included in CFSR.

CMADS is constructed using multiple technologies and scientific methods, including loop nesting of data, projection of resampling models, and bilinear interpolation (www.cmads.org). Data sources for CMADS include nearly 40,000 regional automatic stations under the oversight of China's 2421 national automatic and business assessment centers [29]. CMADS precipitation data use Climate Prediction Center morphing technique (CMORPH) global precipitation products and data from the National Meteorological Information Center of China. The spatial resolution of CMADS V1.1 (www.cmads.org), used in this study, is 0.25°.

2.2.2. Ground Gauge Observations

The daily meteorological observations used to drive the SWAT model include precipitation, maximum and minimum temperatures, solar radiation, wind speed, and relative humidity from 1987 to 2013 at 8 meteorological stations. The data were obtained from China Meteorological Administration. The daily discharge record from 1970 to 2013 was available from Xiangtan station. The locations of these meteorological stations and runoff gauges are shown in Figure 1. An overview of all datasets is given in Table 1.

Table 1. Overview of precipitation datasets.

Datasets	Spatial Resolution	Temporal Resolution	Available Period	Coverage	Source of Data
Gauge	Point	Daily	1987–2013	Xiang River Basin	China Meteorological Administration
CMADS	0.25°	Daily	2008–2016	East Asia	www.cmads.org
3B42V7	0.25°	Daily	1998–present	50° S–50° N	Goddard Space Flight Centre
NCEP-CFSR	38 km	Daily	1979–present	Global	National Centers for Environment precipitation
PERSIANN-CDR	0.25°	Daily	1983–present	60° S–60° N	University of California, Irvine, CA, USA

2.3. Straightforward Comparison

A straightforward comparison was made on two scales: a grid and the whole basin. At the grid scale, only the precipitation estimates for those grids where gauges are located were evaluated. Grid squares were created with the same native resolution for all datasets to identify the grid–gauge pair. Pairwise statistical analyses were conducted between satellite-based/reanalysis precipitation estimates for the grid square and the observations from the gauge located in the grid square [49]. For the whole basin, areal precipitation from precipitation estimates and gauge observations were calculated and compared. The comparisons for both spatial scales were made for daily and monthly timesteps.

Diagnostic Statistics

Seven statistical indexes are used to quantify the accuracy of precipitation predictions: the correlation coefficient (CC), the root mean squared error (RMSE), the mean error (ME), relative bias (BIAS), the probability of detection (POD), the false alarm ratio (FAR), and the critical success index (CSI). The values of the indices are calculated by the following Equations [40,50]:

$$CC = \frac{\sum_{i=1}^{n}(G_i - \bar{G})(S_i - \bar{S})}{\sqrt{\sum_{i=1}^{n}(G_i - \bar{G})^2}\sqrt{\sum_{i=1}^{n}(S_i - \bar{S})^2}} \tag{1}$$

$$RMSE = \sqrt{\frac{1}{n}\sum_{i=1}^{n}(S_i - G_i)^2} \tag{2}$$

$$ME = \frac{1}{n}\sum_{i=1}^{n}(S_i - G_i) \tag{3}$$

$$BIAS = \frac{\sum_{i=1}^{n}(S_i - G_i)}{\sum_{i=1}^{n} G_i} \times 100\% \tag{4}$$

$$POD = \frac{H}{H + M} \tag{5}$$

$$FAR = \frac{F}{H + F} \tag{6}$$

$$CSI = \frac{H}{H + M + F} \tag{7}$$

where G_i is the observed precipitation from gauges, S_i is the precipitation estimates from PERSIANN-CDR, 3B42V7, NCEP-CSFR, and CMADS; H is the observed precipitation correctly detected; M is the observed precipitation not detected; F is the precipitation detected but not observed.

CC reflects the degree of linear correlation, ranging from −1 to 1. The result get the best when the value is equal to 1. ME reflects the average difference between precipitation products and gauge observation. The range of ME is $[0, +\infty)$, with the perfect value of 0. RMSE reflects the average error between precipitation products and gauge observations, imparting bigger weights to larger errors. The range of RMSE is $[0, +\infty)$, and the perfect value of this index is 0. $BIAS$ measures the relative degree of the systematic error of the precipitation estimation, ranging from 0 to positive infinite. The perfect value of BIAS is 0. POD gives the fraction of rain occurrences that are detected. It ranges from 0 to 1, with the perfect value of 1. FAR measures the fraction of rain detections that are wrongly detected. The value field of this index is $[0, 1]$, and the perfect value is 0. CSI gives the fraction of observed and/or detected rain but is correctly detected. The value field of this index is $[0, 1]$, and the perfect value is 1. The precipitation threshold between wet day and dry day is 1 mm in this study.

2.4. Ensemble Bayesian Model Averaging

We used Bayesian model averaging (BMA) to determine which precipitation product is most accurate in simulating streamflow in comparison with streamflow gauge observations by comparing the weights of simulated streamflows predicted by the precipitation products.

The BMA method is as follows. Assume that $f = f_1, \cdots, f_K$ is a set of predictions obtained from K different models, and Δ represents the quantity of interest. In BMA, each ensemble member forecast, $f_k, k = 1, \cdots, K$, is associated with a conditional probability density function (pdf), $g_k(\Delta|f_k)$, which can be interpreted as the conditional pdf of Δ on f_k, given that f_k is the best forecast in the ensemble. The BMA predictive model for dynamic ensemble forecasting can be expressed as a finite mixture model [51]:

$$p(\Delta|f_1, \cdots, f_k) = \sum_{k=1}^{K} w_k g_k(\Delta|f_k) \tag{8}$$

where, w_k denotes the posterior probability of forecast k being the best one. The w_ks are nonnegative and add up to 1. They can be viewed as weights reflecting an individual model's relative contribution to predictive skill over the training period [51]. $g_k(\Delta|f_k)$ of the different ensembles can be approximated by a normal distribution centered at a linear function of the original forecast, $a_k + b_k f_k$ and standard deviation σ.

$$\Delta|f_k \sim N\left(a_k + b_k f_k, \sigma^2\right) \tag{9}$$

The values for a_k and b_k are bias-correction terms derived by linear regression of Δ on f_k for each of the K ensemble members.

The values for $w_k, k = 1, \cdots, K$ and σ^2 in Equations (8) and (9) are estimated by maximum likelihood (ML) from a calibration data set. Assuming the forecast errors in space and time are independent, the log-likelihood function for the BMA predictive model is:

$$l(w_1, \cdots, f_K, \Delta) = \sum_{s,t}^{n} \log\left(\sum_{k=1}^{K} w_k g_k(\Delta_{st}|f_{kst})\right) \tag{10}$$

where, n denotes the total number of measurements in the training data set, s and t denote the number of each dimension of the training data set, Δ_{st} denotes quantities of interest in the training data set, and f_{kst} denotes predictions from K different models in the training data set. However, there are no analytical solutions conveniently maximizing Equation (10). In this study, the DiffeRential Evolution Adaptive Metropolis (DREAM) adaptive Markov chain Monte Carlo (MCMC) algorithm is used to estimate the parameters in Equation (10).

Specific to this study, the ensemble models are the simulated streamflows forced by gauge observations and precipitation estimates from PERSIANN-CDR, 3B42V7, NCEP-CFSR, and CMADS, while the Δ is the observed runoff. According to others' studies, the bias-correction of the value of ensemble models can be ignored ($a_k = 0, b_k = 1$ in Equation (9)) when used in hydrological studies [52]. After maximizing, the probability w_k can denote the relatively applicability of those datasets.

2.5. Model Creation

SWAT was used to create the hydrological model. Details of SWAT, and the model creation are given in Zhu et al. [40].

2.6. Model Calibration and Validation

Two model calibration strategies have mainly been used in previous studies: (1) the SWAT model is calibrated separately for different precipitation datasets; and (2) the best model parameters are obtained from calibrating the SWAT model using observed (gauge) precipitation data and observed streamflow data, and the model is then used for hydrological simulation with other precipitation datasets, such as (in this study) PERSIANN-CDR, NCEP-CFSR, 3B42V7, and CMADS. The first calibration strategy is used in this study because the SWAT model is a semi-distributed model and some sensitive parameters are empirically determined. Model parameters are surrounded by substantial uncertainties because they are inherently non-unique in inverse modeling, and thus it may be that many different sets of parameters will produce the same output signal. In other words, there are no best parameters for a hydrological model because of the inherent uncertainty, but the first calibration strategy guarantees a relatively good simulation result. However, this strategy can confuse because it may result in model predictions that are not comparable. Luckily, some researchers find that there is no obvious difference to the simulated streamflow obtained with the second strategy [40].

The model parameters were calibrated and validated using daily streamflow observed by gauges and simulated streamflow determined by PERSIANN-CDR, NCEP-CFSR, 3B42V7, and CMADS [53,54]. NS, given in Equation (11), and BIAS are used to evaluate the performance of the simulations.

$$\text{NS} = 1 - \frac{\sum_{i=1}^{n} \left(Q_{oi} - Q_{si} \right)^2}{\sum_{i=1}^{n} \left(Q_{oi} - \bar{Q}_o \right)^2} \tag{11}$$

where, Q_{oi} is observed streamflow; Q_{si} is simulated streamflow; and $\bar{Q}_o$ is the mean of observed streamflow.

In view of the overlapping periods of precipitation estimates and streamflow records for the Xiang River basin, the modeling period chosen was from January 2008 to December 2013. The total period was divided into three parts: a warming-up period (1 year from January 2008 to December 2008), a calibration period (calibration; 3 years from January 2009 to December 2011), and a validation period (validation; 2 years from January 2012 to December 2013).

3. Results

3.1. Comparison of Precipitation Estimates

As described in Section 2, precipitation estimates from CMADS, 3B42V7, NCEP-CFSR, and PERSIANN-CDR were compared with daily and monthly gauge observations for the Xiang River basin. The comparisons between precipitation estimates and gauge observations were performed in two spatial scales: grid and the whole basin. At grid scale, the gauge precipitation was compared with the precipitation estimates of the grid where the gauge is located. At the whole basin scale, areal precipitation calculated from precipitation estimates and gauge observations was compared. It should be noted that the comparison method for grid scale used in this paper may introduce errors and uncertainties because the estimate is the average value within a grid while the gauge observation is the value of a point located in the grid.

3.1.1. Spatial Distribution of Annual Precipitation Estimates

The spatial distributions of annual precipitation derived from CMADS, 3B42V7, NCEP-CFSR, PERSIANN-CDR and gauges are shown in Figure 2. The spatial distribution of observed precipitation is interpolated from gauge observations shown in Figure 1 using Kriging method. The results show that the annual precipitation estimates for these four datasets have similar spatial distribution patterns, consistent with observed annual precipitation. The annual precipitation in this area decreases from south to north, and from east to west. As errors and uncertainties may be introduced by interpolation, we did not compare the precipitation magnitudes between estimates and interpolations for each grid in this section. The comparison of precipitation magnitudes was proceeded between stations and the grids where these stations are located in the following sections.

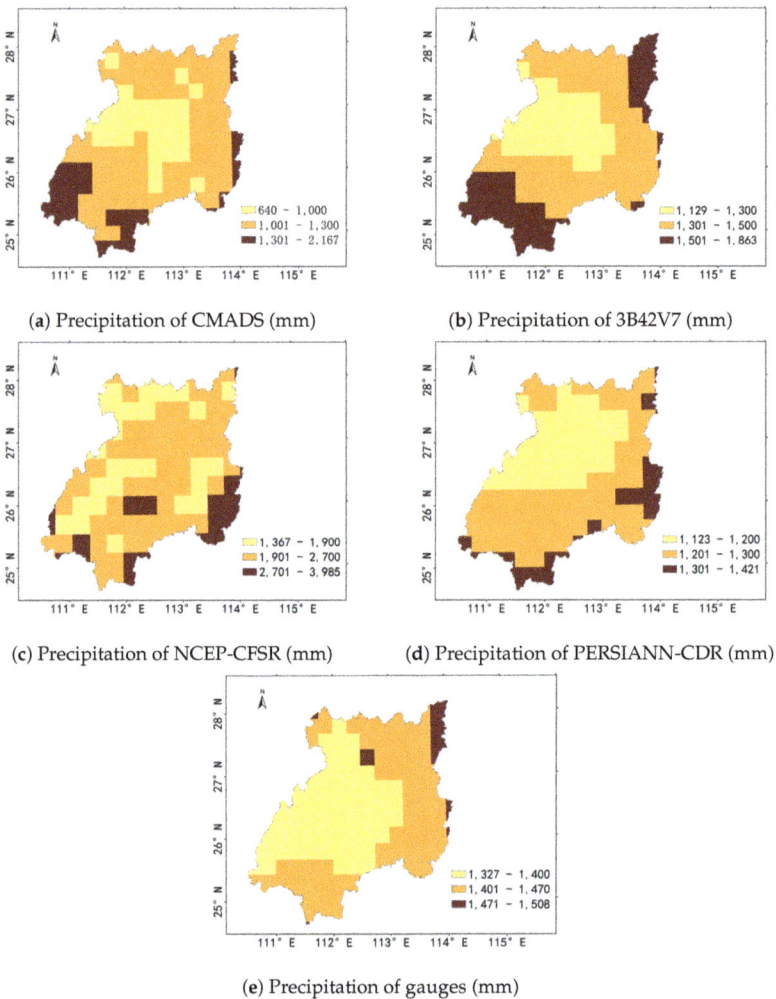

(a) Precipitation of CMADS (mm)   (b) Precipitation of 3B42V7 (mm)

(c) Precipitation of NCEP-CFSR (mm)   (d) Precipitation of PERSIANN-CDR (mm)

(e) Precipitation of gauges (mm)

Figure 2. Spatial distribution of annual precipitation estimates.

3.1.2. Comparison at Monthly Scale

The temporal distribution patterns of all the precipitation datasets and gauge observations for the included grid squares and the whole basin across different months are similar (Figure 3). In the rainy season (from April to September), NCEP-CFSR overestimates the monthly average precipitation, while CMADS underestimates it. Precipitation estimates from PERSIANN-CDR and 3B42V7 are consistent with gauge observations.

Values of CC, RMSE, ME, and BIAS between monthly precipitation estimates and observed (gauge) precipitation for grid station locations and the whole basin are summarized in Table 2. The performance of the datasets in terms of correlation with monthly observed precipitation differs from the daily estimates. There is no obvious correlation between the reanalysis datasets and the satellite-based datasets. CMADS and 3B42V7 perform better than PERSIANN-CDR and NCEP-CFSR in most locations. A comparison of Table 2 with Table 3 shows that precipitation estimates have a greater linear correlation with gauge observations on the monthly scale than the daily scale. This conclusion is consistent with the research of Omranian and Sharif [47]. The monthly BIAS values between precipitation estimates and gauge observation are similar to the daily values. Satellite-based estimates perform better than reanalysis estimates. CMADS tends to underestimate precipitation while NCEP-CFSR tends to overestimate precipitation.

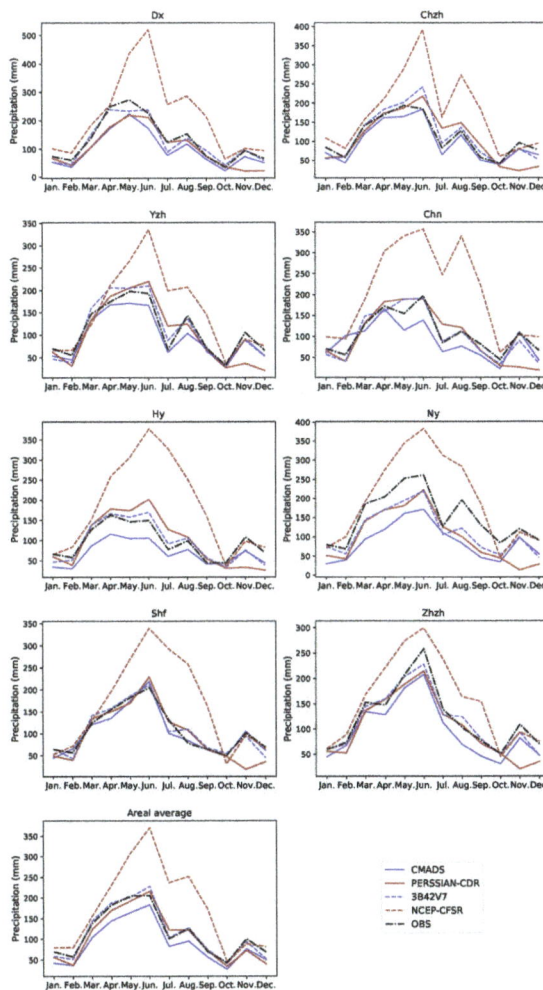

Figure 3. Multi year average monthly precipitation (OBS represents gauge observations).

Table 2. Monthly statistical indexes.

Datasets	CC	RMSE (mm)	ME (mm)	BIAS
CMADS	0.96	49.57	−33.29	−34.24
PERSIANN-CDR	0.79	72.27	−27.75	−27.01
3B42V7	0.93	40.14	−7.27	−5.90
NCEP-CFSR	0.61	159.76	85.86	39.68

(a) Station Dx

Datasets	CC	RMSE (mm)	ME (mm)	BIAS
CMADS	0.97	23.84	−13.21	−13.69
PERSIANN-CDR	0.67	66.34	−2.74	−2.56
3B42V7	0.92	33.24	3.46	3.05
NCEP-CFSR	0.73	101.79	64.97	37.21

(b) Station Chzh

Datasets	CC	RMSE (mm)	ME (mm)	BIAS
CMADS	0.97	27.04	−14.85	−15.54
PERSIANN-CDR	0.76	62.41	−7.48	−7.26
3B42V7	0.94	29.41	0.44	0.40
NCEP-CFSR	0.66	98.07	41.59	27.35

(c) Station Yzh

Datasets	CC	RMSE (mm)	ME (mm)	BIAS
CMADS	0.55	81.06	−17.80	−20.09
PERSIANN-CDR	0.72	62.25	−7.20	−7.26
3B42V7	0.87	38.98	−5.59	−5.55
NCEP-CFSR	0.67	142.98	98.02	47.95

(d) Station Chn

Datasets	CC	RMSE (mm)	ME (mm)	BIAS
CMADS	0.89	41.28	−28.12	−41.37
PERSIANN-CDR	0.65	63.98	1.16	1.19
3B42V7	0.86	36.86	−2.01	−2.14
NCEP-CFSR	0.62	146.09	87.04	47.53

(e) Station Hy

Datasets	CC	RMSE (mm)	ME (mm)	BIAS
CMADS	0.84	82.35	−63.46	−73.29
PERSIANN-CDR	0.73	84.28	−51.63	−52.46
3B42V7	0.92	52.79	−37.18	−32.94
NCEP-CFSR	0.77	105.36	49.44	24.78

(f) Station Ny

Datasets	CC	RMSE (mm)	ME (mm)	BIAS
CMADS	0.93	31.18	−5.11	−5.00
PERSIANN-CDR	0.79	55.38	−9.27	−9.47
3B42V7	0.88	40.13	−0.33	−0.31
NCEP-CFSR	0.74	102.18	57.05	34.74

(g) Station Shf

Datasets	CC	RMSE (mm)	ME (mm)	BIAS
CMADS	0.83	55.42	−24.80	−25.83
PERSIANN-CDR	0.76	60.80	−18.90	−18.55
3B42V7	0.90	37.88	−5.08	−4.39
NCEP-CFSR	0.84	72.31	36.28	23.10

(h) Station Zhzh

Datasets	CC	RMSE (mm)	ME (mm)	BIAS
CMADS	0.98	31.86	−25.41	−28.67
PERSIANN-CDR	0.93	30.07	−9.14	−8.71
3B42V7	0.98	16.78	−0.37	−0.33
NCEP-CFSR	0.81	93.90	61.17	34.91

(i) Areal average

The spatial distributions of monthly precipitation estimates' RMSE are shown in Figure 4. The results show that the spatial distribution patterns of RMSE for the evaluated satellite-based and reanalysis datasets are similar. The gauges located in mountainous regions (Dx and Ny) have much larger RMSE than those located in plain regions, which illustrates that the precipitation estimates derived from both satellite-based datasets and reanalysis datasets perform better in plain regions than in complex orographic areas.

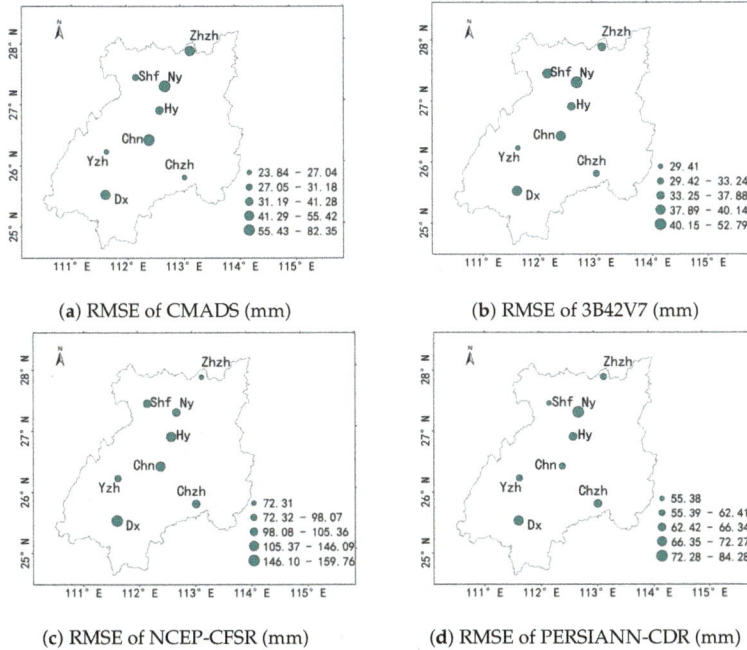

(a) RMSE of CMADS (mm)　　　　　　(b) RMSE of 3B42V7 (mm)

(c) RMSE of NCEP-CFSR (mm)　　　　(d) RMSE of PERSIANN-CDR (mm)

Figure 4. Spatial distribution of the RMSE of monthly precipitation estimates.

3.1.3. Comparison at Daily Scale

Values of the seven diagnostic indexes (CC, RMSE, ME, BIAS, CSI, FAR, and POD) between daily precipitation estimates provided by the reanalysis and satellite datasets and observed (gauged) precipitation for the station locations and for the whole basin are given in Table 3. The results show that reanalysis datasets (CMADS and NCEP-CFSR) are better than satellite-based datasets (PERSIANN-CDR and 3B42V7) in terms of correlation with gauge observations for all evaluated grid squares and for the whole basin. The correlations between precipitation estimates from reanalysis datasets, CMADS and NCEP-CFSR, and observed precipitation are >0.55 for all evaluated grid squares and for the whole basin. The performance of CMADS is similar to NCEP-CFSR in terms of correlation with observed precipitation. The satellite-based dataset precipitations, PERSIANN-CDR and 3B42V7, have relatively low linear correlation with observed precipitation for all grid squares considered. The same results do not hold for the whole basin. The correlation between the 3B42V7 precipitation and observed precipitation over the whole basin is 0.60. It can be deduced from the results that the correlation between dataset precipitation estimates and gauge observations become stronger as the spatial resolution decreases. This conclusion is consistent with the research of Omranian and Sharif [47]. However, the satellite-based datasets (PERSIANN-CDR and 3B42V7) give better estimates than the reanalysis datasets (CMADS and NCEP-CFSR) at most grid locations except for BIAS at station Dx and Ny. All dataset precipitation estimates show much larger BIAS values for this station than for other stations. This result is because station Dx and Ny are located in a mountainous region while other stations are on plains Figure 1. For the whole basin, 3B42V7 shows much lower BIAS than PERSIANN-CDR. CMADS clearly tends to underestimate precipitation. NCEP-CFSR tends to greatly overestimate precipitation at the selected grid squares and for the whole basin because fewer observations were available to initialize the model. NCEP-CFSR is one of a new generation of numerical weather prediction models, and it is very sensitive to the observed data that is used to initialize it [19]. The values of POD, FAR, and CSI show that PERSIANN-CDR cannot accurately detect rainfall events at the grid square level. CMADS and 3B42V7 have a greater probability of detecting rainfall events at the grid scale and for the whole basin but NCEP-CFSR is more accurate in detecting rainfall events.

Table 3. Daily statistical indexes.

Datasets	CC	RMSE (mm)	ME (mm)	BIAS (%)	POD	FAR	CSI
CMADS	0.59	10.13	−1.09	−34.24	0.72	0.26	0.58
PERSIANN-CDR	0.36	12.78	−0.91	−27.01	0.47	0.48	0.33
3B42V7	0.51	11.77	−0.24	−5.90	0.62	0.49	0.39
NCEP-CFSR	0.55	12.76	2.82	39.68	0.47	0.07	0.45

(a) Station Dx

Datasets	CC	RMSE (mm)	ME (mm)	BIAS (%)	POD	FAR	CSI
CMADS	0.60	7.83	−0.43	−13.69	0.74	0.25	0.59
PERSIANN-CDR	0.32	11.15	−0.09	−2.56	0.51	0.45	0.36
3B42V7	0.50	10.25	0.11	3.05	0.65	0.51	0.39
NCEP-CFSR	0.58	9.98	2.13	37.21	0.52	0.09	0.50

(b) Station Chzh

Datasets	CC	RMSE (mm)	ME (mm)	BIAS (%)	POD	FAR	CSI
CMADS	0.62	7.94	−0.49	−15.54	0.69	0.27	0.55
PERSIANN-CDR	0.39	10.97	−0.25	−7.26	0.49	0.45	0.35
3B42V7	0.50	10.33	0.01	0.40	0.62	0.52	0.37
NCEP-CFSR	0.55	9.56	1.37	27.35	0.52	0.10	0.49

(c) Station Yzh

Datasets	CC	RMSE (mm)	ME (mm)	BIAS (%)	POD	FAR	CSI
CMADS	0.55	8.13	−1.20	−53.50	0.69	0.34	0.51
PERSIANN-CDR	0.34	10.97	−0.24	−7.26	0.49	0.46	0.35
3B42V7	0.44	10.43	−0.18	−5.55	0.66	0.55	0.37
NCEP-CFSR	0.56	11.24	3.22	47.95	0.51	0.09	0.48

(d) Station Chn

Datasets	CC	RMSE (mm)	ME (mm)	BIAS (%)	POD	FAR	CSI
CMADS	0.55	7.32	−0.92	−41.37	0.69	0.33	0.51
PERSIANN-CDR	0.31	10.54	0.04	1.19	0.46	0.48	0.32
3B42V7	0.42	9.98	−0.07	−2.14	0.65	0.54	0.37
NCEP-CFSR	0.55	10.32	2.86	47.53	0.49	0.11	0.46

(e) Station Hy

Datasets	CC	RMSE (mm)	ME (mm)	BIAS (%)	POD	FAR	CSI
CMADS	0.58	10.50	−2.08	−73.29	0.71	0.36	0.51
PERSIANN-CDR	0.28	13.63	−1.70	−52.46	0.54	0.47	0.36
3B42V7	0.39	13.29	−1.22	−32.94	0.70	0.55	0.38
NCEP-CFSR	0.60	11.33	1.62	24.78	0.58	0.12	0.54

(f) Station Ny

Datasets	CC	RMSE (mm)	ME (mm)	BIAS (%)	POD	FAR	CSI
CMADS	0.59	9.12	−0.17	−5.00	0.67	0.30	0.53
PERSIANN-CDR	0.31	11.82	−0.30	−9.47	0.48	0.44	0.35
3B42V7	0.43	11.29	−0.01	−0.31	0.65	0.50	0.39
NCEP-CFSR	0.62	9.86	1.87	34.74	0.51	0.11	0.48

(g) Station Shf

Datasets	CC	RMSE (mm)	ME (mm)	BIAS (%)	POD	FAR	CSI
CMADS	0.65	9.33	−0.81	−25.83	0.69	0.26	0.56
PERSIANN-CDR	0.33	12.73	−0.62	−18.55	0.46	0.42	0.34
3B42V7	0.48	12.32	−0.17	−4.39	0.63	0.51	0.38
NCEP-CFSR	0.66	9.95	1.19	23.10	0.56	0.09	0.53

(h) Station Zhzh

Datasets	CC	RMSE (mm)	ME (mm)	BIAS (%)	POD	FAR	CSI
CMADS	0.70	5.77	−0.83	−28.67	0.77	0.21	0.64
PERSIANN-CDR	0.49	7.85	−0.30	−8.71	0.62	0.36	0.46
3B42V7	0.60	7.35	−0.01	−0.33	0.71	0.31	0.54
NCEP-CFSR	0.78	6.08	2.01	34.91	0.62	0.04	0.60

(textbfi) Areal average

The spatial distributions of the *RMSE* of daily precipitation estimates are similar as that of monthly precipitation estimates (Figure 5). Mountainous areas (Dx and Ny) have larger RMSE than plain areas. In addition, CMADS has the minimum RMSE than other datasets, ranging from 7.32 mm to 10.50 mm.

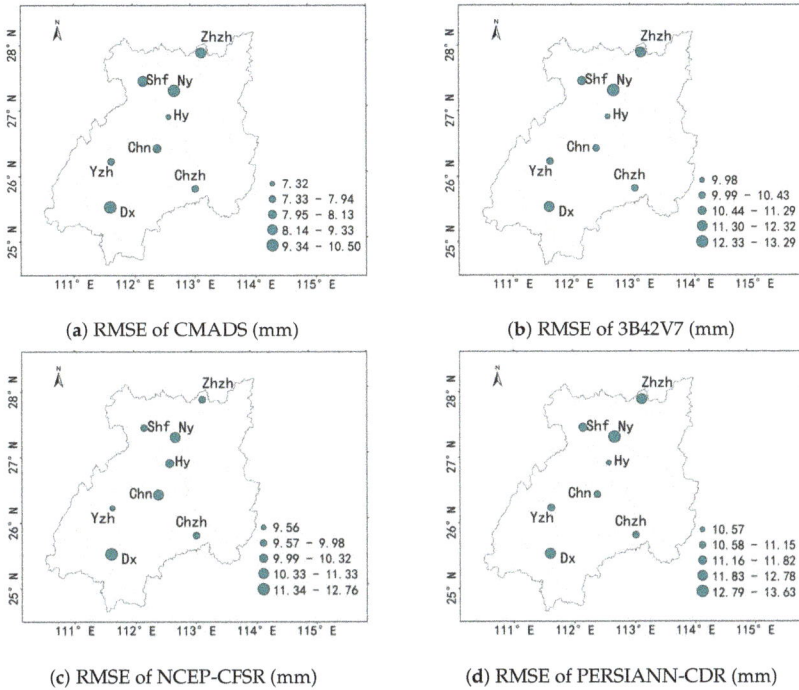

(a) RMSE of CMADS (mm)	(b) RMSE of 3B42V7 (mm)
(c) RMSE of NCEP-CFSR (mm)	(d) RMSE of PERSIANN-CDR (mm)

Figure 5. Spatial distribution of the RMSE of daily precipitation estimates.

3.2. Comparison of Streamflow Simulations

Simulation accuracy in mature hydrological models is mainly determined by meteorological inputs, especially precipitation. Both the total volume and the spatial and temporal distribution of precipitation significantly influence the output of a hydrological model. This section assesses the performance of the different precipitation datasets as drivers (or forcers) of a hydrological model and evaluates the capability of the precipitation datasets to capture the spatial and temporal characteristics of precipitation over the basin as shown by the performance of the hydrological model. The SWAT model is used because it well describes the hydrology of this area [40]. The evaluation is made for both daily and monthly timesteps.

3.2.1. Comparison at Monthly Scale

A comparison of simulated monthly streamflow using the precipitation datasets of observed data, CMADS, 3B42V7, NCEP-CFSR, and PERSIANN-CDR with observed streamflow is shown in Figure 6. The evaluation statistics (NS efficiency coefficient and BIAS) of the monthly simulations are shown in Table 4. As in the daily simulations, NCEP-CFSR greatly overestimates streamflow and is the worst of all the precipitation datasets, having the lowest NS values (−0.12 during calibration and −0.12 during validation) and extremely large BIAS values (36.49% during calibration and 31.31% during validation). Of the other precipitation products, 3B42V7 performs better than others in terms of both NS (0.94 during calibration and 0.88 during validation) and BIAS (−7.20 during calibration and 3.69 during validation). The NS and BIAS for CMADS are 0.92 and −12.06% (calibration) and 0.80 and

2.17% (validation), which indicates that the performance of CMADS is acceptable. PERSIANN-CDR performs well during calibration (NS = 0.89) but not during validation (NS = 0.63). However, the BIAS of PERSIANN-CDR is relatively good, intermediate between CMADS and 3B42V7.

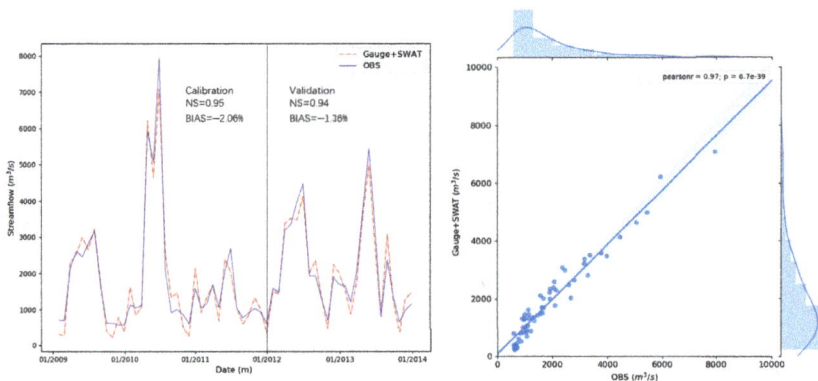

(a) Monthly simulated streamflow with gauged precipitation

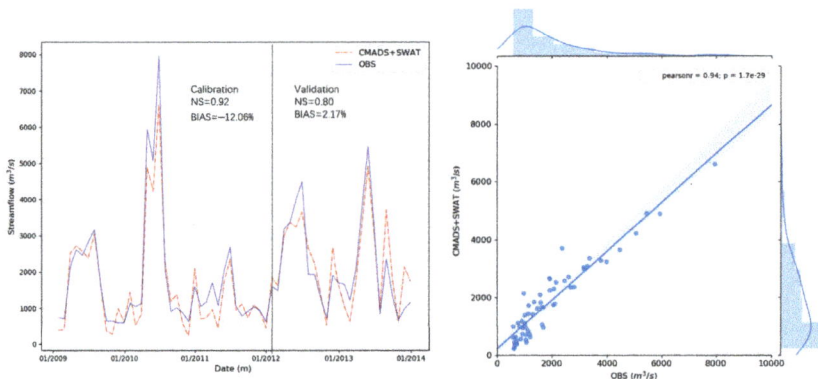

(b) Monthly simulated streamflow with CMADS estimates

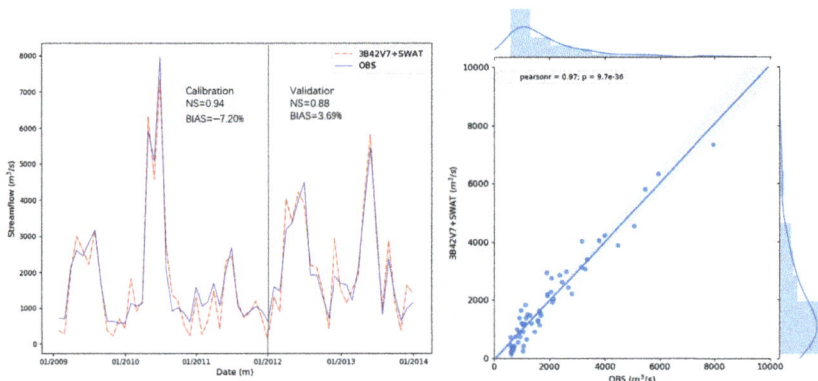

(c) Monthly simulated streamflow with 3B42V7 estimates

Figure 6. *Cont.*

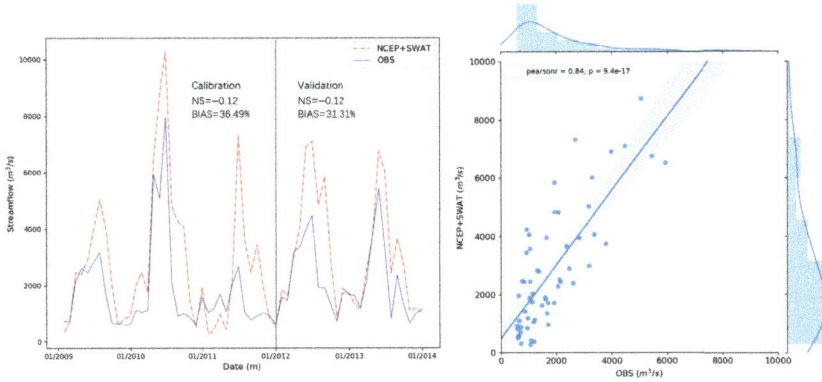

(**d**) Monthly simulated streamflow with NCEP-CFSR estimates

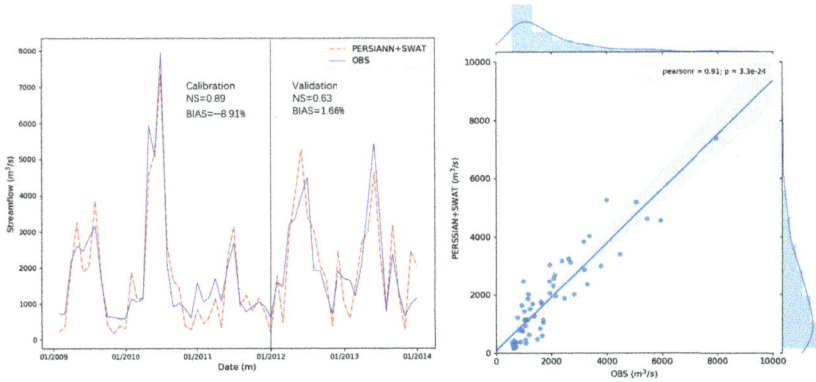

(**e**) Monthly simulated streamflow with PERSIANN-CDR estimates

Figure 6. Monthly simulated flow in Xiangtan station with precipitation from gauge (**a**), PERSIANN-CDR (**b**), NCEP-CFSR (**c**), CMADS (**d**), and 3B42V7 (**e**) (In the scatter plot, pearsonr represents Pearson's correlation coefficient, p represents the significance of paired t-test, and the shade represents the 0.95 confidence interval).

Table 4. NS coefficient and BIAS of monthly simulated streamflow (The number outside and inside the parenthesis is for the calibration period and validation period respectively).

Datasets	NS in Calibration	NS in Validation	BIAS in Calibration (%)	BIAS in Validation (%)
Gauge	0.95	0.94	−2.06	−1.36
CMADS	0.92	0.80	−12.06	2.17
3B42V7	0.94	0.88	−7.20	3.69
NCEP-CFSR	−0.12	−0.12	36.49	31.31
PERSIANN-CDR	0.89	0.63	−8.91	1.66

3.2.2. Comparison at Daily Scale

The simulated flows predicted by the SWAT model, using observed (gauge) data and data from CMADS, 3B42V7, NCEP-CFSR, and PERSIANN-CDR, are shown in Figure 7. The NS efficiency coefficient and the BIAS values of the simulations are shown in Table 5. The calibration and validation periods for the simulations were 1 January 2009–31 December 2011 (calibration) and 1 January 2012–December 2013 (validation). NS and BIAS for the simulation using observed precipitation were 0.86 and −1.58% (calibration) and 0.72 and 3.07% (validation). The flow hydrograph for the simulation using observed precipitation is highly consistent with that for observed streamflow. The results indicate that the SWAT model well predicts daily hydrological response in the Xiang River basin. The NS efficiency coefficients show that the precipitation products, except NCEP-CFSR, predict the streamflow well during calibration. The NS efficiency coefficients for CMADS, 3B42V7, PERSIANN-CDR, and NCEP-CFSR are 0.83, 0.83, 0.71 and −0.46 respectively. The NS efficiency coefficients for PERSIANN-CDR during validation was 0.40, illustrating that the performance of PERSIANN-CDR is not stable during the entire simulation. The NS efficiency coefficients for simulations using observed data (0.72), data from 3B42V7 (0.73), and from CMADS (0.71) during validation are acceptable. The BIAS values show that for simulated streamflow, compared with observed streamflow, observed (gauge) precipitation data gives the best results (−1.58% during calibration and 3.07% during validation), followed by 3B42V7 (−10.84% during calibration and −0.17% during validation) and CMADS (−12.06% during calibration and 2.20% during validation). However, simulation using NCEP-CFSR data greatly overestimates streamflow, with BIAS values of 36.58% (calibration) and 31.40% (validation). The hydrographs (Figure 7) of the simulations show that most datasets, except NCEP-CFSR, which overestimates streamflow during almost the entire simulation period, produce good baseflow predictions during both calibration and validation. However, when streamflow is high, most datasets (observation, PERSIANN-CDR, CMADS, and 3B42V7) underestimate the streamflow to different extents. To eliminate the inherent contribution of the SWAT model to the underestimation of high streamflow, the simulation using observed precipitation is taken as a baseline to judge the performance of the other precipitation datasets in predicting high streamflow (Figure 8). The result shows that NCEP-CFSR overestimates high streamflow substantially, while PERSIANN-CDR overestimates high streamflow slightly. However, there is no obvious overestimation or underestimation for CMADS and 3B42V7 compared with simulation based on gauge precipitation. This implies that CMADS and 3B42V7 perform as well as gauge in capturing temporal features of high streamflow.

NS and BIAS assess the efficiency and water balance predictions of the model and data. Observed data performs best because it has the highest NS efficiency coefficient and lowest BIAS. The differences between NS efficiency coefficients and BIAS values for 3B42V7 and CMADS are insignificant, so it is difficult to discriminate between them in hydrological terms. Bayesian model averaging (BMA) was thus used to distinguish between 3B42V7 and CMADS precipitation. BMA is commonly used to handle conceptual model uncertainty in the analysis of environmental systems and to derive predictive distributions of model output [52]. Comparison of the BMA model weights can show which precipitation dataset performs better in hydrological simulation. A detailed description of BMA analysis was given in Section 2.4. The BMA model weights for simulated flow forced by CMADS and 3B42V7 are shown in Table 6. The results show that CMADS performs better than 3B42V7 when synthetically considering the consistency of simulated and observed streamflow.

Table 5. NS coefficient and BIAS of daily simulated streamflow.

Datasets	NS in Calibration	NS in Validation	BIAS in Calibration (%)	BIAS in Validation (%)
Gauge	0.86	0.72	−1.58	3.07
CMADS	0.83	0.70	−12.06	2.20
3B42V7	0.83	0.73	−10.84	−0.17
NCEP-CFSR	−0.46	−0.46	36.58	31.40
PERSIANN-CDR	0.71	0.40	−16.66	−6.44

Table 6. BMA weights of simulated streamflow forced by different precipitation products.

	3B42V7	CMADS
Weights	0.47	0.53

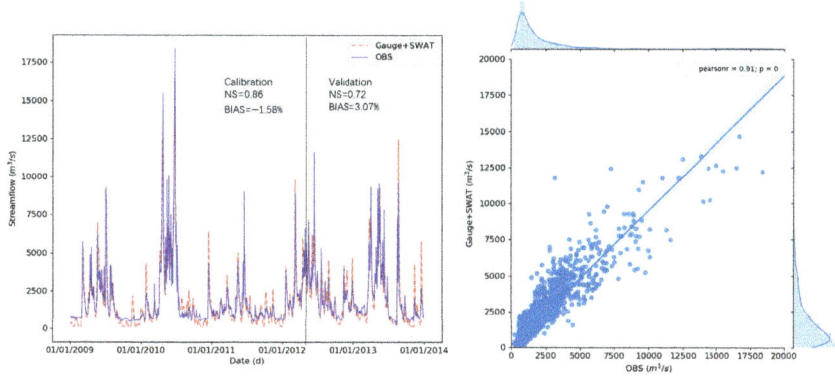

(**a**) Daily simulated streamflow with gauged precipitation

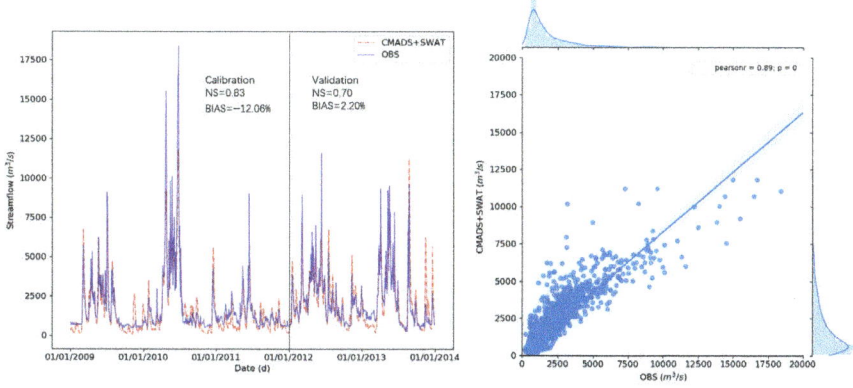

(**b**) Daily simulated streamflow with CMADS estimates

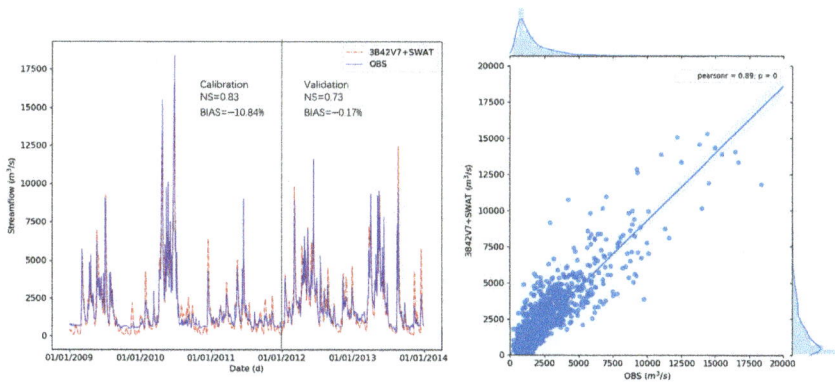

(**c**) Daily simulated streamflow with 3B42V7 estimates

Figure 7. *Cont.*

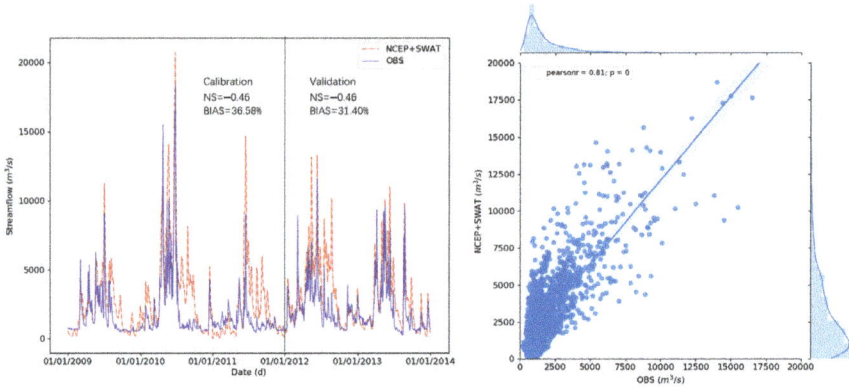

(**d**) Daily simulated streamflow with NCEP-CFSR estimates

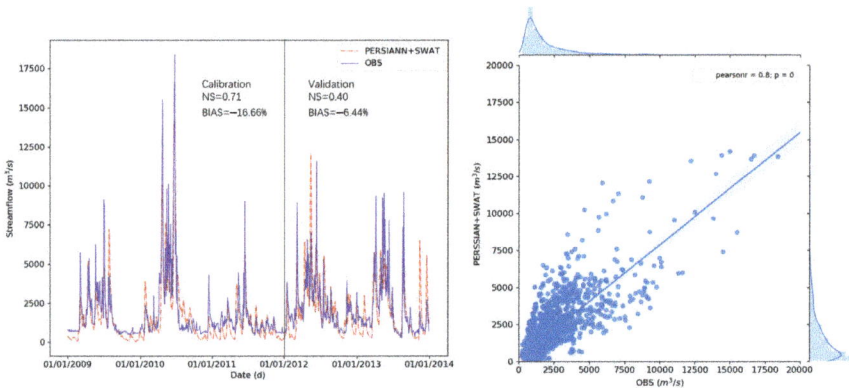

(**e**) Daily simulated streamflow with PERSIANN-CDR estimates

Figure 7. Daily simulated flow in Xiangtan station with precipitation from gauge (**a**), PERSIANN-CDR (**b**), NCEP-CFSR (**c**), CMADS (**d**), and 3B42V7 (**e**) (In the scatter plot, pearsonr represents Pearson's correlation coefficient, p represents the significance of paired t-test, and the shade represents the 0.95 confidence interval).

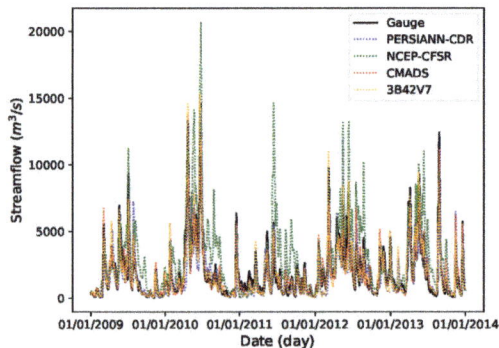

Figure 8. Comparison of simulated streamflow based on PERSIANN-CDR, NCEP-CFSR, CMADS, and 3B42V7 precipitation with that based on gauge precipitation.

3.3. Analysis of Anomalies in Hydrological Simulation

NCEP-CFSR and PERSIANN-CDR are relatively poor in simulating hydrology. The reasons for the poor performances of these two datasets are different. NCEP-CFSR overestimates precipitation during almost the entire simulation period (Table 3). This overestimation is probably the main reason for the extremely low NS values of NCEP-CFSR since calibration with low flows tends to give higher NS values. The cumulative distribution curves of areal precipitation derived from PERSIANN-CDR and that of gauge precipitation for calibration period and validation period are shown in Figure 9. It should be noted that only the precipitation larger than 20 mm is shown in the chart to illustrate the results more clearly. The results show that the relative position of cumulative distribution curves for PERSIANN-CDR and gauge observations is different between calibration period and validation period. This means that the high precipitation characteristics varies between calibration period and validation period. The SWAT model parameters calibrated with precipitation during calibration period are thus not suitable for validation period.

Table 5 shows that relative water volume during validation is greater than during calibration. This is probably caused by the workings of the SWAT model itself. The performance of the SWAT runoff module depends on the calibration. In this study, there was an extreme peak flow during the calibration period. Thus the SWAT parameters were calibrated to fit to high flows. Consequently, the simulated streamflow during validation, using the parameters determined during calibration, tends to be high.

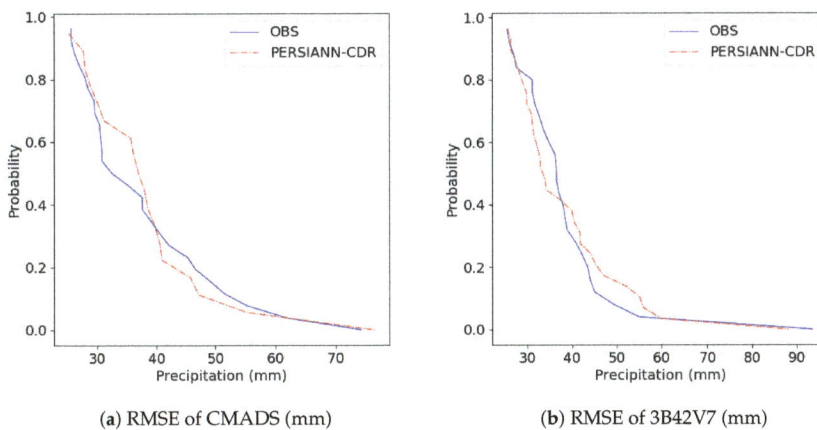

(a) RMSE of CMADS (mm)   (b) RMSE of 3B42V7 (mm)

Figure 9. Cumulative distribution curves of gauge precipitation and PERSIANN-CDR estimates.

4. Discussion

Hydrological simulation is thoroughly influenced by the inputs to the hydrological models. Clearly, there is some linkage between the precipitation estimates and the hydrological simulation. However, a precipitation dataset that shows good linear correlation with gauge observations does not necessarily produce a good hydrological simulation. For example, NCEP-CFSR was best linearly correlated with gauge observations but produced the worst hydrological simulation because of its substantial overestimation and relatively low probability of detecting rainfall events. A comparison of the CC and BIAS values for CMADS and NCEP-CFSR shows that these two precipitation datasets perform similarly. However, CMADS produces a much better hydrological simulation than NCEP-CFSR. This shows that the POD of precipitation estimates has a significant effect on hydrological simulation. Streamflow responds to rainfall events. If a precipitation dataset does not detect most of the rainfall events, it cannot adequately capture streamflow. The influence of FAR on the hydrological simulation cannot be determined from the results; however, it can be analyzed conceptually. If other indexes of precipitation datasets, such as CC, BIAS, and POD, are kept constant, a lower value of FAR will indicate a better hydrological simulation. In addition, the estimates of a precipitation dataset are not always consistent with its hydrological predictions. For example, the BIAS for precipitation estimates and for hydrological simulations are not consistent. The value of BIAS for the CMADS precipitation estimate was −28.67% at the whole basin scale but −12.06% for the simulated streamflow in both daily and monthly timesteps. Many factors may contribute to a difference between precipitation estimates and their corresponding hydrological outputs, such as: (1) areal precipitation is calculated by the Theissen polygon method, which does not consider the impact of topography whereas the SWAT model considers the elevation of precipitation grids; (2) the transformation of precipitation to streamflow is a very complicated nonlinear process, so error will not be transferred from precipitation to streamflow linearly; and (3) there are simplifications in the SWAT model, such as the assumptions in the universal soil loss equation for estimating sediment loss, the assumptions in calculating flow velocity in a river, and the ignoring of some hydrological processes that are considered to have relatively small impact on total hydrology.

The spatial resolution of the areas of comparison (from grids of different sizes to the whole basin) can substantially affect the results of the evaluation. Omranian and Sharif [47] used the GPM IMERG dataset and found that the spatial resolution of the areas compared had a significant effect on the results. The dataset gives better results when the temporal and spatial resolutions are downscaled. However, the spatial resolution of a precipitation dataset has a significant impact on the hydrological simulation. Many studies have shown that in hydrological modeling with high spatial and temporal resolutions, datasets can better characterize streamflow [55]. Thus precipitation datasets with higher spatial and temporal resolution are needed to provide good hydrological simulations. However, as mentioned above, increased spatial and temporal resolution of these datasets worsens the model performance when compared to observation datasets, which can adversely affect the simulation. High-resolution datasets also increase model processing time. From a practical engineering perspective, a more efficient way to combine the input data preparation and the hydrological modeling, that considers both modeling accuracy and modeling efficiency, needs to be further studied [56]. In this study, CMADS performs better in modeling accuracy and is more usable because of its SWAT compatible data structure. Hence, considering the modeling accuracy and modeling efficiency, CMADS are more applicable in practical streamflow simulation.

5. Conclusions

The performance of two reanalysis precipitation datasets (CMADS and NCEP-CFSR) and two satellite-based precipitation datasets (PERSIANN-CDR and 3B42V7) was evaluated at two spatial scales (a grid square and the whole basin) and two timesteps (daily and monthly), and the ability of these datasets to simulate streamflow is assessed for both temporal scales. The results show that: (1) for daily timesteps, the reanalysis datasets perform better than satellite-based datasets in terms of correlation with gauge observations, while satellite-based datasets perform better than reanalysis datasets in most situations in terms of bias. The correlations between reanalysis datasets and gauge observations at both spatial scales are >0.55. The absolute bias values of the two satellite-based datasets are <10% at most grid squares and also for the whole basin. CMADS underestimates precipitation while NCEP-CFSR overestimates it. PERSIANN-CDR cannot accurately detect the spatial distribution of precipitation events compared with other datasets. The POD of PERSIANN-CDR at most grid squares is <0.50; (2) CMADS and 3B42V7 perform better than PERSIANN-CDR and NCEP-CFSR in most situations in terms of correlation with gauge observations and satellite-based datasets perform better than reanalysis datasets in terms of bias; (3) CMADS and 3B42V7 simulate both daily (NS > 0.70) and monthly (NS > 0.80) streamflow well; CMADS performs a little better than 3B42V7 at a daily timestep according to the weights of BMA model, and vice versa for a monthly timestep; NCEP-CFSR performs worst because of its substantial overestimation; PERSIANN-CDR performs badly because of its poor capability to capture the characteristics of streamflow during validation.

Some other studies have shown that precipitation products tend to underestimate flood peaks by directly comparing modeled streamflow driven by precipitation products to observed streamflow. In this study, we eliminated the effects of the model structure on underestimation by comparing modeled streamflow driven by precipitation data from products with streamflow driven by observed precipitation, and we found that there is no obvious underestimation of flood peaks when using precipitation products such as CMADS and 3B42V7 in the Xiang River basin. On the whole, CMADS has great potential in hydrological application in the studied area because that (1) the accuracy of simulated streamflow forced by CMADS is good in the studied area; (2) the dataset is well organized and can be used as inputs of SWAT model directly; (3) as a reanalysis dataset, CMADS can be used in areas with sparse gauges and improved in spatiotemporal resolution in further versions with relatively small cost. (4) Compared with satellite-based datasets, reanalysis datasets such as CMADS usually have much longer time series.

Author Contributions: Conceptualization, X.G. and Z.Y.; Data curation, X.G. and Q.Z.; Formal analysis, X.G.; Funding acquisition, Z.Y.; Investigation, X.G.; Methodology, X.G. and Q.Z.; Project administration, Z.Y.; Supervision, Z.Y.; Visualization, X.G.; Writing—original draft, X.G.; Writing—review & editing, Z.Y. and H.W.

Funding: This study is financially supported by the National Key Research and Development Project (No. 2016YFC0402707 and No. 2016YFA0601503), the Research Fund of the China Institute of Water Resources and Hydropower Research (No. 2017ZY02) and National Natural Science Foundation of China (51709148).

Acknowledgments: Thank the National Meteorological Information Center of China Meteorological Administration for archiving the observed climate data (http://cdc.cma.gov.cn).

Conflicts of Interest: The authors declare no conflict of interest.

References

1. Faurès, J.M.; Goodrich, D.C.; Woolhiser, D.A.; Sorooshian, S. Impact of small-scale spatial rainfall variability on runoff modeling. *J. Hydrol.* **1995**, *173*, 309–326. [CrossRef]
2. Etchevers, P.; Durand, Y.; Habets, F.; Martin, E.; Noilhan, J. Impact of spatial resolution on the hydrological simulation of the Durance high-Alpine catchment, France. *Ann. Glaciol.* **2001**, *32*, 87–92. [CrossRef]
3. Himanshu, S.K.; Pandey, A.; Yadav, B. Assessing the applicability of TMPA-3B42V7 precipitation dataset in wavelet-support vector machine approach for suspended sediment load prediction. *J. Hydrol.* **2017**, *550*, 103–117. [CrossRef]
4. Michelson, D.B. Systematic correction of precipitation gauge observations using analysed metrological variables. *J. Hydrol.* **2004**, *290*, 161–177. [CrossRef]
5. Meng, J.; Li, L.; Hao, Z.; Wang, J.; Shao, Q. Suitability of TRMM satellite rainfall in driving a distributed hydrological model in the source region of Yellow River. *J. Hydrol.* **2014**, *509*, 320–332. [CrossRef]
6. Sorooshian, S.; Hsu, K.L.; Gao, X.; Gupta, H.V.; Imam, B.; Dan, B. Evaluation of PERSIANN System Satellite-Based Estimates of Tropical Rainfall. *Bull. Am. Meteorol. Soc.* **2000**, *81*, 2035–2046. [CrossRef]
7. Joyce, R.J.; Janowiak, J.E.; Arkin, P.A.; Xie, P. CMORPH: A Method That Produces Global Precipitation Estimates from Passive Microwave and Infrared Data at High Spatial and Temporal Resolution. *J. Hydrometeorol.* **2003**, *5*, 287–296. [CrossRef]
8. Kidd, C.; Huffman, G. Global precipitation measurement. *Meteorol. Appl.* **2011**, *18*, 334–353. [CrossRef]
9. Ebert, E.E.; Janowiak, J.E.; Kidd, C. Comparison of near-real-time precipitation estimates from satellite observations and numerical models. *Bull. Am. Meteorol. Soc.* **2007**, *88*, 47–64. [CrossRef]
10. Huffman, G.J.; Bolvin, D.T.; Nelkin, E.J.; Wolff, D.B.; Adler, R.F.; Gu, G.; Hong, Y.; Bowman, K.P.; Stocker, E.F. *The TRMM Multisatellite Precipitation Analysis (TMPA): Quasi-Global, Multiyear, Combined-Sensor Precipitation Estimates at Fine Scales*; Springer: Dordrecht, The Netherlands, 2010; pp. 3–22.
11. Villarini, G.; Krajewski, W.F.; Smith, J.A. New paradigm for statistical validation of satellite precipitation estimates: Application to a large sample of the TMPA 0.25 3-hourly estimates over Oklahoma. *J. Geophys. Res. Atmos.* **2009**, *114*. [CrossRef]

12. Nijssen, B.; Lettenmaier, D.P. Effect of precipitation sampling error on simulated hydrological fluxes and states: Anticipating the Global Precipitation Measurement satellites. *J. Geophys. Res. Atmos.* **2004**, *109*. [CrossRef]

13. Conti, F.L.; Hsu, K.L.; Noto, L.V.; Sorooshian, S. Evaluation and comparison of satellite precipitation estimates with reference to a local area in the Mediterranean Sea. *Atmos. Res.* **2014**, *138*, 189–204. [CrossRef]

14. Seyyedi, H.; Anagnostou, E.N.; Beighley, E.; Mccollum, J. Hydrologic evaluation of satellite and reanalysis precipitation datasets over a mid-latitude basin. *Atmos. Res.* **2015**, *164–165*, 37–48. [CrossRef]

15. Li, C.; Tang, G.; Hong, Y. Cross-evaluation of ground-based, multi-satellite and reanalysis precipitation products: Applicability of the Triple Collocation method across Mainland China. *J. Hydrol.* **2018**, *562*, 71–83. [CrossRef]

16. Inoue, T.; Matsumoto, J. A Comparison of Summer Sea Level Pressure over East Eurasia between NCEP-NCAR Reanalysis and ERA-40 for the Period 1960-99. *J. Meteorol. Soc. Jpn. Ser. II* **2004**, *82*, 951–958. [CrossRef]

17. Marshall, G.J. Trends in Antarctic Geopotential Height and Temperature: A Comparison between Radiosonde and NCEP-NCAR Reanalysis Data. *J. Clim.* **2002**, *15*, 659–674. [CrossRef]

18. Hodges, K.I.; Lee, R.W.; Bengtsson, L. A Comparison of Extratropical Cyclones in Recent Reanalyses ERA-Interim, NASA MERRA, NCEP CFSR, and JRA-25. *J. Clim.* **2011**, *24*, 4888–4906. [CrossRef]

19. Ebisuzaki, W.; Zhang, L. Assessing the performance of the CFSR by an ensemble of analyses. *Clim. Dyn.* **2011**, *37*, 2541–2550. [CrossRef]

20. Dee, D.; Uppala, S.; Simmons, A.J.; Berrisford, P.; Poli, P.; Kobayashi, S.; Andrae, U.; Balmaseda, M.A.; Balsamo, G.; Bauer, P.; et al. The ERA-Interim reanalysis: Configuration and performance of the data assimilation system. *Q. J. R. Meteorol. Soc.* **2011**, *137*, 553–597. [CrossRef]

21. Smith, C.A.; Compo, G.P.; Hooper, D.K. Web-Based Reanalysis Intercomparison Tools (WRIT) for analysis and comparison of reanalyses and other datasets. *Bull. Am. Meteorol. Soc.* **2015**, *95*, 1671–1678. [CrossRef]

22. Meng, X.; Wang, H.; Meng, X.; Wang, H. Significance of the China Meteorological Assimilation Driving Datasets for the SWAT Model (CMADS) of East Asia. *Water* **2017**, *9*, 765. [CrossRef]

23. Zhao, F.; Wu, Y.; Qiu, L.; Sun, Y.; Sun, L.; Li, Q.; Niu, J.; Wang, G. Parameter Uncertainty Analysis of the SWAT Model in a Mountain-Loess Transitional Watershed on the Chinese Loess Plateau. *Water* **2018**, *10*, 690. [CrossRef]

24. Thom, V.; Li, L.; Jun, K.S. Evaluation of Multi-Satellite Precipitation Products for Streamflow Simulations: A Case Study for the Han River Basin in the Korean Peninsula, East Asia. *Water* **2018**, *10*, 642.

25. Meng, X.; Long, A.; Wu, Y.; Yin, G.; Wang, H.; Ji, X. Simulation and spatiotemporal pattern of air temperature and precipitation in Eastern Central Asia using RegCM. *Sci. Rep.* **2018**, *8*, 3639. [CrossRef] [PubMed]

26. Meng, X.Y.; Yu, D.L.; Liu, Z.H. Energy balance-based SWAT model to simulate the mountain snowmelt and runoff—Taking the application in Juntanghu watershed (China) as an example. *J. Mt. Sci.* **2015**, *12*, 368–381. [CrossRef]

27. Meng, X.; Sun, Z.; Zhao, H.; Ji, X.; Wang, H.; Xue, L.; Wu, H.; Zhu, Y. Spring Flood Forecasting Based on the WRF-TSRM Mode. *Tehnički Vjesnik* **2018**, *25*, 27–37.

28. Meng, X.; Wang, H.; Lei, X.; Cai, S.; Wu, H.; Ji, X.; Wang, J. Hydrological modeling in the Manas River Basin using soil and water assessment tool driven by CMADS. *Tehnicki Vjesnik* **2017**, *24*, 525–534.

29. Meng, X.; Wang, H.; Wu, Y.; Long, A.; Wang, J.; Shi, C.; Ji, X. Investigating spatiotemporal changes of the land-surface processes in Xinjiang using high-resolution CLM3.5 and CLDAS: Soil temperature. *Sci. Rep.* **2017**, *7*, 13286. [CrossRef] [PubMed]

30. Liu, J.; Shanguan, D.; Liu, S.; Ding, Y. Evaluation and Hydrological Simulation of CMADS and CFSR Reanalysis Datasets in the Qinghai-Tibet Plateau. *Water* **2018**, *10*, 513. [CrossRef]

31. Pombo, S.; de Oliveira, R.P. Evaluation of extreme precipitation estimates from TRMM in Angola. *J. Hydrol.* **2015**, *523*, 663–679. [CrossRef]

32. Li, D.; Christakos, G.; Ding, X.; Wu, J. Adequacy of TRMM satellite rainfall data in driving the SWAT modeling of Tiaoxi catchment (Taihu lake basin, China). *J. Hydrol.* **2018**, *556*, 1139–1152. [CrossRef]

33. Katiraie-Boroujerdy, P.S.; Asanjan, A.A.; Hsu, K.; Sorooshian, S. Intercomparison of PERSIANN-CDR and TRMM-3B42V7 precipitation estimates at monthly and daily time scales. *Atmos. Res.* **2017**, *193*, 36–49. [CrossRef]

34. Cabrera, J.; Yupanqui, R.T.; Rau, P. Validation of TRMM Daily Precipitation Data for Extreme Events Analysis. The Case of Piura Watershed in Peru. *Procedia Eng.* **2016**, *154*, 154–157. [CrossRef]

35. Ashouri, H.; Hsu, K.L.; Sorooshian, S.; Braithwaite, D.K.; Knapp, K.R.; Cecil, L.D.; Nelson, B.R.; Prat, O.P. PERSIANN-CDR: Daily Precipitation Climate Data Record from Multisatellite Observations for Hydrological and Climate Studies. *Bull. Am. Meteorol. Soc.* **2014**, *96*, 197–210. [CrossRef]

36. Hsu, K.L.; Gao, X.; Sorooshian, S.; Gupta, H.V. Precipitation Estimation from Remotely Sensed Information Using Artificial Neural Networks. *J. Appl. Meteorol.* **2003**, *36*, 1176–1190. [CrossRef]

37. Tan, M.L.; Santo, H. Comparison of GPM IMERG, TMPA 3B42 and PERSIANN-CDR satellite precipitation products over Malaysia. *Atmos. Res.* **2018**, *202*, 63–76. [CrossRef]

38. Yang, X.; Yong, B.; Hong, Y.; Chen, S.; Zhang, X. Error analysis of multi-satellite precipitation estimates with an independent raingauge observation network over a medium-sized humid basin. *Int. Assoc. Sci. Hydrol. Bull.* **2015**, *61*, 1813–1830. [CrossRef]

39. Tan, M.L.; Ibrahim, A.L.; Duan, Z.; Cracknell, A.P.; Chaplot, V. Evaluation of Six High-Resolution Satellite and Ground-Based Precipitation Products over Malaysia. *Remote Sens.* **2015**, *7*, 1504–1528. [CrossRef]

40. Zhu, Q.; Xuan, W.; Liu, L.; Xu, Y. Evaluation and hydrological application of precipitation estimates derived from PERSIANN-CDR, TRMM 3B42V7, and NCEP-CFSR over humid regions in China. *Hydrol. Process.* **2016**, *30*, 3061–3083. [CrossRef]

41. Li, X.; Wan, W.; Yu, Y.; Ren, Z. Yearly variations of the stratospheric tides seen in the CFSR reanalysis data. *Adv. Space Res.* **2015**, *56*, 1822–1832. [CrossRef]

42. Tomy, T.; Sumam, K. Determining the Adequacy of CFSR Data for Rainfall-Runoff Modeling Using SWAT. *Procedia Technol.* **2016**, *24*, 309–316. [CrossRef]

43. Blacutt, L.A.; Herdies, D.L.; de Gonçalves, L.G.G.; Vila, D.A.; Andrade, M. Precipitation comparison for the CFSR, MERRA, TRMM3B42 and Combined Scheme datasets in Bolivia. *Atmos. Res.* **2015**, *163*, 117–131. [CrossRef]

44. Wang, Y.J.; Meng, X.Y.; Liu, Z.H.; Ji, X.N. Snowmelt Runoff Analysis under Generated Climate Change Scenarios for the Juntanghu River Basin, in Xinjiang, China. *Water Sci. Technol.* **2016**, *7*, 41–54.

45. Sorooshian, S.; Aghakouchak, A.; Arkin, P.; Eylander, J.; Foufoulageorgiou, E.; Harmon, R.; Hendrickx, J.M.H.; Imam, B.; Kuligowski, R.; Skahill, B. Advanced Concepts on Remote Sensing of Precipitation at Multiple Scales. *Bull. Am. Meteorol. Soc.* **2011**, *92*, 1353–1357. [CrossRef]

46. Kenawy, A.M.E.; Lopez-Moreno, J.I.; McCabe, M.F.; Vicente-Serrano, S.M. Evaluation of the TMPA-3B42 precipitation product using a high-density rain gauge network over complex terrain in northeastern Iberia. *Glob. Planet. Chang.* **2015**, *133*, 188–200. [CrossRef]

47. Omranian, E.; Sharif, H.O. Evaluation of the Global Precipitation Measurement (GPM) Satellite Rainfall Products over the Lower Colorado River Basin, Texas. *J. Am. Water Resour. Assoc.* **2018**, *54*, 882–898. [CrossRef]

48. Saha, S.; Moorthi, S.; Pan, H.L.; Wu, X.R.; Wang, J.D.; Nadiga, S.; Tripp, P.; Kistler, R.; Woollen, J.; Behringer, D. The NCEP climate forecast system reanalysis. *Bull. Am. Meteorol. Soc.* **2010**, *91*, 1015–1057. [CrossRef]

49. Nkiaka, E.; Nawaz, N.R.; Lovett, J.C. Evaluating global reanalysis precipitation datasets with rain gauge measurements in the Sudano-Sahel region: Case study of the Logone catchment, Lake Chad Basin. *Meteorol. Appl.* **2017**, *24*, 9–18. [CrossRef]

50. Caracciolo, D.; Francipane, A.; Viola, F.; Noto, L.V.; Deidda, R. Performances of GPM satellite precipitation over the two major Mediterranean islands. *Atmos. Res.* **2018**, *213*, 309–322. [CrossRef]

51. Vrugt, J.A.; Diks, C.G.H.; Clark, M.P. Ensemble Bayesian model averaging using Markov Chain Monte Carlo sampling. *Environ. Fluid Mech.* **2008**, *8*, 579–595. [CrossRef]

52. Diks, C.G.H.; Vrugt, J.A. Comparison of point forecast accuracy of model averaging methods in hydrologic applications. *Stoch. Environ. Res. Risk Assess.* **2010**, *24*, 809–820. [CrossRef]

53. Artan, G.; Gadain, H.; Smith, J.L.; Asante, K.; Bandaragoda, C.J.; Verdin, J.P. Adequacy of satellite derived rainfall data for stream flow modeling. *Nat. Hazards* **2007**, *43*, 167–185. [CrossRef]

54. Bitew, M.M.; Gebremichael, M.; Ghebremichael, L.T.; Bayissa, Y.A. Evaluation of High-Resolution Satellite Rainfall Products through Streamflow Simulation in a Hydrological Modeling of a Small Mountainous Watershed in Ethiopia. *J. Hydrometeorol.* **2012**, *13*, 338–350. [CrossRef]

55. Terink, W.; Leijnse, H.; van den Eertwegh, G.; Uijlenhoet, R. Spatial resolutions in areal rainfall estimation and their impact on hydrological simulations of a lowland catchment. *J. Hydrol.* **2018**, *563*, 319–335. [CrossRef]
56. Afshari, S.; Tavakoly, A.A.; Rajib, A.; Zheng, X.; Follum, M.L.; Omranian, E.; Fekete, B.M. Comparison of new generation low-complexity flood inundation mapping tools with a hydrodynamic model. *J. Hydrol.* **2018**, *556*, 539–556. [CrossRef]

Article

Evaluation and Analysis of Grid Precipitation Fusion Products in Jinsha River Basin Based on China Meteorological Assimilation Datasets for the SWAT Model

Dandan Guo [1,2], Hantao Wang [3], Xiaoxiao Zhang [1] and Guodong Liu [1,*]

[1] State Key Laboratory of Hydraulics and Mountain River Engineering, College of Water Resource and Hydropower, Sichuan University, Chengdu 610065, China; jingyugdd@126.com (D.G.); zolazxx@126.com (X.Z.)

[2] School of Civil Engineering, Architecture and Environment, Xihua University, Chengdu 610039, China

[3] Three Gorges Cascade Dispatching & Communication Centre, Yichang 443133, China; hantaow@126.com

* Correspondence: liugd988@163.com; Tel.: +86-138-8180-0968

Received: 17 December 2018; Accepted: 29 January 2019; Published: 1 February 2019

Abstract: Highly accurate and high-quality precipitation products that can act as substitutes for ground precipitation observations have important significance for research development in the meteorology and hydrology of river basins. In this paper, statistical analysis methods were employed to quantitatively assess the usage accuracy of three precipitation products, China Meteorological Assimilation Driving Datasets for the Soil and Water Assessment Tool (SWAT) model (CMADS), next-generation Integrated Multi-satellite Retrievals for Global Precipitation Measurement (IMERG) and Tropical Rainfall Measuring Mission (TRMM) Multi-satellite Precipitation Analysis (TMPA), for the Jinsha River Basin, a region characterized by a large spatial scale and complex terrain. The results of statistical analysis show that the three kinds of data have relatively high accuracy on the average grid scale and the correlation coefficients are all greater than 0.8 (CMADS:0.86, IMERG:0.88 and TMPA:0.81). The performance in the average grid scale is superior than that in grid scale. (CMADS: 0.86(basin), 0.6 (grid); IMERG:0.88 (basin),0.71(grid); TMPA:0.81(basin),0.42(grid)). According to the results of hydrological applicability analysis based on SWAT model, the three kinds of data fail to obtain higher accuracy on hydrological simulation. CMADS performs best (NSE:0.55), followed by TMPA (NSE:0.50) and IMERG (NSE:0.45) in the last. On the whole, the three types of satellite precipitation data have high accuracy on statistical analysis and average accuracy on hydrological simulation in the Jinsha River Basin, which have certain hydrological application potential.

Keywords: CMADS; IMERG; statistical analysis; SWAT hydrological simulation; Jinsha River Basin

1. Introduction

Precipitation is one of the major factors affecting the global water cycle and water balance. Heavy precipitation will cause floods, landslides and other disasters and pose a serious threat to people's safety and national property [1–3]. Precipitation has significant variability in different spatial and temporal scales. Accurate measurement of precipitation distribution can help to is master the spatial distribution characteristics of precipitation and it is of great significance for hydrological and water situation analysis, water resource management, drought and flood disaster prediction and hydrological and ecological simulation [4,5].

At present, the estimation of precipitation based on the data of ground precipitation stations is the main component of quantitative estimation of precipitation. The observation data of ground precipitation stations can provide the real information of a certain point of ground precipitation and have high accuracy. However, the spatial and temporal distribution of precipitation in the basin cannot be accurately reflected due to the limitation of site location and quantity. With the development of remote sensing and computer technology, satellite precipitation products based on data inversion of various sensors are a new source for quantitative estimation of precipitation. Satellite precipitation products have many advantages, such as wide coverage, continuous distribution in space and easy access to data [6,7] and have great application potential in basins without data or lacking data. However, they are not directly measured, so accuracy evaluation and applicability analysis are required for further use [8,9]. Now, the precision evaluation of satellite precipitation products mostly takes the observation data of ground precipitation stations or ground-based radar as the reference value, which can meet the verification requirements to a certain extent. In addition, increasing numbers of researchers turn their attention to the fusion verification based on distributed hydrological models, analyse the hydrological prediction errors caused by satellite precipitation products and test the substitutability of satellite precipitation products to the ground station precipitation in hydrological simulation [10–12].

The China Meteorological Assimilation Driving Datasets for the SWAT model (CMADS) is a public dataset developed by Dr. Xianyong Meng from China Agriculture University (CAU). [13–15]. CMADS provides standard weather model-driven data for SWAT model and other models. CMADSV1.1 used in this paper is an updated version of CMADS V1.0. The TRMM (Tropical Rainfall Measuring Mission Multi-satellite) satellite is mainly used to monitor and study the precipitation in tropical regions [16]. The satellite was launched in Japan in November 1997 and officially retired in June 2015. Presently, the commonly-used TMPA (Tropical Rainfall Measuring Mission Multi-satellite Precipitation Analysis) precipitation data include non-real-time precipitation products 3B42 and 3B43 calibrated by GPCC (the Global Precipitation Climatology Centre) ground stations and real-time precipitation products 3B42RT and so forth [17]. As the follow-up satellite precipitation observation plan of TRMM, GPM (global precipitation measurement) is superior to TRMM in terms of spatial coverage, spatial and temporal resolution and rain and snow data observation. GPM core observation platform (GPM core observatory, GPMCO) was successfully launched in February 2014. Currently, the commonly-used IMERG (Integrated Multi-satellite Retrievals for Global Precipitation Measurement) precipitation data include IMERG-F, a non-real-time precipitation product calibrated by GPCC ground precipitation station, IMERG-E and IMERG-L, quasi-real-time precipitation product [18]. The emergence of these precipitation fusion products provides more basic hydrological and meteorological data for the study of basin water resources.

A lot of researchers evaluate the applicability of the above fusion data, including the applicability of 3B42V7 and IMERG-F satellite precipitation products and their comparative evaluation [19–28]. For example, some researchers evaluate the product quality of IMERG, as well as its diversity and continuity, with the generation of TMPA products within the scope of Mainland China [29]; some evaluate the accuracy and hydrological simulation utility of IMERG-E, IMERG-L, IMERG-F and 3B42V7 product in the Beijiang River Basin in China [30]; some assess the fitness of IMERG-F and 3B42V7 products with the basin and CGDPA, as well as the precipitation drive simulation results under different hydrologic model parameters. Some evaluated the applicability of the above fusion data in Singapore [31], the southern plains [32] and other countries. The verification results show that TMPA and IMERG products have application potential in different studies.

CMADS has received domestic and worldwide attention in the past few years because it is high-resolution and high-quality meteorological data. Research shows that it has shown a good application effect in the basin and region hydrology simulation [33–36]. So far, many achievements have been obtained in the related research in the process of precipitation and runoff through CMADS drive SWAT model. For example, some researchers evaluate the application of CMADS reanalysis data set in China PET [37] and some evaluate the data adaptability on the Qinghai-Tibet plateau of the inland alpine region with sparse distribution weather stations [38]. Some researchers conduct hydrological simulation in South Korea Han River watershed and evaluate the precision of data set [39]. In fact, as basic data, there are few researchers on the statistical accuracy assessment and evaluation of hydrological effect on different climate areas and watersheds and few use TMPA and IMERG in the analysis process of the influence of hydrological model parameters on runoff simulation. Therefore, the study of the application effect on CMADS in more fields has certain significance and can provide more and reliable data set source and theoretical support for the future water resources management, hydrology and meteorology research.

The research contents of the paper are as follows: 1. Statistical analysis part: the accuracy of precipitation data of satellite precipitation fusion products CMADS, TMPA and IMERG in Jinsha River Basin was statistically evaluated by selecting R, RMSE and other statistical analysis indicators and using the hourly precipitation product comparison between CMPA-Hourly (China Hourly Merged Precipitation Analysis combining observations from automatic weather stations with CMORPH) as the reference value 2. Hydrological suitability analysis part: based on SWAT distributed hydrological model, the daily runoff process of the outlet section of Jinsha River Basin was simulated by selecting TMPA, IMERG and CMADS data as precipitation input respectively and the indirect evaluation of precipitation input was realized through hydrological simulation.

This paper is organized as follows: Section 2 introduces the research basin and data; Section 3 introduces the evaluation scheme and corresponding evaluation indexes; Section 4 presents the statistical evaluation of precipitation product CMADS, IMERG and TMPA; Section 5 evaluates the effectiveness of hydrological simulation under the SWAT hydrological model driven by three kinds of precipitation products. Sections 6 and 7 are discussion and summary. The organization flow chart of this paper is shown in Figure 1.

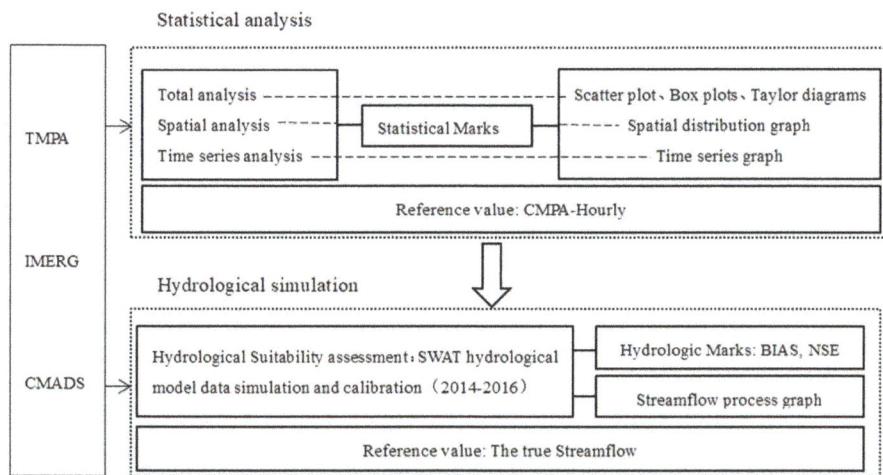

Figure 1. Flowchart of this study.

2. Study Site and Data

2.1. Study Site

The Jinsha River Basin is located at the upper reaches of the Yangtze River and originates from the Tibetan Plateau. The Jinsha is one of the largest rivers in southwest China. The basin is located between 90–105°E and 24–36°N. The length of the river is 2316 km and the area of the river basin is 340,000 km^2. The river begins in Yushu County, Qinghai Province and flows through 5 major terrain and geomorphological units (Tibetan Plateau, West Sichuan Plateau, Hengduan Mountains, Yunnan-Guizhou Plateau and the Mountainous Area of Southwest Sichuan). The river is located in

Qinghai, Tibet, Yunnan and Sichuan. The river is cross-distributed across plateaus, valleys, basins and hills. The basin is covered by plateau alpine meadows, hilly temperate dry river valley shrubs, hilly subtropical arid shrubs, subtropical xerophytic shrubs and subtropical semi humid evergreen broadleaf plants. The climate of the Jinsha River Basin exhibits regional characteristics: The upper and middle reaches experience a southwest monsoon climate with distinct dry and wet seasons (rainy season from May–October and dry season from November–April). The climate includes significant variations with altitude. The lower reaches belong to the central subtropical belt and experience a southeast monsoon climate. In this region, rainfall is characterized by pronounced seasonality and high intensity. In addition, the dry and hot river valley summer season in the lower reaches of the river basin is long and there is no winter [40–44]. Figure 2 shows the study site and its major water systems.

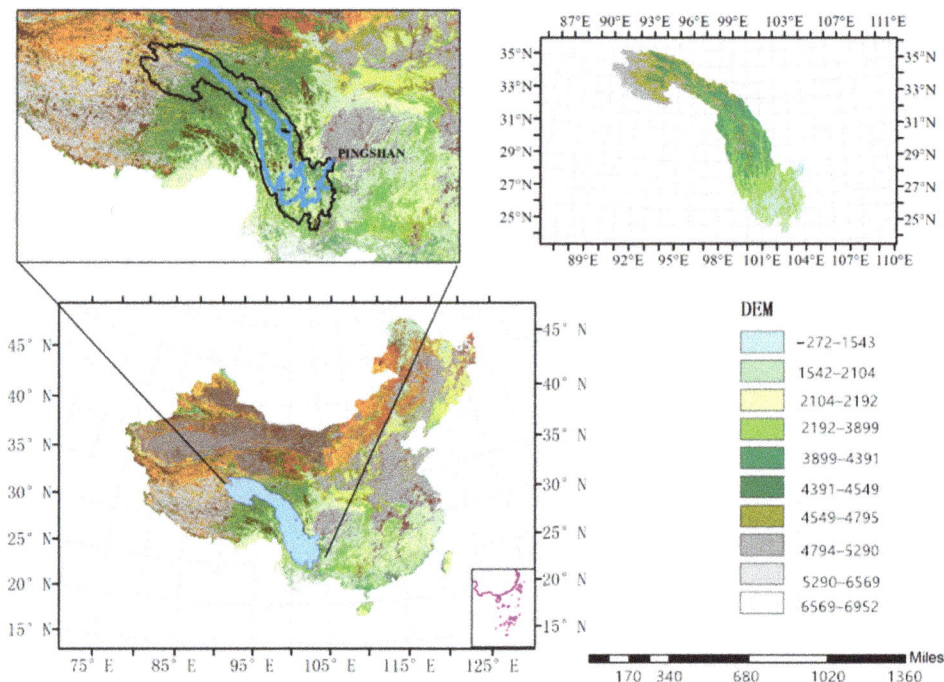

Figure 2. The Jinsha River Basin.

2.2. Study Site Data

In this paper, we assess the suitability of precipitation data from different sources in our research on the Jinsha River Basin. The following paragraphs summarize the spatiotemporal resolution and coverage range of precipitation data types:

1) CMADS

The China Meteorological Assimilation Driving Datasets for the SWAT model (CMADS) are public datasets that were developed by Professor Xianyong Meng from China. The CMADS V1.1 dataset series covers the entire East Asia region (0–65°N, 60–160°E) and has a spatial resolution of 0.25°. The dataset includes precipitation, atmospheric temperature, atmospheric pressure, specific humidity, wind speed and other datasets. These datasets are available in .dbf and .txt formats in order to facilitate analysis and access by researchers from different disciplines [45–47].

The hydrological simulation results of the CMADS + SWAT model can reflect the spatiotemporal distributions of various surface components in different regions and river basins [48]. The CMADS dataset developed by Dr. Xianyong Meng provides 2008–2016 precipitation data to users free of charge and also provides scientific research support. More and more researchers are trusting this dataset and utilizing it to study complex and unique problems [49–52].

2) TMPA

The Tropical Rainfall Measuring Mission (TRMM) was jointly developed and designed by the National Aeronautics and Space Administration (NASA) and the Japan Aerospace Exploration Agency (JAXA). The TRMM tropical satellite precipitation plan covers an area that encircles the globe from 38°N–38°S. The TRMM Multi-satellite Precipitation Analysis (TMPA) is a precipitation product that uses multiple modern satellite precipitation sensors and ground rain gauge networks. TMPA provides 2 standard satellite products, quasi-real-time and non-real-time post-processing. The TRMM plan provides free global 50°N–50°S and multiple time interval precipitation data. The temporal resolution is 3 hours and the spatial resolution is 0.25° [6]. The RMM PR (Ku band) is the world's first space borne precipitation radar and its multiple years of operation have played a major role in improving the understanding of tropical and subtropical precipitation. In this paper, we employed the TMPA 3B42V7 post-real-time processing precipitation product (hereafter termed TMPA).

3) IMERG

The global precipitation measurement (GPM) is a TRMM post-precipitation plan. IMERG represents an improvement of the TRMM in terms of spatial coverage range, spatiotemporal resolution and rain and snow observation data. The GPM DPR (Ku and Ka bands) is the first space borne dual-frequency precipitation radar [53]. As a subsequent satellite of the TRMM, it covers a global range of 65°N–65°S. The Integrated Multi-satellite Retrievals for GPM (IMERG) level-3 product extends the coverage range to the north and south poles and provides a precipitation product with a global temporal resolution of 30 minutes and a spatial resolution of 0.1°. The greater coverage range, higher spatiotemporal resolution, more accurate capture capabilities for trace amounts of precipitation and solid precipitation and overall application accuracy of IMERG have compelled more and more researchers to rely on it. In this paper, we employed the ground station-calibrated non-real-time post-processing satellite precipitation product with an accuracy exceeding that of the quasi real-time/real-time satellite precipitation product, that is, the "final" run IMERG (IMERG-F) product.

3. Statistical Assessment Protocol and Markers

3.1. Statistical Assessment Protocol

CMPA-Hourly (China Hourly Merged Precipitation Analysis combining observations from automatic weather stations with CMORPH) was produced by the meteorological data laboratory of the China Meteorological Data Service Centre [54]. This product is based on the global 30-min and 8-km resolution Climate Prediction Centre MORPHing technique (CMORPH) satellite inversion precipitation product developed by the U.S. Climate Prediction Centre following quality control and correction of hourly precipitation data from 30,000 automated weather station observations in China. This precipitation fusion product effectively combines ground observation data from automated stations in China. Shen et al. validated that an hourly precipitation product with a resolution of 0.1° has an overall error level within 10% and an error level for heavy precipitation and regions with sparse stations within 20%. The fusion product can more accurately capture pronounced precipitation processes in typical regions. This product can provide important input parameters for precipitation in atmospheric and hydrological studies [55].

In the statistical analysis section of this study, the precipitation volume in the Jinsha River Basin from the hourly precipitation 0.1° grid data resulting from the fusion of China's automated stations and CMORPH was used as the observation data to calculate the error markers for TMPA, IMERG precipitation and CMADS precipitation for 910 grid precipitation observations in the Jinsha River Basin. The spatial resolution used for the precipitation calculation was 0.25°.

In this paper, 3 analyses were used to assess the precipitation accuracy of TMPA, IMERG and CMADS in the Jinsha River Basin.

1) Total analysis: Comparative analysis of the error markers of precipitation at all-time points for all 910 grid squares in the river basin, that is, total analysis at the river basin grid scale and the comparative analysis of error markers in precipitation at all-time points after the 910 grid values in the river basin were averaged, that is, total analysis at the average river basin grid scale.

2) Spatial analysis: Comparative analysis of the error markers of precipitation at all-time points for all grids in the river basin, that is, spatial analysis at the river basin grid scale.

3) Time series analysis: Comparative analysis of the error markers of precipitation for all grids at all-time points, that is, analysis of the temporal changes in precipitation in the river basin.

3.2. Assessment Markers

Several widely used assessment markers were selected to quantitatively evaluate the accuracy and error of three precipitation products, including accuracy assessment markers and classification markers. The accuracy assessment markers include the correlation coefficient (R), mean error (ME), relative bias (BIAS), root-mean-square error (RMSE), centred root-mean-squared error (CRMSE), standard deviation (SD) and Nash-Sutcliffe model coefficient of efficiency (NSE). The classification markers include the probability of detection (POD), false alarm ratio (FAR) and critical success index (CSI). Table 1 shows the calculation formulas and optimal values for the various markers.

Table 1. List of accuracy assessment markers.

Evaluation Index	Formula	Value
correlation coefficient (R)	$R = \dfrac{\sum_{i=1}^{n}(D_i - \overline{D})(D_i^* - \tilde{D})}{\sqrt{\sum_{i=1}^{n}(D_i - \overline{D})^2}\sqrt{\sum_{i=1}^{n}(D_i^* - \tilde{D})^2}}$	1
mean error (ME)	$ME = \dfrac{1}{N}\sum_{i=1}^{n} D_i^* - D_i$	0
relative bias (BIAS)	$BIAS = \dfrac{\frac{1}{N}\sum_{i=1}^{n}(D_i^* - D_i)}{\sum_{i=1}^{n} D_i} \times 100\%$	0
root-mean-square error (RMSE)	$RMSE = \sqrt{\dfrac{1}{N}\sum_{i=1}^{n}(D_i^* - D_i)^2}$	0
centred root-mean-squared error (CRMSE)	$CRMSE = \sqrt{\dfrac{1}{N}\sum_{i=1}^{n}\left[(D_i^* - \tilde{D}) - (D_i - \overline{D})\right]^2}$	
standard deviation (SD)	$\sigma/\sigma*$ $\sigma = \sqrt{\dfrac{1}{N}\sum_{i=1}^{n}(D_i - \overline{D})^2}$ $\sigma^* = \sqrt{\dfrac{1}{N}\sum_{i=1}^{n}\left(D_i^* - \tilde{D}\right)^2}$	0
Nash-Sutcliffe model coefficient of efficiency (NSE)	$NSCE = 1 - \dfrac{\sum_{i=1}^{n}(D_i^* - D_i)^2}{\sum_{i=1}^{n}(D_i - \overline{D_i})^2}$	1
probability of detection (POD)	$POD = \dfrac{n_{11}}{n_{11} + n_{01}}$	1
false alarm ratio (FAR)	$FAR = \dfrac{n_{10}}{n_{11} + n_{10}}$	0
critical success index (CSI)	$CSI = \dfrac{n_{11}}{n_{11} + n_{01} + n_{10}}$	1

In the accuracy assessment markers in Table 1, D_i is the observed precipitation data in mm, D_i^* is the test precipitation data in mm, $\overline{D}$ is the mean observed precipitation data in mm, $\tilde{D}$ is the mean test precipitation data in mm and N is sample size. R can be used to assess the consistency between the test precipitation data and the reference precipitation data. The closer R is to 1, the greater the consistency between the test precipitation data and the reference precipitation data. The closer the ME is to 0, the smaller the mean error between the test precipitation data and the observed precipitation data and thus the better the assessment results. The closer the BIAS is to 0, the lower the systematic error between the test precipitation data and the observed precipitation data. The closer the RMSE is to 0, the lower the bias between the test precipitation data and the observed precipitation data and the closer the test precipitation data is to the observed precipitation data. The closer the CRMSE is to 0, the closer the deviation of the test data is to the deviation of the observed data. The closer $\frac{\sigma}{\sigma^*}$ is to 0, the lower the dispersion of the measured precipitation data/observed precipitation dataset. As a hydrological simulation assessment marker, the closer the NSE is to 1, the greater the accuracy of the hydrological simulation.

Among the classification markers, n_{11} represents the precipitation events detected by both the test data and observation data, n_{01} represents precipitation events detected by the observation data but not detected by the test data and n_{10} represents precipitation events detected by the test data but not detected by the observation data. POD is the frequency that precipitation events are detected by both test and observation data and reflects the ability of the test data to identify precipitation events. The closer POD is to 1, the more optimal it is. FAR is the frequency of precipitation events that are detected in test data but not in observation data. The closer FAR is to 0, the more optimal it is. CSI is the probability of accurate judgment when there are precipitation events in either the test data or observation data. A CSI of 1 is the most optimal.

4. Statistical Analysis and Evaluation Results

4.1. Total Analysis

In this paper, the entire hydrological research cycle lasted from November 2014 through October 2016. Table 2 shows the error markers for precipitation.

Table 2. Error markers (the whole study period).

Evaluation Index	Grid 910			Basin (average of 910grids)		
	3B42	IMERG	CMADS	3B42	IMERG	CMADS
R-	0.42	0.71	0.6	0.81	0.88	0.86
RMSE (mm day^{-1})	5.42	3.82	5.18	1.39	1.19	1.17
CRMSE (mm day^{-1})	5.41	3.79	5.17	1.35	1.12	1.13
BIAS (%)	0.00	0.00	0.00	0.03	0.04	0.03
ME (mm day^{-1})	0.32	0.40	0.28	0.32	0.40	0.28
SD$_{GAUGE}$ (mm day^{-1})		4.54			1.91	
SD (mm day^{-1})	5.42	5.24	6.41	2.30	2.36	2.55
POD	0.49	0.70	0.77	0.83	0.91	0.92
FAR	0.58	0.46	0.35	0.32	0.26	0.19
CSI	0.29	0.44	0.54	0.60	0.69	0.76

From Table 2, we can see that in terms of the accuracy assessment markers, the errors after the 910 grid squares were averaged were smaller than the errors from the 910 grid squares in the river basin. This may be due to the fact that the river basin has a large area and averaging the grid squares will eliminate some positive and negative errors. Therefore, the average values of the river basin grid squares exhibited smaller error values. Since the BIAS and ME error values in the river basin grid and the river basin mean range are equivalent, there was no comparison made between the 2. However, the BIAS marker demonstrated satisfactory error calculation results for this river basin. This indicates that the systematic bias between the satellite precipitation data TMPA, IMERG and CMADS datasets and the observed data was low. In terms of classification markers, the mean index error of the river basin still exhibited better performance than the river basin grid scale. For the full cycle error, we referenced the study results of Tang et al. (2016) [56] for R, RMSE, CRMSE, POD, FAR and CSI and found these errors to be acceptable. From the error marker perspective, the 3 precipitation products as well as the observation precipitation product all demonstrated good overall consistency.

In the Jinsha River Basin, the dry season lasts from November through April of the following year, while the rainy season lasts from May through October. In order to better characterize the accuracy performance of the dry season and rainy season data, we divided the study cycle (November 2014–October 2016) into 4 hydrological cycles, with 2014–2015 as the first hydrological cycle (of which November 2014–April 2015 was the first dry season of the hydrological cycle and may 2015–October 2015 was the first rainy season) and 2015–2016 as the second hydrological cycle (of which November 2015–April 2016 was the second dry season of the hydrological cycle and may–October 2016 was the second rainy season). The hourly precipitation grid dataset obtained by the fusion of automated station and CMORPH data was used as observation data for comparative analysis of the corresponding error markers. Table 3 lists the error markers.

We further compared the precipitation data performance of the satellite precipitation data during the 2014–2016 cycle for the full 2-year cycle, first dry season, first rainy season, second dry season and second rainy season. Figure 3 shows the unary linear relationship scatterplot for the TMPA, IMERG, CMADS and observed precipitation data. By combining Figure 3 with Tables 2 and 3, we can see that the fitting line of data (a3,b3 and c3) in the unitary linear scatter plot of flood season is closer to the standard line than the dry season fitting line (a4,b4 and c4),So the data performance for the dry season was weaker than that for the rainy season. Overall, the Jinsha River Basin experiences mainly light rainfall. Moderate to heavy precipitation generally occurs over the middle and lower reaches of the river basin, while the area in which very heavy rain occurs is even smaller [40]. Since the three precipitation products were based on satellite detection or corrections based on satellite detection data, the detection errors for light rain by high-altitude satellites were more affected by other factors. Therefore, the detection efficiency was lower for the dry season than the rainy season since the former is characterized by lighter precipitation. The linear relationship of the full 2-year cycle of precipitation by CMADS and IMERG with the observed precipitation was better than that of the TMPA. The TMPA precipitation product exhibited poor performance in the full cycle and its linear relationship was weaker than that of IMERG and CMADS. One reason for this is due to the fact that the study period experienced instability in the detected precipitation during the 2 dry seasons and some bias occurred when the precipitation was high during the rainy season. Another reason was that the TMPA was calibrated using monthly rain gauges that were relatively sparse, which impacted its overall performance to some extent. IMERG showed improved performances in both dry and rainy seasons compared with the previous generation of TMPA products. Longitudinally, the relative bias of the first dry season was greater than the rainy seasons and the second dry season. The overall performance of CMADS was close to that of IMERG and was relatively good. The bias in the first dry season was greater than that of the other seasons during the study period. The low POD for precipitation during the dry season may stem from the fact that the ability of the satellite to detect solid precipitation is limited by its physical components, thus weakening its detection capability.

Figure 4 shows the box plot of the different error markers. Comparative analysis of the distribution status of various precipitation error markers can be carried out by combining Figure 4 with Tables 2 and 3. Taylor diagrams were plotted by combining SD_{gauge}, SD, R, RMSE and CRMSE (Figure 5). A comparison of the performance of different data at the river basin grid scale and the average river basin grid scale was made by combining Figure 5 with Tables 2 and 3.

Table 3. Error markers (markers related to the dry season and rainy season).

Evaluation Index		Grid 910			Basin (average of 910grids)		
		3B42	IMERG	CMADS	3B42	IMERG	CMADS
R	first dry season	0.43	0.74	0.18	0.64	0.89	0.32
	first rainy season	0.40	0.68	0.74	0.76	0.85	0.89
	second dry season	0.27	0.61	0.82	0.54	0.85	0.95
	second rainy season	0.38	0.70	0.80	0.72	0.82	0.93
RMSE (mm day^{-1})	first dry season	1.83	1.28	8..34	0.55	0.32	1.77
	first rainy season	7.46	5.49	4.58	1.91	1.69	1.12
	second dry season	1.96	1.28	0.90	0.53	0.30	0.15
	second rainy season	7.35	4.95	4.02	1.85	1.62	1.01
CRMSE (mm day^{-1})	first dry season	1.83	1.28	8.34	0.55	0.31	1.76
	first rainy season	7.43	5.43	4.56	1.79	1.48	1.05
	second dry season	1.96	1.28	0.90.	0.53	0.30	0.14
	second rainy season	7.32	4.89	4.00	1.76	1.41	0.90
BIAS (%)	first dry season	0.00	0.00	0.00	-0.04	-0.06	0.33
	first rainy season	0.00	0.00	0.00	0.15	0.19	0.09
	second dry season	0.00	0.00	0.00	0.09	0.08	0.10
	second rainy season	0.00	0.00	0.00	0.13	0.17	0.10
ME (mm day^{-1})	first dry season	−0.03	−0.04	0.23	−0.03	−0.04	0.23
	first rainy season	0.66	0.81	0.39	0.66	0.81	0.39
	second dry season	0.05	0.04	0.06	0.05	0.04	0.06
	second rainy season	0.57	0.79	0.46	0.57	0.79	0.46
SD$_{GAUGE}$ (mm day^{-1})	first dry season		1.65			0.58	
	first rainy season		6.10			2.12	
	second dry season		1.38			0.42	
	second rainy season		5.99			2.31	
SD (mm day^{-1})	first dry season	1.79	1.85	8.48	0.69	0.68	1.86
	first rainy season	7.35	7.19	6.44	2.74	2.76	2.32
	second dry season	1.81	1.50	1.55	0.62	0.55	0.45
	second rainy season	7.03	6.62	6.59	2.38	2.39	2.39
POD	first dry season	0.25	0.40	0.64	0.40	0.80	1.00
	first rainy season	0.51	0.72	0.75	0.83	0.91	0.95
	second dry season	0.23	0.43	0.69	0.00	0.00	1.00
	second rainy season	0.54	0.78	0.82	0.86	0.93	0.95
FAR	first dry season	0.75	0.52	0.49	0.33	0.00	0.38
	first rainy season	0.57	0.48	0.37	0.36	0.26	0.16
	second dry season	0.80	0.61	0.42	1.00	1.00	0.00
	second rainy season	0.53	0.43	0.30	0.25	0.25	0.16
CSI	first dry season	0.14	0.28	0.40	0.33	0.80	0.63
	first rainy season	0.30	0.43	0.53	0.57	0.69	0.80
	second dry season	0.12	0.26	0.46	0.00	0.00	1.00
	second rainy season	0.34	0.49	0.61	0.67	0.70	0.80

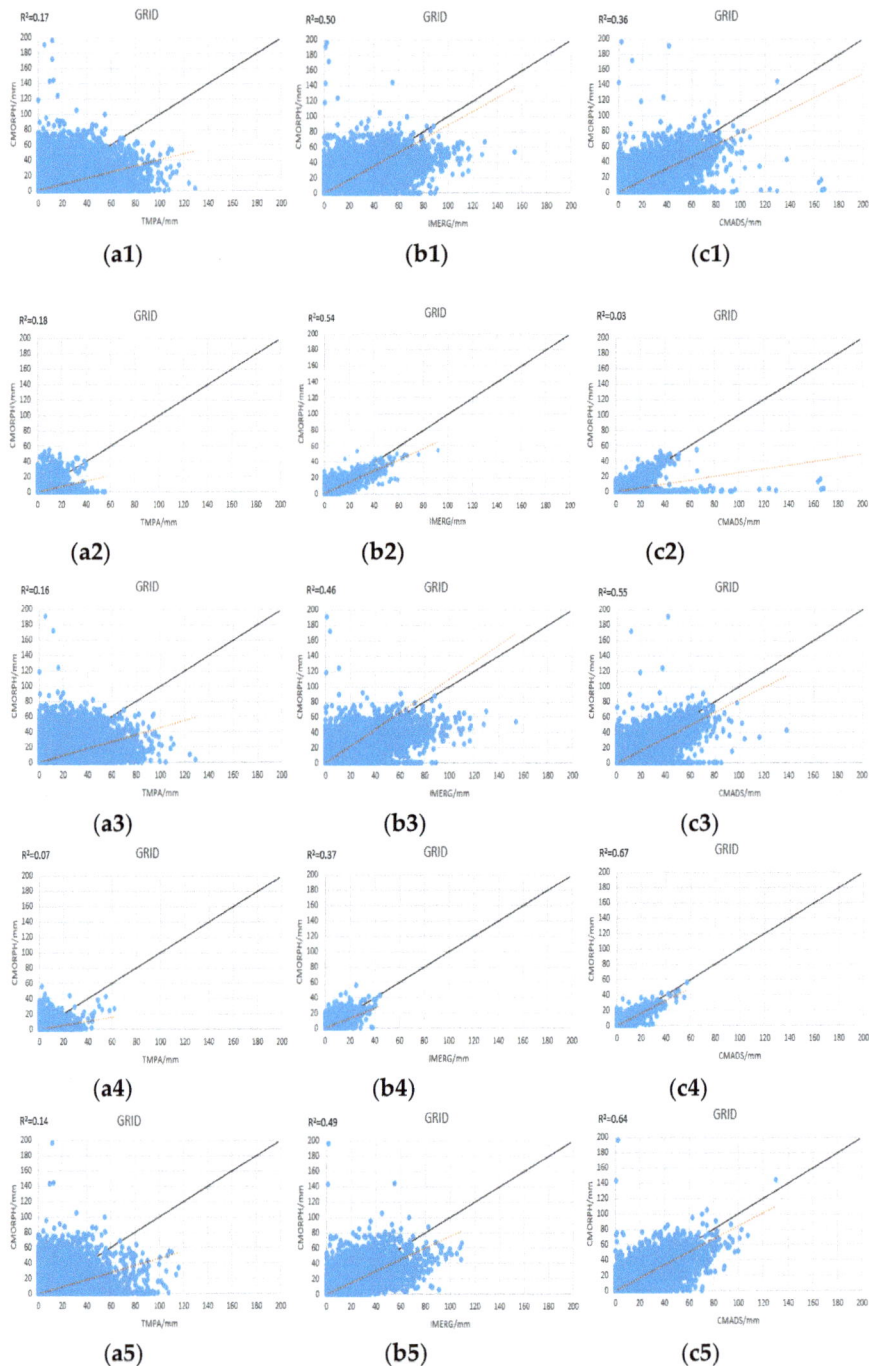

Figure 3. Unary linear scatterplot (**a1**–**a5**), (**b1**–**b5**) and (**c1**–**c5**) represent the full 2014–2016 cycle, first dry season, first rainy season, second dry season and second rainy season for the TMPA, IMERG and CMADS precipitation products.

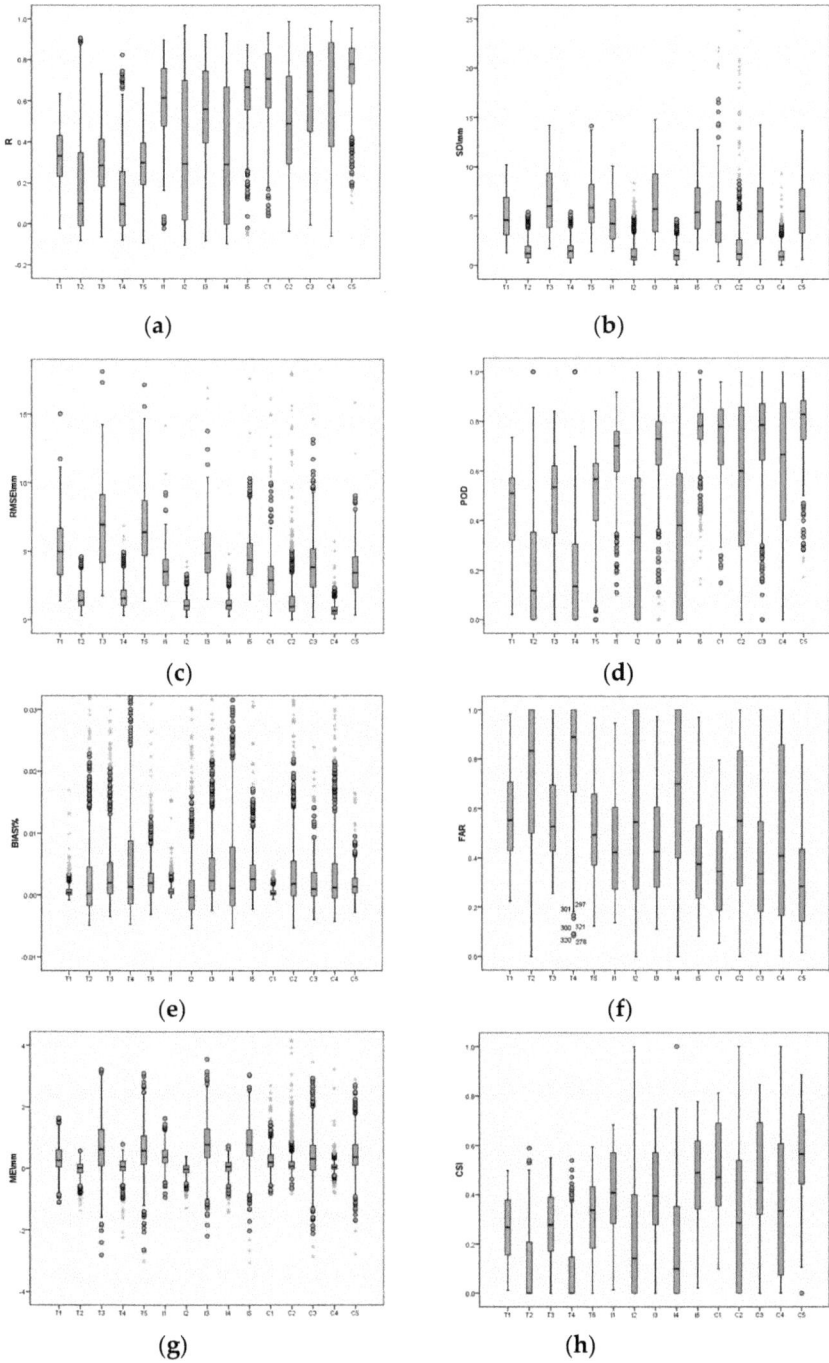

Figure 4. Box plots of assessment markers: T1–T5, I1–I5 and C1–C5 show T (TMPA), I (IMERG) and C (CMADS) in the full 2014–2016 cycle, first dry season, first rainy season, second dry season and second dry season.

If the precipitation product has an R that approaches 1 and SD, RMSE, BIAS and ME that approach 0, this indicates that the error of its precipitation data is small. If the precipitation product has a POD and CSI that approach 1 and a FAR that approaches 0, this indicates that the precipitation data have good detection capabilities for precipitation and non-precipitation events. The solid black lines in the box plot represent the median. The 4 horizontal lines from top to bottom represent the maximum, third quartile, first quartile and minimum. The symbol o represents discrete values and * represents extreme values. From the box plots of the different markers, we can see the median error and the deviation of the integrated error from the median error. From Figure 4d,f,h, we can see that the 3 markers in CMADS are better than those of the TMPA and IMERG precipitation products when the PODs for precipitation events are compared. The frequency of actual precipitation events that were detected by CMADS was optimal. The POD and CSI of the CMADS data were closer to 1 and its FAR was closer to 0, which is optimal. The comparison of the different accuracy assessment markers can be seen in the Taylor diagrams in Figure 5. From the full cycles in Figure 5a–d, we can see that the averaged error markers for the river basin grid were better than those of the river basin scale. The average river basin scale exhibited smaller SD errors. The CMADS of the grid scale and average grid scale for the first dry season all showed lower correlation coefficients and achieved better performances in the second dry season and first rainy season. IMERG is an upgraded precipitation product of the TMPA. From the few comparison cycles, we can see that the errors of IMERG were slightly smaller than those of the TMPA. Using the grid scale and average river basin scale of the first rainy season as an example, we can see that at the grid scale (Figure 5e), $R_{cmads} > R_{imerg} > R_{tmpa}$—that is, the correlation coefficient for CMADS was the greatest and the correlation coefficient of TMPA was the smallest, with that of IMERG falling in between. In addition, $RMSE_{cmads} < RMSE_{imerg} < RMSE_{tmpa}$—that is, the RMSE was the smallest for CMADS, followed by IMERG. The average river basin scale (Figure 5f) produced identical error values as the river basin grid scale, although the mean error was smaller and the correlation coefficient was greater.

In summary, the overall errors for the TMPA, IMERG and CMADS in the river basin in this study were found to be acceptable. These datasets showed better performances at the average grid scale. CMADS demonstrated better performance for the POD marker. The overall error performance of the IMERG precipitation data was better than that of the TMPA precipitation product. CMADS and IMERG exhibited their own strengths and weaknesses in the performances of different markers. The subsequent text will further compare and analyse these 3 types of precipitation data in driving hydrological simulation and actual measured runoff data will be used to further test the performance of the precipitation data.

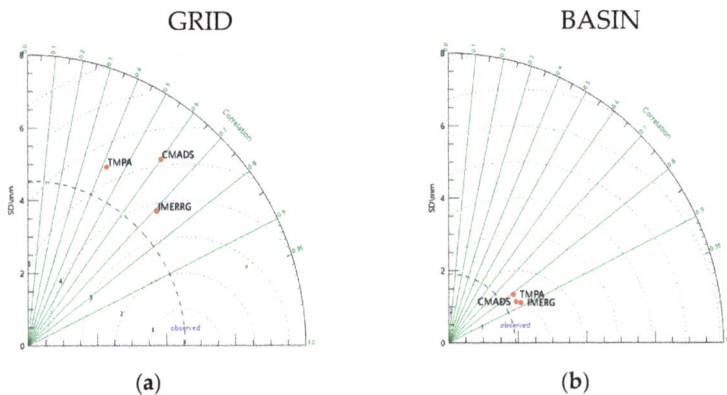

(a) (b)

Figure 5. *Cont.*

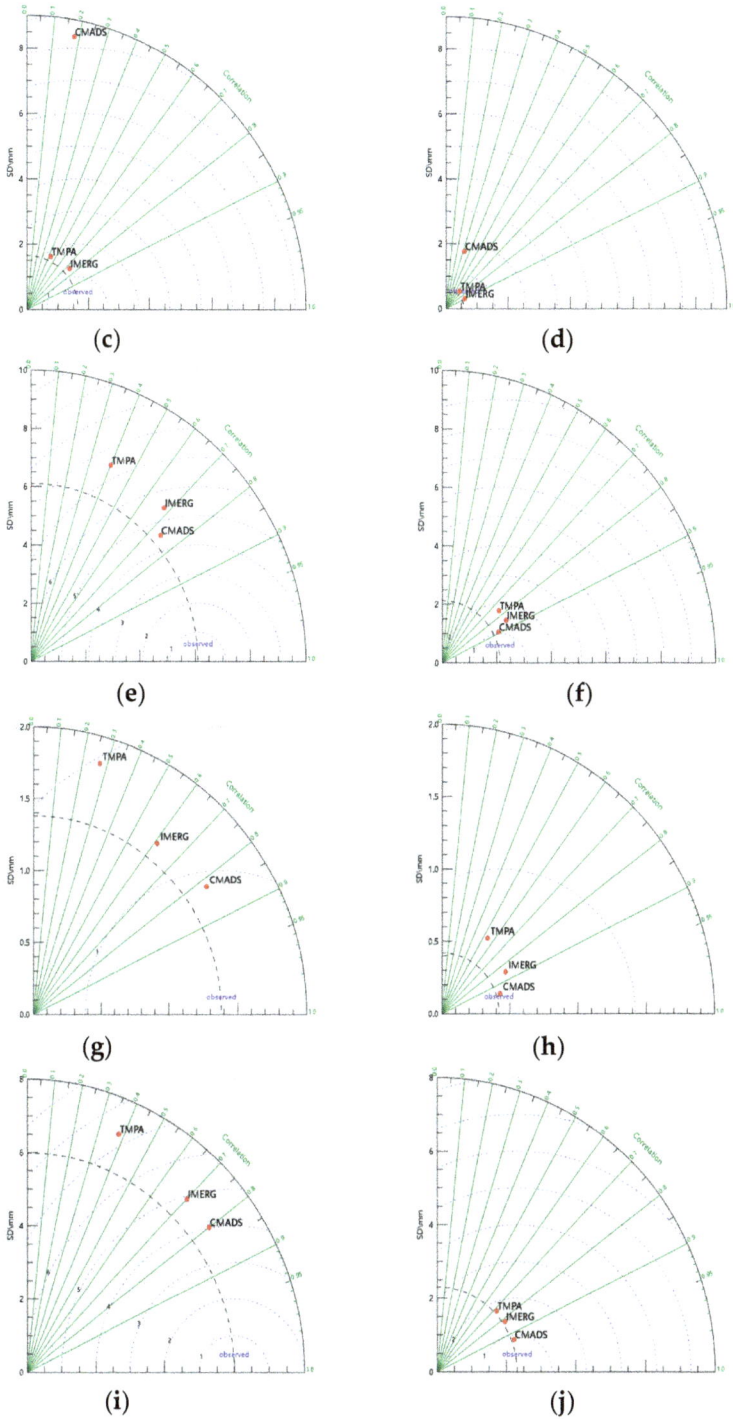

Figure 5. Accuracy assessment Taylor diagrams for satellite precipitation products: (**a**,**b**) full 2014–2016 cycle; (**c**,**d**) first dry season; (**e**,**f**) first rainy season; (**g**,**h**) second dry season; (**i**,**j**) second rainy season.

4.2. Spatial Analysis

The 910 grid squares were used as study objects and the 2014–2016 study period was taken as the entire cycle when analysing the spatial distributions of relevant precipitation markers at the river basin grid scale. Figure 6 shows the spatial distribution status.

From 6a1–c1, we can intuitively see the performances of precipitation-related markers over the entire study period. The spatial distribution of precipitation in the Jinsha River Basin is not uniform. This is because the northwest plateau blocks moist air flow, resulting in less precipitation. In the southeast, the effects of precipitation air flow and oceanic monsoon winds causes increased precipitation. The middle of the basin is a transition zone between these 2 regimes. All 3 precipitation data products demonstrated better performance in the middle and lower reaches of the basin, areas with abundant precipitation. Due to the effects of the plateau westerly winds, the detection correlation of the precipitation data was relatively weak for the comparatively dry middle and upper reaches of the river basin. From the correlation coefficients of the 3 different precipitation products in Figure 6a1–c1, we can see that the correlation coefficients for the middle and lower reaches of the basin were greater than those of the upper reaches and the precipitation data from the middle and lower reaches exhibited better performance(The more blue grids, the better the correlation coefficient). In comparison, CMADS showed more grid correlation coefficients that were close to 1 in the upper and middle reaches. The spatial distribution maps of PODs (Figure 6a4–c4) proved once again that CMADS displayed relatively high POD for precipitation and its POD and CSI for precipitation in the middle and lower reaches were significantly higher than those of the upstream plateau region, while FAR was relatively low. From Figure 6a2–c2, we can see that the RMSE markers for IMERG and CMADS generally fell within the range of 10 mm, while there was 1 downstream site in the TMPA precipitation product that was greater than 10 mm. The MEs (Figure 6a3–c3) of the 3 products all displayed good performance. These values in the grids were generally within the 0.5 range. In addition, the ME error markers for the 3 different precipitation products all tended towards 0. This indicates that when either TMPA, IMERG or CMADS were used in the study site, both the mean error and the observed precipitation error were very small. Thus, the precipitation input can be applied to relevant studies in this river basin within a certain range. From the comparison data, we can see that the overall performances of the spatial distributions of these 3 products were consistent and the error markers and PODs for the plateau region in the upper reaches were all lower than those of the basins and plains in the middle and lower reaches. The overall correlation coefficients showed better performance in the middle and lower reaches than in the upper reaches. The overall mean errors were all within 0.5. IMERG showed slightly better POD, FAR and CSI for detection of precipitation events than TMPA and slightly poorer POD than CMADS.

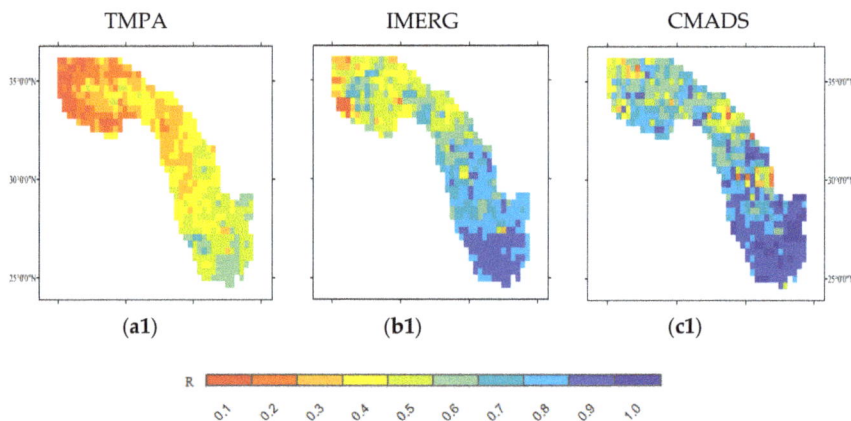

Figure 6. *Cont.*

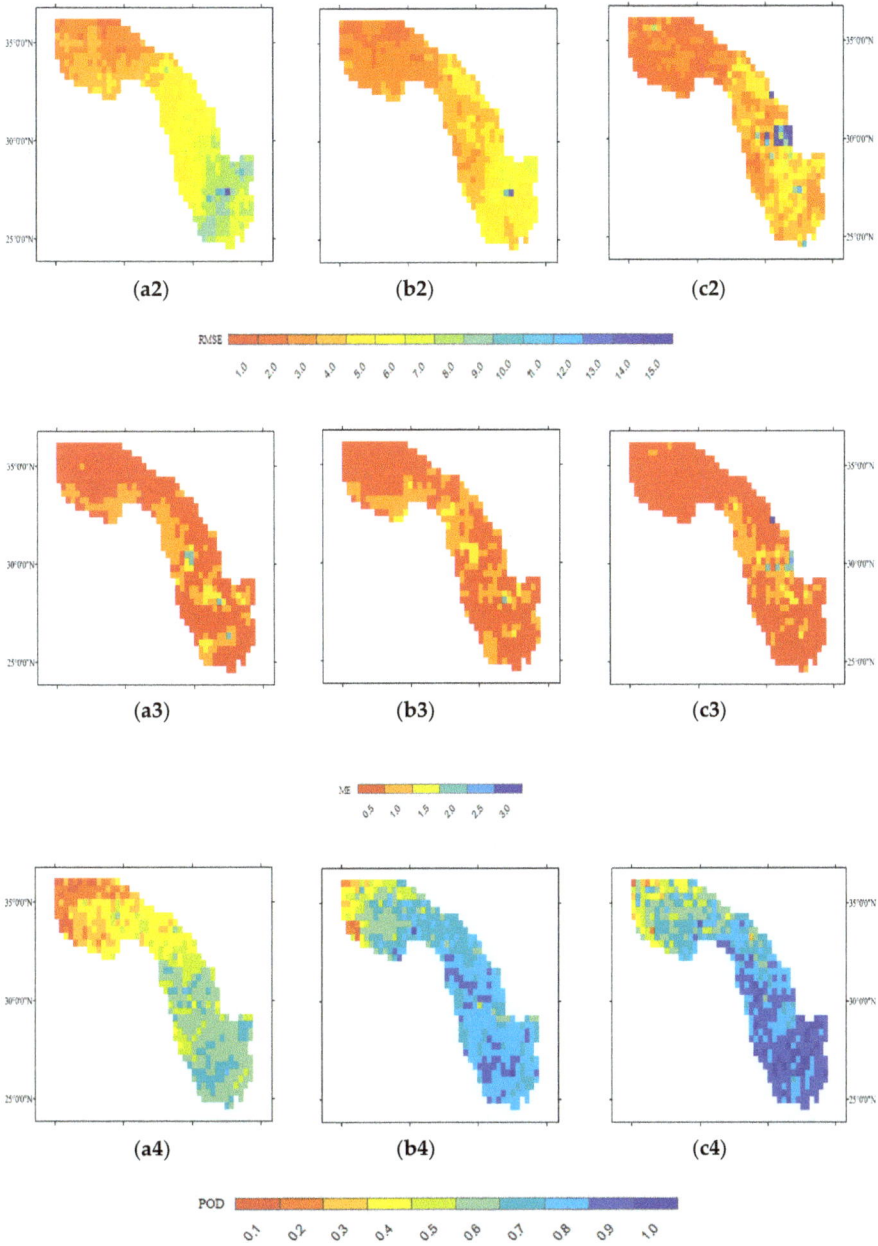

Figure 6. *Cont.*

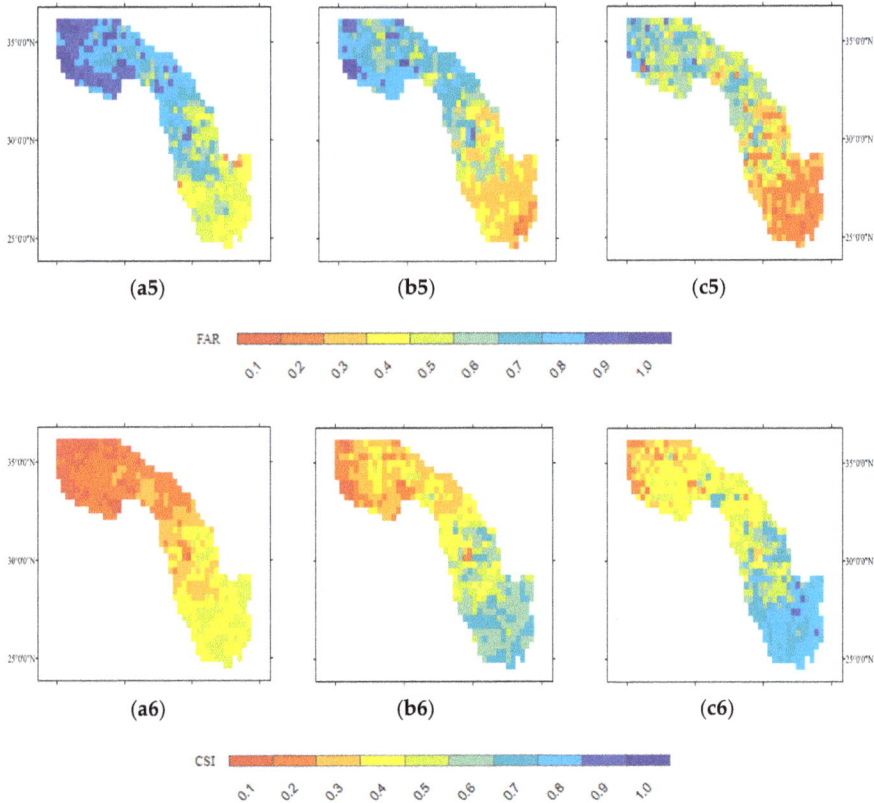

Figure 6. Spatial distribution of precipitation error markers in the Jinsha River Basin: (**a1,b1,c1**) R; (**a2,b2,c2**) RMSE; (**a3,b3,c3**) ME; (**a4,b4,c4**) POD; (**a5,b5,c5**) FAR; (**a6,b6,c6**) CSI.

4.3. Temporal Analysis

Analysis of time series distribution maps:

The 910 grids squares were used as study objects for the analysis calculations in order to obtain the changes of error markers over the full cycle of the 2014–2016 study period (Figure 7).

From Figure 7a, we can see that the correlation coefficients (Rs) of the 3 different precipitation products all tended to approach 1 during the 2 rainy seasons and their RMSE markers exhibited dry-rainy season cyclical changes. The ME showed fluctuation errors around the optimal error value line of 0. The fluctuations of the mean errors for the entire observation data were all within 10 mm.

The correlation coefficients for CMADS and IMERG were better than those of TMPA the majority of the time. In July 2015, the 3 products all had lower correlation coefficient values. This was the month there were more missing values when processing the MERG data of the satellite precipitation data since the TRMM satellite had just entered the atmosphere in June 2015. The subsequent substitute fusion precipitation product, TMPA, may show signs of instability in the first month. During the November 2015–April 2016 dry season, CMADS had a better correlation coefficient than either IMERG or TMPA. As the rainfall station data in China were corrected over the entire coverage range of CMADS, the correction results of the precipitation data from the Jinsha River Basin during this period improved. The mean annual precipitation totals for the rainy season and dry season of the Jinsha River Basin range from 244–886 mm and 12–130 mm, respectively [57]. Since the Jinsha River Basin is characterized by a pronounced vertical gradient, the annual mean precipitation range varies significantly with elevation and climate conditions. The RMSE marker demonstrated significant periodicity during the validation period. Although the deviation of the RMSE marker was greater during the rainy season and its maximum controllable range was 20 mm, comparison of the RMSE of 20 mm with the maximum precipitation of 886 mm during the same period indicated that its relative effects on deviation was not prominent. The precipitation totals detected by TMPA and IMERG from November 2014–April 2015 were less than those detected by CMADS, which is particularly apparent in Figure 7a. Winter precipitation is relatively scarce and plateau temperatures are relatively low, with a mean annual air temperature less than 0 °C. Subsequently, there will be some solid precipitation and plateau radiation that will test the detection capabilities of the satellite. CMADS has sparsely distributed ground station data available for correction, which provides an advantage in the form of relatively small errors.

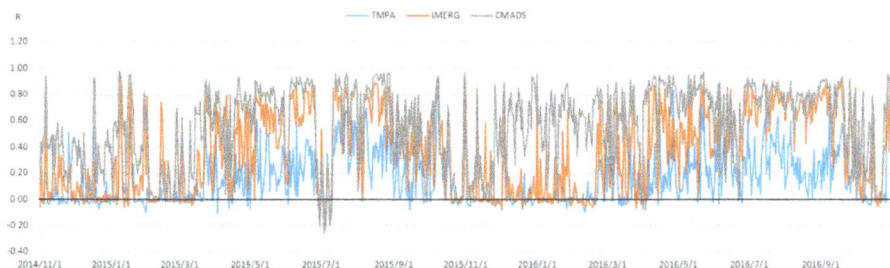

(a)

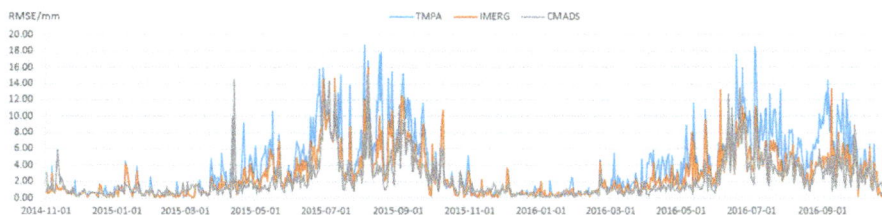

(b)

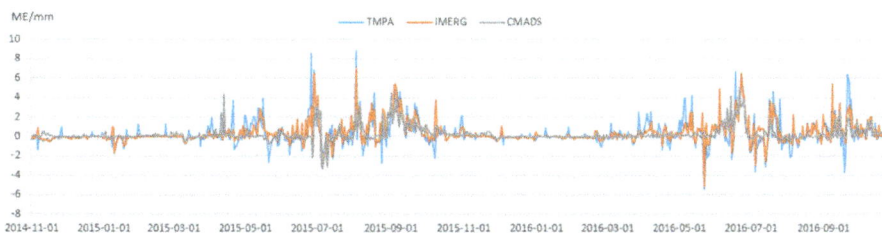

(c)

Figure 7. Time series graph of error markers: (**a**) R; (**b**) RMSE; (**c**) ME.

5. Suitability Assessment for Precipitation Products in the River Basin

5.1. Construction and Calibration of River Basin Hydrological Models

In order to further validate the accuracy and suitability of precipitation data in the study region, we input the precipitation datasets into hydrological models and drove model operation to simulate the variation trends of cross-sectional runoff at the river basin outlet. The model output was then compared with actual runoff data. Comparative analysis was used to validate the results when different precipitation inputs were used to drive runoff at the river basin outlet and to provide theoretical support for applications in water resources and water environment management in this river basin. In this paper, the SWAT distributed hydrological model was used for runoff simulation. This model has demonstrated good simulation results and accuracy for many river basins and its construction is relatively mature.

The Soil and Water Assessment Tool (SWAT) model was developed in 1994 by Dr. Jeff Arnold for the United States Department of Agriculture (USDA) Agriculture Research Service. This is a GIS-based distributed river basin hydrological model. In recent years, the SWAT model has undergone rapid development and application [58]. The physics-based SWAT distributed hydrological model takes into consideration meteorological factors, land use, soil and other elements and divides an entire river basin into different sub-basins. Each sub-basin has different current soil and land use statuses. The model can be used to simulate hydrological processes, soil erosion, chemical processes and biomass changes in a river basin [59]. Its user-friendly interface, open-source program and high simulation accuracy have enabled it to be used in river basin water quality and water quantity simulations worldwide.

5.2. SWAT Hydrological Model Data Simulation and Calibration

In order to accurately simulate the precipitation-runoff process in the Jinsha River Basin and to allow for the effects of terrain in the river basin on surface runoff, the model first extracted the catchment area, elevation, gradient and other information based on digital elevation model (DEM) data. Next, the soil type, land use type and gradient levels in the sub-basin were determined. The sub-basin was further divided into hydrologic response units (HRUs). After this was completed, the compiled meteorological data (including precipitation, temperature, wind speed, air pressure, solar radiation and so on) were uploaded into the model. Following that, the runoff volume at the outlet of this sub-basin was added. SWAT Calibration and Uncertainty Procedures (SWAT-CUP) were used for parameter sensitivity analysis and SUF12 was used to calibrate the accuracy of model parameters. The parameters selected for sensitivity analysis were CN2, ALPHA_BF, GW_DELAY, GE_REAVP, ESCO, CH_N2, CH_K2, ALPHA_BNK, SOL_AWC, SOL_K, SOL_BD and SFTMP.

In this paper, 2014–2016 was the study period. The Pingshan hydrological station was used as the river basin control outlet. The 3 different precipitation products were used as precipitation input to simulate the daily mean runoff volume of this river basin. Based on CMADS datasets, the simulation period was used for parameter calibration and data fitting. During the validation period, we compared and analysed the performances of these 3 types of precipitation data in driving runoff and further validated the suitability of different precipitation data for the study site of this paper.

5.3. Hydrological Model Simulation Results and Analysis

In this study, the runoff simulation data during the 2014–2016 validation cycle was used for error analysis. BIAS and NSE were used as error markers to assess the runoff simulation results and to validate the runoff simulation results. Table 4 shows the error markers for the full cycle of the study site. Figure 8 is the comparison chart for the simulated runoff and actual runoff in the full cycle of the study site.

Table 4. Comparison of hydrological model simulation accuracy.

Time	Evaluation Index	Precipitation Product		
		TMPA	IMERG	CMADS
November 2014–December 2016	BIAS (%)	−0.02	−0.02	−0.03
	NSE	0.39	0.35	0.53
May 2015–December 2016	BIAS (%)	−0.01	0.00	−0.03
	NSE	0.50	0.45	0.55

(a)

(b)

(c)

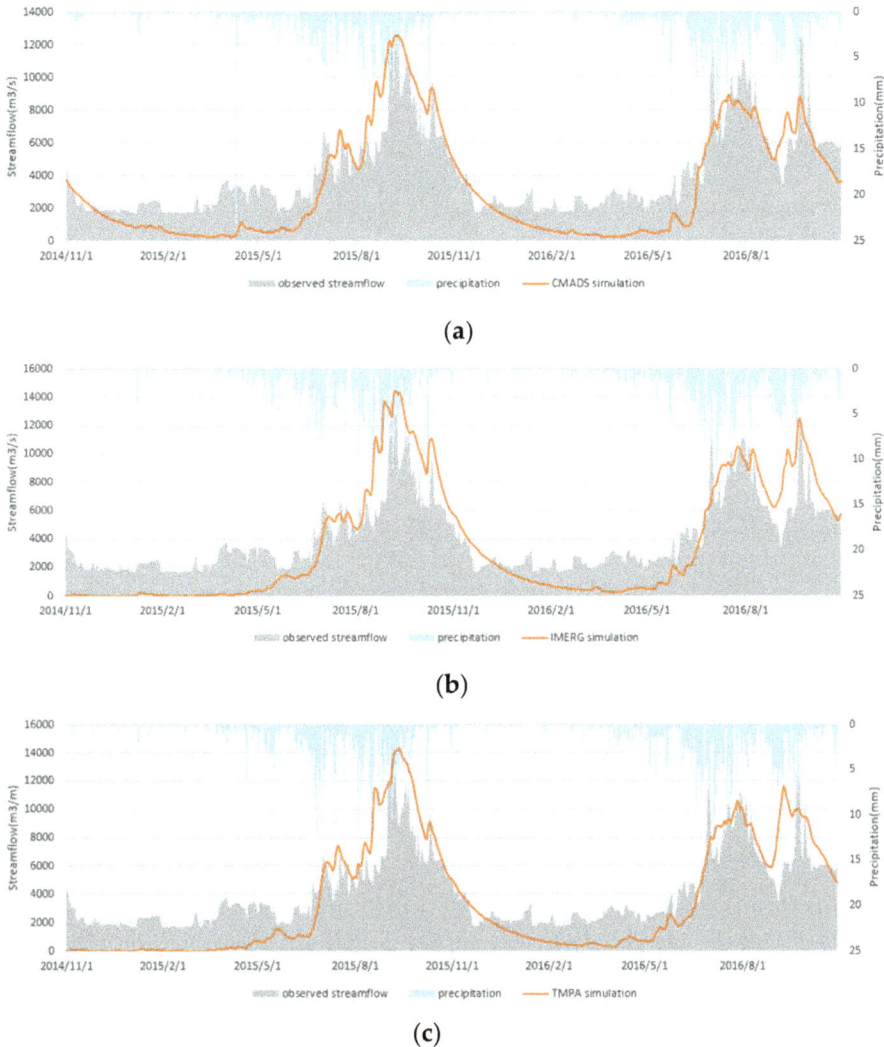

Figure 8. Comparison of runoff results from hydrological model simulation of the Jinsha River Basin from November 2014–October 2016: (**a**) TMPA; (**b**) IMERG; (**c**) CMADS.

By combining Figure 8 with Table 4, we can see from the runoff volume calculation results and runoff simulation error analysis at all timepoints in the validation cycle for the SWAT model simulation that good overall variation trends, as well as timing for trough values, peak values and flood peaks for runoff, can be simulated using the 3 products as model precipitation input. IMERG had a relatively large bias in the first dry season. This indicates that the first dry season from November 2014–April 2015 did not capture the actual runoff. The TMPA product showed a bias that was similar in magnitude to that of IMERG during the first dry season. Both the TMPA and IMERG (which is a subsequently improved version of the TMPA) precipitation products exhibited unstable performances in data quality during the study period and did not capture maximum flow at the same time. This may be due to the presence of inherent defects in the sensors of the TRMM and GPM satellites used to detect trace precipitation. The detection capabilities of these sensors require further improvements. The CMADS-driven simulated runoff captured the flow volume during this period although there was still considerable bias. On the one hand, this bias may be associated with the generation of the CMADS data, which was based on raw satellite data during the cold and dry winter months and may contain some satellite detection biases. These biases can be confirmed by the performances of the aforementioned TMPA and IMERG data. On the other hand, this bias may be due to the low total precipitation volume in the river basin during the dry season when the sensitivity of the river basin outlet simulation results to precipitation accumulation at the upper reaches was magnified. However, the ground stations in the plateau at the upper reaches of the basin that are used to correct the precipitation data are sparsely distributed. This in itself results in poor data quality. Even though there was considerable bias, the use of CMADS during this time period as precipitation input still captured the runoff volume to some extent and exhibited the best performance among the 3 products. In September 2015, the 3 different precipitation products all successfully captured the flood peak volume. The good flood peak capture performance of the 3 products during the first rainy season proved that the SWAT model has good runoff simulation accuracy for the Jinsha River Basin during the rainy season. This is particularly true for the capture of flood peak volume. IMERG did successfully capture the 2 flood peak volumes in September 2015 and October 2015. During these 2 periods, the performance of TMPA was better than that of IMERG. The relatively good hydrological performances of TMPA and IMERG proved that both of these products have application potential for future short-term flood forecasting for this river basin. During this study cycle, the NSE of CMADS was closer to 1, showing that although the capture abilities of TMPA and IMERG for some flood peaks were better than CMADS, the CMADS demonstrated better overall hydrological simulation accuracy during the entire cycle. The lowest biases for the hydrological simulations of the 3 products were TMPA: -0.02%, IMERG: -0.02% and CMADS: -0.03%. This further verified that satellite precipitation fusion products produce better simulation and detection results when precipitation is abundant.

Considering that there is no transition of parameters and data in the model at the beginning of the verification period, which may affect the accuracy of simulation. The study also took the dry seasons from November 2014 to April 2015 as the transition time of the model. Then we compared the prediction performance of runoff with the same parameters from May 2015 to October 2016. From the error markers, we can see that the relative biases for runoff simulations of the study site by the 3 different precipitation products over the entire validation period were relatively close. Excluding the simulation adaptation process during the transition period, IMERG and TMPA precipitation products were also able to approximate the simulation results of CMADS during the study period. In this article, the Nash coefficients of transitional runoff simulation are excluded to be 0.50 (TMPA), 0.45 (IMERG) and 0.55 (CMADS), respectively. According to the simulation results of Hydrological Information Forecasting Specification, it is in C program. The Nash efficiency coefficient does not reach the A and B schemes. The large reason for that is the large watershed area in the survey area (340,000 km^2) and the amount of runoff change captured in the outlet section of the basin has certain simulation accuracy limits of the global parameters. However, refer to the relevant large watershed parameters, according to He [60] and the research results of runoff Nash coefficient (0.43 (gauge), 0.49 (tmpa), 0.53 (imerg)) in the larger watershed upstream of the Lancang River Basin (150,000 km^2). The hydrological simulation accuracy of the basin is still acceptable in this paper.

6. Discussion

Precipitation data studies have demonstrated that the generation of the TMPA precipitation data has provided a new avenue for hydrological and meteorological researchers to obtain accurate precipitation data. Researchers can download grid precipitation data for their study site free of charge and are no longer limited to the data acquisition from national hydrological and meteorological stations. This promotes hydrological and meteorological research in different research domains. The usage accuracy of TMPA in many regions and river basins has been verified by relevant studies. Assessments of the hydrological performance of IMERG, the subsequent precipitation product of

TMPA, are currently underway. Therefore, in this paper, analysis in combination with observation data determined that the accuracy of the IMERG precipitation data in a river basin having a large area and complex terrain (in this case the 340,000 km^2 Jinsha River Basin) can generally satisfy the hydrological simulation requirements and demonstrated improvements compared with TMPA in certain areas, as well as good accuracy for rainy seasons. The free CMADS dataset that was compiled by Chinese researchers also demonstrated good usage simulation results for the Jinsha River Basin.

For the November 2014–October 2016 time frame, we can see that the runoff simulation during the validation period produced good results, with relative biases of −0.02%, −0.02% and −0.03% for the TMPA, IMERG and CMADS products, respectively. This demonstrated the operability of the models. The SWAT model was used for runoff simulation prediction for this river basin, while the satellite precipitation data and the CMADS precipitation dataset were used to drive the SWAT model for hydrological simulation, which demonstrated good suitability for this river basin. Our simulation results indicated that when the CMADS, which was constructed by Dr. Meng for East Asia, was used as the meteorological data input to drive the SWAT model, the variation patterns and accuracy of runoff for the cross-section of the river basin outlet were both feasible, thus providing a theoretical reference for future research.

The Jinsha River Basin outlet—Pingshan hydrological station—is located at the stone ladder in Jinping Village, Pingshan County, Sichuan Province (28 km upstream from the Xiangjiaba Dam). The river segment that is measured by the Pingshan station is straight; there are 500 m upstream and 2000 m downstream of the measured segment which are bends in the river. There are no major tributaries that flow into the river near its upper or lower reaches. The station was built in August 1937. Due to the reservoir and hydropower generation of the Xiangjiaba Dam, the Pingshan station was changed to a hydrological station on 20 June 2012. The Xiangjiaba Dam is located at the Lianhuachi Campsite, town of Anbian, Yibing County, Sichuan Province. The test section is located 2 km downstream of the Xiangjiaba hydropower station, was completed in June 2008 and belongs to the Pingshan station (which is located 30 km upstream of the Xiangjiaba hydrological station). The Yangtze River Hydrological Bureau manages the operation of the dam. In 2012, the Pingshan station was removed and replaced by the Xiangjiaba hydrological station. Given this change, there will be errors in cross-sectional measurements for flow volume at the outlet of the Jinsha River Basin in long-series runoff simulations and the relevant effects can be verified by researchers in the future.

7. Summary and Conclusions

In this study, the Jinsha River Basin was used as an example for quantitative analysis and assessment of the general accuracy and hydrological accuracy of three precipitation products: TMPA, IMERG and CMADS. The following paragraphs are a summary of this paper. The major study conclusions are as follows:

(1) At the river basin grid scale, the three different precipitation products all yielded small relative errors. The next-generation satellite precipitation product IMERG demonstrated performance comparable to that of the previous generation TMPA satellite precipitation product. The overall RMSE for IMERG was only 3.82 mm /day, indicating that the data had small precipitation errors. On the other hand, the mean error for CMADS was only 0.28 mm /day. For the average river basin grid scale, the overall assessment markers were better than those of the river basin grid scale. Since the Jinsha River Basin has a large area with numerous grid squares, the average river basin scale will eliminate some positive and negative grid biases. This is also one of the reasons the average river basin scale exhibited better error performance.

(2) In terms of the detection capabilities for precipitation events, the three types of precipitation data all demonstrated better performance at the middle and lower reaches of the Jinsha River Basin than the upper reaches. The upper reaches of the Jinsha River are in the Tibetan Plateau and the differences between the upper and lower reaches are great due to the effects of melting glaciers. The relevant precipitation data may cause inferior relative error performance for solid precipitation

detection capabilities. For the precipitation detection capability markers, we found that CMADS > IMERG > TMPA. This proves that IMERG has better precipitation detection capabilities than the previous generation satellite precipitation product, TMPA. The CMADS precipitation dataset that was designed for the East Asian region was corrected using ground precipitation measurement data. Therefore, it now displays outstanding performance in terms of precipitation detection.

(3) From the hydrological simulation results, we found that the three open-source free precipitation products could all accurately detect the flood peak volume during the first rainy season. Although there were varying degrees of overestimation, the overall trends were good. TMPA and IMERG showed good capture performance for flood peak volume during the two rainy seasons. However, the performance of TMPA and IMERG in precipitation POD for the first dry season was relatively poor, when no flow was detected at all. During data processing, this portion of the data had more missing values, which affected the accuracy of hydrological simulations. We analysed the reasons for this and determined that the detection capabilities by satellite precipitation data for trace amounts of precipitation during the dry season are still weak and the corresponding fusion product is still affected by the sparse distribution of ground rainfall observation stations. In the future, improved processing methods may become available, resulting in more accurate detection of small precipitation events. CMADS data successfully captured maximum flow values in the runoff simulation for the first dry season and its performance in this regard was better than that of either IMERG or TMPA. In addition, the CMADS dataset contains precipitation, atmospheric temperature, air pressure, humidity and other data. This not only makes data usage convenient but also yielded good results for the Jinsha River Basin in this study.

(4) In this paper, SWAT model simulation and sensitivity analysis of measured runoff data from the Pingshan station in 2014–2016 was applied during the study period, leading to good hydrological simulation results. The relative NSE for the full validation cycle reaches 0.55. The issue of numerous missing precipitation data values during the first dry season was the primary factor affecting the SWAT model's simulation results and caused the bias for the first dry season to be larger, affecting the simulation accuracy for overall runoff. However, performance for the rainy season was good. In the future, researchers should continue attempting to validate the effects and errors of the Pingshan station migration on hydrological simulation and water resource management in this river basin through long-series runoff simulations.

In summary, the TMPA, IMERG and CMADS precipitation products were used for the Jinsha River Basin and displayed pros and cons in terms of precipitation error markers, precipitation POD and hydrological simulation runoff analysis markers. Their applications in river basin precipitation-runoff studies have shown acceptable ranges of error. These results demonstrate the suitability of these datasets for the Jinsha River Basin study site. TMPA, IMERG and CMADS datasets exhibited good performance in terms of predictions for the study site and can be applied in both hydrological and meteorological management for the Jinsha River Basin when the long-series product becomes available. Future researchers may investigate new precipitation products from different sources in order to analyse their application potential and to identify data and theoretical support for use in the water resource management of river basins.

Author Contributions: Data curation, D.G. and H.W.; Investigation, D.G.; Methodology, D.G. and X.Z.; Supervision, G.L.; Writing—original draft, D.G.

Conflicts of Interest: The authors declare no conflict of interest.

References

1. Meng, X.Y.; Wang, H.; Lei, X.H.; Cai, S.Y.; Wu, H.J. Hydrological Modeling in the Manas River Basin Using Soil and Water Assessment Tool Driven by CMADS. *Teh. Vjesn.* **2017**, *24*, 525–534.
2. Hsun-Chuan, C.; Po-An, C.; Jung-Tai, L. Rainfall-Induced Landslide Susceptibility Using a Rainfall-Runoff Model and Logistic Regression. *Water* **2018**, *10*, 1354. [CrossRef]

3.   De Luca, D.L.; Biondi, D. Bivariate Return Period for Design Hyetograph and Relationship with T-Year Design Flood Peak. *Water* **2017**, *9*, 673. [CrossRef]

4.   Shi, Y. The development trend of natural disasters in China under the influence of global warming. *J. Nat. Disasters* **1996**, *2*, 106–121. (In Chinese)

5.   Zhang, X.; Liu, G.; Wang, H.; Li, X. Application of a Hybrid Interpolation Method Basedon Support Vector Machine in the Precipitation Spatial Interpolation of Basins. *Water* **2017**, *9*, 760. [CrossRef]

6.   Tang, G.; Wan, W.; Zeng, Z.; Guo, X.; Li, N.; Long, D.; Hong, Y. An overview of the Global Precipitaition Measurement (GPM) mission and it's latest development. *Remote Sens. Technol. Appl.* **2015**, *30*, 607–615. (In Chinese)

7.   Li, Z.; Yang, D.; Tian, F. Flood forecast for Three Gorges region of the Yangtze based on ground-observed rainfall. *J. Hydroelectr. Eng.* **2013**, *32*, 44–49, 62. (In Chinese)

8.   Li, N. The Hydrometeorological Evaluation of Multiple Quantitative Precipitation Estimation and Radar-Based Nowcasting Precipitation: Ganjiang River Baisn. Master's Thesis, Chinese Academy of Metrorological Sciences, Beijing, China, 2016. (In Chinese)

9.   Shi, C.X.; Xie, Z.H.; Qian, H.; Liang, M.L.; Yang, X.C. China land soil moisture EnKF data assimilation based on satellite remote sensing data. *Sci. China Earth Sci.* **2011**, *54*, 1430–1440. [CrossRef]

10. Tang, G.; Li, Z.; Xue, X.; Hu, Q.; Yong, B.; Hong, Y. A study of substitutability of TRMM remote sensing precipitation for gauge-based observation in Ganjiang River basin. *Adv. Water Sci.* **2015**, *26*, 340–346. (In Chinese)

11. Biondi, D.; De Luca, D.L. Rainfall-runoff model parameter conditioning on regional hydrological signatures: Application to ungauged basins in southern Italy. *Hydrol. Res.* **2016**, *48*, 714–725. [CrossRef]

12. Zhao, F.; Wu, Y. Parameter Uncertainty Analysis of the SWAT Model in a Mountain-Loess Transitional Watershed on the Chinese Loess Plateau. *Water* **2018**, *10*, 690. [CrossRef]

13. Meng, X.; Wang, H.; Shi, C.; Wu, Y.; Ji, X. Establishment and Evaluation of the China Meteorological Assimilation Driving Datasets for the SWAT Model (CMADS). *Water* **2018**, *10*, 1555. [CrossRef]

14. Meng, X.; Wang, H.; Wu, Y.; Long, A.; Wang, J.; Shi, C.; Ji, X. Investigating spatiotemporal changes of the land-surface processes in Xinjiang using high-resolution CLM3.5 and CLDAS: Soil temperature. *Sci. Rep.* **2017**, *7*, 13286. [CrossRef] [PubMed]

15. Meng, X.; Wang, H. Significance of the China Meteorological Assimilation Driving Datasets for the SWAT Model (CMADS) of East Asia. *Water* **2017**, *9*, 765. [CrossRef]

16. Liu, S.; Yan, D.; Wang, H.; Li, C.; Qin, T.; Weng, B.; Xing, Z. Evaluation of the quality of TRMM precipitation in the watershed regions of mainland China. *Adv. Water Sci.* **2016**, *27*, 639–651. (In Chinese)

17. NASA (NATIONAL AERONAUTICS AND SPACE ADMINISTRATION) PRECIPITATION MEASURE-MENT MISSIONS. Available online: http://precip.gsfc.nasa.gov/pub/trmmdocs/3B42_3B43_doc.pdf (accessed on 1 February 2019).

18. Li, N.; Tang, G.; Zhao, P.; Hong, Y.; Gou, Y.; Yang, K. Statistical assessment and hydrological utility of the latest Multi-satellite precipitation analysis IMERG in Ganjiang River basin. *Atmos. Res.* **2017**, *183*, 212–223. [CrossRef]

19. Satya, P.; Ashis, K.; Imranali, M. Comparison of TMPA-3B42 Versions 6 and 7 Precipitation Products with Gauge-Based Data over India for the Southwest Monsoon Period. *J. Hydrometeorol.* **2015**, *16*, 346–362.

20. Jiang, S.; Zhou, M.; Ren, L.; Cheng, X.; Zhang, P. Evaluation of latest TMPA and CMORPH satellite precipitation products over Yellow River Basin. *Water Sci. Eng.* **2016**, *9*, 87–96. [CrossRef]

21. Zhao, H.; Yang, B.; Yang, S.; Huang, Y.; Dong, G.; Bai, J.; Wang, Z. Systematical estimation of GPM-based global satellite mapping of precipitation products over China. *Atmos. Res.* **2018**, *201*, 206–217. [CrossRef]

22. Kong, Y. Evaluation of the Accuracy of GPM/IMERG over the Mainland of China. Master's Thesis, Nanjng Uniwersity of information Science & Technology, Nanjing, China, 2017. (In Chinese)

23. Wang, Z.; Zhong, R.; Lai, C.; Chen, J. Evaluation of the GPM IMERG satellite-based precipitation products and the hydrological utility. *Atmos. Res.* **2017**, *196*, 151–163. [CrossRef]

24. Chen, X.; Zhong, R.; Wang, Z.; Lai, C.; Chen, J. Precision and hydrological utility evaluation of new generation GPM IMERG remote sensing precipitation data in southern China. *J. Water Conserve.* **2017**, *48*, 1147–1156. (In Chinese)

25. Jin, X.; Shao, H.; Zhang, C.; Yan, Y. Applicability analysis of GPM satellite precipitation data in Tianshan Mountains. *J. Nat. Resour.* **2016**, *31*, 2074–2085. (In Chinese)

26. Wu, Q.; Yang, M.; Dou, F. Analysis of snow detection capability by GPM two-frequency precipitation observation radar. *Meteorol. Mon.* **2017**, *43*, 348–353. (In Chinese)

27. Tang, G.; Zeng, Z.; Long, D.; Guo, X. Statistical and Hydrological Comparisons between TRMM and GPM Level-3 Products over a Midlatitude Basin: Is Day-1 IMERG a Good Successor for TMPA 3B42V7. *J. Hydrometeorol.* **2016**, *17*, 121–137. [CrossRef]

28. Wang, W.; Lu, H. Evaluation and comparison of newest GPM and TRMM products over Mekong River Basin at daily scale. In Proceedings of the 2016 IEEE International Geoscience and Remote Sensing Symposium (IGARSS), Beijing, China, 10–15 July 2016.

29. Liu, Z. Comparison of Integrated Multisatellite Retrievals for GPM (IMERG) and TRMM Multisatellite Precipitation Analysis (TMPA) Monthly Precipitation Products: Initial Results. *J. Hydrometeorol.* **2016**, *17*, 777–790. [CrossRef]

30. Tang, G.; Ma, Y.; Long, D.; Zhong, L.; Hong, Y. Evaluation of GPM Day-1 IMERG and TMPA Version-7 legacy products over Mainland China at multiple spatiotemporal scales. *J. Hydrol.* **2016**, *533*, 152–167. [CrossRef]

31. Tan, M.; Duan, Z. Assessment of GPM and TRMM Precipitation Products over Singapore. *Remote Sens.* **2017**, *9*, 720. [CrossRef]

32. Qiao, L.; Hong, Y.; Sheng, C.; Zou, C.; Gourley, J. Performance assessment of the successive Version 6 and Version 7 TMPA products over the climate-transitional zone in the southern Great Plains, USA. *J. Hydrol.* **2014**, *513*, 446–456. [CrossRef]

33. Cao, Y.; Zhang, J.; Yang, M. Application of SWAT Model with CMADS Data to Estimate Hydrological Elements and Parameter Uncertainty Based on SUFI-2 Algorithm in the Lijiang River Basin, China. *Water* **2018**, *10*, 742. [CrossRef]

34. Shao, G.; Guan, Y.; Zhang, D.; Yu, B.; Zhu, J. The Impacts of Climate Variability and Land Use Change on Streamflow in the Hailiutu River Basin. *Water* **2018**, *10*, 814. [CrossRef]

35. Zhou, S.; Wang, Y.; Chang, J.; Guo, A.; Li, Z. Investigating the Dynamic Influence of Hydrological Model Parameters on Runoff Simulation Using Sequential Uncertainty Fitting-2-Based Multilevel-Factorial-Analysis Method. *Water* **2018**, *10*, 1177. [CrossRef]

36. Gao, X.; Zhu, Q.; Yang, Z.; Wang, H. Evaluation and Hydrological Application of CMADS against TRMM 3B42V7, PERSIANN-CDR, NCEP-CFSR, and Gauge-Based Datasets in Xiang River Basin of China. *Water* **2018**, *10*, 1225. [CrossRef]

37. Tian, Y.; Zhang, K.; Xu, Y.; Gao, X.; Wang, J. Evaluation of Potential Evapotranspiration Based on CMADS Reanalysis Dataset over China. *Water* **2018**, *10*, 1126. [CrossRef]

38. Liu, J.; Shanguan, D.; Liu, S.; Ding, Y. Evaluation and Hydrological Simulation of CMADS and CFSR Reanalysis Datasets in the Qinghai-Tibet Plateau. *Water* **2018**, *10*, 513. [CrossRef]

39. Vu, T.; Li, L.; Jun, K. Evaluation of Multi-Satellite Precipitation Products for Streamflow Simulations: A Case Study for the Han River Basin in the Korean Peninsula, East Asia. *Water* **2018**, *10*, 642. [CrossRef]

40. Ren, F.; Zhang, C.; Chen, X.; Xu, P.; Chen, J. Analysis of the spatial heterogeneity of the Jinsha River trunk stream vegetation and its impact on ecological restoration. *J. Yangtze River Sci. Res. Inst.* **2016**, *33*, 24–30. (In Chinese)

41. Deng, M.; Han, S. Study on development mode of healthy tourism in Jinsha River basin. *Tour. Manag. Stud.* **2012**, *24*, 18–19. (In Chinese)

42. Liu, X. A Preliminary Study of Meteorological Factors and Runoff in the Jinsha River Basin. Master's Thesis, Chinese Academy of Metrorological Sciences, Wuhan, China, 2016. (In Chinese)

43. Chen, Y.; Wang, W.; Wang, G.; Wang, S. Analysis of temperature and precipitation change characteristics in Jinsha River basin. *Plateau Mt. Meteorol. Res.* **2010**, *30*, 51–56. (In Chinese)

44. Cen, S.; Qin, N.; Li, Y. Analysis of climatic characteristics of runoff flow change in Jinsha River basin during flood season. *Resour. Sci.* **2012**, *34*, 1538–1544. (In Chinese)

45. Meng, X.; Dan, L.; Liu, Z. Energy balance-based SWAT model to simulate the mountain snowmelt and runoff—Taking the application in Juntanghu watershed (China) as an example. *Mt. Sci.* **2015**, *12*, 368–381. [CrossRef]

46. Wang, Y.; Meng, X. Snowmelt runoff analysis under generated climate change scenarios for the Juntanghu River basin in Xinjiang, China. *Tecnol. Cienc. Agua* **2016**, *7*, 41–54.

47. Meng, X. Simulation and spatiotemporal pattern of air temperature and precipitation in Eastern Central Asia using RegCM. *Sci. Rep.* **2018**, *8*, 3639. [CrossRef]

48. Meng, X.; Wang, H.; Cai, S.; Zhang, X.; Leng, G.; Lei, X.; Shi, C.; Liu, S.; Shang, Y. The China Meteorological Assimilation Driving Datasets for the SWAT Model (CMADS) Application in China: A Case Study in Heihe River Basin. *PearlRiver* **2016**, *37*, 1–19.

49. Meng, X. Spring Flood Forecasting Based on the WRF-TSRM mode. *Teh. Vjesn.* **2018**, *25*, 27–37.

50. Qin, G.; Liu, J.; Wang, T.; Xu, S.; Su, G. An Integrated Methodology to Analyze the Total Nitrogen Accumulation in a Drinking Water Reservoir Based on the SWAT Model Driven by CMADS: A Case Study of the Biliuhe Reservoir in Northeast China. *Water* **2018**, *10*, 1535. [CrossRef]

51. Guo, B.; Zhang, J.; Xu, T.; Croke, B.; Jakeman, A.; Song, Y.; Yang, Q.; Lei, X.; Liao, W. Applicability Assessment and Uncertainty Analysis of Multi-Precipitation Datasets for the Simulation of Hydrologic Models. *Water* **2018**, *10*, 1611. [CrossRef]

52. Dong, N.; Yang, M.; Meng, X.; Liu, X.; Wang, Z.; Wang, H.; Yang, C. CMADS-Driven Simulation and Analysis of Reservoir Impacts on the Streamflow with a Simple Statistical Approach. *Water* **2018**, *11*, 178. [CrossRef]

53. Xie, P.; Xiong, A. A conceptual model for constructing high-resolution gauge-satellite merged precipitation analyses. *J. Geophys. Res. Atmos.* **2011**, *116*, 116. [CrossRef]

54. Shen, Y.; Xiong, A. Validation and comparison of a new gauge-based precipitation analysis over mainland China. *Int. J. Climatol.* **2016**, *36*, 252–265. [CrossRef]

55. Shen, Y.; Xiong, A.; Wang, Y.; Xie, P. Performance of high-resolution satellite precipitation products over China. *J. Geophys. Res.* **2010**, *115*, D02114. [CrossRef]

56. Tang, G.; Wen, Y.; Gao, J.; Long, D.; Ma, Y.; Wan, W.; Hong, Y. Similarities and differences between three coexisting spaceborne radars in global rainfall and snowfall estimation. *Water Resour. Res.* **2017**, *10*, 1002. [CrossRef]

57. Zeng, X.; Ye, L.; Zhai, J.; Zhang, H. Spatiotemporal evolution characteristics of precipitation in Jinsha River basin in 1961_2010. *Resour. Environ. Yangtze Basin* **2015**, *24*, 402–407. (In Chinese)

58. Feng, Z. Application of SWAT Model in Daxi River Basin. Master's Thesis, Chinese Academy of Metrorological Sciences, Xi'an, China, 2015. (In Chinese)

59. Yuan, J.; Su, B.; Li, H.; Lu, Y. Runoff simulation study of Chaihe reservoir basin based on SWAT model. *J. Beijing Norm. Univ. (Nat. Sci.)* **2010**, *3*, 361–365. (In Chinese)

60. He, Z.; Yang, L.; Hou, A.; Lu, H. Intercomparisons of Rainfall Estimates from TRMM and GPM Multisatellite Products over the Upper Mekong River Basin. *J. Hydrometeorol.* **2017**, *18*, 413–430. [CrossRef]

Article

Evaluation and Hydrological Simulation of CMADS and CFSR Reanalysis Datasets in the Qinghai-Tibet Plateau

Jun Liu [1,2], Donghui Shanguan [1,*], Shiyin Liu [3] and Yongjian Ding [1]

[1] State Key Laboratory of Cryospheric Science, Northwest Institute of Eco-Environment and Resources, Chinese Academy of Sciences, Lanzhou 730000, China; Liujun16@lzb.ac.cn (J.L.); dyj@lzb.ac.cn (Y.D.)

[2] University of Chinese Academy of Sciences, Beijing 100049, China

[3] Institute of International Rivers and Eco-Security, Yunnan University, Kunming 650500, China; shiyin.liu@ynu.edu.cn

* Correspondence: dhguan@lzb.ac.cn; Tel.: +86-13919104740

Received: 11 March 2018; Accepted: 10 April 2018; Published: 20 April 2018

Abstract: Multisource reanalysis datasets provide an effective way to help us understand hydrological processes in inland alpine regions with sparsely distributed weather stations. The accuracy and quality of two widely used datasets, the China Meteorological Assimilation Driving Datasets to force the SWAT model (CMADS), and the Climate Forecast System Reanalysis (CFSR) in the Qinghai-Tibet Plateau (TP), were evaluated in this paper. The accuracy of daily precipitation, max/min temperature, relative humidity and wind speed from CMADS and CFSR are firstly evaluated by comparing them with results obtained from 131 meteorological stations in the TP. Statistical results show that most elements of CMADS are superior to those of CFSR. The average correlation coefficient (R) between the maximum temperature and the minimum temperature of CMADS and CFSR ranged from 0.93 to 0.97. The root mean square error (RMSE) for CMADS and CFSR ranged from 3.16 to 3.18 °C, and ranged from 5.19 °C to 8.14 °C respectively. The average R of precipitation, relative humidity, and wind speed for CMADS are 0.46; 0.88 and 0.64 respectively, while they are 0.43, 0.52, and 0.37 for CFSR. Gridded observation data is obtained using the professional interpolation software, ANUSPLIN. Meteorological elements from three gridded data have a similar overall distribution but have a different partial distribution. The Soil and Water Assessment Tool (SWAT) is used to simulate hydrological processes in the Yellow River Source Basin of the TP. The Nash Sutcliffe coefficients (NSE) of CMADS+SWAT in calibration and validation period are 0.78 and 0.68 for the monthly scale respectively, which are better than those of CFSR+SWAT and OBS+SWAT in the Yellow River Source Basin. The relationship between snowmelt and other variables is measured by GeoDetector. Air temperature, soil moisture, and soil temperature at 1.038 m has a greater influence on snowmelt than others.

Keywords: CMADS; Qinghai-Tibet Plateau (TP); SWAT; CFSR

---

1. Introduction

The Qinghai-Tibet Plateau (TP, 26°–40° N, 73°–104° E) is the largest and highest plateau in the world, with an average altitude higher than 4000 m over an area of about 2.5 million km^2 [1,2]. Complex orographic and harsh weather conditions make it difficult to install and maintain synoptic stations in the TP. Almost all of the existing stations are located in the east and south of TP, and 70% of the stations are located below 4000 m. Scarcity and low elevation of the existing weather stations cannot accurately present the meteorological status of the TP. Likewise, scarcity and weak representation of meteorological data is fruitless for developing hydrological models [3,4]. Distributed models scientifically delineate water cycles in basin-scales, but also need high quality weather data as input [5]. Therefore, the poorly observed network is one of the reasons for the slowly progress of hydrological simulation and analysis in the basins of the TP.

Reanalysis datasets are based on remote sensing products and climate model outputs, and some are corrected by observed data and are an important surrogate for observations [6]. They will play a major role in the development of models [7], showing climate change trends, [8] un-gauged regions [9]. However, in view of uncertainties which exist in the process of data acquisition and assimilation, most studies are still predicated on the evaluation of reanalysis [10]. The accuracy of reanalysis is mainly assessed in one of the following two ways: (1) comparison of reanalysis with corresponding observed data [11,12]; (2) using reanalysis as input data to drive hydrological models and then comparing the hydrological features of model output with the observed [13,14].

The first way is always used over a large-scale, with a certain number of weather stations. Many statistic indexes are utilized to measure the quality of reanalysis datasets, such as: correlation coefficient (R), relative bias (BIAS), root mean square error (RMSE) et al. Evaluation of average temperature and precipitation from reanalysis are more frequently used than other meteorological variables. Wang et al. [15] compared two types of ERA-Interim datasets with gridded observation datasets. Results showed that after topographic correction, temperature distribution of reanalysis closely reproduces the temperature conditions of the TP, and that the increased trend is similar to observed data. Likewise, achievements of Gao et al. [16] showed that ERA-Interim temperature in the TP works well; R of temperature in the monthly scale ranges from 0.973 to 0.999 when compared with 75 stations' data above 3000 m. Song et al. [17] compared precipitation from eight gridded datasets with station observations in Asian high mountains; the result indicated that gauge-based or multi-source datasets showed better performance, and that merged datasets are of potential use in modeling water cycles. You et al. [18] compared multisource datasets with gridded precipitation observations over the TP; most datasets can capture the precipitation distribution and identity varieties of mean monthly precipitation. Wang et al. [19] compared precipitation, temperature, radiation, wind speed and surface pressure from six multi-reanalysis products with observed data. Results indicate that different products have different abilities in calculating meteorological elements. For example; ERA-Interim performance is good with temperature, whereas the Global Land Data Assimilation Systems (GLDAS) shows the best performance with precipitation. In conclusion, reanalysis datasets can display the broad distribution of meteorological elements of the TP, but corrections using observed data are essential to minimize errors.

With long continuous time series and high spatial resolution, reanalysis datasets are suitable to create hydrological models, especially in regions that have few weather stations. High quality temporal and spatial resolution meteorological input data for distributed models largely determines the result of model output. Much research evaluates reanalysis at the watershed-scale by using hydrological models. Thomas et al. [20] evaluated ten satellite and reanalysis datasets in six, different sized watersheds in West Africa. Gilles et al. [21] analyzed the impact of combining different reanalysis and weather station data on the accuracy of discharge modeling in Canada and the USA. Both concluded that reanalysis datasets can be an alternative for observed data. Some reanalysis datasets do well in runoff simulation, and NSE are satisfactory, especially in the reanalysis datasets bias-corrected by weather station data. Kan et al. [22] evaluated "the Climate Prediction Center Morphing Technique (CMORPH)", "Tropical Rain Measurement Mission Multi-satellite Precipitation Analysis (TRMM 3B42 V6)", "China Meteorological Forcing Dataset (CMFD)" and "Asian Precipitation-Highly Resolved Observational Data Integration Towards Evaluation Of Water Resource (APHRODITE)" in the upper Yarkant River. Results indicate that datasets of distribution of precipitation from CMFD are more appropriate because they are consistent with the distribution of glaciers, and CMORPH based on satellite data, gets better results in forcing the Variable Infiltration Capacity (VIC) model. Guo et al. [23] compared two kind of multisource reanalysis data in hydrological simulation in the Lasa River Basin: the NSE is above 0.7 in the daily scale, and 0.8 in the monthly scale based on the HIMS model.

Gao et al. [24] analyzed the application of CFSR, ERA-Interim in driving the VIC model in the Kash River Basins, and results indicate ERA-Interim is superior to CFSR. Hence, a set of reliable datasets can be a substitute for observed data in basins with sparsely distributed weather stations.

Previous studies, whether they use the first or the second method, always focus upon precipitation and average temperature. However, a complete set of data for distributed or semi-distributed hydrological models needs precipitation and average temperature, but also max/min temperatures, relative humidity, atmospheric pressure, wind speeds, solar radiation etc. For example, the SWAT model, as one of the most popular models, is extensively applied in runoff simulation and prediction, sediment transition etc. [25], and requires not only daily precipitation and temperature, but also relative humidity, atmospheric pressure, and wind speeds as input weather data, to obtain evapotranspiration [26]. Relative humidity, wind speed and max/min-temperatures also have great value in research. Relative humidity reflects the saturation of moisture in the atmosphere and has an impact on surface water, energy budgets, formation of aerosols, growth of plants and animals, etc. [27,28]. Wind speed depicts the movement of atmosphere and its influence affects other weather phenomena like precipitation, smog [29,30]. Maximum and minimum temperatures are more responsive to extreme weather events [31,32].

CMADS and CFSR, as two more complete datasets, contain several meteorological elements and are recommended by the SWAT official website (https://swat.tamu.edu/). CFSR has been widely used around the world. Dile et al. [33] and Abeyou et al. [34] used CFSR to drive three different hydrological models in the Blue Nile River Basin, and their results indicate that CFSR has the ability in forcing hydrological models; its simulation results were the same as, or better than, those forced by weather station data. In China, CFSR was used in the Bahe River Basin [35], Kaidu River Basin [36], Kash River Basins [24] etc. CMADS, built by Dr. Xianyong Meng from China Institute of Water Resources and Hydropower Research (IWHR), and bias-correction by observed data has been used in several basins including China's Juntanghu watershed [37–39] and the Manas River Basin [40]; the results are satisfactory. However, comprehensive evaluation and application of these two datasets in the TP is scarce, especially CMADS. Thus, precipitation, max/min-temperatures, relative humidity and wind speed from CMADS and CFSR were evaluated using data from 131 weather stations in this paper. The Yellow River Source Basin was also selected for hydrological simulation and analysis.

2. Study Area

Located in south central Eurasia, affected by high elevation and far from the ocean, the TP forms a complex plateau climate system. The average annual temperature ranges from 20 °C in the southeast to −6 °C in the northwest, and precipitation declines from 2000 to 50 mm, correspondingly [41]. The TP is composed of a series of plateaus, mountains and valleys. The Yellow River Source Basin was selected to analyze the ability of two reanalysis in forcing hydrological models. The Yellow River Source Basin is located in the northeastern part of the TP (Figure 1), and refers to the basin above the Tangnaihai hydrological station (100°09′ E,35°30′ N, 2546 m) [42]. The catchment area is about 122 thousand square kilometers and the elevation ranges from 2676 to 6254 m (Figure 1). Permafrost is widely distributed within the Yellow River Source Basin, and most of it is seasonally frozen. The Yellow River Source Basin is rich in water resources and there are a large number of plateau lakes and wetlands. The Zaling and Erling lakes are the highest freshwater lakes in China [43].

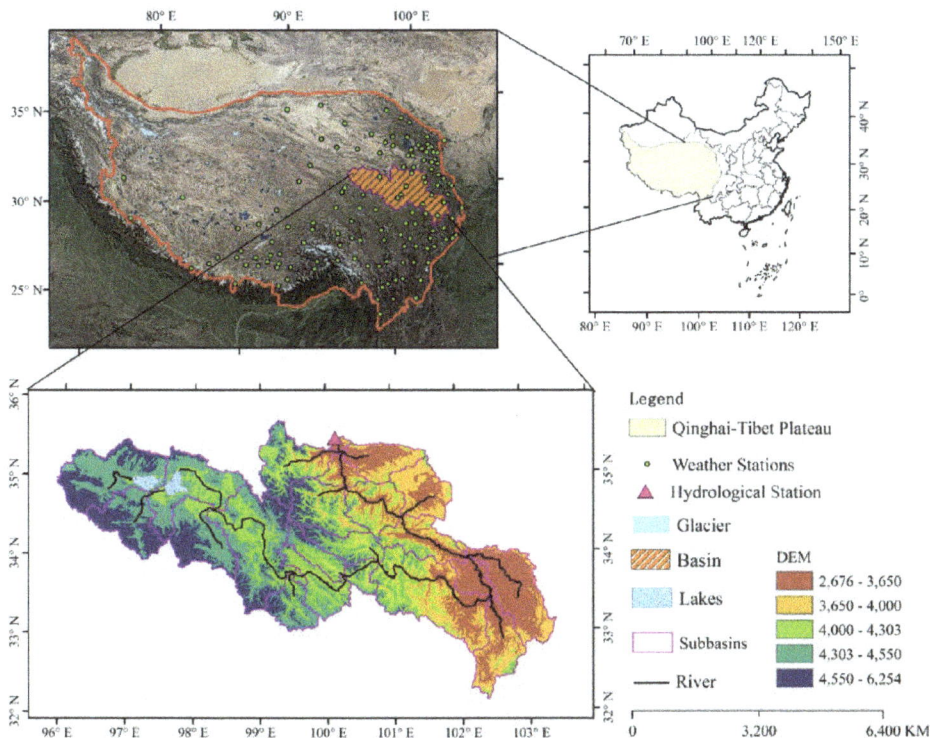

Figure 1. The Locations of TP and the Digital Elevation Model of Yellow River Source Basin.

3. Data and Methods

3.1. Data

The China Meteorological Assimilation Driving Datasets for the SWAT model version 1.0 (CMADS V1.0) was developed by Dr. Xianyong Meng using STMAS assimilation techniques [44]. Temperature, atmospheric pressure, specific humidity and wind speed of CMADS is based on The National Center for Environmental Prediction Global Forecast System (NECP/GFS), and is corrected by observed data. The background field for precipitation is CMORPH, and this is adjusted by observed precipitation data [44]. CMADS V1.0 provides: daily maximum/average/minimum temperatures, cumulative 24 h-precipitation, average solar radiation, air pressure, relative humidity, and average wind speed from 2008 to 2016. Ten layers of soil temperature from CMADS-ST are also used in this paper [45,46]. The depth from the first to the tenth are 0.007 m, 0.028 m, 0.062 m, 0.119 m, 0.212 m, 0.366 m, 0.62 m, 1.038 m, 1.727 m and 2.865 m. Climate and soil temperature data of CMADS can be downloaded CMADS official website (http://www.cmads.org/).

The Climate Forecast System Reanalysis datasets (CFSR) is developed by The National Center for Environmental Prediction (NCEP) and is derived from the Global Forecast System [47]. With high spatial resolution, reliability and long time series, CFSR is widely used in climate analysis and hydrological simulation. The SWAT official website provides data from a 36-year period (from 1979 to 2014) in the format requested by the SWAT model, with elements including: precipitation, max/min temperatures, relative humidity, wind speed and solar radiation [33]. For comparison purposes, we selected the period from 2008 to 2014; the CFSR dataset was freely accessible from the SWAT official website (https://globalweather.tamu.edu/).

We also collected measured data from 131 weather stations (Figure 1); meteorological elements included: mean/max/min temperatures, precipitation, wind speed, and relative humidity in daily scale, provided by the China Meteorological Administration Meteorological Data Center. Among the 131 weather stations, elevation of 9 stations are less than 2000 m, 42 stations are between 2000 and 3000 m, 55 stations range from 3000 to 4000 m, and 25 stations are above 4000 m, with highest at an altitude of 4800 m. Seventy four percent of sites were at an elevation of between 2000 and 4000 m.

Geographical and hydrological data includes: DEM (digital elevation model), soil, land cover, the 90 m DEM was downloaded from CGIAR-CSI (http://srtm.csi.cgiar.org/). Soil and land use data was provided by the Cold and Arid Regions Sciences Data Center at Lanzhou (http://westdc.westgis.ac.cn/).

3.2. Hydrological Models

SWAT was developed by the US Department of Agriculture in the 1990s and plays an important role in runoff simulation, sediment movement, and non-point source modeling [48]. According to elevation, a watershed will be divided into several sub-basins which will be further divided into hydrological response units (HRUs) based on land use, soil type and slope. Water balance will be calculated in each HRU. Soil Convention Service (SCS) runoff curve and Penman-Monteith methods are used to model surface runoff process and evapotranspiration. Precipitation will be divided into rain or snow, according to critical temperature. Snowfall is stored as snow on the surface, and the process of addition (P_{day}), ablation ($SNOW_{mlt}$) and sublimation E_{sub} will be calculated by the snow mass conservation Equation (1). Degree day method is used to simulate snow melt ($SNOW_{mlt}$), the snow temperature (T_{snow}), daily maximum temperature of the basin (T_{max}) and snowmelt threshold temperature (T_{mlt}) combined with snow cover area ($SNOW_{cov}$) and degree-day factor (b_{mlt}). These parameters pertain to snowmelt, and their relationship is Equation (2). Lakes and reservoirs belong to the river network water cycle calculation. Wetland belongs to corresponding sub-basins, and the change is also based on the corresponding water balance equation. Therefore, SWAT has the ability to simulate the complicated hydrological process of the Yellow River Source Basin, which is a snow dominated watershed with wetland, lakes etc.

$$SNOW = SNOW + P_{day} + E_{sub} + SNOW_{mlt}, \tag{1}$$

$$SNOW_{mlt} = b_{mlt} \times SNOW_{cov} \times [\frac{T_{snow} + T_{max}}{2} - T_{mlt}], \tag{2}$$

3.3. Spatial Analysis Methods

Observed meteorological data from 131 weather stations are used to interpolate though ANUSPLIN, which is a professional interpolation software based on the thin plate smooth spline technique. Wahba proposed the thin plate smoothing spline surface fitting technique in 1979; the theoretical model formula is as follows,

$$Z_i = f(x_i) + b^T y_i + e_i, \tag{3}$$

Z_i is a dependent variable, x_i is independent variable, f is unknown smooth function, y_i is independent covariate, b is coefficient and e_i is random error.

Bates, Eblen, Hutchinson et al. updated this spatial interpolation method and eventually formed ANUSPLIN [49]. ANUSPLIN is convenient and has been widely used in Australia, Europe, the United States etc. [50]; more detailed information about each module is described by Liu et al. [51].

GeoDetector is an effective way to measure spatially stratified heterogeneity of variables, and to test the connection between variables according to the consistency of their spatial distributions [52]. It is widely used in the field of health to detect the correlation of distribution of disease incidence and their impact factors. However, Zhao et al. [53] use this model to analysis the impacts of terrestrial environmental factors on precipitation variation over the Beibu Gulf Economic Zone in Coastal Southwest China. Foroogh et al. [54] used this method to analyze the relationship between air temperature and land use, elevation, latitude et al. Therefore, we use this detector model for quantitative analysis of the relationship between snow melt and related factors, like soil temperature, soil humidity, or topographic parameters. The parameter to measure the degree of correlation between variables is *q*-stastic, and the formula is as follows,

$$q(Y|h) = 1 - \frac{1}{N\sigma^2} \sum_{h=1}^{L} N_h \sigma_h{}^2, \tag{4}$$

σ^2 stands for the variance of Y; N is the number of units of Y; Y is composed of L strata ($h = 1, 2, \ldots, L$). It should be noted that $q \in [0,1]$, and $q = 0$ means there is no association between Y and X; $q = 1$ indicates Y is completely determined by X.

3.4. Evaluation Index

The correlation coefficient (*R*), relative bias (BIAS), root mean square error (RMSE) and ratio of standard deviation (σ/σ_{obs}) are used to measure the accuracy of reanalysis datasets compared to observed data in both daily and monthly scale. *R* is Pearson correlation coefficient (*R*) and its square is coefficient of determination (R^2). They are used to measure the correlation between variables. The range of R and R2 is [0,1]. If R = 0, there is no correlation between two variables. If R = 1, the two variables are linearly related. BIAS and RMSE are used to measure the deviation between variables; ranges are $[-\infty,+\infty]$ and $[0,+\infty]$ respectively. σ/σ_{obs} is used to measure the simulated value compared to observed data.

$$R = \frac{\sum_i (X_i - \overline{X})(Y_i - \overline{Y})}{\sqrt{\sum_i (X_i - \overline{X})^2}\sqrt{\sum_i (Y_i - \overline{Y})^2}}, \tag{5}$$

$$BIAS = \sum_i \frac{X_i - Y_{i,obs}}{Y_{i,obs}}, \tag{6}$$

$$RMSE = \sqrt{\frac{1}{N}\sum_i (X_i - Y_i)^2}, \tag{7}$$

$$\sigma/\sigma_{obs} = \frac{\sqrt{\sum_i (X_i - \overline{X})}}{\sqrt{\sum_i (Y_i - \overline{Y})}}, \tag{8}$$

Nash-Sutcliffe Efficiency (NSE) and the coefficient of determination (R^2) are used to evaluate the simulation effect of hydrological models on runoff, the range is $(-\infty,1)$. $0.75 < \text{NSE} \leq 1$ means that the simulation results are excellent, $0.65 < \text{NSE} \leq 0.75$ means the simulation results are good. $0.5 < \text{NSE} \leq 0.65$ means the simulation results are acceptable. When NSE < 0.5, the simulation filed and the results is unacceptable [55]. NSE and R^2 are calculated as follows,

$$NSE = 1 - \frac{\sum (Q_{sim} - Q_{obs})^2}{\sum (Q_{obs} - \overline{Q_{obs}})^2}, \tag{9}$$

$$R^2 = \frac{\left(\sum (Q_{obs} - \overline{Q_{obs}})(Q_{sim} - \overline{Q_{sim}})\right)^2}{\sum (Q_{obs} - \overline{Q_{obs}})^2 \sum (Q_{sim} - \overline{Q_{sim}})^2}, \tag{10}$$

4. Results

4.1. Comparison of CMADS and CFSR with Observation Data

Precipitation is affected by multi-factors and has large spatial heterogeneity in alpine regions, which is very difficult to capture accurately (Figure 2). Mean R of CMADS precipitation is within 0.16–0.66, with an average value of 0.46; Sixty-four percent of stations drop to 0.4–0.6 (Table 1). Range of BIAS is from −0.64 to 3.76, with an average value of 0.08, among which 56% weather stations have a positive value, 44% have a negative bias, and three stations have abnormal BIAS values beyond 2. RMSE ranges from 0.54 to 6.78 mm, and 80% stations are located in the 3–5 mm range with an average value of 3.77 mm. σ/σ_{obs} stands for the ratio of deviation used to measure the dissociation of two time series. σ/σ_{obs} of CMADS precipitation is within 0.28–2.12, with an average value of 1.07. Among the 131 stations, just one station is beyond 2, meaning that the degree of deviation is twice that of the observed data in this station. For CFSR, the evaluation results are still not grounds for optimism. Mean R of precipitation is in the range of 0.13–0.6, with a mean value of 0.43, approaching the result of CMADS; Sixty-five percent of stations are within 0.4–0.5 (Figure 2). BIAS shows that 77% stations overestimate precipitation, and one station presents unusually beyond 10. RMSE of CFSR precipitation ranges between 1.38 and 13.67 mm, with an average value of 4.5 mm; 22% stations are greater than 5 mm. Compared with observed data, σ/σ_{obs} of 63% of stations is larger than observations, and 9 stations are double. Precipitation of CMADS uses CMORPH as the background field and assimilates more than 30,000 mobile observation stations in China. CMORPH is derived from low orbiter satellite microwave observations, whose features are transported via spatial propagation information that is obtained entirely from geostationary satellite IR data. However, precipitation observed by satellites always underestimate light rainfall events, and tend to fail over snow- and ice-coved surfaces [56]; these system biases results in underestimation of precipitation provided by CMADS. CFSR is derived from the global forecast system and this dataset always overestimates precipitation in northwest China, which have been obtained in many studies [24,36,57,58].

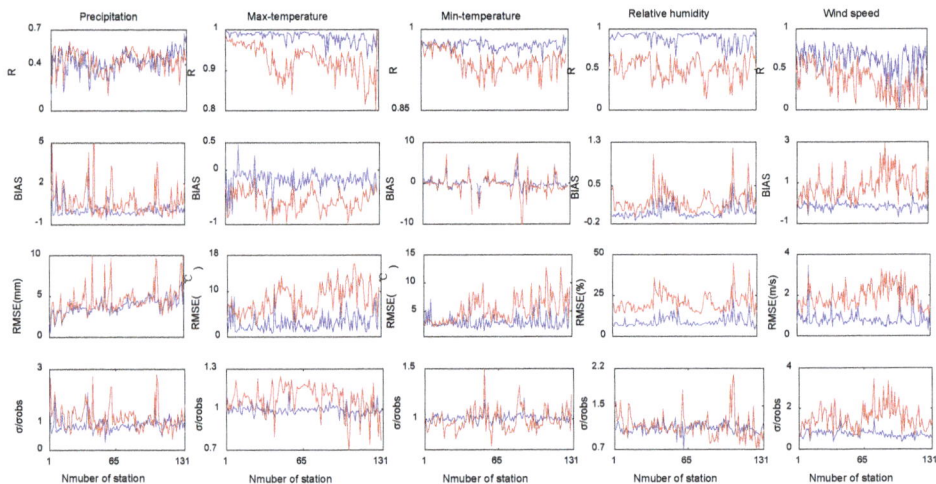

Figure 2. Statistical factors map from CMADS, CFSR compared to 131 observations stations from 2008 to 2013 (Red line is CFSR, blue line is CMADS).

Table 1. Average statistic value of CMADS and CFSR.

Datasets	Elements	R	BIAS	RMSE	σ / σ_{obs}
CMADS	Precipitation	0.46	0.08	3.77	0.92
	Max-temperature	0.98	−0.18	2.99	0.99
	Min-temperature	0.97	−0.26	3.06	1.01
	Humidity	0.88	0.01	8.92	1.09
	Wind speed	0.64	−0.15	0.83	0.74
CFSR	Precipitation	0.43	0.76	4.50	1.19
	Max-temperature	0.93	−0.56	8.22	1.07
	Min-temperature	0.94	−0.95	5.21	0.99
	Humidity	0.52	0.22	20.63	1.13
	Wind speed	0.37	0.83	1.93	1.53

Evaluation of results of max/min temperatures improved significantly compared to precipitation: R of CMADS max/min-temperatures are close to 1 and CFSR is within 0.78–0.98 (Table 1). However, CMADS underrates max/min-temperatures since the BIAS of max-temperature of 94% stations is less than zero and the corresponding value of min-temperature is 52%. Meanwhile, all 131 BIAS of CFSR max-temperatures have a cold value with the mean value of −0.56. This result improves in min-temperature, although 60% stations have a cold value. The RMSE is not optimal; this is also the case with Gao et al. [16], where RMSE is within 0.78–16.9 °C of CMADS maximum temperature and 1.4–10 °C of min-temperature, and the range for CFSR maximum and minimum temperature are 2.4–16 °C and 2.2–12.8 °C respectively. Two stations of CMADS maximum temperature have RMSE value 16.9 °C and 14.2 °C, and others lower than 10 °C. Forty-one stations have a RMSE of max-temperature of CFSR higher than 10 °C. The σ / σ_{obs}, of the maximum minimum temperature performs better and most stations of CMADS and CFSR are close to 1.

Evaluation results of relative humidity still have a gap between CMADS and CFSR (Figure 2). R of CMADS relative humidity is within 0.66–0.95, with an average value of 0.88, and 86% stations record above 0.8. Average value of BIAS is 0.01, and 62% have a negative value. RMSE is in the range of 3.99–22.4%, with a mean value of 8.9%, and 68% stations are within 10%. σ / σ_{obs} is within 0.7–1.33, with average value 1.09. R of CFSR relative humidity is within 0.13–0.79 with an average value of 0.51; average BIAS is 0.22. RMSE is within the range of 11.9–44.7%, and the average value is 20.63%. The wind speed of CFSR is worse, with an average R of 0.36; BIAS is 0.83, the average RMSE is 1.93 m/s, and average σ / σ_{obs} is 1.5 times that of observation. Thirty-two percent stations have a BIAS greater than 1, and 7 stations have a negative value; 40% show RMSE over 2 m/s. As for CMADS, BIAS of 91% stations has a negative value, which indicates that CMADS underestimates wind speed in the TP. The average value of R, BIAS and RMSE improved significantly compared with CFSR but a small number of stations still underperformed. Temperature, humidity and wind speed of CMADS are based on NCEP/GFS; they assimilate 2421 national automatic stations and 39,439 regional climate stations, so the results are an improvement compared to CFSR. Thus, observed climate data plays a significant role in the process of developing a high-quality reanalysis dataset.

As is shown in Figure 3, the RMSE distribution of precipitation of CMADS decreases from southeast to northwest, which can be divided into four grades. The first gradient is the worst: RMSE ranges from 4.91 to 6.78 mm. All of these stations are located in the southeastern margin of the TP. Because of the complex orographic features and effects of the Pacific and Indian Ocean monsoons, this region's precipitation exhibits large spatial heterogeneity. The second gradient surrounds the first gradient, mainly in the eastern and southern parts of the TP, and presents a RMSE range from 3.27 to 4.91 mm. The remaining stations are mainly located in the Qaidam Basin, which is flat and has a relatively stable weather pattern. This phenomenon is also reflected in the precipitation in the CFSR, which further corroborates the difficulty in describing precipitation in complex orographic regions. The abnormal values of relative humidity from both CMADS and CFSR are mainly located in the south of the TP. In the case of wind speed, three stations with large RMSE values are located in the east of the TP; this level of CFSR is mainly located in the south and east of the TP.

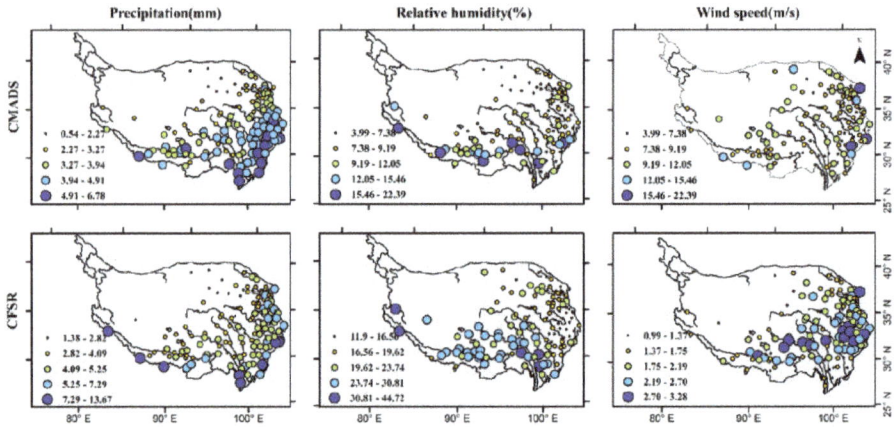

Figure 3. RMSE distribution of precipitation, relative humidity and wind speed.

4.2. Distribution of Observed Data, CMADS and CFSR

We use ANUSPLIN to obtain gridded distribution of meteorological elements including precipitation, max/min temperatures, relative humidity and wind speed based on 131 observation stations with a spatial resolution of 0.3° (which approaches the resolution of CMADS V1.0 (1/3°) and CFSR (0.313°)). The distribution diagram displays the annual average values from 2008 to 2013.

Precipitation can be divided into three regions: abundant regions, relative rainy areas and arid regions (Figure 4). In the southeast margin of the TP, water vapor from Pacific and Indian Ocean brings abundant rainfall, so these are called abundant regions. Precipitation in this region is over 800 mm and, in some regions, greater than 1200 mm, as is shown by OBS (stands for ANUSPLIN interpolation results hereinafter). Precipitation distribution of CMADS in this region is not as high as OBS; some areas show precipitation within 800–1200 mm, others less than 800 mm, and a fraction show higher than 1200 mm. However, the precipitation of CFSR has abnormal characteristics: some regions show over 3000 mm, and individual sites even show over 10,000 mm. In addition, the annual average precipitation of CFSR in the southeast margin of the TP shows over 1200 mm. The second region is a relatively rainy area, that is, mainly around abundant regions and gradually decreases to the northwest. This is due to the increase in elevation, distance from the ocean, and a decrease in the presence of water vapor. This region shows a similarity between OBS and CMADS: precipitation is within 400–800 mm. The vast northwest is basically an arid region and precipitation is below 400 mm. Although precipitation of CFSR in this region still overestimated, it displays an obviously increasing trend from low altitudes to high mountains. This is important since precipitation in high mountainous areas occupies a vital status in runoff. In general, CMADS overestimated the amount of precipitation and CFSR widely overestimated it, compared with OBS.

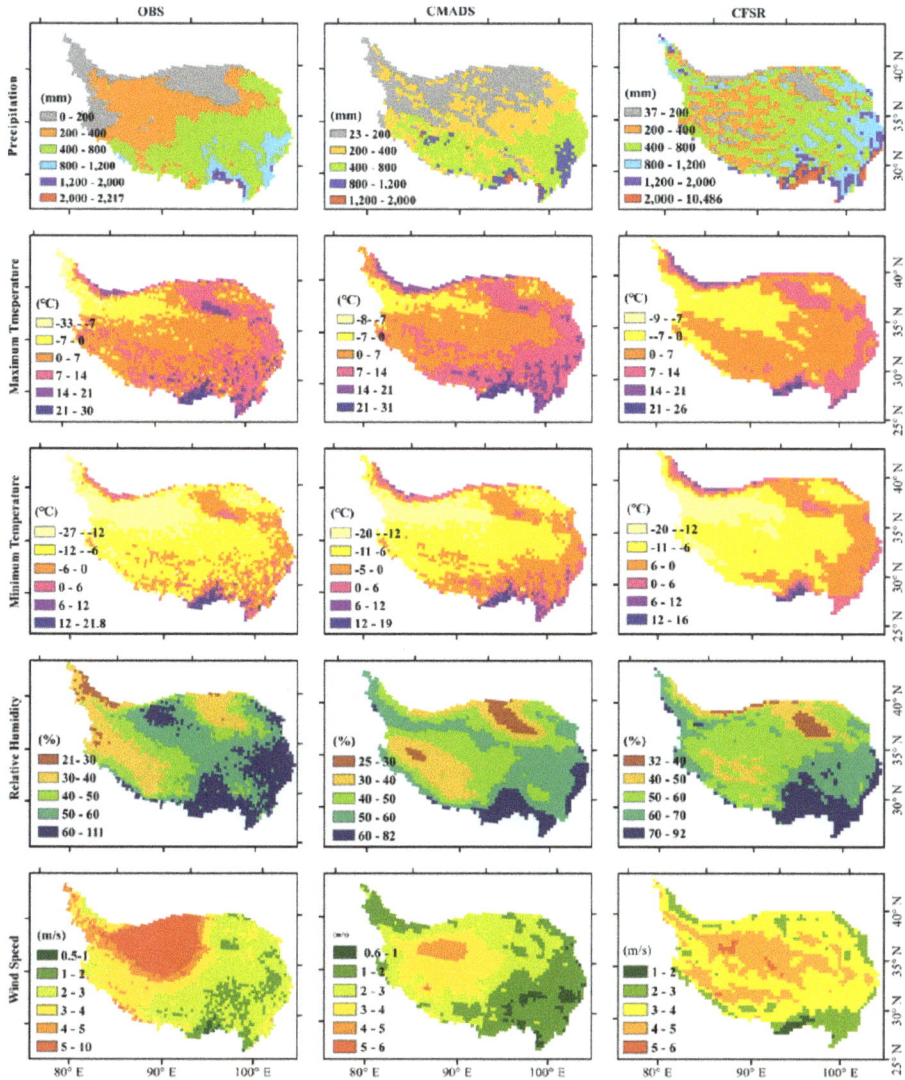

Figure 4. Distribution characteristics of five meteorological factors of observation (ANUSPLIN), CMADS and CFSR.

Distribution of max/min-temperatures from OBS, CMADS and CFSR are more consistent when compared with precipitation (Figure 4). Factors which influence temperature are mainly (a) elevation, and (b) latitude. In the east of the TP, temperature increases from north to south with the corresponding decrease of latitude. Likewise, temperature decreases from east to west with elevation increase at the same latitude. In the southeast margin of the TP with low latitude and elevation, max/min-temperatures are high, and in the north-west high mountains, the opposite is true. CFSR underestimates maximum temperature in the south and east of the TP. Annual average Max-temperatures of OBS and CMADS mainly lie between 7 and 14 °C in the east and south of the TP, but in corresponding regions of CFSR records, they obviously go down.

Relative humidity is primarily affected by precipitation, and the distribution is consistent with precipitation (Figure 4). Relative humidity of CFSR is higher than in the observed data and CMADS in most part of the TP. For example, relative humidity of gridded observation and CMADS is within 30–40% in the south-central TP, but CFSR shows that it is within 40–60% in this region. In the hinterland, relative humidity of gridded observation and CMADS is within 30–40%, and the range of CFSR is 40–60%. Wind speed from gridded observation, CMADS and CFSR also show wide differences. Gridded observation reveals that annual average wind speed is in excess of 5 m/s in the northwest, but CMADS and CFSR do not show this characteristic, and the annual average wind speed is below 5 m/s. In the south east of the TP, wind speed of CFSR is overestimated at within 2–4 m/s, compared to 1–3 m/s, 1–2 m/s of gridded observation and CMADS respectively.

The TP is divided into three parts including: Tibet (I), Qinghai Provence (II) and the remaining regions include parts of Gansu Provence, Sichuan Provence and Yunnan Provence (III). Precipitation of CMADS is very close to observed data, but CFSR overestimates in all three regions (Figure 5). CMADS and CFSR underestimate maximum and minimum temperatures in most regions; only the comparison result of the minimum temperature of Qinghai Provence is satisfactory, and CFSR is substantially undervalued. Relative humidity and wind speed of CMADS and CFSR left little room for optimism: CFSR overestimates relative humidity in January to April and November to December in Qinghai Provence and each month is overvalued in Tibet and other regions. CMADS is consistent with observed data, except for a little overestimation in April to August in Tibet. Wind speed of CFSR still overestimated heavily in all three basins. Wind speed calculation of CMADS is more satisfactory in Tibet, but it underestimates in Qinghai and other regions. CMADS and CFSR calculate greater wind speed in winter and spring in comparison with summer and autumn, and this seasonal distribution is similar to the observed, gridded data (Figure 4).

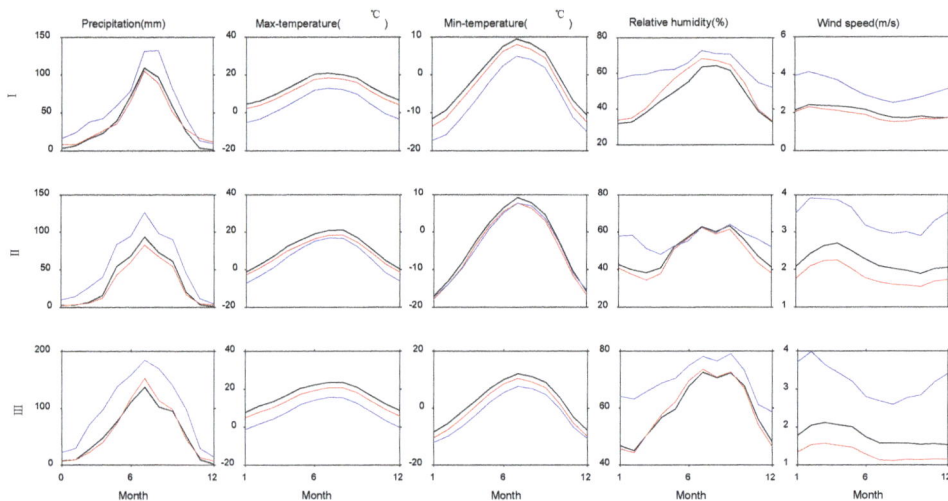

Figure 5. Monthly average value in Tibet (I), Qinghai Provence (II), and other areas(III). (Bleak line is observed, Red line is CMADS, blue line is CFSR).

4.3. Runoff Simulation in the Yellow River Source Basin

DEM is used in watershed delineation to generate stream networks and divide sub-basins; 25 sub-basins are divided in the Yellow River Source Basins. 2455 HRUs are generated by land use, soil and slope. Eleven observed stations, 107 CMADS grid, and 118 CFSR grid points are used to force SWAT in the Yellow River Source Basin. A weather generator which comes from the SWAT model is used to make up for the factors which are lacking from observed data. CMADS and CFSR provide all climate input data. Twelve sensitive parameters are selected to be calibrated. SWAT forced by CMADS (CMADS+SWAT) and CFSR (CFSR+SWAT) are better than observed data (OBS+SWAT) overall (Table 2 and Figure 6). In the monthly scale, NSE for CMADS+SWAT and CFSR+SWAT range from 0.42 to 0.68, which are improvements compared to OBS+SWAT. OBS+SWAT underestimate runoff: the NSE is −0.8 and −0.72 in calibration and validation over the period of a monthly scale. CFSR+SWAT overestimate runoff in most years and seriously overestimated the summer runoff in 2009, 2011 and 2013. NSE of CFSR+SWAT is 0.59 and 0.42 in calibration and validation over the period of a monthly scale. CMADS+SWAT do well in forcing the hydrological model and the simulation results are better, NSE of CMADS+SWAT is 0.68 to 0.58 in the monthly scale. In the daily scale, simulation results are not as good as in the monthly scale for all three datasets, but NSE of CMADS+SWAT is still above 0.5 (Table 2, Figure 7).

Table 2. Evaluation for simulation result of monthly scale.

Forcing Data	Monthly				Daily			
	Calibration		Validation		Calibration		Validation	
	R^2	NSE	R^2	NSE	R^2	NSE	R^2	NSE
OBS+SWAT	0.56	−0.8	0.49	−0.72	0.46	−0.72	0.43	−0.91
CMADS+SWAT	0.91	0.78	0.86	0.68	0.83	0.63	0.71	0.59
CFSR+SWAT	0.89	0.69	0.75	0.52	0.65	0.42	0.58	0.47

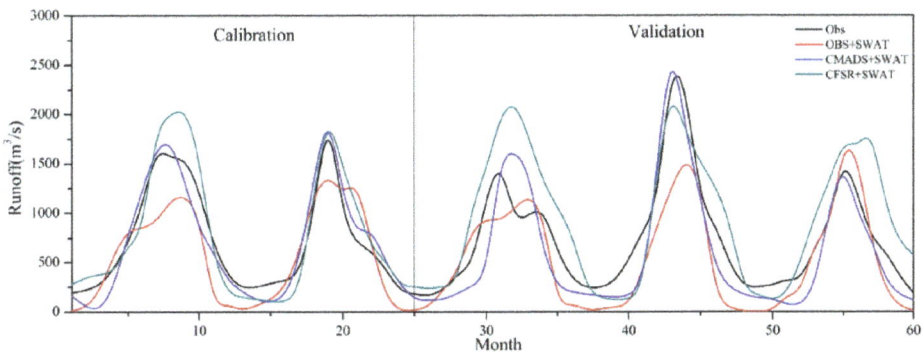

Figure 6. Simulation result of OBS+SWAT, CMADS+SWAT and CFSR+SWAT in Yellow River Source Basin in monthly scale (2009–2013).

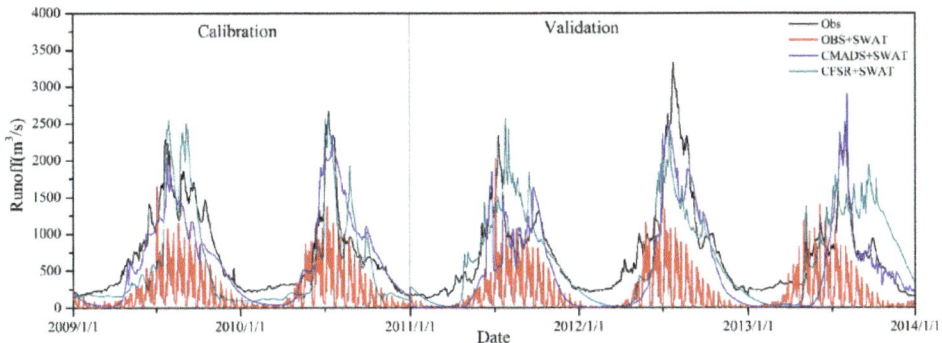

Figure 7. Simulation result of CMADS+SWAT in Yellow River Source Basin in daily scale (2009–2013).

Different results of runoff simulation are mainly caused by the different climate forcing data. The SWAT output results show that annual average precipitation of CMADS is 560.1 mm in 2009–2013, and the corresponding value of OBS and CFSR are 536.5 mm, 815.4 mm respectively. This is consistent with the evaluation results. Precipitation of CFSR is overestimated in the TP (Table 1, Figure 2). The different amounts of precipitation result in the deviation simulation of runoff and streamflow of OBS+SWAT, CMADS+SWAT and CFSR+SWAT and are 171.9 mm, 203.4 mm, 284.7 mm respectively. However, evapotranspiration CMADS+SWAT and CFSR+SWAT is 314.8 mm, 526.4 mm. The difference between humidity, wind speed and max/min temperatures between CMADS and CFSR is the main reason for the deviation, according to Penman-Monteith.

4.4. Analysis of Snowmelt and Related variables

The hydrological process of CMADS+SWAT in March and April 2010 is used to find the relationship between snowmelt (Y) and other variables (X), including: precipitation, 2 m air temperature, solar radiation, soil relative humidity, potential evapotranspiration and soil temperature in different depths. q-statistic measures the association between Y and X variables, both linearly or nonlinearly. X are explanatory variables of Y, and the higher the q-statistic, the more contact between Y and X. Results of q value show that 2 m temperature and soil temperature at a depth of 1.038 m have greater effect on snowmelt; q-statistic values are 0.72 and 0.75 respectively. This is followed by soil relative humidity and soil temperature at depth of ST2, ST3, and ST9 (Table 3). Different q-statistics of different soil layer results from the complicated soil hydrothermal process during the snowmelt period [59]. Shallow soil temperatures are mainly affected by air temperature and have a high correlation with snowmelt results in q-statistics (which are high in ST2 and ST3 (Table 3)). Deep soil temperatures are protected by upper soil layers and have a smaller change. However, when the snow and frozen layers start to melt, this recharges the deeper layers and so the temperature will increase. Therefore, q-statistic of ST8 and ST9 is high.

Table 3. Results of q-statistic (Pre, precipitation; Tem, temperature; PET, potential evapotranspiration; SW, soil moisture; Ele, elevation; STi, soil temperature of layer i).

Factors	Pre	Tem	Wind	Solar	PET	SW	Ele	Slope	ST1
q statistic	0.32	0.72	0.24	0.57	0.34	0.63	0.04	0.04	0.49
Factors	**ST2**	**ST3**	**ST4**	**ST5**	**ST6**	**ST7**	**ST8**	**ST9**	**ST10**
q statistic	0.63	0.66	0.42	0.47	0.47	0.50	0.75	0.65	0.27

The eighth sub-basin has a large amount of snowmelt and is selected for further analysis. Snowmelt increased significantly after March 16, accompanied by a rapid increase in soil moisture (Figure 8a). This process is caused by the increase of temperature. 2 m air temperature increased significantly around 16 March, and two depths soil temperature also show the same clear increase trend (Figure 8b). The change of snowmelt is stable before snowmelt, and soil moisture increased significantly as the snowmelt process occurs. Soil moisture supplementation by the melt of snow and frozen soil, and the large area of wetlands make this process more obvious. After snowmelt, evaporation rates increased and more soil water is evaporated. Deep soil temperature is less affected by air temperature, as shown in Figure 8. Before snowmelt, air temperature and soil temperature rises fluctuate, but as air temperature rises, there is an obvious increase in snowmelt and soil temperature.

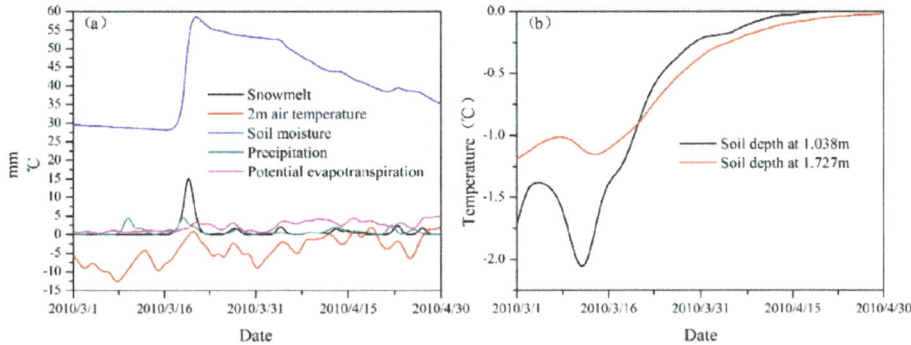

Figure 8. Time series of hydrological factors from CMADS+SWAT in March and April 2010 (**a**) and the soil temperature during corresponding period (**b**).

5. Conclusions

Reanalysis datasets are an important alternative to observed data, especially for regions with few weather stations. They can provide several meteorological factors with higher-resolution data, which is profitable for hydrological simulation. However, various reanalysis datasets still have differences in sources, bias-corrected methods, resolution and temporal coverage et al. CMADS and CFSR are evaluated in this paper and the results show that bias-correction by observed weather data is important for reanalysis. CMADS assimilates nearly 40,000 regional automatic stations under China's 2421 National Automatic and Business Assessment Centers, so that data accuracy is considerably improved. A complete set of data should contain as many climate elements as possible. Accuracy of relative humidity, wind speed, solar radiation etc. should receive more attention, not only due to their meteorological significance, but because they are also important to hydrological, ecological and erosion research etc. Evaluation results of CMADS and CFSR indicate that relative humidity and wind speed still have room for improvement (Table 1, Figures 2 and 5). Besides, long-term series are more representative. CMADS just covers 9 years, compared to 35 years of CFSR, and is therefore too short. In overview, with good manifestation in meteorological elements and forcing hydrological models, it is hoped that authors expand the time series so as to provide convenience in assessing hydrological changes in a long-term context.

In this paper, precipitation, max/min temperatures, relative humidity and wind speed from CMADS and CFSR are evaluated. Discrepancies between these two datasets are fully demonstrated and main results are displayed as follows:

Compared with 131 metrological stations, daily precipitation is more difficult to simulate accurately. The average R for CMADS precipitation is 0.46, which is similar to CFSR ($R = 0.43$). R of CMADS and the CFSR max/min-temperatures is better, and the range is within 0.93–0.98. CMADS and CFSR both underestimate max/min-temperatures and average BIAS is cold. The average RMSE of max/min-temperatures is within 2.99–8.22 °C. Deviation of CMADS and CFSR temperature time series is close to observed data. Relative humidity and wind speed for CMADS is superior to those of CFSR according to various indexes.

The professional interpolation software ANUSPLIN is used to obtain the spatial distribution of annual average precipitation, max/min-temperatures, relative humidity and wind speed, based on the data from weather stations. Distribution of the three kinds of data is generally similar, but the local differences are more obvious. Precipitation of CFSR is overestimated in the whole TP, and unusually large values appeared in the southeast. Precipitation of CMADS is similar with observed data in distribution and in amount. As for the maximum and minimum temperatures, all three datasets have better consistency. Distribution of relative humidity of observed data shows that it is moister in the

Water **2018**, *10*, 513

southeast and drier in the west, and this is different to what CMADS and CFSR present. A difference of distribution of wind speed is obvious in the northwest between observation and reanalysis.

CMADS has unique advantages in hydrological simulations compared with observed data and CFSR. Runoff simulations have achieved satisfactory results in the Yellow River Source Basin. NSE of CMADS+SWAT is 0.78 and 0.68 in calibration and validation, NSE of CFSR+SWAT is 0.69 and 0.52 in the Yellow River Source Basin and OBS+SWAT is unsatisfactory (NSE < 0). Obvious snow melting processes appeared in March and the temperature and soil moisture increased significantly around this time period. There are only eleven weather stations located in the Yellow River Source Basin, and these are located in the lower elevation areas of the eastern region, which means they are not representative. It is therefore difficult to achieve satisfactory simulation results only through adjustment of parameters in SWAT. Simulation results of runoff in the watershed are improved by two reanalysis datasets, due to their high resolution and quality, though CFSR overestimates precipitation in the Yellow River Source Basin, and results in excessive runoff. 2 m air temperature, soil moisture and 1.038 m depth soil temperature contribute more to snowmelt as shown, when measured by GeoDetector. Climate forcing data is important, deviation of precipitation (Table 1, Figure 2) results in the different amounts of runoff (Figure 6), and the temperature, humidity and wind speed, etc. also play an important role in calculating evapotranspiration. Evaluation of various reanalyses before forcing hydrological models is essential.

Acknowledgments: This work was financially supported by the National Natural Science Foundation of China (Grant nos. 41671066 & 41730751 & 41401084) and KJF-STS-ZDTP-015 & SKLCS-ZZ-2017.

Author Contributions: Donghui Shangguan, Shiyin Liu and Yongjian Ding conceived and designed the experiments; Jun Liu performed the experiments and wrote the paper.

Conflicts of Interest: The authors declare no conflict of interest.

References

1. Qiu, J. The Third Pole. *Nature* **2008**, *454*, 393–396. [CrossRef] [PubMed]
2. Zhang, Q.; Kong, D.; Shi, P.; Singh, V.P.; Sun, P. Vegetation phenology on the Qinghai-Tibetan Plateau and its response to climate change (1982–2013). *Agric. For. Meteorol.* **2018**, *248*, 408–417. [CrossRef]
3. Zhou, J.; Pomeroy, J.W.; Zhang, W.; Cheng, G.; Wang, G.; Chen, C. Simulating cold regions hydrological processes using a modular model in the west of China. *J. Hydrol.* **2014**, *509*, 13–24. [CrossRef]
4. Silberstein, R.P. Hydrological models are so good, do we still need data? *Environ. Model. Softw.* **2006**, *21*, 1340–1352. [CrossRef]
5. Sun, W.; Ishidaira, H.; Bastola, S.; Yu, J. Estimating daily time series of streamflow using hydrological model calibrated based on satellite observations of river water surface width: Toward real world applications. *Environ. Res.* **2015**, *139*, 36–45. [CrossRef] [PubMed]
6. Ma, Y.; Yang, Y.; Han, Z.; Tang, G.; Maguire, L.; Chu, Z.; Hong, Y. Comprehensive evaluation of Ensemble Multi-Satellite Precipitation Dataset using the Dynamic Bayesian Model Averaging scheme over the Tibetan plateau. *J. Hydrol.* **2018**, *556*, 634–644. [CrossRef]
7. Huang, D.; Gao, S. Impact of different reanalysis data on WRF dynamical downscaling over China. *Atmos. Res.* **2018**, *200*, 25–35. [CrossRef]
8. Agutu, N.O.; Awange, J.L.; Zerihun, A.; Ndehedehe, C.E.; Kuhn, M.; Fukuda, Y. Assessing multi-satellite remote sensing, reanalysis, and land surface models' products in characterizing agricultural drought in East Africa. *Remote Sens. Environ.* **2017**, *194*, 287–302. [CrossRef]
9. Moalafhi, D.B.; Sharma, A.; Evans, J.P. Reconstructing hydro-climatological data using dynamical downscaling of reanalysis products in data-sparse regions—Application to the Limpopo catchment in southern Africa. *J. Hydrol. Reg. Stud.* **2017**, *12*, 378–395. [CrossRef]
10. Stopa, J.E.; Cheung, K.F. Intercomparison of wind and wave data from the ECMWF Reanalysis Interim and the NCEP Climate Forecast System Reanalysis. *Ocean Model.* **2014**, *75*, 65–83. [CrossRef]

11. Sharp, E.; Dodds, P.; Barrett, M.; Spataru, C. Evaluating the accuracy of CFSR reanalysis hourly wind speed forecasts for the UK, using in situ measurements and geographical information. *Renew. Energy* **2015**, *77*, 527–538. [CrossRef]

12. Zeng, J.; Li, Z.; Chen, Q.; Bi, H.; Qiu, J.; Zou, P. Evaluation of remotely sensed and reanalysis soil moisture products over the Tibetan Plateau using in-situ observations. *Remote Sens. Environ.* **2015**, *163*, 91–110. [CrossRef]

13. Xu, H.; Xu, C.-Y.; Chen, S.; Chen, H. Similarity and difference of global reanalysis datasets (WFD and APHRODITE) in driving lumped and distributed hydrological models in a humid region of China. *J. Hydrol.* **2016**, *542*, 343–356. [CrossRef]

14. Tomy, T.; Sumam, K.S. Determining the Adequacy of CFSR Data for Rainfall-Runoff Modeling Using SWAT. *Procedia Technol.* **2016**, *24*, 309–316. [CrossRef]

15. Wang, X.; Pang, G.; Yang, M.; Zhao, G. Evaluation of climate on the Tibetan Plateau using ERA-Interim reanalysis and gridded observations during the period 1979–2012. *Quat. Int.* **2017**, *444*, 76–86. [CrossRef]

16. Gao, L.; Hao, L.; Chen, X.-W. Evaluation of ERA-interim monthly temperature data over the Tibetan Plateau. *J. Mt. Sci.* **2014**, *11*, 1154–1168. [CrossRef]

17. Song, C.; Huang, B.; Ke, L.; Ye, Q. Precipitation variability in High Mountain Asia from multiple datasets and implication for water balance analysis in large lake basins. *Glob. Planet. Chang.* **2016**, *145*, 20–29. [CrossRef]

18. You, Q.; Min, J.; Zhang, W.; Pepin, N.; Kang, S. Comparison of multiple datasets with gridded precipitation observations over the Tibetan Plateau. *Clim. Dyn.* **2014**, *45*, 791–806. [CrossRef]

19. Wang, A.; Zeng, X. Evaluation of multireanalysis products with in situ observations over the Tibetan Plateau. *J. Geophys. Res. Atmos.* **2012**, *117*. [CrossRef]

20. Poméon, T.; Jackisch, D.; Diekkrüger, B. Evaluating the performance of remotely sensed and reanalysed precipitation data over West Africa using HBV light. *J. Hydrol.* **2017**, *547*, 222–235. [CrossRef]

21. Essou, G.R.C.; Brissette, F.; Lucas-Picher, P. Impacts of combining reanalyses and weather station data on the accuracy of discharge modelling. *J. Hydrol.* **2017**, *545*, 120–131. [CrossRef]

22. Kan, B.Y.; Su, F.G.; Tong, K.; Zhang, L.L. Analysis of the Applicability of Four Precipitation Datasets in the Upper Reaches of the Yarkant River, the Karakorum. *J. Glaciol. Geocryol.* **2013**, *35*, 710–722.

23. Guo, Y.; Wang, Z.; Wu, Y. Comparison of applications of different reanalyzed precipitation data in the Lhasa River Basin based on HIMS model. *Prog. Geogr.* **2017**, *36*, 1033–1039.

24. Gao, R.; Mu, Z.; Peng, L.; Z, Y.; Yin, Z.; Tang, R. Application of CFSR and ERA-Interim Reanalysis Data in Runoff Simulation in High Cold Alpine Area. *Water Resour. Power* **2017**, *35*, 8–12.

25. Malago, A.; Bouraoui, F.; Vigiak, O.; Grizzetti, B.; Pastori, M. Modelling water and nutrient fluxes in the Danube River Basin with SWAT. *Sci. Total Environ.* **2017**, *603–604*, 196–218. [CrossRef] [PubMed]

26. Patil, A.; Ramsankaran, R. Improving streamflow simulations and forecasting performance of SWAT model by assimilating remotely sensed soil moisture observations. *J. Hydrol.* **2017**, *555*, 683–696. [CrossRef]

27. Vellei, M.; Herrera, M.; Fosas, D.; Natarajan, S. The influence of relative humidity on adaptive thermal comfort. *Build. Environ.* **2017**, *124*, 171–185. [CrossRef]

28. Xiong, Y.; Meng, Q.-S.; Gao, J.; Tang, X.-F.; Zhang, H.-F. Effects of relative humidity on animal health and welfare. *J. Integr. Agric.* **2017**, *16*, 1653–1658. [CrossRef]

29. Wang, Y.; Ma, H.; Wang, D.; Wang, G.; Wu, J.; Bian, J.; Liu, J. A new method for wind speed forecasting based on copula theory. *Environ. Res.* **2018**, *160*, 365–371. [CrossRef] [PubMed]

30. Sedaghat, A.; Hassanzadeh, A.; Jamali, J.; Mostafaeipour, A.; Chen, W.-H. Determination of rated wind speed for maximum annual energy production of variable speed wind turbines. *Appl. Energy* **2017**, *205*, 781–789. [CrossRef]

31. Li, T.; Li, J. A 564-year annual minimum temperature reconstruction for the east central Tibetan Plateau from tree rings. *Glob. Planet. Chang.* **2017**, *157*, 165–173. [CrossRef]

32. Villarini, G.; Khouakhi, A.; Cunningham, E. On the impacts of computing daily temperatures as the average of the daily minimum and maximum temperatures. *Atmos. Res.* **2017**, *198*, 145–150. [CrossRef]

33. Dile, Y.T.; Srinivasan, R. Evaluation of CFSR climate data for hydrologic prediction in data-scarce watersheds: An application in the Blue Nile River Basin. *JAWRA J. Am. Water Resour. Assoc.* **2014**, *50*, 1226–1241. [CrossRef]

34. Worqlul, A.W.; Yen, H.; Collick, A.S.; Tilahun, S.A.; Langan, S.; Steenhuis, T.S. Evaluation of CFSR, TMPA 3B42 and ground-based rainfall data as input for hydrological models, in data-scarce regions: The upper Blue Nile Basin, Ethiopia. *Catena* **2017**, *152*, 242–251. [CrossRef]

35. Hu, S.; Qiu, H.; Yang, D.; Cao, M.; Song, J.; Wu, J.; Huang, C.; Gao, Y. Evaluation of the applicability of climate forecast system reanalysis weather data for hydrologic simulation: A case study in the Bahe River Basin of the Qinling Mountains, China. *J. Geogr. Sci.* **2017**, *27*, 546–564. [CrossRef]

36. Tian, L.; Liu, T.; Bao, A.M.; Huang, Y. Application of CFSR Precipitation Dataset inHydrological model for Arid Mountains Area: A Case Study in the Kaidu River Basin. *Arid Zone Res.* **2017**, *34*, 755–761.

37. Meng, X.Y.; Dan, L.Y.; Liu, Z.-H. Energy balance-based SWAT model to simulate the mountain snowmelt and runoff—Taking the application in Juntanghu watershed (China) as an example. *J. Mt. Sci.* **2015**, *12*, 368–381. [CrossRef]

38. Wang, Y.J.; Meng, X.Y.; Liu, Z.H.; Ji, X.N. Snowmelt Runoff Analysis under Generated Climate Change Scenarios for the Juntanghu River Basin, in Xinjiang, China. *Tecnología y Ciencias del Agua* **2016**, *7*, 41–54.

39. Meng, X.Y. Spring Flood Forecasting Based on the WRF-TSRM Mode. *Tehnički Vjesnik* **2018**, *25*, 141–151.

40. Meng, X.; Wang, H.; Lei, X.; Cai, S.; Wu, H.; Ji, X.; Wang, J. Hydrological Modeling in the Manas River Basin Using Soil and Water Assessment Tool Driven by CMADS. *Tehnički Vjesnik* **2017**, *24*, 525–534.

41. Zhu, X.; Wu, T.; Li, R.; Wang, S.; Hu, G.; Wang, W.; Qin, Y.; Yang, S. Characteristics of the ratios of snow, rain and sleet to precipitation on the Qinghai-Tibet Plateau during 1961–2014. *Quat. Int.* **2017**, *444*, 137–150. [CrossRef]

42. Qin, Y.; Yang, D.; Gao, B.; Wang, T.; Chen, J.; Chen, Y.; Wang, Y.; Zheng, G. Impacts of climate warming on the frozen ground and eco-hydrology in the Yellow River source region, China. *Sci. Total Environ.* **2017**, *605–606*, 830–841. [CrossRef] [PubMed]

43. Nicoll, T.; Brierley, G.; Yu, G.A. A broad overview of landscape diversity of the Yellow River source zone. *J. Geogr. Sci.* **2013**, *23*, 793–816. [CrossRef]

44. Meng, X.; Wang, H. Significance of the China Meteorological Assimilation Driving Datasets for the SWAT Model (CMADS) of East Asia. *Water* **2017**, *9*, 765. [CrossRef]

45. Meng, X.; Wang, H.; Wu, Y.; Long, A.; Wang, J.; Shi, C.; Ji, X. Investigating spatiotemporal changes of the land-surface processes in Xinjiang using high-resolution CLM3.5 and CLDAS: Soil temperature. *Sci. Rep.* **2017**, *7*, 13286. [CrossRef] [PubMed]

46. Shi, C.X.; Xie, Z.H.; Qian, H.; Liang, M.L.; Yang, X.C. China land soil moisture EnKF data assimilation based on satellite remote sensing data. *Sci. China Earth Sci.* **2011**, *54*, 1430–1440. [CrossRef]

47. Fuka, D.R.; Walter, M.T.; MacAlister, C.; Degaetano, A.T.; Steenhuis, T.S.; Easton, Z.M. Using the Climate Forecast System Reanalysis as weather input data for watershed models. *Hydrol. Process.* **2014**, *28*, 5613–5623. [CrossRef]

48. Grusson, Y.; Anctil, F.; Sauvage, S.; Sánchez Pérez, J. Testing the SWAT Model with Gridded Weather Data of Different Spatial Resolutions. *Water* **2017**, *9*, 54. [CrossRef]

49. Hutchinson, M.F.; Gessler, P.E. Splines—More than just a smooth interpolator. *Geoderma* **1994**, *62*, 45–67. [CrossRef]

50. McKenney, D.W.; Pedlar, J.H.; Papadopol, P.; Hutchinson, M.F. The development of 1901–2000 historical monthly climate models for Canada and the United States. *Agric. For. Meteorol.* **2006**, *138*, 69–81. [CrossRef]

51. Liu, Z.H.; Li, L.T.; McVicar, T.R.; Van Niel, T.G.; Yang, Q.K.; Li, R. Introduction of the Professional Interpolation Software for Meteorology Data: ANUSPLINN. *Meteorol. Mon.* **2008**, *34*, 92–100.

52. Wang, J.F.; Li, X.H.; Christakos, G.; Liao, Y.L.; Zhang, T.; Gu, X.; Zheng, X.Y. Geographical Detectors-Based Health Risk Assessment and its Application in the Neural Tube Defects Study of the Heshun Region, China. *Int. J. Geogr. Inf. Sci.* **2010**, *24*, 107–127. [CrossRef]

53. Zhao, Y.; Deng, Q.; Lin, Q.; Cai, C. Quantitative analysis of the impacts of terrestrial environmental factors on precipitation variation over the Beibu Gulf Economic Zone in Coastal Southwest China. *Sci. Rep.* **2017**, *7*, 44412. [CrossRef] [PubMed]

54. Golkar, F.; Sabziparvar, A.A.; Khanbilvardi, R.; Nazemosadat, M.J.; Parsa, S.Z.; Rezaei, Y. Estimation of instantaneous air temperature using remote sensing data. *Int. J. Remote Sens.* **2018**, *39*, 258–275. [CrossRef]

55. Moriasi, D.N.; Arnold, J.G.; Liew, M.W.V.; Bingner, R.L.; Harmel, R.D.; Veith, T.L. Model Evaluation Guidelines for Systematic Quantification of Accuracy in Watershed Simulations. *Trans. Asabe* **2007**, *50*, 885–900. [CrossRef]

56. Beck, H.E.; van Dijk, A.I.J.M.; Levizzani, V.; Schellekens, J.; Miralles, D.G.; Martens, B.; de Roo, A. MSWEP: 3-hourly 0.25°global gridded precipitation (1979–2015) by merging gauge, satellite, and reanalysis data. *Hydrol. Earth Syst. Sci.* **2017**, *21*, 589–615. [CrossRef]

57. Sheng, H.U.; Cao, M.; Qiu, H.; Song, J.; Jiang, W.U.; Yu, G.; Jingzhong, L.I.; Sun, K. Applicability evaluation of CFSR climate data for hydrologic simulation: A case study in the Bahe River Basin. *Acta Geogr. Sin.* **2016**, *71*, 1571–1587.

58. Zeng-Yun, H.U.; Yong-Yong, N.I.; Shao, H.; Yin, G.; Yan, Y.; Jia, C.J. Applicability study of CFSR, ERA-Interim and MERRA precipitation estimates in Central Asia. *Arid Land Geogr.* **2013**, *36*, 700–708.

59. Wang, J.Z.; Liu, Z.H.; Tiyip, T.; Wang, L.; Zhang, B. Thawing Process of Seasonal Frozen Soil on Northern Slope of the Tianshan Mountains During Snowmelt Period. *Arid Zone Res.* **2017**, *34*, 282–291.

Article

Evaluation of Multi-Satellite Precipitation Products for Streamflow Simulations: A Case Study for the Han River Basin in the Korean Peninsula, East Asia

Thom Thi Vu, Li Li and Kyung Soo Jun *

Graduate School of Water Resources, Sungkyunkwan University, Suwon 16419, Korea;
vuthom.khtn@gmail.com (T.T.V.); lili0809@skku.edu (L.L.)
* Correspondence: ksjun@skku.edu; Tel.: +82-31-290-7515

Received: 27 March 2018; Accepted: 13 May 2018; Published: 16 May 2018

Abstract: The accuracy and sufficiency of precipitation data play a key role in environmental research and hydrological models. They have a significant effect on the simulation results of hydrological models; therefore, reliable hydrological simulation in data-scarce areas is a challenging task. Advanced techniques can be utilized to improve the accuracy of satellite-derived rainfall data, which can be used to overcome the problem of data scarcity. Our study aims to (1) assess the accuracy of different satellite precipitation products such as Tropical Rainfall Measuring Mission (TRMM 3B42 V7), Precipitation Estimation from Remotely Sensed Information using Artificial Neural Networks (PERSIANN), PERSIANN-Climate Data Record (PERSIANN-CDR), and China Meteorological Assimilation Driving Datasets for the SWAT Model (CMADS) by comparing them with gauged rainfall data; and (2) apply them for runoff simulations for the Han River Basin in South Korea using the SWAT model. Based on the statistical measures, that is, the proportion correct (PC), the probability of detection (POD), the frequency bias index (FBI), the index of agreement (IOA), the root-mean-square-error (RMSE), the mean absolute error (MAE), the coefficient of determination (R^2), and the bias, the rainfall data of the TRMM and CMADS show a better accuracy than those of PERSIANN and PERSIANN-CDR when compared to rain gauge measurements. The TRMM and CMADS data capture the spatial rainfall patterns in mountainous areas as well. The streamflow simulated by the SWAT model using ground-based rainfall data agrees well with the observed streamflow with an average Nash-Sutcliffe efficiency (NSE) of 0.68. The four satellite rainfall products were used as inputs in the SWAT model for streamflow simulation and the results were compared. The average R^2, NSE, and percent bias (PBIAS) show that hydrological models using TRMM ($R^2 = 0.54$, NSE = 0.49, PBIAS = [-52.70–28.30%]) and CMADS ($R^2 = 0.44$, NSE = 0.42, PBIAS = [-29.30–41.80%]) data perform better than those utilizing PERSIANN ($R^2 = 0.29$, NSE = 0.13, PBIAS = [38.10–83.20%]) and PERSIANN-CDR ($R^2 = 0.25$, NSE = 0.16, PBIAS = [12.70–71.20%]) data. Overall, the results of this study are satisfactory, given that rainfall data obtained from TRMM and CMADS can be used to simulate the streamflow of the Han River Basin with acceptable accuracy. Based on these results, TRMM and CMADS rainfall data play important roles in hydrological simulations and water resource management in the Han River Basin and in other regions with similar climate and topographical characteristics.

Keywords: TRMM; PERSIANN; PERSIANN-CDR; CMADS; satellite-derived rainfall; streamflow simulation; SWAT; Han River

1. Introduction

Precipitation is one of the most essential components of the hydrological cycle [1]. The quantity and quality of the precipitation data used as the principal input to hydrological models affect the accuracy of the simulation results [2,3]. Rain gauges provide direct precipitation measures; however, scarcity and irregularity problems with respect to the gauge network considerably influence the data reliability [4,5]. In addition, an effective spatial coverage of precipitation over a large area is difficult. Compared with rain gauges that provide rainfall data by accumulating rainfall over a time interval, weather radar systems provide an instantaneous spatial measure of precipitation and thus, produce rapid climate information [6]. However, Westrick et al. [7] investigated the limitations of the radar network for quantitative precipitation measurement and showed that the radar-derived precipitation estimates could not represent the regional precipitation since radar coverage is limited to lowland areas. The drawbacks of radar-derived data, such as coverage area limitations, costly infrastructure construction, and inaccuracy under complex atmospheric conditions, result in the poor performance of hydrological models [8]. Currently, visible and thermal infrared sensors onboard the geostationary Earth-orbiting satellites and passive microwave sensors onboard the low-Earth-orbiting satellites provide more accurate rainfall estimates at a higher measurement frequency. Based on the advancements of these techniques, several satellite-based precipitation products with global high-resolution (up to 0.25°) are now available such as those derived from the Tropical Rainfall Measuring Mission (TRMM), Precipitation Estimation from Remotely Sensed Information using Artificial Neural Networks (PERSIANN), and Climate Prediction Center morphing technique (CMORPH) [9]. Those global and near-real-time rainfall estimates are extremely attractive for hydrological and weather studies [10–12]. The rainfall estimates from PERSIANN, CMORPH, and TRMM-based Multi-satellite Precipitation Analysis (TMPA), which combines satellite data from different sensors and ground station data from the Global Precipitation Climatology Centre, have been widely applied in numerous studies [13–21]. Recently, the China Meteorological Assimilation Driving Datasets for the SWAT Model (CMADS), which consist of reanalyzed data based on assimilation techniques, provided important basic data (that is, rainfall, maximum and minimum temperature, solar radiation, relative humidity, and wind speed) that are extremely useful to analyze climate–water cycles and "macro" energy balances in hydrological studies [22,23]. The CMADS was developed by Dr. Xianyong Meng from the China Institute of Water Resources and Hydropower Research (IWHR) and has received worldwide attention [22]. Based on the full coverage of East Asia and an improved accuracy, CMADS promises to be one of the most useful satellite-derived weather datasets for meteorological and hydrological research.

Distributed hydrological models have been widely applied in water resource management and hydrological research [24]. They include the Hydrologic Simulation Program-Fortran (HSPF) [25], MIKE SHE [26], the Hydrologic Modeling System (HEC-HMS) [27], and the Soil and Water Assessment Tool (SWAT) [28]. Those models reduce the dependency on specific precipitation inputs and fully use satellite-based hydrometeorological data [29]. The SWAT has been widely applied because many studies showed that the SWAT model can simulate streamflow in regions with limited data well [30–32]. Studies of SWAT applications in South Korea include those by Kang et al. [33], Kim et al. [34], Bae et al. [35], Kim et al. [36], Shope et al. [37], and Cho et al. [38]. Kim et al. [34] suggested the integrated SWAT-MODFLOW model which can simulate the interaction between the river flow and the saturated aquifer. Kim et al. [36] proposed a method for the evaluation of the flow regulation effects by dams on river flow using the SWAT model for the Han River Basin. In those studies, the SWAT model was successfully applied to mountainous areas and river basins with various sizes; however, a SWAT model using satellite-based rainfall data has not been considered.

Satellite-derived rainfall estimates with high spatial resolution contribute to the water resources management especially in areas where the ground-based climate data are limited. The hydrologic performance of different satellite rainfall products varies regionally because of several factors such as instrument characteristics and retrieval algorithms. Kim et al. [39] compared four satellite precipitation products for the hydrological utility at a mountainous basin in South Korea and found that TMPAv6 and TMPAv7 products were closer to ground-based rainfall than CMORPH and global satellite mapping of precipitation (GSMaP). In the streamflow simulation, TMPAv6 and TMPAv7 performed well while CMORPH and GSMaP resulted in a large underestimation.

Although there are several flood control dams in the river, large floods have occurred in the downstream area of the Han River Basin, causing severe damages. The use of advanced techniques to predict precipitation that causes floods has attracted a large interest in recent years [40]. The application of satellite precipitation data with high accuracy and resolution has been widely studied to be able to respond quickly in real-world situations. Numerous studies successfully applied the SWAT model or used satellite rainfall products for various regions of South Korea. However, combining satellite rainfall products with the SWAT model was not considered. This study investigates the hydrologic application of different satellite rainfall products in the Han River Basin of South Korea by applying the SWAT model. The CMADS, a newly developed dataset with the full coverage of East Asia, was evaluated in comparison with other satellite rainfall products. This is a case study for a specific period from 2008 to 2013 as the CMADS is only available from 2008. This study aimed to (1) compare four satellite rainfall products (TRMM 3B42 V7, PERSIANN, PERSIANN-CDR, and CMADS) with gauge rainfall data; and (2) evaluate the accuracy and suitability of the four satellite precipitation products as inputs for streamflow simulations in the Han River Basin. The results of the study can provide information on the performance of different satellite rainfall products in hydrologic modeling for the Han River Basin. In addition, this study contributes to enriching the scientific database on hydrologic applications of different satellite precipitation datasets, especially for data scarce regions.

2. Materials and Methods

2.1. Study Area

This study focused on the Han River Basin; the river originates from Mt. Taebaek and flows into the Yellow Sea (Figure 1). The Han River Basin is the largest river basin (26,219 km^2) in South Korea and occupies approximately 27% of the country's area [41]. The river has a total length of 5417 km and comprises of two major branches, that is, the North Han River (NHR; 10,652 km^2) and South Han River (SHR; 12,514 km^2). These branches converge at the immediate upstream of the Paldang Lake, which is the major source of water supply to the Seoul metropolitan area and forms the main stream (Figure 1).

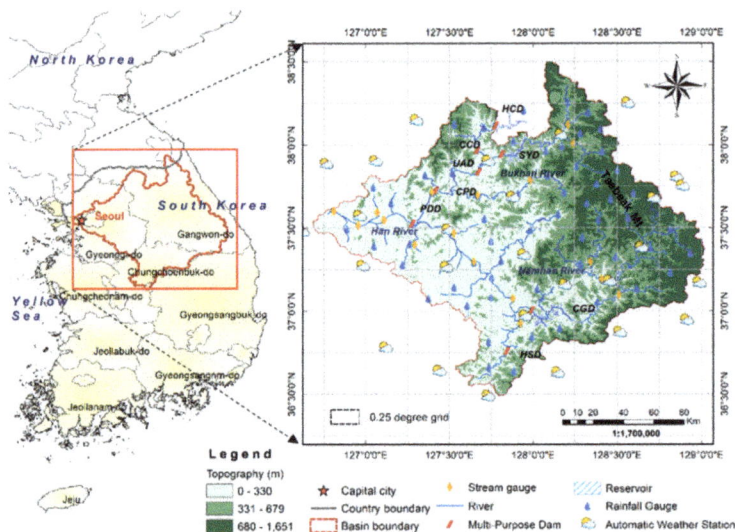

Figure 1. The Han River basin.

The Han River Basin has four distinct seasons. The monthly average temperature varies from −6 °C to 3 °C during winter and from 23 °C to 25 °C during summer. The average precipitation during the rainy season is 1272.5 mm, which accounts for 70% of the average annual precipitation. A large amount of precipitation is concentrated in the mountainous area in the eastern part of the basin, while less precipitation occurs in the lowlands of the western part due to the influence of western North Pacific typhoons and summer monsoon precipitation [42,43].

Both precipitation and groundwater significantly affect the water resources of the Han River [44]. The surface runoff mainly originates from local precipitation during the summer monsoon season (June to September) and is maintained by groundwater seepage during the remainder of the year [44]. Several hydraulic structures, such as flood control dams, significantly affect the river flow (Figure 1). However, the flow of the Han River generally varies depending on the season, that is, it ranges from 150 m^3/s in January to 4300 m^3/s in August [44].

2.2. Precipitation Data from Rain Gauges

The ground-based precipitation data used in this study consist of daily records from 89 rain gauges from 2007 to 2013 (Figure 1). The daily precipitation data were gathered from an extensive ground-based data network consisting of 60 synoptic stations of the Water Management Information System and 29 automatic weather stations (AWS) of the Korea Meteorological Administration. At the AWS sites, rain is first detected by the sensor of the rain detector and then by the sensor of the tipping bucket rain gauge, with an accuracy of 0.5 mm [45]. Homogenization and outlier processing were used for the quality control of the gauged data [46].

In the Han River Basin, the precipitation data were collected from 89 gauge stations that could be considered as a large number of stations. For a precise climatological/hydrological study, however, 89 stations may not be sufficient for a basin with the area of 26,219 km^2. Observations with more homogeneous distribution and fine spatial and temporal resolution could produce more accurate results. The limitation of ground-based observations can be compensated by well-validated satellite-derived estimates. Moreover, since various satellite-derived data are freely available, it could be a cost-effective way to collect data, especially for data-scarce areas.

2.3. Satellite-Derived Precipitation Products

2.3.1. TRMM 3B42 Precipitation Products

The TMPA generated precipitation data for a global coverage of 50° N–S and spatial resolution of 0.25° × 0.25° based on various meteorological satellites [47]. The TRMM 3B42 V7 includes two types of datasets, that is, the 3-hourly (corresponding to 3-hour intervals per day, that is, UTC 00:00, 03:00, 06:00, 09:00, 12:00, 15:00, 18:00, and 21:00) and the daily precipitation products. The datasets were obtained by the TRMM Algorithm 3B42, which was calibrated by multiple independent precipitation estimates from the optimal combination of 2B-31, 2A-12, Special Sensor Microwave Imager, Special Sensor Microwave Imager/Sounder, Advanced Microwave Sounding Unit, Microwave Humidity Sounder, and microwave-adjusted merged geo-infrared (IR). In this study, TRMM 3B42 V7 satellite rainfall data with a daily resolution were used for the period of 2008–2013. The data can be downloaded from the Goddard Space Flight Center website (https://mirador.gsfc.nasa.gov).

2.3.2. PERSIANN and PERSIANN-CDR Products

The PERSIANN data were produced by using the artificial neural network algorithm to estimate the rainfall rate based on longwave IR images from geostationary Earth-orbiting satellites. Rainfall data with a spatial coverage of 60° S–N and spatial resolution of 0.25° × 0.25° are available for March 2000 to the present [48]. The PERSIANN-CDR rainfall data were generated by the PERSIANN algorithm using Gridsat-B1 IR satellite data and the bias was adjusted using monthly products from the Global Precipitation Climatology Project. The spatial resolution and coverage of this rainfall product are consistent with that of the PERSIANN dataset; data are available for January 1983 to April 2017. In this study, both the PERSIANN and PERSIANN-CDR products on a daily time scale were obtained from the Center for Hydrometeorology and Remote Sensing (http://chrsdata.eng.uci.edu/) from 2008 to 2013.

2.3.3. CMADS Precipitation Products

The CMADS is an atmospheric reanalysis dataset, which is obtained by using various assimilation techniques and multi-source data. The Space and Time Multiscale Analysis System (STMAS) assimilation technique combined with big data projection and processing methods was used to produce the CMADS climate dataset. Multi-source data were obtained from the National Centers for Environmental Prediction/National Center for the Atmospheric Research NCEP/NCAR-R1 reanalysis dataset, the National Centers for Environmental Prediction-Department of Energy (NCEP-DOE)-(R2) reanalysis dataset, the Climate Forecast System Reanalysis (CFSR) by NCEP, the European Centre for Medium-Range Weather Forecasts 15-year Reanalysis (ECMWF ERA-15), the ECMWF Reanalysis (ERA-40), the ECMWF Reanalysis-Interim (ERA-Interim), the Japanese 25-year Reanalysis project (JRA-25), and the CMA Land Data Assimilation System (CLDAS) [22]. The CMADS rainfall data provided by the China National Meteorological Information Center were produced using CMORPH's integrated precipitation products and validated with gauged rainfall data. In this study, CMADS V1.1 data (2008–2013) with a spatial coverage of 60° E–160° E longitude and 0° N–65° N latitude, daily resolution, and spatial resolution of 0.25° × 0.25° were used. The detailed description of different satellite rainfall datasets is shown in Table 1.

Table 1. The description of satellite-based rainfall datasets used in this study.

Dataset	Version	Spatial/Temporal Resolution	Areal Coverage	Time Coverage	Sources
TRMM	3B42 V7	0.250/daily	Near Global	1998–present	Huffman et al. [47]
PERSIANN	-	0.250/daily	Near Global	2000–present	Sorooshian et al. [48]
PERSIANN-CDR	CDR	0.250/daily	Near Global	1983–2017	Ashouri et al. [49]
CMADS	V1.1	0.250/daily	East Asia	2008–2014	Meng [22]

2.4. SWAT Model

The SWAT is a conceptual and semi-distributed model that simulates, based on a daily time step, various variables related to hydrology, weather, soil erosion, soil temperature, plant growth, nutrients, pesticides, and land management [28]. The SWAT model uses the SCS curve number and modified rational methods to calculate the surface runoff and estimate the peak discharge, respectively. The Penman–Monteith and Muskingum methods are used to simulate the evapotranspiration and channel routing, respectively [50]. Based on a delineation process, the basin is divided into sub-basins and hydrologic response units utilizing three essential inputs including the digital elevation model, land use map, and soil map.

Daily rainfall data from 89 stations and other climate data including air temperature, relative humidity, wind speed, and solar radiation from 29 meteorology stations were used as inputs to the model. The outflow of eight reservoirs (Figure 1) obtained from the Korea Water Resources Management Information System were also used as model input. A detailed description of the input data for the SWAT model is shown in Table 2. The SWAT-CUP sequential uncertainty fitting (SUFI-2) program was used for the autocalibration and sensitivity analysis of the SWAT model [51]. In the SUFI-2 program, the calibration objective was to maximize the NSE [52,53]. The SUFI-2 algorithm has been widely used for the calibration of the SWAT model. Wu and Chen [54] and Khoi et al. [55] showed that the SWAT model calibrated with the SUFI-2 algorithm makes better and reasonable predictions than models calibrated with other auto-calibrated methods. To construct a homogeneous set of optimal parameters for the model, a calibration was performed using the streamflow (2008–2010) measured at 16 gauging stations in the main branches. The calibrated parameters were then used for the model validation (2011–2013). The data from 2007 were used for the model warm-up.

Table 2. The description of the input data used in the Soil and Water Assessment Tool (SWAT).

Data Type	Data Description	Scale	Data Source
Topography map	Digital elevation map (DEM)	90 m	USGS-HydroSHEDS
Land-use/Land cover map	Land use/Land cover classification 2010	1:1,250,000	Korea Ministry of Environment
Soil map	Soil types (2007)	10 km	Food and Agriculture Organization
Meteorology	Daily precipitation, Minimum and maximum temperature, Solar radiation, Relative humidity, Wind speed	1990–2013	Korea Meteorological Administration and Water Resources Management Information System
Hydrological data	Discharge, Dam operation, Reservoir characteristics	2008–2013	Water Resources Management Information System

2.5. Statistical Measures for Precipitation and Runoff

The spatiotemporal variability of satellite precipitation products was compared to the ground-based data by pixel-to-point comparison. To categorically evaluate and compare the daily satellite-derived precipitation data with ground-based precipitation data, three statistical measures were calculated, including the proportion correct (PC), probability of detection (POD), and frequency bias index (FBI) [56,57]. Those statistical measures indicate the detection capability of satellite rainfall data to estimate the possibility of precipitation events. A precipitation event represents a precipitation day when the daily rainfall is greater than 1 mm/day. The POD, PC, and FBI are defined as follows:

$$PC = (a + d)/n \tag{1}$$

$$POD = a/(a + c) \tag{2}$$

$$FBI = (a + b)/(a + c), \tag{3}$$

where a, b, c, and d are the numbers of precipitation events (a: satellite yes, observation yes; b: satellite yes, observation no; c: satellite no, observation yes; and d: satellite no, observation no) and n is the total number of satellite observation pairs.

The POD, which is also called the hit rate, determines the likelihood of detected rainfall data and the PC represents the accuracy of detected rainfall data. The POD and PC values range from 0 to 1 and can be used to assess the level of agreement between satellite-based rainfall and the gauged values. The perfect POD and PC score is 1. The FBI ranges from 0 to infinity, with a perfect score of 1. These three statistical indicators were calculated using a 2 × 2 contingency table (Table 3).

Table 3. The contingency table for the satellite and gauged precipitations with a threshold of 1.0 mm.

Satellite Event	Observation Event		Marginal Total
	Yes ($p \geq 1.0$ mm)	No ($p < 1.0$ mm)	
Yes ($p \geq 1.0$ mm)	a	b	$a + b$
No ($p < 1.0$ mm)	c	d	$c + d$
Marginal total	$a + c$	$b + d$	$n = a + b + c + d$

To evaluate the accuracy of satellite-derived rainfall data by comparing them with gauged precipitation, five statistical indicators were adopted, including the index of agreement (IOA), the root-mean-square-error (RMSE), the mean absolute error (MAE), the coefficient of determination (R^2), and the bias [57]. The IOA measures additive and proportional differences in the observed and satellite-derived means and variances. The RMSE measures the average magnitude of the errors. Since the errors are squared before averaged, RMSE gives relatively high weights to large errors. The MAE is a linear score which means that all the individual differences are weighted equally in the average. The R^2 is a measure of the proportion of variance, and bias measures the difference between gauged and satellite-derived data. These indicators are defined as follows:

$$IOA = 1 - \frac{\sum_{i=1}^{n} (M_i - O_i)^2}{\sum_{i=1}^{n} (|M_i - \overline{O}| + |O_i - \overline{O}|)^2} \tag{4}$$

$$RMSE = \sqrt{\frac{1}{n}\sum_{i=1}^{n} (M_i - O_i)^2} \tag{5}$$

$$MAE = \frac{1}{n}\sum_{i=1}^{n} |M_i - O_i| \tag{6}$$

$$R^2 = \frac{\sum\limits_{i=1}^{n} (O_i - \overline{O}) \times (M_i - \overline{M})}{\sqrt{\sum\limits_{i=1}^{n} (O_i - \overline{O})^2 \times (M_i - \overline{M})^2}} \tag{7}$$

$$Bias = \frac{1}{n}\sum_{i=1}^{n} (M_i - O_i), \tag{8}$$

where M_i is the estimated grid-scale precipitation from satellite products (that is, TRMM, PERSIANN, PERSIANN-CDR, and CMADS); O_i are ground-based measurement data; $\overline{O}$ and $\overline{M}$ are the average values of ground-based measurements and satellite precipitation data, respectively; and n is the total number of data.

The suitability of satellite-derived rainfall data for the streamflow simulation was evaluated using the Nash–Sutcliffe efficiency (NSE), R^2, and percent bias (PBIAS), which are widely used performance measures in hydrologic studies [58]. NSE is a normalized statistic that determines the relative magnitude of the residual variance compared to the measured data variance, and PBIAS measures the average tendency of the estimated data to be larger or smaller than their observed counterparts.

$$NSE = 1 - \left[\frac{\sum\limits_{i=1}^{n} (O_i - P_i)^2}{\sum\limits_{i=1}^{n} (O_i - \overline{O})^2} \right] \tag{9}$$

$$R^2 = \frac{\sum\limits_{i=1}^{n} (O - \overline{O}) \times (P - \overline{P})}{\sqrt{\sum\limits_{i=1}^{n} (O - \overline{O})^2 \times (P - \overline{P})^2}} \tag{10}$$

$$PBIAS = 100. \frac{\sum\limits_{i=1}^{n} (O_i - P_i)}{\sum\limits_{i=1}^{n} O_i} \tag{11}$$

where O_i is the *i*th observation; P_i is the *i*th predicted value; $\overline{O}$ and $\overline{P}$ are the mean observed and predicted values, respectively; and n is the total number of observations.

3. Results

3.1. Evaluation of Different Satellite-Derived Precipitation Data

In this study, the applicability of four satellite-derived precipitation datasets (2008–2013), that is, from TRMM 3B42 V7, PERSIANN, PERSIANN-CDR, and CMADS, were evaluated.

First, these satellite-based rainfall data were estimated and compared with gauged rainfall data. The contingency measures of the POD, PC, and FBI were used for categorical data analysis. Unlike gauged rainfall data that can be measured in a quantitative manner, it is difficult to determine the threshold for the precipitation and non-precipitation events of satellite-based rainfall data. Based on the studies by Dai [59] and Dinku et al. [60], a minimum threshold of 1.0 mm per day was used to discriminate between rain ($\geq$1.0 mm/day) and no rain (<1.0 mm/day). Figure 2 shows the statistical measures calculated for the four satellite-based precipitation datasets. With respect to the POD values, TRMM has the largest POD with an average of 0.73 (ranging from 0.21 to 0.88), which is closest to the perfect score of 1. The CMADS has a slightly higher POD than PERSIANN, with a mean of 0.62 (ranging from 0.47 to 0.76), while the PERSIANN mean is 0.58 (ranging from 0.44 to 0.75). The PERSIANN-CDR has the smallest POD with a mean of 0.52 (ranging from 0.21 to 0.75; Figure 2a). With respect to the PC values, TRMM shows the best performance, with an average of 0.78 (ranging from 0.71 to 0.89) and CMADS has the second-best performance, with an average of 0.70 (ranging from 0.56 to 0.82). The mean PERSIANN and PERSIANN-CDR values are 0.60 (ranging from 0.40 to 0.77) and 0.50 (ranging from 0.24 to 0.69), respectively (Figure 2b). With respect to the FBI values, TRMM has an average value of 2.0 (ranging from 1.50 to 3.0), which is the smallest value and closest to the perfect score of 1. The average FBI of CMADS is 2.75, ranging from 1.25 to 4.25. The PERSIANN and PERSIANN-CDR averages are 2.75 (ranging from 1.50 to 5.50) and 4.50 (ranging from 1.85 to 5.54), respectively (Figure 2c). These results indicate that the TRMM data show the best agreement with the gauged data, while the PERSIANN-CDR data display the biggest difference from the gauged data.

The TRMM has an average IOA value of 0.30 (ranging from 0.25 to 0.37). The CMADS has a slightly bigger average value of 0.32 (ranging from 0.21 to 0.34). The PERSIANN and PERSIANN-CDR have relatively lower mean values of 0.24 (ranging from 0.20 to 0.26) and 0.23 (ranging from 0.19 to 0.26), respectively (Figure 2d). Therefore, the average IOA of CMADS is the closest to the perfect value of 1. The average RMSEs of the TRMM, CMADS, PERSIANN, and PERSIANN-CDR are 11.22 mm/day (ranging from 2.15 mm/day to 21.29 mm/day), 10.72 mm/day (ranging from 2.05 mm/day to 17.47 mm/day), 11.61 mm/day (ranging from 1.81 mm/day to 22.53 mm/day), and 11.35 mm/day (ranging from 1.04 mm/day to 22.46 mm/day), respectively (Figure 2e). This shows that the four satellite-based rainfall datasets have similar RMSE averages and ranges. The average MAE of the CMADS is 3.53 mm/day (ranging from 1.52 mm/day to 4.80 mm/day), which is smaller than that of the TRMM, PERSIANN, and PERSIANN-CDR, with average values of 3.96 mm/day (ranging from 2.75 mm/day to 5.0 mm/day), 3.80 mm/day (ranging from 2.01 mm/day to 5.25 mm/day), and 3.84 mm/day (ranging from 2.0 mm/day to 5.30 mm/day), respectively (Figure 2f). In Figure 2, it is difficult to see whether the RMSEs (Figure 2e) and MAE (Figure 2f) of each satellite rainfall data are significantly different. In order to determine if there are significant differences among the mean RMSEs and the mean MAEs of the four satellite rainfall data, one-way Analysis of Variance (ANOVA) was carried out. The results of ANOVA showed that the mean RMSEs of the four satellites' rainfall data are not significantly different (p-value = 0.692) at a confidence level of 0.05. However, the p-value for MAE was 0.013, which is smaller than 0.05, and the multiple comparisons showed that the mean MAE of CMADS is significantly greater than that of TRMM. Based on the average IOA, RMSE, and MAE values, the CMADS data show the best accuracy when compared with the gauged data.

The spatial correlation and bias patterns of the four satellite-derived rainfall datasets for the quantitative verification of precipitation are illustrated in Figures 3 and 4. Overall, the TRMM shows a better correlation (average of 0.58, ranging from 0.16 to 0.80) with the gauged data and a smaller bias pattern (ranging from −5.1 mm/day to 4.0 mm/day) for most gauges, which indicates that the TRMM data are similar to the gauge observations. The average R^2 of the CMADS data is 0.52, which is slightly smaller than that of the TRMM data. The bias of the CMADS data varies between −5.3 mm/day and 6.0 mm/day. The average R^2 of the PERSIANN and PERSIANN-CDR data is 0.41 (ranging from 0.20 to 0.58) and 0.46 (ranging from 0.22 to 0.61), respectively, and the bias varies between −5.3 mm/da and 7.1 mm/day, and between −5.5 mm/day and 7.9 mm/day, respectively.

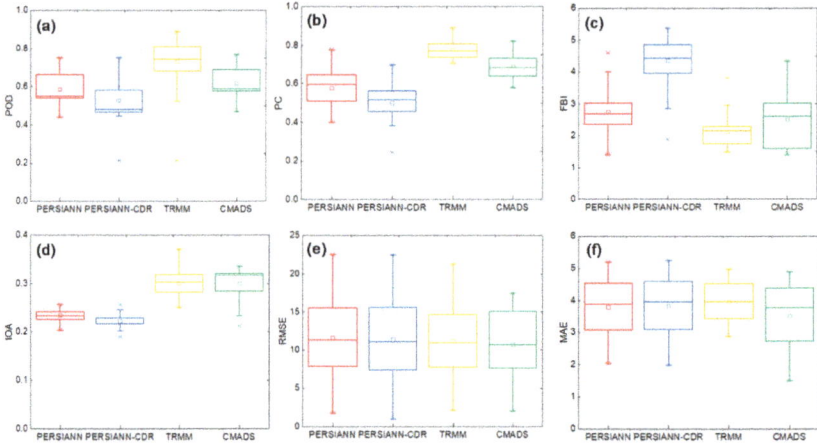

Figure 2. The box plots for the statistical measures of (**a**) probability of detection (POD), (**b**) proportion correct (PC), (**c**) frequency bias index (FBI), (**d**) index of agreement (IOA), (**e**) root mean square error (RMSE), and (**f**) mean absolute error (MAE). The square symbol represents the mean value. The median is presented by the middle line in the box. Each box ranges from the lower (25th) to upper quartile (75th).

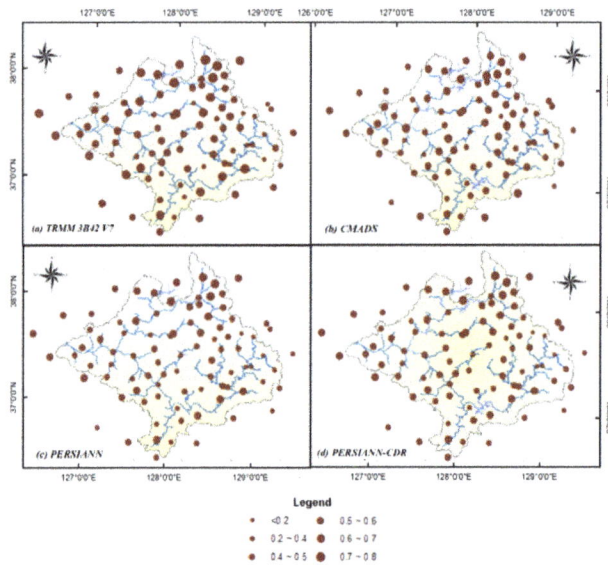

Figure 3. The spatial correlation pattern for ground-based and satellite-derived rainfall during 2008–2013. The circles represent the gauge stations. (**a**) TRMM 3B42 V7, (**b**) CMADS, (**c**) PERSIANN, (**d**) PERSIANN-CDR.

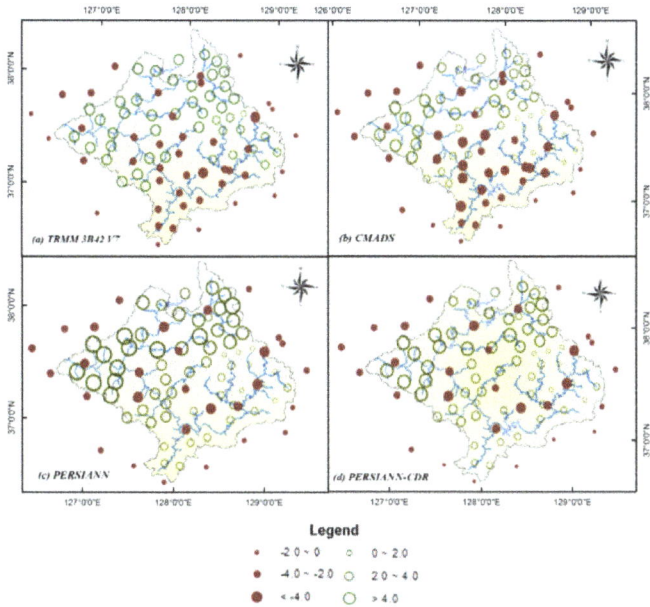

Figure 4. The spatial bias pattern for ground-based and satellite-derived rainfall during 2008–2013. The circles represent the gauge stations. (**a**) TRMM 3B42 V7, (**b**) CMADS, (**c**) PERSIANN, (**d**) PERSIANN-CDR.

3.2. SWAT Calibration and Validation

The SWAT model was calibrated and validated using gauged precipitation as model input. The simulated streamflow was then compared to the data observed at 16 stream gauges that are distributed near homogeneously and have no missing data. A sensitivity analysis was used to identify the key parameters required for model calibration [61]. Six parameters, that is, CN2, ALPHA_BF, CH_K2, CH_N2, CANMX, and CH_N1, were selected to calibrate the model. The initial and calibrated values of those six parameters are shown in Table 4. An automated baseflow separation technique was used [31] to obtain a more reasonable ALPHA BF (ratio of surface runoff to baseflow) value for the Han River Basin. Based on the obtained values of ALPHA_BF at different streamflow stations, the initial range of ALPHA_BF was set for the SWAT model calibration. The calibration (2008–2010) and validation (2011–2013) were performed using the SUFI-2 algorithm. The objective function of the SWAT model calibration is to maximize the NSE. The distribution of behavioral model parameters (1000 runs) for the model calibration is shown in Figure 5. Figure 5 shows that most of the identified parameters are distributed with the NSE ranging from 0.4 to 0.6. The statistical measures for model performances computed using daily streamflow observations are listed in Table 5. The results show an overall good agreement between the observed and simulated streamflow; the NSE, R^2, and PBIAS values vary in the ranges of 0.50 to 0.94, 0.51 to 0.95, and −14.10% to 24.30%, respectively, during the calibration and from 0.50 to 0.90, 0.51 to 0.90, and −38.80% to 21.47%, respectively, during the validation. The NSE, R^2, and PBIAS values at the outlet of the basin (Haengjudaegyo Station) are 0.58%, 0.59%, and −10.00%, respectively, during the model calibration and 0.77%, 0.81%, and −38.80%, respectively, during the model validation.

Table 4. The initial and calibrated parameters selected for the SWAT model.

Parameters	Parameter Description	Initial Range	Calibrated Range	Best Value
ͬ CN2	Initial SCS CN II value	−0.20 to 0.20	−0.07 to 0.79	0.18
ᵛ ALPHA_BF	Baseflow alpha factor	0.0035 to 0.80	0.28 to 0.80	0.72
ᵛ CH_K2	Effective hydraulic conductivity of the main channel	−0.01 to 500	−0.01 to 268	4.65
ᵛ CH_N2	Manning's value for main channels	−0.01 to 0.30	−0.10 to 0.17	0.03
ᵛ CANMX	Maximum canopy storage	0 to 100	0 to 55.40	55.19
ᵛ CH_N1	Manning's value for tributary channels	0.01 to 30	10 to 30	28.46

ͬ The parameter was multiplied by one plus a given value; ᵛ The parameter was replaced by a given value.

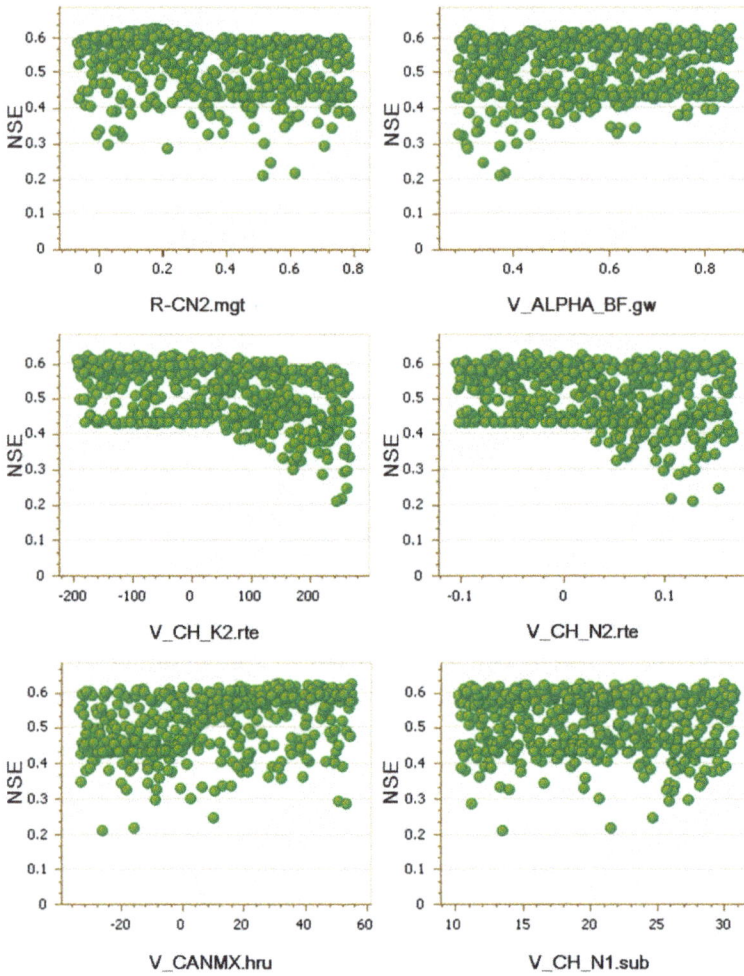

Figure 5. The distribution of behavioral model parameters.

Table 5. The statistical measurements of the model performance for the streamflow simulation.

Code	Station	Calibration (2008–2010)			Validation (2011–2013)		
		NSE	R^2	PBIAS (%)	NSE	R^2	PBIAS (%)
SG6	PanUn	0.53	0.54	−7.50	-	-	-
SG8	YeongWeol1	0.57	0.59	14.70	0.64	0.72	−33.41
SG9	YeongChun	0.59	0.59	3.50	0.61	0.71	−26.99
SG10	DalCheon	0.63	0.67	24.30	0.75	0.78	−17.19
SG11	Mokgyegyo	0.94	0.95	14.70	0.61	0.73	−10.10
SG15	Yeojudaegyo	0.82	0.82	−0.20	-	-	-
SG17	Heukcheongyo	0.51	0.53	20.00	0.61	0.61	−7.73
SG19	WeonTong	0.53	0.67	11.30	0.50	0.51	21.47
SG20	NaeLinCheon	0.69	0.70	17.10	0.58	0.66	−12.44
SG23	Jueumchigyo	0.50	0.51	16.90	0.56	0.56	18.90
SG25	Bangokgyo	0.67	0.71	24.20	0.58	0.63	−6.78
SG26	Daeseongri	0.71	0.73	0.60	0.61	0.73	−14.71
SG28	Sumthlgyo	0.63	0.67	20.50	0.75	0.78	20.77
SG31	Gwangjingyo	0.56	0.63	−14.10	0.56	0.77	−24.96
SG33	Jungranggyo	0.56	0.58	18.40	0.90	0.90	−11.80
SG37 (outlet)	Haengjudaegyo	0.58	0.59	−13.00	0.77	0.81	−38.80

3.3. Streamflow Simulation Using Four Satellite-Derived Rainfall Datasets

The SWAT model calibrated using ground-based rainfall data was used for the evaluation of the hydrologic performance of four satellite-derived rainfall products. Propagation of the uncertainties in the parameters leads to uncertainties in the model output, that is, the discharge of the streamflow, which are expressed as the 95% probability distributions. These are calculated at the 2.5% and 97.5% levels of the cumulative distribution of the discharge. This is referred to as the 95% prediction uncertainty, or 95PPU. Figures 6–10 illustrate the prediction uncertainty at the outlet (Haengjudaegyo station) of the Han River Basin from May to October 2008. These figures show that the TRMM rainfall data have a better performance than other satellite rainfall data in streamflow simulations.

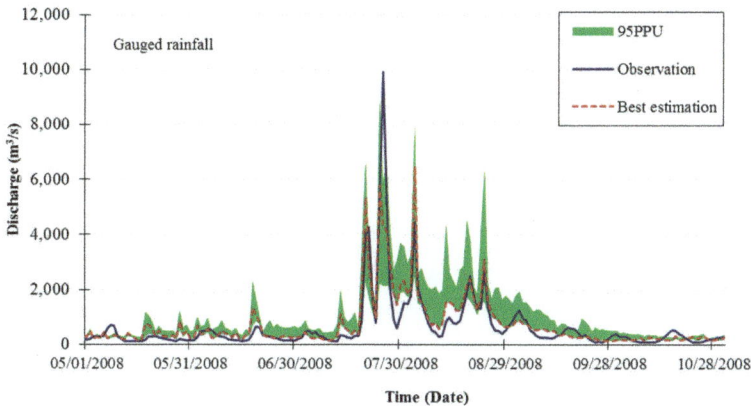

Figure 6. The 95% prediction uncertainty (95PPU) for streamflow simulations using gauged rainfall from May to October (2008).

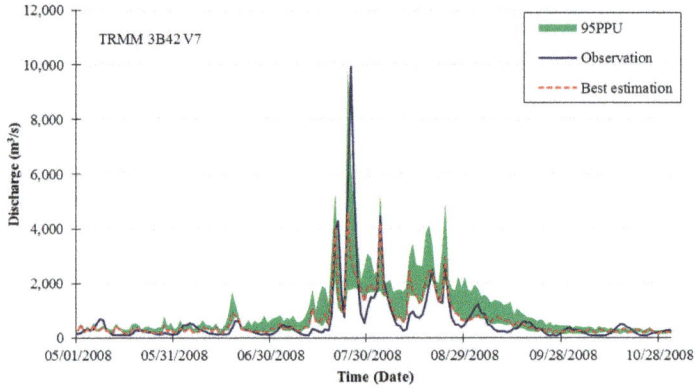

Figure 7. The 95PPU for streamflow simulations using Tropical Rainfall Measuring Mission (TRMM 3B42 V7) rainfall estimates from May to October (2008).

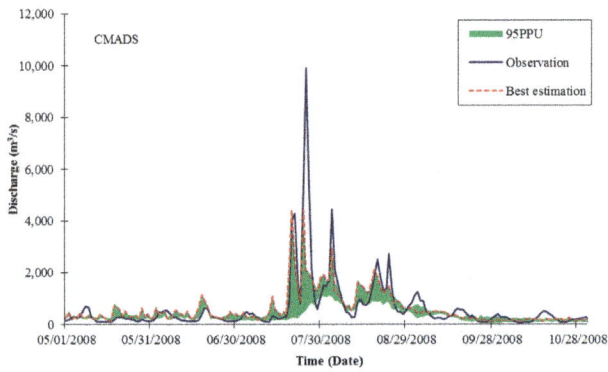

Figure 8. The 95PPU for streamflow simulations using China Meteorological Assimilation Driving Datasets for the SWAT Model (CMADS) rainfall estimates from May to October (2008).

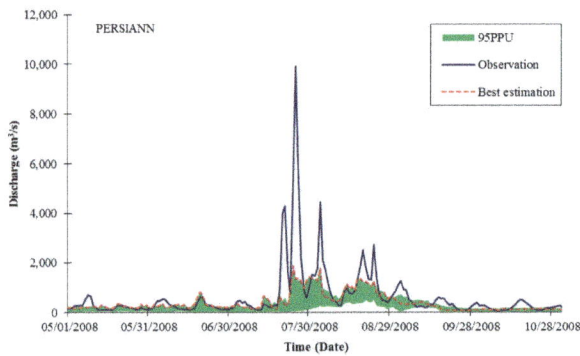

Figure 9. The 95PPU for streamflow simulations using Precipitation Estimation from Remotely Sensed Information using Artificial Neural Networks (PERSIANN) rainfall estimates from May to October (2008).

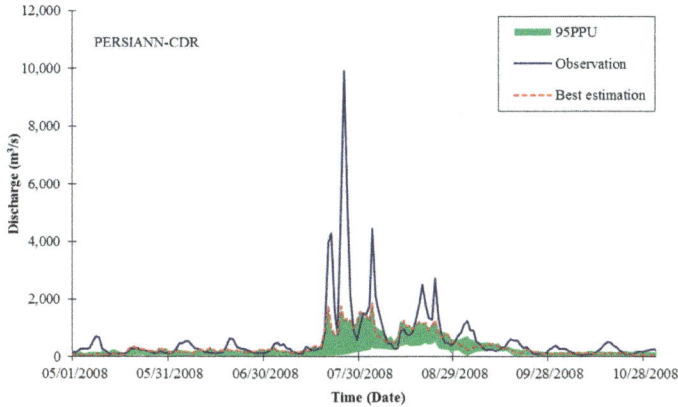

Figure 10. The 95PPU for streamflow simulations using PERSIANN-Climate Data Record (PERSIANN-CDR) rainfall estimates from May to October (2008).

To quantify the fit between the simulation results, expressed as 95PPU, and the observations, the P-factor and R-factor were calculated. The P-factor is the percentage of observed data enveloped by the out modeling result, the 95PPU. The R-factor is the thickness of the 95PPU envelope. A P-factor of 1 and an R-factor of zero is a simulation that exactly corresponds to the measured data. The two statistics for simulations using gauged rainfall and satellite rainfall estimates are shown in Table 6.

Table 6. The P-factor and R-factor due to different rainfall datasets.

Statistics	Gauged Rainfall (Calibration)	Gauged Rainfall (Validation)	TRMM	CMADS	PERSIANN	PERSIANN-CDR
P-factor	0.54	0.47	0.51	0.40	0.39	0.33
R-factor	0.47	0.56	0.42	0.42	0.37	0.43

Table 7 shows that the daily streamflow simulation using TRMM data, with an average R^2 of 0.54 (ranging from 0.29 to 0.81), average NSE of 0.49 (ranging from 0.27 to 0.79), and the PBIAS ranging from −52.70% to 28.30%, performs better than the simulations using the other three satellite-derived rainfall datasets. The streamflow simulation using the CMADS data shows the second-best performance, with an average R^2 of 0.44 (ranging from 0.22 to 0.70), average NSE of 0.42 (ranging from 0.3 to 0.62), and the PBIAS ranging from −29.3% to 41.8%. The average R^2, NSE, and PBIAS show that models using the PERSIANN and PERSIANN-CDR perform relatively poor.

Table 7. The statistical indicators for the streamflow simulations using different rainfall datasets.

Code	Station	Product	R^2	NSE	PBIAS (%)
SG6	PanUn	Rain gauge	0.54	0.53	−7.50
		PERSIANN	0.44	0.23	55.70
		PERSIANN-CDR	0.21	0.19	35.00
		TRMM 3B42 V7	0.48	0.42	−52.70
		CMADS	0.39	0.37	11.20
SG8	YeongWeol1	Rain gauge	0.66	0.61	−9.36
		PERSIANN	0.55	0.21	67.20
		PERSIANN-CDR	0.55	0.21	67.20
		TRMM 3B42 V7	0.46	0.45	−18.40
		CMADS	0.49	0.43	23.20

Table 7. *Cont.*

Code	Station	Product	R^2	NSE	PBIAS (%)
SG9	YeongChun	Rain gauge	0.65	0.60	−11.75
		PERSIANN	0.49	0.25	59.50
		PERSIANN-CDR	0.28	0.25	31.30
		TRMM 3B42 V7	0.59	0.54	−33.00
		CMADS	0.44	0.42	20.90
SG10	DalCheon	Rain gauge	0.73	0.69	3.56
		PERSIANN	0.41	0.15	68.90
		PERSIANN-CDR	0.29	0.23	46.30
		TRMM 3B42 V7	0.33	0.33	−14.80
		CMADS	0.36	0.33	10.10
SG11	Mokgyegyo	Rain gauge	0.84	0.78	2.30
		PERSIANN	0.61	0.18	71.10
		PERSIANN-CDR	0.60	0.42	50.10
		TRMM 3B42 V7	0.81	0.79	−7.30
		CMADS	0.70	0.62	35.50
SG15	Yeojudaegyo	Rain gauge	0.82	0.82	−0.20
		PERSIANN	0.06	0.04	40.50
		PERSIANN-CDR	0.05	0.03	31.20
		TRMM 3B42 V7	0.60	0.43	−13.57
		CMADS	0.31	0.31	−29.30
SG17	Heukcheongyo	Rain gauge	0.57	0.56	6.14
		PERSIANN	0.08	0.03	69.90
		PERSIANN-CDR	0.06	0.03	58.80
		TRMM 3B42 V7	0.48	0.42	12.70
		CMADS	0.30	0.38	41.80
SG19	WeonTong	Rain gauge	0.59	0.52	16.39
		PERSIANN	0.40	0.04	83.20
		PERSIANN-CDR	0.44	0.17	71.20
		TRMM 3B42 V7	0.59	0.58	28.30
		CMADS	0.57	0.49	27.10
SG20	NaeLinCheon	Rain gauge	0.68	0.64	2.33
		PERSIANN	0.32	0.12	72.60
		PERSIANN-CDR	0.23	0.17	54.10
		TRMM 3B42 V7	0.59	0.58	−4.50
		CMADS	0.31	0.45	19.50
SG23	Jueumchigyo	Rain gauge	0.52	0.51	17.90
		PERSIANN	0.04	0.02	73.80
		PERSIANN-CDR	0.04	0.02	59.90
		TRMM 3B42 V7	0.58	0.42	11.30
		CMADS	0.50	0.43	22.90
SG25	Bangokgyo	Rain gauge	0.67	0.63	8.71
		PERSIANN	0.14	0.02	73.70
		PERSIANN-CDR	0.08	0.02	62.00
		TRMM 3B42 V7	0.56	0.55	16.00
		CMADS	0.47	0.41	20.70
SG26	Daeseongri	Rain gauge	0.73	0.66	−7.06
		PERSIANN	0.45	0.35	48.90
		PERSIANN-CDR	0.50	0.44	36.00
		TRMM 3B42 V7	0.55	0.51	−8.90
		CMADS	0.53	0.51	24.30

Table 7. *Cont.*

Code	Station	Product	R^2	NSE	PBIAS (%)
SG28	Sumthlgyo	Rain gauge	0.73	0.69	20.64
		PERSIANN	0.13	0.20	75.30
		PERSIANN-CDR	0.07	0.10	67.40
		TRMM 3B42 V7	0.62	0.57	25.10
		CMADS	0.59	0.55	23.90
SG31	Gwangjingyo	Rain gauge	0.70	0.56	−14.53
		PERSIANN	0.08	0.05	38.10
		PERSIANN-CDR	0.06	0.01	12.70
		TRMM 3B42 V7	0.57	0.44	−51.20
		CMADS	0.49	0.42	−10.70
SG33	Jungranggyo	Rain gauge	0.74	0.73	3.30
		PERSIANN	0.19	0.04	75.80
		PERSIANN-CDR	0.30	0.15	62.50
		TRMM 3B42 V7	0.29	0.27	13.40
		CMADS	0.39	0.30	21.80
SG37 (Outlet)	Haengjudaegyo	Rain gauge	0.70	0.68	−25.90
		PERSIANN	0.23	0.16	52.60
		PERSIANN-CDR	0.18	0.16	32.90
		TRMM 3B42 V7	0.46	0.47	−32.50
		CMADS	0.22	0.32	14.60
Average		Rain gauge	0.68	0.64	0.31
		PERSIANN	0.29	0.13	64.18
		PERSIANN-CDR	0.25	0.16	48.66
		TRMM 3B42 V7	0.54	0.49	−8.13
		CMADS	0.44	0.42	17.34

Figure 11 shows the daily streamflow simulated by using the gauged rainfall and different satellite-derived rainfall datasets as inputs. Based on Figure 11, the models using the four satellite-based rainfall datasets capture the behavior of the streamflow in most cases by characterizing the rising and falling of flow. At SG11, SG26, SG31, and SG37, the hydrological models driven by the TRMM and CMADS data simulate the streamflow very well. However, the models using the four satellite-derived rainfall datasets tend to underestimate the peak flow of most flood events.

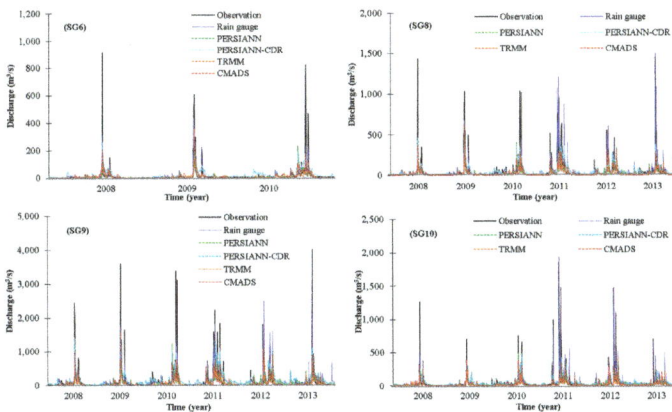

Figure 11. *Cont.*

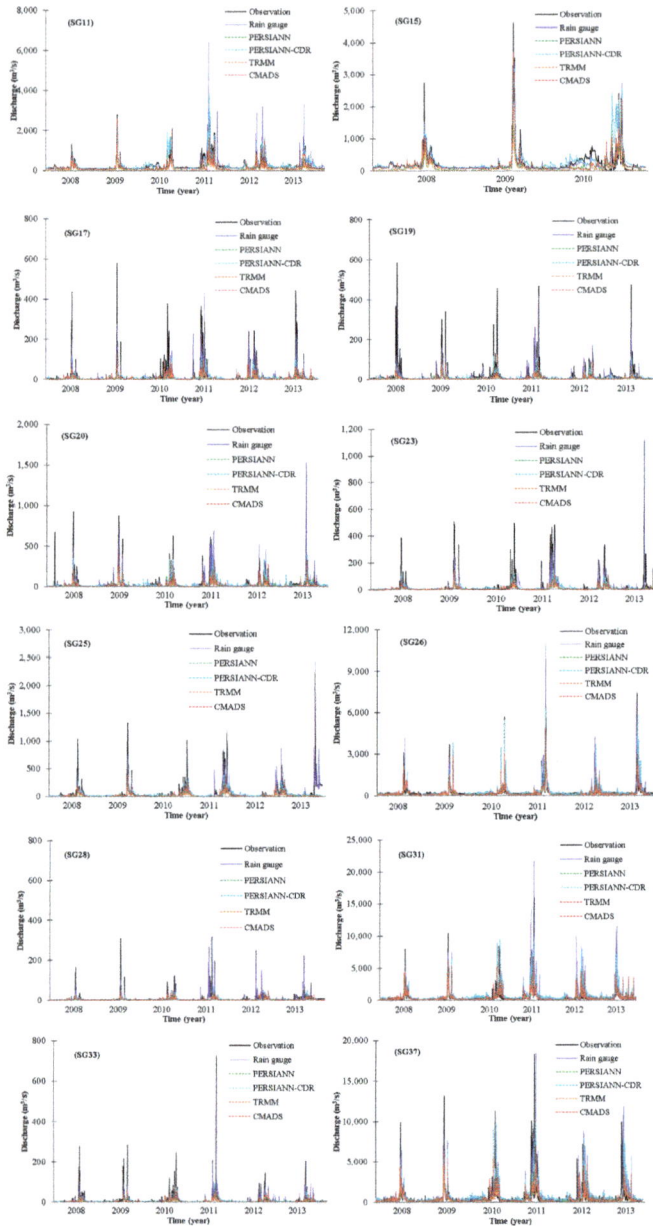

Figure 11. The comparison of the observed and computed flows obtained by using ground-based rainfall and the four satellite-derived rainfall datasets.

4. Discussion

Our study indicates that the TRMM and CMADS rainfall data show higher accuracy than the PERSIANN and PERSIANN-CDR data. Additionally, the models using the TRMM and CMADS data show a better performance than those using the PERSIANN and PERSIANN-CDR data for streamflow simulation in the Han River Basin. The results show that the PERSIANN-CDR data, which are bias-adjusted products, and the PERSIANN data have a similar accuracy. The hydrologic models using these two rainfall datasets as inputs show similar performances. Contrary to the results of this study, Ashouri et al. [49] evaluated an extreme weather event (Hurricane Katrina) in the United States and showed that PERSIANN-CDR data have a higher correlation with gauged rainfall than the TMPA data.

While the satellite rainfall data either overestimate or underestimate the gauged rainfall data at different stations due to the spatiotemporal uncertainty of satellite rainfall products, the streamflow simulation results show that the SWAT model using different satellite rainfall datasets mostly underestimates the peak flow. Figures 12–15 show the comparison of spatially averaged gauged rainfall and satellite-derived rainfall estimates, and Figure 16 shows the comparison of spatially averaged annual maximum daily rainfall. These figures show that except TRMM, the other three satellites' data tend to underestimate the spatially averaged gauged data. Therefore, the underestimation of streamflow could be attributed to the overall underestimation of satellite rainfall data. Such a tendency may or may not be only for the Han River Basin and/or for the specific period from 2008 to 2013. Hromadka and McCuen [62], Maskey et al. [63], and Jones et al. [64] indicated that one of the main sources of uncertainty in streamflow simulations using rainfall–runoff models is the spatiotemporal uncertainty of the catchment rainfall. This means that the satellite-based rainfall estimates have a significant effect on the streamflow simulation [65]. The uncertainty in satellite-based rainfall estimates can have several causes. Gebregiorgis and Hossain [66] characterized the errors of satellite-derived rainfall data based on climate type and topography (elevation). They found that the uncertainty of satellite rainfall data depends more on the topography of the region than the regional climate. Xu et al. [67] found that a large amount of uncertainty in a satellite precipitation dataset can be explained by the normalized difference vegetation index, digital elevation model, and land surface temperature. Bitew and Gebremichael [68] evaluated the performance of satellite rainfall products in streamflow simulations (that is, CMORPH, TMPA 3B42RT, TMPA 3B42, and PERSIANN) for two different watersheds of the Ethiopian highlands and showed that the application of different satellite rainfall products is related to the watershed area. All these factors can explain the errors in streamflow simulations using satellite-derived rainfall data. An error propagation from satellite-derived rainfall to streamflow simulation of hydrological models is inevitable. Thus, the improvement of the accuracy of satellite-derived rainfall data is very important and a systematic error correction of satellite data for hydrological applications should be considered. Several studies showed that the use of satellite data for the calibration of hydrological models results in a better streamflow prediction [18,69]. However, such a recalibration may fail due to errors in the satellite data, which are more difficult to control than those of ground-based data.

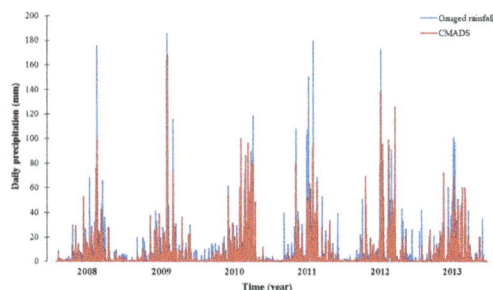

Figure 12. The comparison of spatially averaged gauged rainfall and CMADS.

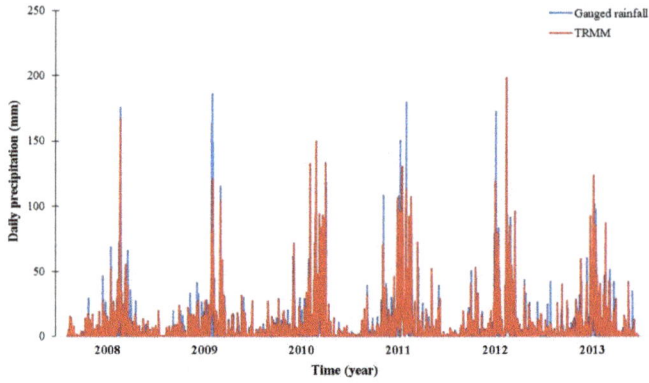

Figure 13. The comparison of spatially averaged gauged rainfall and TRMM.

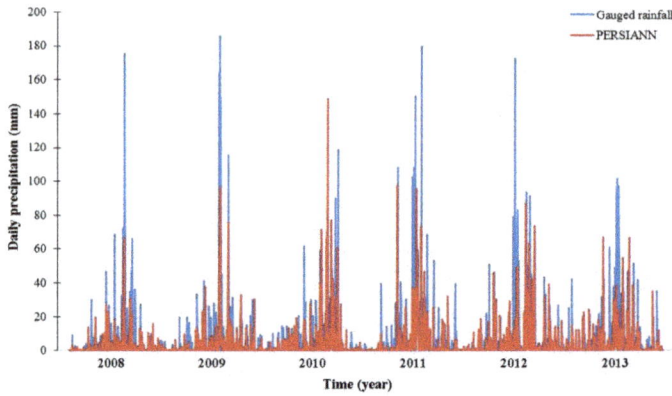

Figure 14. The comparison of spatially averaged gauged rainfall and PERSIANN.

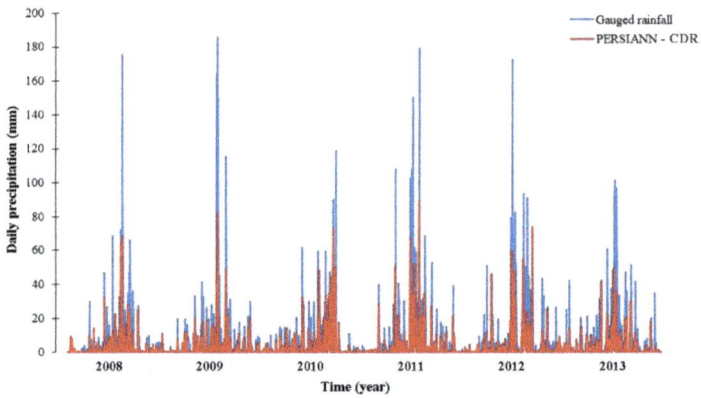

Figure 15. The comparison of spatially averaged gauged rainfall and PERSIANN-CDR.

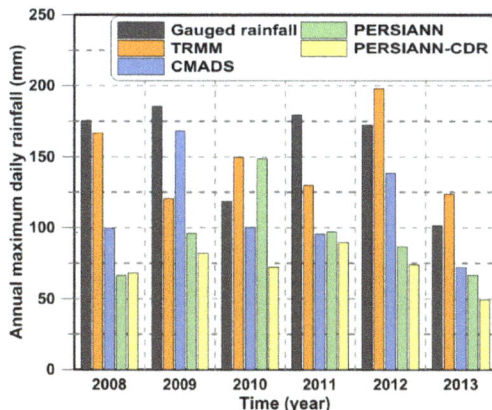

Figure 16. The comparison of annual maximum daily rainfall of different datasets.

5. Conclusions

The main purpose of this study was to evaluate the utility of satellite-derived precipitation data in a hydrological model for streamflow simulation in the Han River Basin. Four different satellite-derived rainfall datasets were compared with ground-based rainfall data at a daily time step. The TRMM and CMADS rainfall data have relatively higher accuracy. The runoff simulations indicate that the use of ground-based rainfall data in the SWAT model leads to an overall good agreement between observed and computed streamflow. Based on the comparison of the use of the four satellite rainfall products for streamflow simulation in the Han River Basin, hydrological models using the TRMM and CMADS rainfall data perform better than those using PERSIANN and PERSIANN-CDR data. The results of this study indicate that the TRMM and CMADS rainfall data can play significant roles in water resource management and flood control in the Han River Basin. This study also indicates that the CMADS will provide important basic data with acceptable accuracy for hydrological research in East Asia, especially in areas with scarce data.

In the study, a pixel-to-point comparison was made between satellite rainfall data and gauged data. However, a point-to-pixel comparison using some statistical interpolation methods such as multiple linear regression, optimal interpolation or Kriging should be further discussed in the future study.

One of the main purposes of the study is to evaluate the newly developed CMADS in comparison with gauged data and other satellite rainfall products. In this study, data for a limited period (2008–2013) were used since the CMADS is only available from 2008. However, from a climatological point of view, using data for only 6 years may not lead to robust results since it is too short to include all the different characteristics of the rainfall regime of the area. Using data from different periods could show different results in the rainfall data comparison as well as in the SWAT model simulation. In addition, 3-year data (2008–2010) was used for the SWAT model calibration; however, data from different periods may also produce different results. Therefore, the findings and the conclusions of this study may not be generalized for different time periods. Further studies using a longer period of data would produce more robust results.

Author Contributions: T.T.V. designed the framework and analyzed the data of this study; T.T.V. and L.L. collected the data and wrote the paper; K.S.J. provided significant suggestions on the methodology and structure of the manuscript. All authors read and approved the final manuscript.

Funding: This research was funded by Ministry of Land, Infrastructure and Transport of Advanced Water Management Research Program (13AWMP-B066744-01).

Conflicts of Interest: The authors declare no conflict of interest.

References

1. González-Rouco, J.F.; Jiménez, J.L.; Quesada, V.; Valero, F. Quality control and homogeneity of precipitation data in the southwest of Europe. *J. Clim.* **2001**, *14*, 964–978. [CrossRef]
2. Price, K.; Purucker, S.T.; Kraemer, S.R.; Babendreier, J.E.; Knightes, C.D. Comparison of radar and gauge precipitation data in watershed models across varying spatial and temporal scales. *Hydrol. Process.* **2014**, *28*, 3505–3520. [CrossRef]
3. Sikorska, A.; Seibert, J. Importance of precipitation data quality for streamflow predictions. *Geophys. Res. Abstr.* **2015**, *17*, 13369.
4. Buarque, D.C.; De Paiva, R.C.D.; Clarke, R.T.; Mendes, C.A.B. A comparison of Amazon rainfall characteristics derived from TRMM, CMORPH and the Brazilian national rain gauge network. *J. Geophys. Res. Atmos.* **2011**, *116*. [CrossRef]
5. Kidd, C.; Bauer, P.; Turk, J.; Huffman, G.J.; Joyce, R.; Hsu, K.L.; Braithwaite, D. Intercomparison of high-resolution precipitation products over northwest Europe. *J. Hydrometeorol.* **2012**, *13*, 67–83. [CrossRef]
6. Thorndahl, S.; Einfalt, T.; Willems, P.; Nielsen, J.E.; Veldhuis, M.C.; Arnbjerg-Nielsen, K.; Rasmussen, M.R.; Molnar, P. Weather radar rainfall data in urban hydrology. *Hydrol. Earth Syst. Sci.* **2017**, *21*, 1359–1380. [CrossRef]
7. Westrick, K.J.; Mass, C.F.; Colle, B.A. The limitation of the WSR-88D radar network for quantitative precipitation measurement over the coastal western United States. *Bull. Am. Meteorol. Soc.* **1999**, *80*, 2289–2298. [CrossRef]
8. Rico-Ramirez, M.A.; Liguori, S.; Schellart, A.N.A. Quantifying radar-rainfall uncertainties in urban drainage flow modeling. *J. Hydrol.* **2015**, *528*, 17–28. [CrossRef]
9. Joyce, R.J.; Janowiak, J.E.; Arkin, P.A.; Xie, P. CMORPH: A method that produces global precipitation estimates from passive microwave and infrared data at high spatial and temporal resolution. *J. Hydrometeorol.* **2004**, *5*, 487–503. [CrossRef]
10. Kidd, C.; Huffman, G. Global precipitation measurement. *Meteorol. Appl.* **2011**, *18*, 334–353. [CrossRef]
11. Kidd, C.; Levizzani, V. Status of satellite precipitation retrievals. *Hydrol. Earth Syst. Sci.* **2011**, *15*, 1109–1116. [CrossRef]
12. Hou, A.Y.; Kakar, R.K.; Neeck, S.; Azarbarzin, A.A.; Kummerow, C.D.; Kojima, M.; Oki, R.; Nakamura, K.; Iguchi, T. The global precipitation measurement mission. *Bull. Am. Meteorol. Soc.* **2014**, *95*, 701–722. [CrossRef]
13. Su, F.; Hong, Y.; Lettenmaier, D.P. Evaluation of TRMM multisatellite precipitation analysis (TMPA) and its utility in hydrologic prediction in the La Plata basin. *J. Hydrometeorol.* **2008**, *9*, 622–639. [CrossRef]
14. Collischonn, B.; Collischonn, W.; Tucci, C.E.M. Daily hydrological modeling in the Amazon basin using TRMM rainfall estimates. *J. Hydrol.* **2008**, *360*, 207–216. [CrossRef]
15. Scheel, M.L.M.; Rohrer, M.; Huggel, C.; Villar, D.S.; Huffman, G.J. Evaluation of TRMM multi-satellite precipitation analysis (TMPA) performance in the Central Andes region and its dependency on spatial and temporal resolution. *Hydrol. Earth Syst. Sci.* **2011**, *15*, 2649–2663. [CrossRef]
16. Ouma, Y.O.; Owiti, T.; Kibiiy, J.; Ouma, Y.O.; Tateishi, R.; Kipkorir, E. Multitemporal comparative analysis of TRMM-3B42 satellite-estimated rainfall with surface gauge data at basin scales: Daily, decadal and monthly evaluations. *Int. J. Remote Sens.* **2012**, *33*, 7662–7684. [CrossRef]
17. Xue, X.; Hong, Y.; Limaye, A.S.; Gourley, J.J.; Huffman, G.J.; Khan, S.I.; Dorji, C.; Chen, S. Statistical and hydrological evaluation of TRMM-based multi-satellite precipitation analysis over the Wangchu Basin of Bhutan: Are the latest satellite precipitation products ready for use in ungauged basins? *J. Hydrol.* **2013**, *499*, 91–99. [CrossRef]
18. Stisen, S.; Sandholt, I. Evaluation of remote-sensing-based rainfall products through predictive capability in hydrological runoff modelling. *Hydrol. Process.* **2010**, *24*, 879–891. [CrossRef]
19. Behrangi, A.; Khakbaz, B.; Jaw, T.C.; AghaKouchak, A.; Hsu, K.; Sorooshian, S. Hydrologic evaluation of satellite precipitation products over a mid-size basin. *J. Hydrol.* **2011**, *397*, 225–237. [CrossRef]
20. Shen, Y.; Xiong, A.; Wang, Y.; Ha, P. Performance of high-resolution satellite precipitation products over China. *J. Geophys. Res. Atmos.* **2010**, *115*. [CrossRef]
21. Hirpa, F.A.; Gebremichael, M.; Hopson, T. Evaluation of high-resolution satellite precipitation products over very complex terrain in Ethiopia. *J. Appl. Meteorol. Climatol.* **2010**, *49*, 1044–1051. [CrossRef]
22. Meng, X.; Wang, H. Significance of the China meteorological assimilation driving datasets for the SWAT model (CMADS) of East Asia. *Water (Switzerland)* **2017**, *9*, 765. [CrossRef]

23. Meng, X.Y.; Wang, H.; Lei, X.H.; Cai, S.Y.; Wu, H.J. Hydrological modeling in the Manas River Basin using Soil and Water Assessment Tool driven by CMADS. *Teh. Vjesn.* **2017**, *24*, 525–534.

24. Setegn, S.G.; Donoso, M.C. *Sustanability of Intergrated Water Resources Management: Water Governace, Climate and Ecohydrology*; Springer: Basel, Switzerland, 2015.

25. Bicknell, B.R.; Imhoff, J.C.; Kittle, J.L.; Donigian, A.S.; Johanson, R.C. *Hydrologic Simulation Program—Fortran: User's Manual for Release 11*; Environmental Research Laboratory, Office of Research and Development, U.S. Environmental Protection Agency: Athens, GA, USA, 1996.

26. Refsgaard, J.C.; Storm, B. MIKE SHE. In *Computer Models of Watershed Hydrology*; Sing, V.P., Ed.; Water Resource Publications: Highland Ranch, CO, USA, 1995; pp. 806–846.

27. USACE. *HEC-5 Simulation of Flood Control and Conservation System*; US Army Corps of Engineers, Hydrologic Engineering Center: Davis, CA, USA, 1998.

28. Arnold, J.G.; Srinivasan, R.; Muttiah, R.S.; Williams, J.R. Large area hydrologic modeling and assessment part I: Model development. *J. Am. Water Resour. Assoc.* **1998**, *34*, 73–89. [CrossRef]

29. Sivapalan, M.; Takeuchi, K.; Franks, S.W.; Gupta, V.K.; Karambiri, H.; Lakashmi, V.; Liang, X.; McDonnell, J.J.; Mendiondo, E.M.; O'connell, P.E.; et al. IAHS Decade on Predictions in Ungauged Basins (PUB), 2003–2012: Shaping an exciting future for the hydrological sciences. *Hydrol. Sci. J.* **2003**, *48*, 857–880. [CrossRef]

30. Eckhardt, K.; Arnold, J.G. Automatic calibration of a distributed catchment model. *J. Hydrol.* **2001**, *251*, 103–109. [CrossRef]

31. Arnold, J.G.; Allen, P.M. Automated methods for estimating baseflow and ground water recharge from streamflow records. *J. Am. Water Resour. Assoc.* **1999**, *35*, 411–424. [CrossRef]

32. White, K.L.; Chaubey, I. Sensitivity analysis, calibration, and validations for a multisite and multivariable SWAT model. *J. Am. Water Resour. Assoc.* **2005**, *41*, 1077–1089. [CrossRef]

33. Kang, M.S.; Park, S.W.; Lee, J.J.; Yoo, K.H. Applying SWAT for TMDL programs to a small watershed containing rice paddy fields. *Agric. Water Manag.* **2006**, *79*, 72–92. [CrossRef]

34. Kim, N.W.; Chung, I.M.; Won, Y.S.; Arnold, J.G. Development and application of the integrated SWAT-MODFLOW model. *J. Hydrol.* **2008**, *356*, 1–16. [CrossRef]

35. Bae, D.H.; Jung, I.W.; Lettenmaier, D.P. Hydrologic uncertainties in climate change from IPCC AR4 GCM simulations of the Chungju Basin, Korea. *J. Hydrol.* **2011**, *401*, 90–105. [CrossRef]

36. Kim, N.W.; Lee, J.E.; Kim, J.T. Assessment of flow regulation effects by dams in the Han River, Korea, on the downstream flow regimes using SWAT. *J. Water Resour. Plan. Manag.* **2012**, *131*, 24–35. [CrossRef]

37. Shope, C.L.; Maharjan, G.R.; Tenhunen, J.; Seo, B.; Kim, K.; Riley, J.; Arnhold, S.; Koellner, T.; Ok, Y.S.; Peiffer, S.; et al. Using the SWAT model to improve process descriptions and define hydrologic partitioning in South Korea. *Hydrol. Earth Syst. Sci.* **2014**, *18*, 539–557. [CrossRef]

38. Cho, K.H.; Pachepsky, Y.A.; Kim, M.; Pyo, J.; Park, M.H.; Kim, Y.M.; Kim, J.W.; Kim, J.H. Modeling seasonal variability of fecal coliform in natural surface waters using the modified SWAT. *J. Hydrol.* **2016**, *535*, 377–385. [CrossRef]

39. Kim, J.P.; Jung, I.; Park, K.W.; Yoon, S.K.; Lee, D. Hydrological utility and uncertainty of multi-satellite precipitation products in the mountainous region of South Korea. *Remote Sens.* **2016**, *8*, 608. [CrossRef]

40. Smith, P.J.; Panziera, L.; Beven, K.J. Forecasting flash floods using data-based mechanistic models and NORA radar rainfall forecasts. *Hydrol. Sci. J.* **2014**, *59*, 1343–1357. [CrossRef]

41. KOWACO. *The Pre-Investigation Report for Groundwater Resources*; Korea Water Resources Corporation: Daejeon, Korea, 1993; p. 340.

42. Korea Meteorological Administration. *Annual Report 2016*; Korea Meteorological Administration: Seoul, Korea, 2016; p. 49.

43. Kim, J.S.; Jain, S.; Yoon, S.K. Warm season streamflow variability in the Korean Han River Basin: Links with atmospheric teleconnections. *Int. J. Climatol.* **2012**, *32*, 635–640. [CrossRef]

44. Lee, K.S.; Bong, Y.S.; Lee, D.; Kim, Y.; Kim, K. Tracing the sources of nitrate in the Han River watershed in Korea, using δ^{15}N-NO$_3$ and δ^{18}O-NO$_3$ values. *Sci. Total Environ.* **2008**, *395*, 117–124. [CrossRef] [PubMed]

45. Heo, B.-H.; Kim, K.-E.; Kang, S.-G. Removals of noises from automatic weather station data and radial velocity data of doppler weather radar using modified median filter. *J. Korean Meteorol. Soc.* **1999**, *35*, 127–135.

46. Li, D.; Christakos, G.; Ding, X.; Wu, J. Adequacy of TRMM satellite rainfall data in driving the SWAT modeling of Tiaoxi catchment (Taihu lake basin, China). *J. Hydrol.* **2018**, *556*, 1139–1152. [CrossRef]

47. Huffman, G.J.; Adler, R.F.; Bolvin, D.T.; Gu, G.; Nelkin, E.J.; Bowman, K.P.; Hong, Y.; Stocker, E.F.; Wolff, D.B. The TRMM multi-satellite precipitation analysis (TMPA): Quasi-global, multi-year, combined-sensor precipitation estimates at fine scale. *J. Hydrometeorol.* **2007**, *8*, 38–55. [CrossRef]

48. Sorooshian, S.; Hsu, K.L.; Gao, X.; Gupta, H.V.; Imam, B.; Braithwaite, D. Evaluation of PERSIANN system satellite-based estimates of tropical rainfall. *Bull. Am. Meteorol. Soc.* **2000**, *81*, 2035–2046. [CrossRef]

49. Ashouri, H.; Hsu, K.L.; Sorooshian, S.; Braithwaite, D.K.; Knapp, K.R.; Cecil, L.D.; Nelson, B.R.; Prat, O.P. PERSIANN-CDR: Daily precipitation climate data record from multisatellite observations for hydrological and climate studies. *Bull. Am. Meteorol. Soc.* **2015**, *96*, 69–83. [CrossRef]

50. Neitsch, S.L.; Arnold, J.G.; Kiniry, J.R.; Williams, J.R. *Soil and Water Assessment Tool Theoretical Documentation Version 2009*; Texas Water Resources Institute: College Station, TX, USA, 2011.

51. Abbaspour, K.C.; Johnson, C.A.; van Genuchten, M.T. Estimating uncertain flow and transport parameters using a sequential uncertainty fitting procedure. *Vadose Zone J.* **2004**, *3*, 1340–1352. [CrossRef]

52. Yang, J.; Reichert, P.; Abbaspour, K.C.; Xia, J.; Yang, H. Comparing uncertainty analysis techniques for a SWAT application to the Chaohe Basin in China. *J. Hydrol.* **2008**, *358*, 1–23. [CrossRef]

53. Narsimlu, B.; Gosain, A.K.; Chahar, B.R.; Singh, S.K.; Srivastava, P.K. SWAT model calibration and uncertainty analysis for streamflow prediction in the Kunwari River Basin, India, using sequential uncertainty fitting. *Environ. Process.* **2015**, *2*, 79–95. [CrossRef]

54. Wu, H.; Chen, B. Evaluating uncertainty estimates in distributed hydrological modeling for the Wenjing River watershed in China by GLUE, SUFI-2, and ParaSol methods. *Ecol. Eng.* **2015**, *76*, 110–121. [CrossRef]

55. Khoi, D.N.; Thom, V.T. Parameter uncertainty analysis for simulating streamflow in a river catchment of Vietnam. *Glob. Ecol. Conserv.* **2015**, *4*, 538–548. [CrossRef]

56. Willmott, C.J.; Matsuura, K. Advantages of the mean absolute error (MAE) over the root mean square error (RMSE) in assessing average model performance. *Clim. Res.* **2005**, *30*, 79–82. [CrossRef]

57. Baik, J.; Choi, M. Spatio-temporal variability of remotely sensed precipitation data from COMS and TRMM: Case study of Korean peninsula in East Asia. *Adv. Space Res.* **2015**, *56*, 125–1138. [CrossRef]

58. Moriasi, D.N.; Arnold, J.G.; Van Liew, M.W.; Bingner, R.L.; Harmel, R.D.; Veith, T.L. Model evaluation guidelines for systematic quantification of accuracy in watershed simulations. *Trans. ASABE* **2007**, *50*, 885–900. [CrossRef]

59. Dai, A. Precipitation characteristics in eighteen coupled climate models. *J. Clim.* **2006**, *19*, 4605–4630. [CrossRef]

60. Dinku, T.; Chidzambwa, S.; Ceccato, P.; Connor, S.J.; Ropelewski, C.F. Validation of high-resolution satellite rainfall products over complex terrain. *Int. J. Remote Sens.* **2008**, *29*, 4097–4110. [CrossRef]

61. Ma, L.; Ascough Ii, J.C.; Ahuja, L.R.; Shaffer, M.J.; Hanson, J.D.; Rojas, K.W. Root Zone Water Quality Model sensitivity analysis using Monte Carlo simulation. *Trans. ASAE* **2000**, *43*, 883–895. [CrossRef]

62. Hromadka, T.V.; McCuen, R.H. Uncertainty estimates for surface runoff models. *Adv. Water Resour.* **1998**, *11*, 2–14. [CrossRef]

63. Maskey, S.; Guinot, V.; Price, R.K. Treatment of precipitation in rainfall-runoff modelling: A fuzzy set approach. *Adv. Water Resour.* **2004**, *27*, 889–898. [CrossRef]

64. Jones, P.D.; Lister, D.H.; Wilby, R.L.; Kostopoulou, E. Extended riverflow reconstructions for England and Wales, 1865–2002. *Int. J. Climatol.* **2006**, *25*, 219–231. [CrossRef]

65. Andreassian, V.; Perrin, C.; Michel, C.; Usart-Sanchez, I.; Lavabre, J. Impact of imperfect rainfall knowledge on the efficency and the parameters of watershed models. *J. Hydrol.* **2001**, *250*, 206–223. [CrossRef]

66. Gebregiorgis, A.S.; Hossain, F. Understanding the dependence of satellite rainfall uncertainty on topography and climate for hydrologic model simulation. *IEEE Trans. Geosci. Remote Sens.* **2013**, *51*, 704–718. [CrossRef]

67. Xu, S.G.; Niu, Z.; Shen, Y. Understanding the dependence of the uncertainty in a satellite precipitation data set on the underlying surface and a correction method based on geogrphically weighted reggression. *Int. J. Remote Sens.* **2014**, *35*, 6508–6521. [CrossRef]

68. Bitew, M.M.; Gebremichael, M. Assessment of satellite rainfall products for streamflow simulation in medium watersheds of the Ethiopian highlands. *Hydrol. Earth Syst. Sci.* **2011**, *15*, 1147–1155. [CrossRef]

69. Artan, G.; Gadain, H.; Smith, J.L.; Asante, K.; Bandaragoda, C.J.; Verdin, J.P. Adequacy of satellite derived rainfall data for streamflow modeling. *Nat. Hazards* **2007**, *43*, 167–185. [CrossRef]

water

MDPI

Article

CMADS-Driven Simulation and Analysis of Reservoir Impacts on the Streamflow with a Simple Statistical Approach

Ningpeng Dong [1], Mingxiang Yang [2,*], Xianyong Meng [3,4,*], Xuan Liu [5], Zhaokai Wang [6], Hao Wang [2] and Chuanguo Yang [1]

[1] State Key Laboratory of Hydrology-Water Resources and Hydraulic Engineering, Hohai University, Nanjing 210098, China; dongningpeng@hhu.edu.cn (N.D.); cgyang@hhu.edu.cn (C.Y.)
[2] Department of Water Resources, China Institute of Water Resource and Hydropower Research, Beijing 100044, China; wanghao@iwhr.com
[3] College of Resources and Environmental Science, China Agricultural University (CAU), Beijing 100094, China
[4] Department of Civil Engineering, The University of Hong Kong (HKU), Pokfulam 999077, Hong Kong, China
[5] College of Civil Engineering, Tianjin University, Tianjin 300072, China; sandaliuxuan@163.com
[6] College of Hydrology and Water Resources, Hohai University, Nanjing 210098, China; wzkhhu@163.com
* Correspondence: yangmx@iwhr.com (M.Y.); xymeng@cau.edu.cn (X.M.)

Received: 10 December 2018; Accepted: 17 January 2019; Published: 21 January 2019

Abstract: The reservoir operation is a notable source of uncertainty in the natural streamflow and it should be represented in hydrological modelling to quantify the reservoir impact for more effective hydrological forecasting. While many researches focused on the effect of large reservoirs only, this study developed an online reservoir module where the small reservoirs were aggregated into one representative reservoir by employing a statistical approach. The module was then integrated into the coupled Noah Land Surface Model and Hydrologic Model System (Noah LSM-HMS) for a quantitative assessment of the impact of both large and small reservoirs on the streamflow in the upper Gan river basin, China. The Noah LSM-HMS was driven by the China Meteorological Assimilation Driving Datasets for the Soil and Water Assessment Tool (SWAT) model (CMADS) with a very good performance and a Nash-Sutcliffe coefficient of efficiency (NSE) of 0.89, which proved to be more effective than the reanalysis data from the National Centers for Environmental Prediction (NCEP) over China. The simulation results of the integrated model indicate that the proposed reservoir module can acceptably depict the temporal variation in the water storage of both large and small reservoirs. Simulation results indicate that streamflow is increased in dry seasons and decreased in wet seasons, and large and small reservoirs can have equally large effects on the streamflow. With the integration of the reservoir module, the performance of the original model is improved at a significant level of 5%.

Keywords: reservoirs; operation rule; Noah LSM-HMS; capacity distribution; aggregated reservoir; CMADS

1. Introduction

Since decades ago, intensive human activities have brought about a growing challenge to a sustainable watershed management [1–3]. In particular, the construction and the management of numerous reservoirs are regarded as a major source of variability and uncertainty in the flow regime [4]. As compared to larger ones, small-scale reservoirs have recently become a preferred choice in the design of hydraulic projects given its cost-effectiveness and its less sociopolitical and environmental consequence. [5,6]. While small reservoirs may be able to provide additional economic benefits, they, together with larger reservoirs, also substantially divide the river basin into small segments, thereby disrupting the natural hydrological processes to a point where conventional hydrologic models without reservoirs are no longer desirable [7].

Hydrological modelling with the consideration of reservoirs have been reported by multiple studies, especially with respect to relatively large-sized reservoirs [8,9]. Hanasaki et al. [10] proposed a generalized reservoir scheme for global river routing models, where reservoirs are regulated using a monthly operation rule for both non-irrigation and irrigation reservoirs. Mateo et al. [4] employed the H08 Water Resources Model and the CaMa-Flood River Routing Model to assess the impact of reservoir operation on the flood propagation. Deng et al. [5] integrated an offline single-reservoir module to Xinanjiang model, a semi-distributed hydrologic model, and made an accurate prediction on the storage and water level change of the studied reservoir. Zhao et al. [11] integrated a reservoir operation module into the Distributed Hydrology-Soil-Vegetation Model (DHSVM) where they employed generalized reservoir operation rules to determine the outflow of reservoir and achieved satisfactory results. All of these studies emphasized the effect of large-scale reservoirs and disregarded the effect of smaller reservoirs.

However, smaller reservoirs should not be excluded in the hydrological modelling in many basins. A survey of Poyang Lake basin in China reveals that there are over 10,000 small reservoirs in the basin accounting for over 1/5 of the combined capacity of all reservoirs. In smaller sub-basins, this value can go up to over a half. Studies involving small reservoirs, however, are relatively scarce but they give some directions. Güntner et al. [12] employed a process-oriented semi-distributed hydrological model to quantify the effect of large and small reservoirs in a semi-arid area in Brazil by aggregating small reservoirs into one large reservoir. However, it does not account for the heterogeneity of water storage or runoff among reservoirs. Cao et al. [13] employed remote sensing techniques to detect the variation in the water surface of small reservoirs before and after the floods to quantify the flood detention effect of the small reservoirs. Deitch et al. [14] developed a Geographic Information System (GIS) based watershed model in which the fine digital elevation model (DEM) and mass-balance equation are employed to assess the cumulative effect of small reservoirs on the downstream flow. It is noted that the function of reservoir to supply water is not considered in their study, which may not be a negligible process in a medium- or long-term hydrological simulation [15]. Besides, most of these studies were based on offline reservoir modules, which can hardly depict the interacting effect of reservoirs on the water cycle.

Finally, meteorological forcing is a crucial impacting factor of the accuracy of hydrological modelling [16]. While rain gauges and meteorological stations traditionally played a significant role in providing forcing data for hydrological models, the recorded data of many stations is hardly accessible for researchers. Additionally, they are often sparsely distributed, especially in mountainous and arid areas [17], and thus are considered less desirable for a fine-scale modelling. The China Meteorological Assimilation Driving Datasets for the SWAT model (CMADS), which was developed by Dr. Xianyong Meng from China Agricultural University, integrates multi-satellite meteorological products with ground observation gauges and has been proven more effective than many of the counterparts since published [18–25]. For instance, Meng et al. [26] applied CMADS to Heihe basin for streamflow simulation and achieved comparatively better performance than other datasets. Similar results were also obtained by Gao et al. [27], who applied CMADS to Xiang River basin. However, the CMADS were mostly applied in the SWAT model so far, whether or not it can desirably drive other land surface models, such as Noah Land Surface Model (Noah LSM) or hydrological models, such as the Hydrologic Model System (HMS), remains a question.

Hence, the first objective of this paper is to develop an online reservoir module in the consideration of particularly small reservoirs and fully integrate this module into a coupled land surface-hydrological model, Noah LSM-HMS. The proposed module aggregated a group of small reservoirs into one representative reservoir using a statistical approach to depict the heterogeneity of storage capacity.

Secondly, to have a better understanding of the reservoir impact for more efficient hydrological forecasting for water agencies, the integrated model was applied to the upper Gan River basin, China to depict in detail the effect of both large and small reservoirs on the streamflow. Meanwhile, a comparative study was conducted between the CMADS and the reanalysis data from the National Centers for Environmental Prediction (NCEP) to evaluate the reliability and effectiveness of CMADS, which demonstrates that the CMADS can serve as a reliable dataset to drive hydrological models in China. Simulation results quantitatively indicate that the streamflow is increased in dry seasons and decreased in wet seasons, and large and small reservoirs have equally large effects on the streamflow. With the reservoir module, the performance of the model is also improved.

2. Methods

2.1. The Coupled Land Surface-Hydrological Model System (Noah LSM-HMS)

Noah LSM-HMS is a two-way coupled land surface-hydrological model system developed by Yuan et al. [28] as part of a joint Sino-German research program (hereinafter 'LSM-HMS'), with its structure being illustrated in Figure 1. Noah LSM is a land surface scheme of the Weather Research and Forecasting Model (WRF) and it is characterized by four soil temperature and moisture layers with canopy moisture and snow cover prediction [29]. It solves the Richards equation to obtain the soil moisture content and the vertical flow, and it employs the water balance and energy balance equations to derive streamflow, evapotranspiration, and recharge, which are then passed on to the hydrological model (HMS). HMS is a spatially distributed hydrological model and it solves the surface and subsurface flow using two-dimensional hydrodynamic equations [30]. HMS also calculates other variables, such as groundwater discharge and vadose-zone soil moisture, and it returns these variables to the LSM. The LSM-HMS is intended to provide a closed description of the water cycle between the land surface and the subsurface and is able to produce reasonable simulations on a range of hydrological variables in a mesoscale basin [29,31–33].

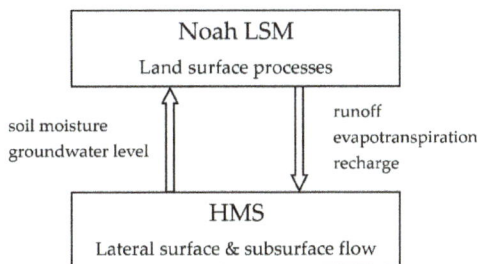

Figure 1. A schematic diagram of the coupled land surface-hydrological model system.

2.2. Reservoir Modelling

A typical basin, depending on its area, can normally contain up to hundreds or thousands of reservoirs with different capacities. These reservoirs connect each other to form a reservoir system and segment the basin into pieces. To account for the difference of reservoirs in size and importance, a reasonable classification of reservoirs in terms of their capacity is often necessary in the reservoir modelling [12]. In this study, reservoirs with a storage capacity of more than 1×10^7 m^3 are categorized as large-sized reservoirs, those with a capacity between 1×10^7 m^3 and 1×10^5 m^3 are categorized as small-sized. Those with a capacity less than 1×10^5 m^3 are regarded as earth dams and pools instead of reservoirs, therefore they are not included in this study.

In view of the scale and the considerable hydrological impact, large reservoirs are directly integrated into grid cells of their actual locations. These reservoirs often function differently with respect to the amount of the incoming flow and the water volume they stored at a certain time.

In this study, a simplified operation rule that achieved good performance in Lake Whitney and Lake Aquilla, Texas was employed to estimate the outflow of reservoirs in a daily or monthly scale—see Zhao et al. [11] for more details.

While there are probably not too many of large reservoirs in a typically mesoscale basin, small reservoirs can amount to several hundreds and even thousands, which are dotted everywhere, and most of them do not have larger ones as detailed data. It is thus impractical to quantify the effect of each of them in an analogous way to large reservoirs. A basic idea is to group them to one reservoir, as in the study of Güntner et al. [12] and Malveira et al. [34]. This idea also applies to this study, with the basin of interest divided into a few sub-basins and the small reservoirs in each sub-basin aggregated to a representative reservoir (hereinafter 'aggregated reservoir'), and the parameters of the aggregated reservoir (e.g., storage capacity) are derived from the summation of all small reservoirs combined. Based on the above idea, the information of heterogeneity in terms of the capacity of each small reservoir is lost. To preliminarily overcome this problem, a capacity cumulative distribution function for small reservoirs in a certain basin is introduced in the form of:

$$V_i = f(\alpha) \tag{1}$$

where V_i is the storage capacity of the ith small reservoir sorted in an ascending order and α is the portion of small reservoirs that has a capacity less than V_i. It should be noted that the integral of this function on its domain is equal to the mean capacity of the small reservoirs, or the capacity of the aggregated reservoir divided by the number of small reservoirs, N, in the sub-basin. A preliminary curve-fitting study on a few basins of Southeast China suggests that their cumulative distribution function follows a similar pattern, i.e.,

$$V_i = k_1 e^{k_2 \alpha} + k_3 e^{k_4 \alpha} \tag{2}$$

where k_i, $i = 1,2,3,4$ are the regression coefficients related to basins. Whether or not this distribution pattern suits all basins requires future investigation.

The capacity cumulative distribution function is employed to determine the release of the aggregated reservoirs. To be specific, the mean water storage of all small reservoirs in a sub-basin, V_{mean} (m³), is first calculated from the water storage of the aggregated reservoir divided by the number of small reservoirs. Subsequently, as illustrated in Figure 2, the percentage of small reservoirs with a water storage of V_{mean} smaller than and larger than their storage capacity V_i (i.e., unfilled and filled reservoirs) can be respectively calculated with the mean water storage and the distribution function, and is used to determine the outflow of the aggregated reservoir below.

Given that small reservoirs are mostly used for water supply instead of flood control because of a small capacity, the operation rule is simplified such that, when a reservoir is 'filled', all of the water above the capacity will be released immediately and that, when a reservoir is 'unfilled', the outflow is determined as the human water demand. The operation rule is then arranged in the following form to determine the reservoir outflow with the consideration of the heterogeneity of capacity:

$$Q_t = \alpha_1 \times U + \alpha_2 \frac{V_t - V}{\Delta t} \tag{3}$$

where Q_t (m³/s) and V_t (m³) are the outflow and the water storage of the aggregated reservoir at time t, respectively. U (m³/s) is the human water demand. α_1 and α_2 are the proportion of the combined water storage of unfilled and filled reservoirs, respectively.

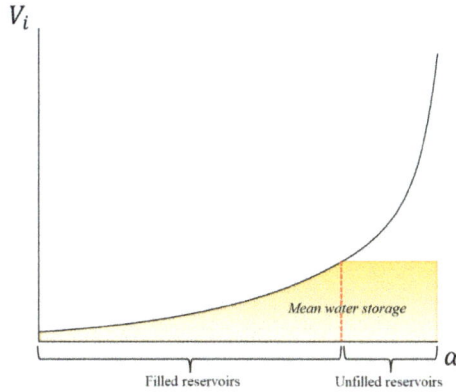

Figure 2. A schematic illustration of the capacity cumulative distribution function (black curve). The shaded area is the mean water storage of small reservoirs in a sub-basin and can be used to determine the proportion of unfilled and filled reservoirs in this sub-basin.

After the release of reservoirs is determined, the water balance equation is then employed to calculate V_t (m³), the reservoir water storage at time t:

$$V_t = V_{t-1} + \Delta t \times (Q_{in} - Q_{out} + A_t \times P - A_t \times E - A_t \times D) \tag{4}$$

where V_{t-1} (m³) is the water storage of the reservoir for the previous timestep. The initial water storage of reservoir is based on the soil moisture of the local grid. Δt (s) is the timestep. A_t (m²) is the current surface area of the reservoir. Q_{in} and Q_{out} (m³/s) are the inflow and outflow of the reservoir, respectively. P (m/s) and E (m/s) are the precipitation and evaporation on the reservoir surface. D (m/s) is the two-way flux between the reservoir and the vadose zone or groundwater. It is calculated as a part of the channel-groundwater interaction module in HMS, i.e.,

$$D = K_d \times \left(h_t - h_g \right) \tag{5}$$

where K_d (s⁻¹) is the hydraulic conductivity between channel and subsurface. h_t and h_g (m) are respectively the water level of the reservoir and groundwater level.

2.3. Module Integration

By integrating the reservoir module, the model is able to depict the interaction between the reservoir regulation and the subsequent hydrological processes. Firstly, the inflow and release difference due to reservoirs will change the surface water height in the two-dimensional diffusive wave equation and therefore alter the streamflow routing [35]. Secondly, the flux between reservoirs and groundwater will also affect the groundwater depth and the groundwater routing, which is realized by adding an additional source term to the two-dimensional Boussinesq equation. While the surface and subsurface flow routing can directly adjust the streamflow, it can also result in the variation of the hydrological conditions around the reservoirs, e.g., infiltration, river recharge, and groundwater table, which in turn gives feedback to the reservoir operation.

2.4. Performance Indexes

Three indexes are employed for a quantitative evaluation of the downstream streamflow simulation with or without the reservoir module, namely the water balance index (WBI), the Pearson product-moment correlation coefficient (R), and the Nash–Sutcliffe coefficient of efficiency (NSE), for the evaluation of water balance, data correlation, and flood peak simulation, respectively [32,35]:

$$\text{WBI} = \frac{\sum S_i}{\sum O_i} \tag{6}$$

$$R = \frac{\sum (S_i - \overline{S})(O_i - \overline{O})}{\left[\sum (S_i - \overline{S})^2 \sum (O_i - \overline{O})^2\right]^{0.5}} \tag{7}$$

$$\text{NSE} = 1 - \frac{\sum (S_i - O_i)^2}{\sum (O_i - \overline{O})^2} \tag{8}$$

where S_i and O_i are the simulated and observed streamflow for each timestep, respectively. The overbar symbolizes average.

According to the model evaluation guidelines [36], for a monthly timestep, the model simulation can be considered very good if NSE > 0.75, R > 0.70, and WBI < 0.1.

3. Case Study and Data

3.1. Study Area

The upper Gan River basin, as part of the Yangtze River basin, has an area of around 18,000 km^2 (Figure 3). The average annual precipitation is over 1300 mm and it exhibits an uneven distribution within a year. The wet season normally lasts for seven months from March to September, whereas the period from October to February is denoted as dry seasons. The reservoir amount and capacity were stable during the study period of 2008–2015, where there were, in total, about 453 reservoirs in this area, including eight large reservoirs with a combined capacity of 3.78 × 10^8 m^3. The largest reservoir, Tuanjie Reservoir, has a storage capacity of 1.46 × 10^8 m^3. The other 445 small reservoirs have a total storage capacity of 4.25 × 10^8 m^3, with nearly no reservoir newly built or under construction. The small reservoirs account for 98% of the reservoir amount and 53% of the reservoir storage in this basin so that the small reservoirs are supposed to have a considerable cumulative effect on the local hydrologic cycle.

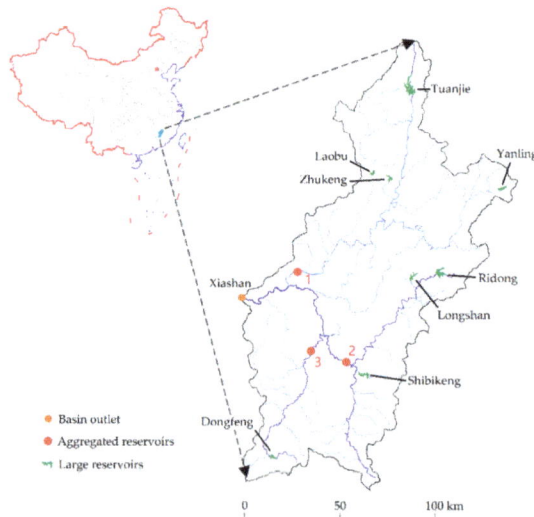

Figure 3. The upper Gan River basin. Eight large reservoirs are marked in green. Three red round markers with number 1, 2, 3 indicate aggregated reservoirs, which divide the basin into three sub-basins.

3.2. Model Setup

LSM-HMS: The computational timestep for LSM and HMS were determined to be half an hour and one day, respectively, with a spatial resolution of 10×10 km. While the study period is 2008–2015, the simulation period starts one year earlier such that the year 2007 is included as part of the model spin-up process.

Reservoir module: The computational domain of LSM-HMS is fully discretized in the unit of grid cell. For large reservoirs, they are directly integrated into the computational grids corresponding to their actual location. However, the integration of aggregated reservoirs is more complicated because they do not really have an actual location or actual amount. Two considerations for the location and number of aggregated reservoirs in this study are presented, as follows:

1. The aggregated reservoir can be placed in the proximity of the convergence point between the mainstream and the tributary or between two tributaries so that each tributary is a sub-basin and most of the small reservoirs in the entire basin can be included.
2. The number and location of aggregated reservoirs or sub-basins should be in conformity to data availability, so that the sum of the reservoir capacity for each sub-basin can be known.

With the two guidelines above, the detailed configuration of each aggregated reservoir is determined and presented in Table 1. The aggregated reservoirs are generally located downstream three largest tributaries and are around the demarcation of local administrative regions (i.e., counties), such that the reservoir data for each county, e.g., the sum of storage capacity and the number of reservoirs in each sub-basin, are well collected by the local administrations. The study area is therefore divided into three sub-basins with one aggregated reservoir placed at the outlet of each sub-basin, as illustrated in Figure 3.

Table 1. Configuration of aggregated reservoirs.

Aggregated Reservoir	Storage Capacity (10^8 m^3)	Average Water Demand	
		Non-Irrigation Period (m^3/s)	Irrigation Period (m^3/s)
1	2.52	1.7	3.2
2	0.93	0.9	1.8
3	0.80	0.8	1.5

3.3. Data Input

The land surface scheme of the model is driven by a set of meteorological forcing. The precipitation, surface air temperature, surface pressure, solar radiation, humidity, and wind speed were obtained from the China Meteorological Assimilation Driving Datasets for the SWAT model (CMADS) [37]. Downward longwave radiation is also needed for the calculation of evapotranspiration and it is obtained from NCEP reanalysis data [38]. The land use and soil data were collected from the Moderate Resolution Imaging Spectroradiometer (MODIS) 1km data and the Harmonized World Soil Database (HWSD), which were both accessed from the Cold and Arid Regions Science Data Center at Lanzhou. The terrain data were obtained from HYDRO1K DEM established by the United States Geological Survey (USGS) [39]. Daily discharge data of Xiashan between 2008–2015 were available for the model calibration and validation.

The yearly water use from reservoirs from 2008–2015 was collected from annual reports that were published by the Department of Water Resources of Jiangxi Province. The irrigation water demand generally accounted for 65% of the total water demand, and it is then evenly downscaled to monthly values throughout the wet seasons. The monthly values are further partitioned to daily values for reservoir input. While the irrigation use occurs during only the irrigation period or wet seasons, other water uses are evenly partitioned from annual to daily values. The water demand of the basin is then allocated to reservoirs in proportion to the storage capacity. Other human activities, including the consumptive use of water and groundwater/river extraction, are not included in this study, since the study focuses on the reservoir effect.

The capacity cumulative distribution function for small reservoirs is fitted using 275 small reservoirs with known capacity in the study area and in the larger Poyang Lake basin, in the form of $V_i = 185850e^{2.274\alpha} + 0.00031e^{23.82\alpha}$, and it is assumed to be applicable to the entire basin in this study.

The relationship between water level, surface area and water storage is available for all large reservoirs. For small reservoirs, these correlations are estimated using a linear fitting of the data points from the combination of the large reservoirs.

4. Results and Discussions

In this study, the model was mainly calibrated against the downstream streamflow data series of Xiashan. The calibration and validation period are 2012–2015 and 2008–2011, respectively. The reservoir module was calibrated against the water storage of Tuanjie reservoir from 2012–2015. For conciseness, simulations without the reservoir module, with only large reservoirs and with all reservoirs are hereinafter denoted as LSM-HMS, LH-L, and LH-A, respectively.

4.1. Calibration and Evaluation of the Model

4.1.1. Reservoir Module

To quantify the effect of large reservoirs, the water storage of Tuanjie Reservoir during 2012 and 2015 was employed for the calibration of the operation rule [11]. Subject to the data scarcity, the calibration results were applied to all of the large reservoirs in this study. With respect to Tuanjie Reservoir, the largest reservoir in the basin, the monthly simulation of its storage over the study period and the observation data of its water storage during 2008 and 2015 are illustrated in Figure 4. Most of the simulated water storage follows an increase and decrease alteration within a year, indicating that the reservoir functions well according to the generalized operation rule. The simulation of water storage from 2008 to 2015 generally matches the observation, and the monthly WBI, R, and NSE for the study period are 1.03, 0.84, and 0.65, respectively. A major source of error can be the inaccuracy of inflow computed in LSM-HMS, and the difference between the generalized regulation rule and the reservoir operation in reality is also considered to have greatly contributed to the error. It is especially noted that the water storage of Tuanjie Reservoir is much overestimated, especially in some dry seasons, indicating that the operation rule in reality is very flexible such that the downstream water demand can be satisfied in dryer years. However, this is not presented in the generalized operation rule.

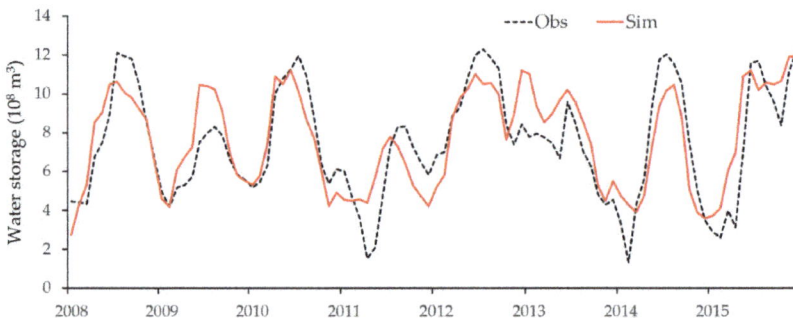

Figure 4. Monthly simulated and observed water storage of Tuanjie Reservoir during 2008–2015.

4.1.2. LSM-HMS and the Integrated Model

To investigate the impact of the reservoir module on the downstream discharge, the LSM-HMS was calibrated against the monthly observed streamflow discharge in Xiashan for the period of 2012–2015, and then validated in 2008–2011. In all, six parameters were included in the calibration, namely streambed conductivity, Manning's roughness, saturated hydraulic conductivity, porosity, wilting point, and aquifer thickness. The last four parameters were initially collected from the HWSD database, but it can also be adjusted within a limited range (i.e., ±50%). The calibration results were presented in Table 2.

Table 2. Calibration results of land surface-hydrological model (LSM-HMS).

Parameters	Input Value	Parameters	Input Value
Streambed conductivity	$0.90\ \mathrm{s}^{-1}$	Porosity	×1.0
Manning's roughness	0.07	Wilting point	×1.0
Saturated hydraulic conductivity	×1.0	Aquifer thickness	×1.0

In terms of the LSM-HMS, the monthly WBI, R, and NSE are, respectively, 1.03, 0.97, and 0.91 over the calibration period, 1.15, 0.92, and 0.86 over the validation period, and 1.08, 0.95, and 0.89 over the entire study period (see Figure 5). According to the model evaluation guidelines [36], it can be generally regarded as a very good simulation result, indicating that the CMADS-driven LSM-HMS can serve as a reasonable tool to investigate basin-scale hydrological variations and that the CMADS can successfully drive the coupled land surface-hydrological model for an accurate simulation. The error can be attributed to the size of grid cell in this study, as a 10 km grid cell can be too large for a better simulation result in view of the size of the basin. Besides, the spatial heterogeneity of some hydrogeological parameters, such as the wilting point and the Manning coefficient, were not considered, i.e., the entire basin employs a same value because of the data scarcity and model complexity. In addition, the accuracy of CMADS forcing data, after all, is subject to its own spatial resolution and it cannot fully reveal the fine-scale spatial distribution of each meteorological variable. Also, the lack of human activities can also be a source of error in this case.

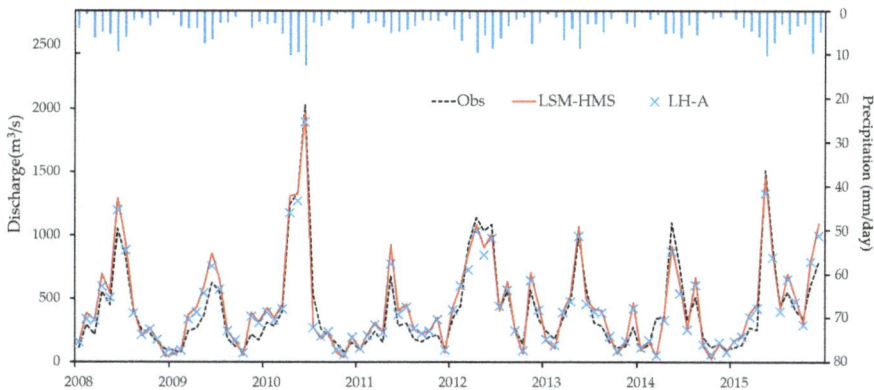

Figure 5. Monthly observed discharge and simulated discharge of China Meteorological Assimilation Driving Datasets for the SWAT model (CMADS)-driven LSM-HMS and all-reservoir condition (LH-A) in Xiashan during 2008–2015.

By incorporating the effect of large reservoirs, LH-L considers the effect of relatively large reservoirs, thus improving the NSE to 0.90 (1.1%) as compared to LSM-HMS. With the WBI of LH-L decreasing by 2.8% to 1.05, the loss of water is presumably a result of reservoir evaporation and infiltration. With small reservoirs included, it can be seen in Figure 5 that LH-A further improves the simulation. The WBI of LH-A reduces to 1.04 (1.90%) as compared to LH-L for similar reasons. The R and NSE also improve to 0.96 (1.0%) and 0.91 (1.1%) as compared to LH-L, respectively. It is noted that the improvement of LH-A is basically in a same order of magnitude as that of LH-L by comparing the change in WBI, R, and NSE, indicating a large group of small reservoirs can have an effect as considerable as large reservoirs on the streamflow discharge simulation. This effect can be enlarged, especially in this case study where the large reservoirs are all located very upstream with a relatively small capacity than in other typical basins, whereas the small reservoirs are mostly dotted downstream with a larger combined capacity.

A paired Student's *t*-Test was also performed to evaluate the magnitude of improvement. The result demonstrates that, for all three sets of comparison, the simulation of downstream streamflow sees a significant improvement at a significant level of 5%, indicating that the proposed reservoir module can effectively reduce the error of the original simulation. The *p*-values of the *t*-Test and the performance of each simulation are summarized in Table 3.

Table 3. Comparison of monthly streamflow simulations.

Index	LSM-HMS	LH-L	LH-A	Difference (%)		
				LH-L/LSM-HMS	LH-A/LH-L	LH-A/LSM-HMS
WBI	1.08	1.05	1.03	−2.8	−1.9	−4.6
R	0.95	0.95	0.96	0	1.0	1.0
NSE	0.89	0.90	0.91	1.1	1.1	2.2
Probability of the paired *t*-Test				<0.01 *	<0.01 *	<0.01 *

* Indicates *t*-Test is significant at the 0.05 significance level.

4.2. Evaluation of CMADS against NCEP Database

Although the model performs well with the CMADS database, it is necessary to compare the accuracy and the efficiency of CMADS with other meteorological databases in driving LSM-HMS. As the existing literature has reported the superiority of CMADS over multiple meteorological data sources, such as the Precipitation Estimation from Remotely Sensed Information using Artificial Neural Networks–Climate Data Record (PERSIANN-CDR) in driving SWAT [25,27], the meteorological reanalysis dataset from NCEP was employed in this section to further evaluate the efficiency of CMADS in LSM-HMS. The NCEP reanalysis data covers the globe in T62 grids and it is widely used in macroscale and mesoscale hydrological modelling. Therefore, the precipitation, air temperature, air pressure, relative humidity, longwave radiation, downward solar radiation, and wind speed were processed and employed in this study to drive LSM-HMS for a comparative study. In general, the NCEP precipitation is 12% larger than the CMADS precipitation for the period of 2008–2015. This is in accordance with the finding of Yang [31] that NCEP tends to overestimate the precipitation in China. The NCEP-driven simulated discharge of Xiashan for 2008–2015 was therefore overestimated by 17% (see Figure 6b), and the WBI, R, and NSE over the study period are 1.17, 0.81, and 0.56, respectively. As also indicated in Figure 6a, the simulation result of NCEP-driven LSM-HMS is much less accurate than that of CMADS-driven LSM-HMS. It suggests that LSM-HMS is sensitive to the input forcing and, when compared to NCEP data, CMADS can serve as a much more reliable source of meteorological forcing for the hydrological modelling in China.

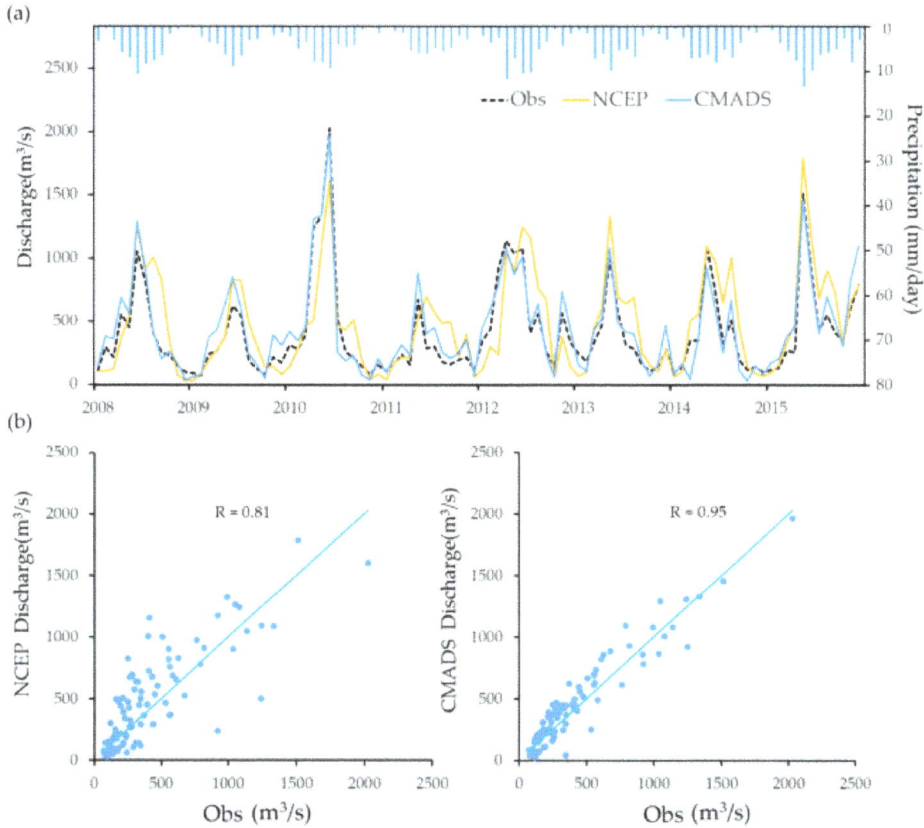

Figure 6. (**a**) Monthly observed discharge and simulated discharge of National Centers for Environmental Prediction (NCEP)-driven and China Meteorological Assimilation Driving Datasets for the SWAT model (CMADS)-driven LSM-HMS in Xiashan during 2008–2015, with the NCEP precipitation shown at the top, (**b**) the scatter plot of observed discharge against NCEP-driven simulation (left) and CMADS-driven simulation (right).

4.3. Effects of Reservoirs on Streamflow

In the previous section, it was found that the introduction of reservoirs can improve the performance of Noah LSM-HMS at a significant level of 5%. The temporal distribution of the difference between all-reservoir condition (LH-A) and no-reservoir condition (LSM-HMS) was presented in Figure 7. It can be seen that the improvement in LH-A as compared to LSM-HMS can be mostly attributed to the increase of streamflow in dry seasons and the decrease of streamflow in wet seasons, since reservoirs can be regulated to mitigate the flood in wet seasons and to supply water in dry seasons. To be specific, the streamflow sees a 5.1 m³/s (2.0%) increase in dry seasons and a 45.4 m³/s decrease (7.8%) in wet seasons. In light of the different modelling methods between large and small reservoirs, their effects are separately discussed, as follows.

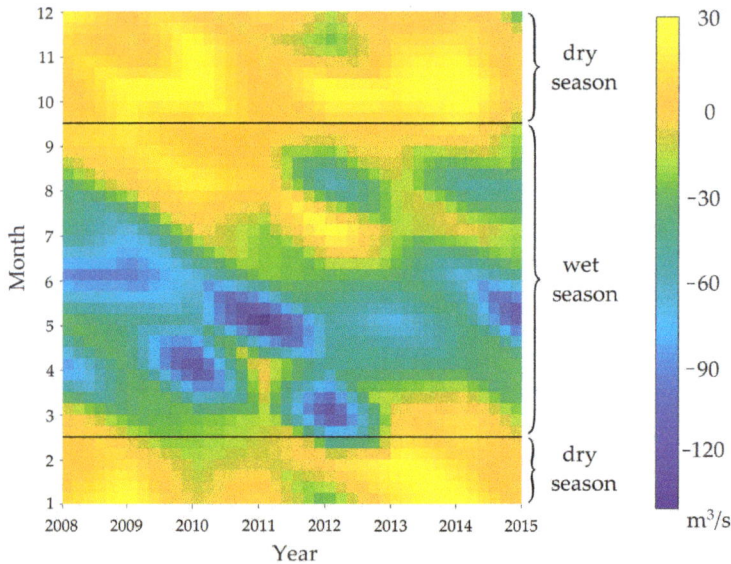

Figure 7. Temporal distribution of reservoir effect on the downstream discharge at Xiashan for 2008–2015. Yellow indicates the streamflow is increased with the consideration of reservoirs whereas blue indicates otherwise (see the color bar).

Large reservoirs: Large reservoirs are generally intended to mitigate floods in wet seasons and supply more water in dry seasons. To quantitatively explain the simulated effects of the large reservoirs on extreme climate conditions, the averaged maximum monthly inflow (high flow) and minimum monthly inflow (low flow), and the release of each reservoir were selected and presented in Table 4. It is noted that, for all large reservoirs, the maximum monthly inflow is reduced, with the largest reduction of 53.5% for the Shibikeng Reservoir. On the other hand, the monthly low flow is maintained or increased for all of the reservoirs except Ridong Reservoir. These results indicate that, with the generalized operation rule, the large reservoirs effectively mitigate floods and, in most cases, supply more water in dry seasons.

Table 4. Effects of reservoirs on the monthly maximum and minimum flow.

Reservoir	Averaged Monthly Maximum			Averaged Monthly Minimum		
	Inflow (m^3/s)	Outflow (m^3/s)	Difference (%)	Inflow (m^3/s)	Outflow (m^3/s)	Difference (%)
Tuanjie	100.0	87.4	−12.6	1.6	2.5	56.3
Yanling	47.0	45.0	−4.3	1.2	1.3	8.3
Ridong	33.8	27.9	−17.5	1.2	0.7	−41.7
Longshan	13.9	11.5	−17.3	0.1	0.1	0.0
Shibikeng	25.4	11.8	−53.5	0.3	0.3	0.0
Zhukeng	15.4	12.7	−17.5	0.2	0.3	50.0
Laobu	15.9	14.1	−11.3	0.1	0.2	100.0
Dongfeng	24.6	22.7	−7.7	0.4	0.7	75.0
AR[†] 1	1323.2	1289.6	−2.5	20.3	20.4	0.0
AR[†] 2	402.8	390.7	−3.0	6.2	7.0	12.9
AR[†] 3	210.1	208.8	−0.7	2.3	2.8	21.7

AR[†] indicates aggregated reservoir.

Small reservoirs: While most small reservoirs are not primarily intended for flood mitigation because of the limited capacity, a large group of small reservoirs can indirectly reduce the local floods because there is always a portion of small reservoirs that are not filled enough to release all of the incoming flood. For example, the simulated inflow and outflow of small reservoirs represented by aggregated reservoir 3 for the first half of 2010 are illustrated in Figure 8b, where the flood peaks are mitigated by different minor values. As the wet season arrives, the proportion of unfilled and filled small reservoirs in the sub-basin is constantly changing in terms of the magnitude of the incoming flow (see Figure 8a). An increasing number of small reservoirs become dry towards the end of dry season because of the human water demand. When the flood season comes, these reservoirs start to fill up as the flood peaks are somewhat reduced. It is also noted that, due to the limited capacity, most of the small reservoirs are filled during a significant flood where almost all of the flood is released immediately.

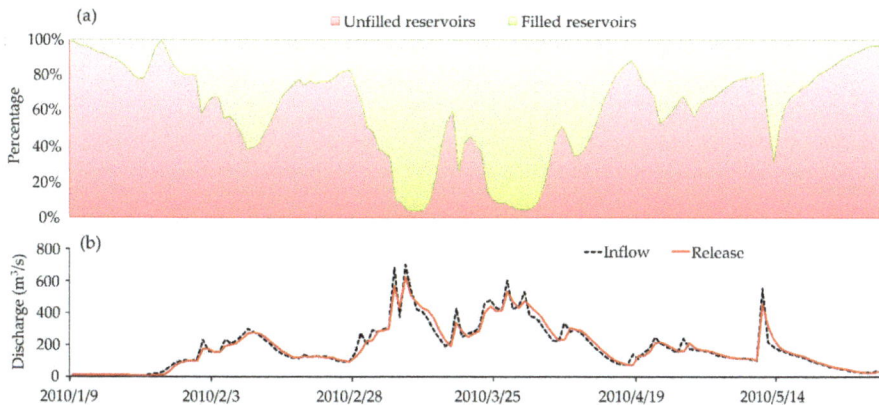

Figure 8. For small reservoirs represented by aggregated reservoir 3, (**a**) the daily simulated percentage of unfilled and filled small reservoirs, and (**b**) the corresponding simulated inflow and release of the aggregated reservoir 3 during January and May 2010.

The averaged maximum and minimum monthly inflow and outflow of each aggregated reservoir are presented in Table 4. It can be seen the relative flood mitigation effect of aggregated reservoirs is minor (0.7%–2.5%) as compared to large reservoirs, but the absolute value of mitigation is not necessarily smaller than large reservoirs. In a monthly level, the small reservoirs can increase the local runoff by 0%–21% in dry seasons, indicating that small reservoirs can serve as an effective source of human water use in dry seasons.

For a more direct view of the effect of reservoirs, three flood events are selected for comparison between LSM-HMS and LH-A. It can be seen from Figure 9 that the flood peaks are mitigated by different values. For instance, with respect to the floods on January 2008, May 2009, and April 2013, the peak discharges simulated by LH-A are, respectively, 95 m^3/s, 120 m^3/s, and 108 m^3/s lower than those simulated by LSM-HMS alone, making the simulated values closer to the observed ones. Similarly, a flood detention effect can be observed in some flood events, i.e., the flood tends to rise and fall less drastically in the simulation of LH-A (see also Figure 9). This effect is more conspicuous, especially for the first flood after a long relatively dry period, presumably because the reservoir group needs to fill themselves first before releasing the flood.

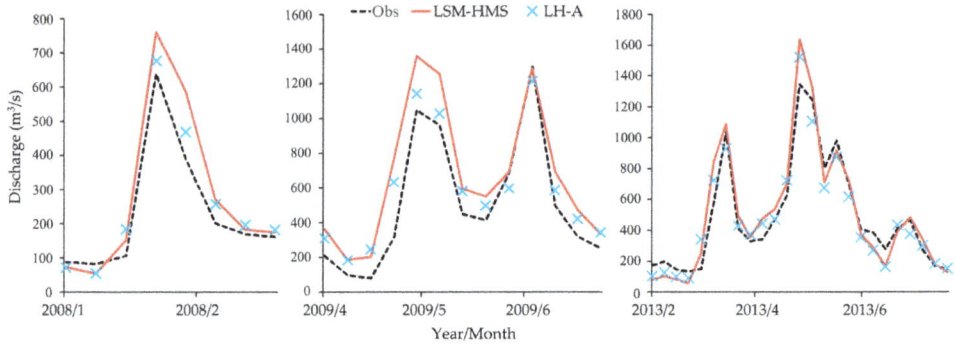

Figure 9. Comparison of weekly mean observed downstream streamflow in Xiashan and simulated streamflow of LH-A and LSM-HMS for three selected periods.

5. Conclusions

Reservoir operation can result in notable uncertainty in terms of hydrological modelling and it is an important aspect that should be handled to gain better knowledge of the reservoir impact and to support a more effective hydrological forecasting for water agencies. While carrying out detailed investigations on the regulation scheme of each reservoir is time-consuming and impractical, especially for small reservoirs, this paper aggregated small reservoirs into one representative reservoir with the use of the capacity cumulative distribution function, which was then integrated into a coupled land surface-hydrological model, Noah LSM-HMS. Large important reservoirs were also represented in the model using a set of generalized operation rules. With the application of the integrated model to a case study in the upper Gan river basin, the following conclusions are made:

- CMADS can serve as a high-quality meteorological database for the coupled land surface-hydrological model. CMADS-driven LSM-HMS generally have a much better performance than NCEP-driven LSM-HMS.

- The reservoir module can depict the annual and interannual variation in the water storage well for both large and small reservoirs. The integrated model yields improved simulation results at a significant level with the incorporation of reservoirs.

- Both large reservoirs and small reservoirs have a similar effect in reducing the floods in wet seasons and increasing the flow in dry seasons. Although small reservoirs are not primarily intended for flood mitigation, a large group of small reservoirs can indirectly reduce the local floods by up to 2.5% in a monthly level.

- The error of LSM-HMS is related to the input data and grid resolution as well as input parameter error. With a finer modelling resolution, the error is expected to be reduced. The simplification of the reservoir representation and the operation rule is also considered to be a source of error.

The idea using a statistical distribution and an aggregated reservoir to represent small reservoirs in this study serves as a compromise between the convenience and model accuracy. It saves time from investigating and integrating each of the small reservoirs, especially in a basin where there are too many small reservoirs to consider one by one. However, this idea of introducing aggregated reservoirs, after all, is based on a lumped hydrologic concept rather than a distributed concept to be used in a distributed hydrological model. While one can employ a more distributed method by, for example, distributing one or more reservoirs into each of the grid, their interconnection in the model needs to be considered with care to be an appropriate representation of the reality, and further work is expected on this aspect.

The idea using a statistical distribution and an aggregated reservoir to represent small reservoirs in this study serves as a compromise between the convenience and model accuracy. It saves time from investigating and integrating each of the small reservoirs, especially in a basin where there are too many small reservoirs to consider one by one. However, this idea of introducing aggregated reservoirs, after all, is based on a lumped hydrologic concept rather than a distributed concept to be used in a distributed hydrological model. While one can employ a more distributed method by, for example, distributing one or more reservoirs into each of the grid, their interconnection in the model needs to be considered with care to be an appropriate representation of the reality, and further work is expected on this aspect.

Author Contributions: Conceptualization: M.Y., X.M., H.W. and C.Y., Investigation: X.L., X.M., C.Y. and Z.W., Methodology: N.D., X.M., H.W. and M.Y., Formal analysis: N.D., X.L. and Z.W., Writing: N.D. All the authors have approved of the submission of this manuscript.

Funding: Financial support for this study was provided by the National Key Research and Development Project (2016YFC0402201), the National Science Foundation for Young Scientists of China (Grant No.51709271), the Young Elite Scientists Sponsorship Program by CAST (2017QNRC001) and the National Natural Science Foundation of China (41761134090; 41323001; 41471016).

Acknowledgments: Special thanks to Jianhui Wei and Qing Zhao for providing instructions and suggestions on this work.

Conflicts of Interest: The authors declare no conflict of interest.

References

1. Zhang, L.; Karthikeyan, R.; Bai, Z.; Wang, J. Spatial and temporal variability of temperature, precipitation, and streamflow in upper sang-kan basin, china. *Hydrol. Process.* **2018**, *31*, 279–295. [CrossRef]
2. Yang, L.; Feng, Q.; Yin, Z.; Wen, X.; Si, J.; Li, C.; Deo, R.C. Identifying separate impacts of climate and land use/cover change on hydrological processes in upper stream of Heihe River, Northwest China. *Hydrol. Process.* **2017**, *31*, 1100–1112. [CrossRef]
3. Marhaento, H.; Booij, M.J.; Rientjes, T.H.M.; Hoekstra, A.Y. Attribution of changes in the water balance of a tropical catchment to land use change using the SWAT model. *Hydrol. Process.* **2017**, *31*, 2029–2040. [CrossRef]
4. Mateo, C.M.; Hanasaki, N.; Komori, D.; Tanaka, K.; Kiguchi, M.; Champathong, A.; Sukhapunnaphan, T.; Dai Yamazaki, D.; Oki, T. Assessing the impacts of reservoir operation to floodplain inundation by combining hydrological, reservoir management, and hydrodynamic models. *Water. Resour. Res.* **2015**, *50*, 7245–7266. [CrossRef]
5. Deng, C.; Liu, P.; Liu, Y.; Wu, Z.; Wang, D. Integrated hydrologic and reservoir routing model for real-time water level forecasts. *J. Hydrol. Eng.* **2014**, *20*, 05014032. [CrossRef]
6. Mushtaq, S.; Dawe, D.; Hafeez, M. Economic evaluation of small multi-purpose ponds in the Zhanghe irrigation system, China. *Agric. Water. Manag.* **2007**, *91*, 61–70. [CrossRef]
7. Potter, K.W. Small-scale, spatially distributed water management practices: Implications for research in the hydrologic sciences. *Water. Resour. Res.* **2006**, *42*. [CrossRef]
8. Christensen, N.S.; Lettenmaier, D.P. A multimodel ensemble approach to assessment of climate change impacts on the hydrology and water resources of the Colorado River Basin. *Hydrol. Earth Syst. Sci. Discuss.* **2006**, *3*, 3727–3770. [CrossRef]
9. VanRheenen, N.T.; Wood, A.W.; Palmer, R.N.; Lettenmaier, D.P. Potential implications of PCM climate change scenarios for Sacramento–San Joaquin River Basin hydrology and water resources. *Clim. Chang.* **2004**, *62*, 257–281. [CrossRef]
10. Hanasaki, N.; Kanae, S.; Oki, T. A reservoir operation scheme for global river routing models. *J. Hydrol.* **2006**, *327*, 22–41. [CrossRef]
11. Zhao, G.; Gao, H.; Naz, B.S.; Kao, S.C.; Voisin, N. Integrating a reservoir regulation scheme into a spatially distributed hydrological model. *Adv. Water. Resour.* **2016**, *98*, 16–31. [CrossRef]
12. Güntner, A.; Krol, M.S.; Araújo, J.C.D.; Bronstert, A. Simple water balance modelling of surface reservoir systems in a large data-scarce semiarid region/Modélisation simple du bilan hydrologique de systèmes de réservoirs de surface dans une grande région semi-aride pauvre en données. *Hydrol. Sci. J.* **2014**, *49*. [CrossRef]

13. Cao, M.; Zhou, H.; Zhang, C.; Zhang, A.; Li, H.; Yang, Y. Research and application of flood detention modeling for ponds and small reservoirs based on remote sensing data. *Sci. China Tech. Sci.* **2011**, *54*, 2138–2144. [CrossRef]

14. Deitch, M.J.; Merenlender, A.M.; Feirer, S. Cumulative effects of small reservoirs on streamflow in Northern Coastal California catchments. *Water Resour. Manag.* **2013**, *27*, 5101–5118. [CrossRef]

15. Magilligan, F.J.; Nislow, K.H. Changes in hydrologic regime by dams. *Geomorphology* **2005**, *71*, 61–78. [CrossRef]

16. Salamon, P.; Feyen, L. Assessing parameter, precipitation, and predictive uncertainty in a distributed hydrological model using sequential data assimilation with the particle filter. *J. Hydrol.* **2009**, *376*, 428–442. [CrossRef]

17. Guo, B.; Zhang, J.; Xu, T.; Croke, B.; Jakeman, A.; Song, Y.; Yang, Q.; Lei, X.; Liao, W. Applicability Assessment and Uncertainty Analysis of Multi-Precipitation Datasets for the Simulation of Hydrologic Models. *Water* **2018**, *10*, 1611. [CrossRef]

18. Cao, Y.; Zhang, J.; Yang, M. Application of SWAT Model with CMADS Data to Estimate Hydrological Elements and Parameter Uncertainty Based on SUFI-2 Algorithm in the Lijiang River Basin, China. *Water* **2018**, *10*, 742. [CrossRef]

19. Meng, X.; Wang, H.; Shi, C.; Wu, Y.; Ji, X. Establishment and Evaluation of the China Meteorological Assimilation Driving Datasets for the SWAT Model (CMADS). *Water* **2018**, *10*, 1555. [CrossRef]

20. Meng, X.; Wang, H.; Wu, Y.; Long, A.; Wang, J.; Shi, C.; Ji, X. Investigating spatiotemporal changes of the land-surface processes in Xinjiang using high-resolution CLM3.5 and CLDAS: Soil temperature. *Sci. Rep.* **2017**, *7*, 13286. [CrossRef]

21. Meng, X.; Wang, H. Significance of the China Meteorological Assimilation Driving Datasets for the SWAT Model (CMADS) of East Asia. *Water* **2017**, *9*, 765. [CrossRef]

22. Meng, X.; Dan, L.; Liu, Z. Energy balance-based SWAT model to simulate the mountain snowmelt and runoff—Taking the application in Juntanghu watershed (China) as an example. *J. Mt. Sci.* **2015**, *12*, 368–381. [CrossRef]

23. Meng, X.; Wang, H.; Lei, X.; Cai, S.; Wu, H. Hydrological Modeling in the Manas River Basin Using Soil and Water Assessment Tool Driven by CMADS. *Teh. Vjesn.* **2017**, *24*, 525–534. [CrossRef]

24. Zhao, F.; Wu, Y. Parameter Uncertainty Analysis of the SWAT Model in a Mountain Loess Transitional Watershed on the Chinese Loess Plateau. *Water* **2018**, *10*, 690. [CrossRef]

25. Liu, J.; Shanguan, D.; Liu, S.; Ding, Y. Evaluation and Hydrological Simulation of CMADS and CFSR Reanalysis Datasets in the Qinghai-Tibet Plateau. *Water* **2018**, *10*, 513. [CrossRef]

26. Meng, X.; Wang, H.; Cai, S.; Zhang, X.; Leng, G.; Lei, X.; Shi, C.; Liu, S.; Shang, Y. The China Meteorological Assimilation Driving Datasets for the SWAT Model (CMADS) Application in China: A Case Study in Heihe River Basin. *Pearl River* **2016**, *37*, 1–9.

27. Gao, X.; Zhu, Q.; Yang, Z.; Wang, H. Evaluation and Hydrological Application of CMADS against TRMM 3B42V7, PERSIANN-CDR, NCEP-CFSR, and Gauge-Based Datasets in Xiang River Basin of China. *Water* **2018**, *10*, 1225. [CrossRef]

28. Yuan, F.; Kunstmann, H.; Yang, C.; Yu, Z.; Ren, L.; Fersch, B.; Xie, Z. Development of a coupled land-surface and hydrology model system for mesoscale hydrometeorological simulations. In *New Approaches to Hydrological Prediction in Data-Sparse Regions, Proceedings of Symposium HS.2 at the Joint Convention of the International Association of Hydrological Sciences (IAHS) and The International Association of Hydrogeologists (IAH), Hyderabad, India, 6–12 September 2009*; IAHS Press: Wallingford, UK, 2009.

29. Wagner, S.; Fersch, B.; Yuan, F.; Yu, Z.; Kunstmann, H. Fully coupled atmospheric-hydrological modeling at regional and long-term scales: Development, application, and analysis of WRF-HMS. *Water Resour. Res.* **2016**, *52*, 3187–3211. [CrossRef]

30. Yu, Z.; Pollard, D.; Cheng, L. On continental-scale hydrologic simulations with a coupled hydrologic model. *J. Hydrol.* **2006**, *331*, 110–124. [CrossRef]

31. Yang, C. Research on Coupling Land Surface-Hydrology Model and Application. Ph.D. Thesis, Hohai University, Nanjing, China, 2009.

32. Yang, C.; Lin, Z.; Yu, Z.; Hao, Z.; Liu, S. Analysis and simulation of human activity impact on streamflow in the Huaihe River basin with a large-scale hydrologic model. *J. Hydrometeorol.* **2010**, *11*, 810–821. [CrossRef]

33. Yang, C.; Yu, Z.; Hao, Z.; Zhang, J.; Zhu, J. Impact of climate change on flood and drought events in Huaihe River Basin, China. *Hydrol. Res.* **2012**, *43*, 14–22. [CrossRef]

34. Malveira, V.T.C.; Araújo, J.C.D.; Güntner, A. Hydrological impact of a high-density reservoir network in semiarid northeastern Brazil. *J. Hydrol. Eng.* **2011**, *17*, 109–117. [CrossRef]

35. Lv, M.; Hao, Z.; Lin, Z.; Ma, Z.; Lv, M.; Wang, J. Reservoir operation with feedback in a coupled land surface and hydrologic model: A case study of the Huai River Basin, China. *J. Am. Water Resour. Assoc.* **2016**, *52*, 168–183. [CrossRef]

36. Moriasi, D.N.; Arnold, J.G.; Liew, M.W.V.; Bingner, R.L.; Harmel, R.D.; Veith, T.L. Model evaluation guidelines for systematic quantification of accuracy in watershed simulations. *Trans. ASABE* **2007**, *50*, 885–900. [CrossRef]

37. The China Meteorological Assimilation Driving Datasets for the SWAT model (CMADS). Available online: http://www.cmads.org (accessed on 25 October 2018).

38. NOAA Earth System Research Laboratory. Available online: https://www.esrl.noaa.gov (accessed on 26 October 2018).

39. USGS EROS Archive-Digital ElevationHYDRO1K. Available online: https://www.usgs.gov/centers/eros/science/usgs-eros-archive-digital-elevation-hydro1k (accessed on 1 October 2018).

water

MDPI

Article

Assessing the Impact of Reservoir Parameters on Runoff in the Yalong River Basin using the SWAT Model

Xuan Liu [1], Mingxiang Yang [2,*], Xianyong Meng [3,4,*], Fan Wen [5] and Guangdong Sun [2]

1   College of Civil Engineering, Tianjin University, Tianjin 300072, China; liuxuantju@163.com
2   Department of Water Resources, China Institute of Water Resources and Hydropower Research,
    Beijing 100038, China; sungd@webmail.hzau.edu.cn
3   College of Resources and Environmental Science, China Agricultural University (CAU),
    Beijing 100094, China
4   Department of Civil Engineering, The University of Hong Kong (HKU),
    Pokfulam 999077, Hong Kong, China
5   School of Hydropower & Information Engineering, Huazhong University of Science and Technology,
    Wuhan 430074, China; wenfan@hust.edu.cn
*   Correspondence: yangmx@iwhr.com (M.Y.); xymeng@cau.edu.cn or xymeng@hku.hk (X.M.)

Received: 30 January 2019; Accepted: 23 March 2019; Published: 27 March 2019

Abstract: The construction and operation of cascade reservoirs has changed the natural hydrological cycle in the Yalong River Basin, and reduced the accuracy of hydrological forecasting. The impact of cascade reservoir operation on the runoff of the Yalong River Basin is assessed, providing a theoretical reference for the construction and joint operation of reservoirs. In this paper, eight scenarios were set up, by changing the reservoir capacity, operating location, and relative location in the case of two reservoirs. The aim of this study is to explore the impact of the capacity and location of a single reservoir on runoff processes, and the effect of the relative location in the case of joint operation of reservoirs. The results show that: (1) the reservoir has a delay and reduction effect on the flood during the flood season, and has a replenishment effect on the runoff during the dry season; (2) the impact of the reservoir on runoff processes and changes in runoff distribution during the year increases with the reservoir capacity; (3) the mitigation of flooding is more obvious at the river basin outlet control station when the reservoir is further downstream; (4) an arrangement with the smaller reservoir located upstream and the larger reservoir located downstream can maximize the benefits of the reservoirs in flood control.

Keywords: reservoir parameters; runoff; CMADS; SWAT; Yalong River

1. Introduction

Climate change and human activities are the two major factors affecting the hydrological cycle in basin [1]. Human activities make hydrological processes more complicated by interfering with the transmission and distribution of runoff, sediments, etc. The influence of human activities on the hydrological characteristics of basins is deepening [2]. Reservoirs were built for multiple purposes, including hydropower generation, flood control, irrigation, water supply, and navigation [3,4]. However, with the increasing number of reservoirs in the basin, the characteristics of the basin have been changed significantly, which affects the runoff generation in basin and leads to many difficulties in research on watershed hydrological forecasting, water resources planning, and hydrological analysis and calculation [5].

The long-term trend of interannual variation in runoff downstream of the reservoir does not appear to be related to long-term climate change, as there was no correlation (or very weak correlation) between the flow downstream of the reservoir and the climate variables (precipitation and temperature) [6]. The results of Vicente-Serrano et al. [7] showed that the gradual increase of reservoir capacity in the basin complicated the non-linear correlation between precipitation and runoff, and that the reservoir could lead to a significant decline in runoff downstream and also to significant changes to the natural river conditions. One of the main hydrological effects of the reservoirs on runoff was a reduction of runoff variation, and uniformity of flow, which was due to the decrease in peak flow and the enhancement of minimum flow [8]. Döll and Aus Der Beek [9,10] pointed out that this change is due to the temporal mismatch between water supply and demand; the reservoir stores water during the rainy season, and supplies irrigated fields and urban areas during the dry season. Research on Caoe River [11] found that the average annual flow after the construction of a reservoir was slightly lower than that before, and that the annual distribution of flow tends to be even. Biemans et al. [12] quantitatively estimated the effects of reservoirs on runoff and irrigation water supply in the 20th century at global, continental, and basin scales. The combined effect of reservoir operation and irrigation reduced the annual average flow into the ocean, compared to natural conditions, and significantly changed the time lag before discharge into the sea.

The occurrence of floods is often affected by human activities, as well as natural factors, such as climate and land cover. As the most important approach for flood control, hydraulic engineering has an increasing impact on floods, and even plays a decisive effect in some cases [13]. Zhang et al. [14] used multiple linear regression, an artificial neural network, and support vector machine to simulate the monthly runoff processes of the basin before the construction of the hydraulic engineering, and selected a neural network model to simulate the monthly runoff process after the reservoirs were built. Comparing the observed and simulated monthly runoff after the reservoirs were built, they analysed the influence of the reservoirs on the monthly runoff. In terms of hydrological analysis and calculation, in order to improve the rationality of reservoir design, Yigzaw et al. [15] studied the impact of reservoir size on the probable maximum rainfall and probable maximum flood in the basin. Chen et al. [16] analyzed the impact of cascade reservoir regulation for flood control on flooding downstream of hydropower stations in different typical years. The operation of two large reservoirs reduces summer runoff by 10% to 50%, and winter runoff by 45% to 85%, in the upstream area of the Yenisei River [17].

Finally, meteorological data is a key factor influencing the accuracy of hydrological models. Usually, data from meteorological stations is scarce and hardly accessible. Reanalysis datasets, such as CMADS (China Meteorological Assimilation Driving Datasets for the SWAT model), are playing an increasingly important role. Liu et al. [18] and Meng et al. [19] applied, respectively, the CMADS in Qinghai-Tibet Plateau and Heihe Basin and achieved better performance than other datasets. Dong et al. [20] indicated that a coupled land surface-hydrological model was driven by the CMADS with good performance. A series of researches showed that the CMADS can provide the necessary meteorological data for hydrological modes' simulations and support parameter calibration [21–24]. However, the CMADS has not been applied in the Yalong River Basin, and whether or not it supports research on the impacts of reservoir parameters on runoff remains a question.

The basic purpose of the studies above is to analyse the impact of reservoir operation on runoff. However, there are few studies investigating the impact of the reservoir parameters, including the capacity, location of a single reservoir, and joint operation of multiple reservoirs, on the runoff processes. The objective of this paper is to assess the impact of reservoir parameters on runoff.

If the construction scheme of the cascade reservoir group is reasonably planned, it will be possible to make use of its flood control measures to regulate flooding. It is a common goal to use hydraulic engineering to exert greater flood control, increasing benefits and minimizing losses [25]. However, the natural circulation processes of the basin are interfered with to a certain extent by the construction and operation of the cascade reservoirs, resulting in fragmented flow generation and concentration. Furthermore, the accuracy and effectiveness of hydrological forecasting have been affected greatly. It is therefore essential to carry out research on the impact of the construction and operation of the reservoir group on the runoff processes in the Yalong River Basin.

This paper takes the Yalong River Basin as its study area, where several hydropower stations (Jinping I, Jinping II, Guandi, Lianghekou, Tongzilin, and Ertan) have been constructed [26]. In this paper, eight scenarios are discussed, by changing the capacity, location, and number of reservoirs in the Yalong River Basin, and then analysing the annual maximum peak flow, peak time, and maximum three-day flood volume at the outlet control station, before and after the construction of the reservoirs. At the same time, the variations in the monthly average flow are analysed during the flood season and the non-flood season. Then, the influences of the reservoir parameters on the runoff processes of the river basin are defined. This paper provides a theoretical reference for the joint operation of reservoirs, flood forecasting, and flood control.

2. Study Area

The Yalong River Basin is located in the eastern part of the Qinghai-Tibet Plateau, west of Sichuan Province, between Jinsha River and Dadu River. The geographical location is between 96°52'-102°48' E and 26°32'-33°58' N. Grassland covers approximately 51% of the catchment, and forest covers approximately 35% [27]. The total length of the main stream, which is the longest tributary of the Yangtze River, is about 1500 km, and the drainage area is about 130,000 km^2. There are many tributaries of the Yalong River, and the water system is well developed.

The Yalong River Basin belongs to the climatic region of the Western Sichuan Plateau, with a clear wet and dry season. The dry season is from November to April, and the rainy season is from May to October. The precipitation in the rainy season accounts for 90–95% of the annual precipitation, and rainy days account for about 80% of the whole year. The average annual precipitation in the Yalong River Basin is 500–2470 mm, and it has a growth trend from north to south. Watershed runoff is formed by precipitation, groundwater, and snowmelt (ice) water.

Due to the extremely rich water resources, the hydropower capacity in the Yalong River Basin is listed in the top ten in the country. Jinping I Hydropower Station, which started operation in 2013, has an installed capacity of 3600 MW, with an annual electricity production of 17.41 billion KW·h. The Ertan Hydropower Station started operation in 1998, with an installed capacity of 3300 MW and an annual electricity production of 17 billion KW·h. Guandi Hydropower Station started operation in 2012, with an installed capacity of 2400 MW and an average annual power generation of about 11.87 billion KW·h. Observation data for 2008–2011 are available for a total of 7 hydrological stations: Yajiang, Madilong, Liewa, Jinping, Huning, Daluo, and Ertan. The distributions of hydropower stations and hydrological stations are shown in Figure 1.

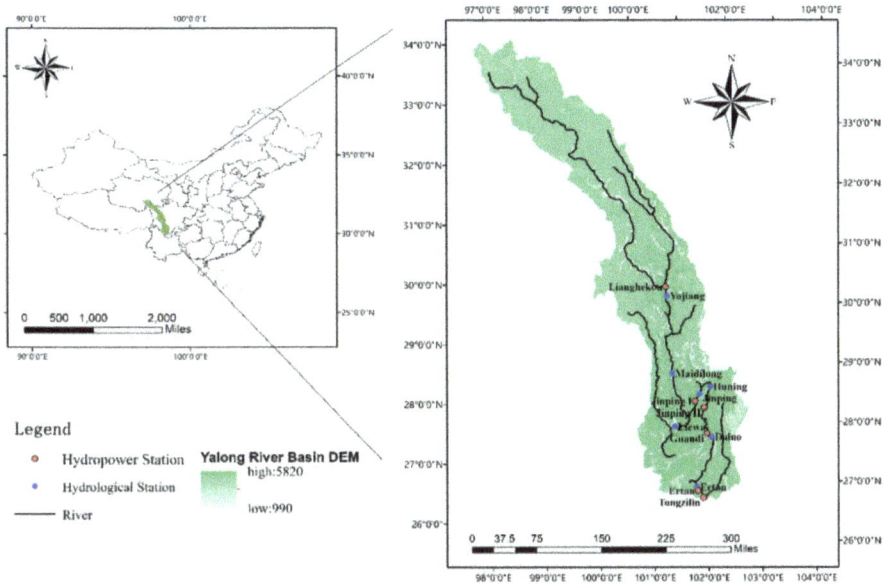

Figure 1. The Yalong River basin.

3. Data and Methods

3.1. Hydrological Model

The distributed watershed hydrological model is an important tool for simulating and analysing watershed hydrological processes. The SWAT (Soil and Water Assessment Tools) model is a typical semi-distributed hydrological model that can take natural factors, such as topography, climate, soil, land use, and land cover, into account, so that it has a strong physical foundation [28]. At present, many scholars have used this model to study the water quality of surface and underground runoff, sediment and nutrient transport, and point source and non-point source pollution [29–31]. This study uses the SWAT model at a daily scale to simulate the impact of reservoir operations on floods.

The SWAT model provides three methods to simulate the reservoir outflow [32]: (1) computing the outflow with an average annual release rate for an uncontrolled reservoir; (2) with pre-defined monthly/daily outflow; and (3) simulating the outflow with generalized operation rules based on the target reservoir storage. This study employs the third method, which simplifies the reservoir release process into two types: principal spillway and emergency spillway. Based on the corresponding reservoir storage of the two types of spillways, the target reservoir storage for the flood season and dry season is established. The advantage is that less data is required, and the high flow and low flow can be reasonably reconstructed.

For the target release approach, the principal spillway volume corresponds to the storage with a maximum flood control capacity, while the emergency spillway volume corresponds to the storage with no flood control capacity. In the non-flood season, the target storage is set at the emergency spillway volume. During the flood season, the flood control reservation for wet ground conditions is set at the maximum, and for dry ground conditions, it is set at 50% of the maximum. The target storage is calculated as [33]:

$$V_{targ} = V_{em} \text{ if } mon_{fld,beg} < mon < mon_{fld,end} \tag{1}$$

$$V_{targ} = V_{pr} + \frac{(1 - \min[\frac{SW}{FC}, 1])}{2} \cdot (V_{em} - V_{pr}) \text{ if}$$
$$mon \leq mon_{fld,beg} \text{ or } mon \geq mon_{fld,end} \tag{2}$$

where V_{targ} is the target reservoir storage for a given day (m^3), V_{em} is the storage of the reservoir when filled to the emergency spillway (m^3), V_{pr} is the storage of the reservoir when filled to the principal spillway (m^3), SW is the average soil water content in the sub-basin (m^3), FC is the water content of the sub-basin soil at field capacity (m^3), mon is the month of the year, $mon_{fld,beg}$ is the first month of the flood season, and $mon_{fld,end}$ is the last month of the flood season.

3.2. Data

The data used as inputs to the SWAT model include meteorology, a digital elevation model (DEM), and soil and land use datasets. Among these, the meteorological data are critical. The CMADS is a public dataset developed by Professor Xianyong Meng from China. The CMADS incorporates the technologies of LAPS (Local Analysis and Prediction System)/STMAS (Space-Time Multiscale Analysis System), and was constructed using multiple technologies and scientific methods, including loop nesting of data, projection of resampling models, and bilinear interpolation [34–36]. CMADS V1. 0 (spatial range 0°N to 65°N, 60°E to 160°E; spatial resolution 1/3°; temporal range 2008–2016) spatially divides the whole of East Asia into 300 × 195 grid points, a total of 58,500 sites; each site contains elements for daily average temperature, daily high/low temperature, daily precipitation, daily average solar radiation, daily average air pressure, daily specific humidity, and daily average wind speed [23,37].

Soil properties and land use types determine the runoff generation and confluence characteristics of different hydrological units, and they are also the basis for the definition of hydrological response units (HRU) in the SWAT model. The land use data (GLC2000 Data) is obtained from the Western China Environmental and Ecological Science Data Centre [38]. The soil data selected for this study is the China Soil Dataset, based on the Harmonized World Soil Database (HWSD) [39]. The Digital Elevation Model (DEM) is SRTM (Shuttle Radar Topography Mission) data, with a resolution of 90 m, from the International Scientific & Technical Data Mirror Site, Computer Network Information Centre, Chinese Academy of Sciences. A description of the data is given in Table 1.

This paper is a study of the impact of reservoirs on runoff processes. Information on the reservoirs used in this study is shown in Table 2. Guandi is an annual regulation reservoir, with a regulation period of one year; the excess water in the wet season is stored to increase the water supply in the dry season. Jinping is a daily regulation reservoir, with a regulation period of one day; it regulates the water demand that varies within the course of a single day. This study required the establishment of a seasonal regulation reservoir to explore the impact of the storage capacity. The ratio of regulated storage capacity (V) to the annual average runoff (W) is the coefficient of storage (β). β of Jinping is 0.13 and β of Guandi is 0.0027. For a seasonal regulation reservoir, β is between 0.02 and 0.08; for this study, it was β = 0.05. Because we know the annual average runoff and β, we can calculate the regulated storage capacity, and then also calculate other factors. In the chart, we use "Seasonal" to represent a seasonal regulation reservoir created by a coefficient of storage.

Table 1. A description of the data used for this study.

Dataset	Description of Dataset	Data Source	Time Scale
DEM	Shuttle Radar Topography Mission (SRTM) DEM 90M	http://www.gscloud.cn/	2007
Soil	China Soil Dataset based on the Harmonized World Soil Database (HWSD)	http://westdc.westgis.ac.cn	2009
Land use	European Union Joint Research Centre (JRC) Space Applications Institute (SAI) 2000 Global Land Cover Data Products (GLC2000)	http://westdc.westgis.ac.cn	2000
CMADS V1.0	Assimilation driving datasets applied to better reflect the spatial distribution of precipitation and meteorology	http://www.cmads.org	2008–2016

Table 2. Information on the reservoirs used for this study.

Name	Start Year	Upper Water Level for Flood Control			Flood Control Level		
		Water Level (m)	Volume (10^4 m^3)	Area (ha)	Water Level (m)	Volume (10^8 m^3)	Area (ha)
Guandi	2013	1330	72,920	1469	1328	70,070	1436
Jinping I	2012	1880	776,500	8255	1859.06	616,230	7064.9
Seasonal	2012	1838.09	479,334	6011.7	1827.06	416,038	5456.9

3.3. Model Building

Based on watershed delineation, the basin is divided into 34 sub-basins and 694 hydrologic response units. The CMADS, soil, and land use were used as inputs to the model. Details of the input data for the SWAT model are shown in Table 1. The surface runoff is calculated by the SCS runoff curve method; the potential evapotranspiration is calculated by the Penman/Monteith method, and the flow concentration is calculated by the variable storage method.

With the construction and operation of the Ertan Hydropower Station in 1998, the basin above it remained a natural watershed. Therefore, the Ertan Hydropower Station was selected as the basin outlet. In all, seven observation stations were employed separately in the calibration, with one optimal parameter set obtained for each station. The observed flow for 2009–2011 (taking 2009 as a warm-up period) was used for calibration. Firstly, taking the sub-basins above the Yajiang station as a whole region, we used a set of parameters in this region. Adjusting this set of parameters enables the NSE (Nash-Sutcliffe Efficiency) and R^2 (Coefficient of Determination) to be optimized. Then, the set of optimal parameters was then brought back to the SWAT model. The same method was used to calibrate the next station, Maidilong. The calibration was not finished until the parameters of the outlet control station, Ertan, were determined. Then, the calibration parameters were used for model validation for 2008–2009 (taking 2008 as a warm-up period).

3.4. Sensitivity Analysis and Calibration Methodology

The physics of the SWAT model is controlled by multiple parameters, and each of them can affect the simulation results to some extent. Therefore, it is necessary to perform a parameter sensitivity analysis using the SWAT-CUP (SWAT Calibration and Uncertainty Programs) to improve the calculation efficiency and the accuracy of the simulation results [40].

SWAT-CUP is a computer program for the calibration of SWAT models, which links SUFI2 (Sequential Uncertainty Fitting Version 2), PSO (Particle Swarm Optimization), GLUE (Generalized Likelihood Uncertainty Estimation), ParaSol (Parameter Solution), and MCMC (Markov Chain Monte Carlo) procedures to the SWAT model. It enables sensitivity analysis, calibration, validation, and uncertainty analysis of SWAT models [41]. The SUFI-2 (Sequential Uncertainty Fitting ver. 2) algorithm has been widely used in model calibration and uncertainty analysis in recent years.

Dao et al. [42] used four uncertainty algorithms to compare the runoff simulations of a SWAT model in a basin in Vietnam. The results show that SUFI-2 can obtain the best simulation results and uncertainty confidence intervals with the least number of simulations. In SUFI-2, parameter uncertainty accounts for all sources of uncertainties, such as the uncertainties in the driving variables (e.g., rainfall), conceptual model, parameters, and measured data [41]. The uncertainty is quantified by a measure referred to as the *p*-factor, which is the percentage of measured data bracketed by the 95% prediction uncertainty (95PPU). The 95PPU is calculated at the 2.5% and 97.5% levels of the cumulative distribution of an output variable obtained through Latin hypercube sampling [29].

3.5. Statistical Criteria for Evaluation

Different statistical criteria were used to evaluate the performance of the SWAT model during calibration and validation. According to Zhang et al. [43], the evaluation coefficients for runoff simulation effects include the coefficient of determination (R^2) and the Nash-Sutcliffe efficiency (NSE).
R^2 is calculated as

$$R^2 = \left[\sum_i (Q_{obs} - \overline{Q_{obs}})(Q_{sim} - \overline{Q_{sim}})\right]^2 / \left[\sum_i (Q_{obs} - \overline{Q_{obs}})^2\right]\left[\sum_i (Q_{sim} - \overline{Q_{sim}})^2\right] \qquad (3)$$

where Q_{obs} and Q_{sim} represent the observed and simulated flow at Ertan, respectively. $\overline{Q_{obs}}$ is the mean of the observed flow for the entire time period of the evaluation, and $\overline{Q_{sim}}$ is the mean of the simulated flow for the entire time period of the evaluation. R^2 represents the proportion of the total variance in the observation data that can be explained by the model. R^2 ranges between 0.0 and 1.0, with higher values indicating a better performance [43].
NSE is calculated as

$$NSE = 1.0 - \sum_i (Q_{obs} - Q_{sim})^2 / \sum_i (Q_{obs} - \overline{Q_{obs}})^2 \qquad (4)$$

The meaning of the symbols is the same as above. The NSE indicates how well the observation values fit the simulated values, ranging from $-\infty$ to 1 [44]. When the NSE is closer to 1, the simulation performance is better. A result of 1 means the observation values are completely consistent with the simulated values.

It is generally believed that a result of NSE > 0.5 and R^2 > 0.6 indicates a satisfactory performance [45]. According to the evaluation criteria of Moriasi [46], the NSE of the model is acceptable between 0.5 and 0.65, good between 0.65 and 0.75, and excellent between 0.75 and 1.0.

3.6. Scenarios

Both the capacity and operating location of the reservoir can have an impact on the downstream hydrological regime. Therefore, the runoff processes were simulated in eight scenarios (Table 3) with the reservoir added to the SWAT model with different capacities, operating locations (of a single reservoir), and relative locations (of two reservoirs). The simulations were then conducted by comparing the natural flow of the Ertan Control Station with the flow under different scenarios.

Table 3. The scenarios for this study.

Scenario	Description	Objective	Reservoir capacity	Location
Scenario 1	Natural conditions	For comparison with other scenarios.	\	\
Scenario 2	One annual regulation reservoir	The impact of reservoir capacity on runoff.	Annual regulation reservoir	Sub-basin 19
Scenario 3	One annual regulation reservoir		Seasonal regulation reservoir	Sub-basin 19
Scenario 4	One daily regulation reservoir		Daily regulation reservoir	Sub-basin 19
Scenario 5	One annual regulation reservoir	The impact of reservoir location on runoff.	Annual regulation reservoir	Sub-basin 9
Scenario 6	One annual regulation reservoir		Annual regulation reservoir	Sub-basin 22
Scenario 7	Two reservoirs, with the annual regulation reservoir upstream	The impact of joint operation of reservoirs on runoff.	Annual regulation reservoir & daily regulation reservoir	Sub-basins 19 & 24
Scenario 8	Two reservoirs, with the daily regulation reservoir upstream		Daily regulation reservoir & annual regulation reservoir	Sub-basins 19 & 24

4. Results and Discussion

4.1. SWAT Calibration and Validation

According to the relevant literature on SWAT calibration and the characteristics of the Yalong River Basin, 17 sensitive parameters were selected for this study (shown in Table 4). These 17 parameters were calibrated and validated using observation data from seven stations. The model calibration period was from 2009 to 2011, and 2009 was the model warm-up period. The model validation period was 2008 to 2009, and 2008 was the model warm-up period. Calibration efforts focused on improving the model performance at the main gauging stations. Since the basin upstream of Ertan Station is a natural watershed, Ertan Station is the focus of this study, and is used as an export control station. Table 4 shows the parameter sensitivity ranking and the optimum values of the station.

Table 4. The calibrated parameter values selected for the SWAT model at Ertan station.

Parameter	Description	Type	Best Value	Sensitivity Ranking
CN2	SCS runoff curve number factor	r	0.06	1
ALPHA_BNK	Baseflow alpha factor for bank storage	v	0.84	2
CH_N2	Manning's "n" value for the main channel	v	0.14	8
CH_K2	Effective hydraulic conductivity in main channel alluvium	v	304.5	12
ALPHA_BF	Baseflow alpha factor (days)	v	0.34	14
GW_DELAY	Groundwater delay (days)	v	270.5	6
GWQMN	Threshold depth of water in the shallow aquifer for "revap" to occur (mm)	v	50.81	7
REVAPMN	Threshold depth of water in the shallow aquifer for "revap" to occur (mm)	v	262.98	5
ESCO	Soil evaporation compensation factor	v	0.94	9
EPCO	Plant uptake compensation factor	v	0.12	3
SLSUBBSN	Average slope length	r	−0.25	4
SOL_AWC(1)	Available water capacity of the soil layer	r	3.23	13
SOL_K(1)	Saturated hydraulic conductivity	r	0.10	15
SOL_BD(1)	Moist bulk density	r	0.21	11
TLAPS	Temperature lapse rate	v	−4.42	10
SFTMP	Snowfall temperature	v	1.734	17
SMFMN	Minimum melt rate for snow during the year	v	10	16

On the daily scale, the SWAT model simulation results achieved satisfactory results in seven control stations in the Yalong River basin (Table 5). For the model calibration period, the R^2 and NSE values of all stations were greater than 0.78 and 0.73, respectively. We also found that the model performance was better with the station further downstream. These evaluation criteria indicate that the results of the stations in the Yalong River Basin are very good on the daily scale. For the validation period, the R^2 value of each station in the basin was greater than 0.8, and even reached 0.9 for Jinping, Huning, and Ertan. The NSE of each station was greater than 0.77, which indicates that the runoff simulation results in the Yalong River basin were in good agreement with the daily observation values. To sum up, this model is applicable to the Yalong River basin. The results of the runoff at the Ertan station during the calibration and validation period are shown in Figure 2. It can be seen from the figure that the simulation is generally consistent with the observations, and the performance of the SWAT model is satisfactory.

Table 5. Evaluation indexes for the simulation result at the daily scale in Yalong River Basin.

Forcing Data	Indexes	Stations						
		Yajiang	Maidilong	Liewa	Jinping	Huning	Daluo	Ertan
Calibration	R^2	0.84	0.84	0.78	0.85	0.85	0.85	0.87
(2010–2011)	NSE	0.73	0.76	0.76	0.75	0.78	0.77	0.80
Validation	R^2	0.85	0.86	0.80	0.90	0.91	0.92	0.93
(2009)	NSE	0.77	0.82	0.79	0.86	0.89	0.85	0.89

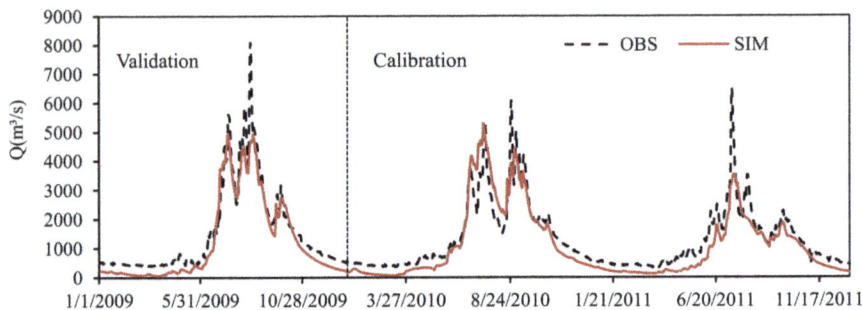

Figure 2. The daily simulation results for runoff (2009–2011) at Ertan control station. (OBS represents the observation data, and SIM represents the simulation data.).

4.2. The Impact of Reservoir Capacity on Runoff

Four scenarios (Scenarios 1, 2, 3, and 4) are considered in this section. The parameters of the reservoirs are shown in Table 2, and the location information is shown in Figure 3(a).

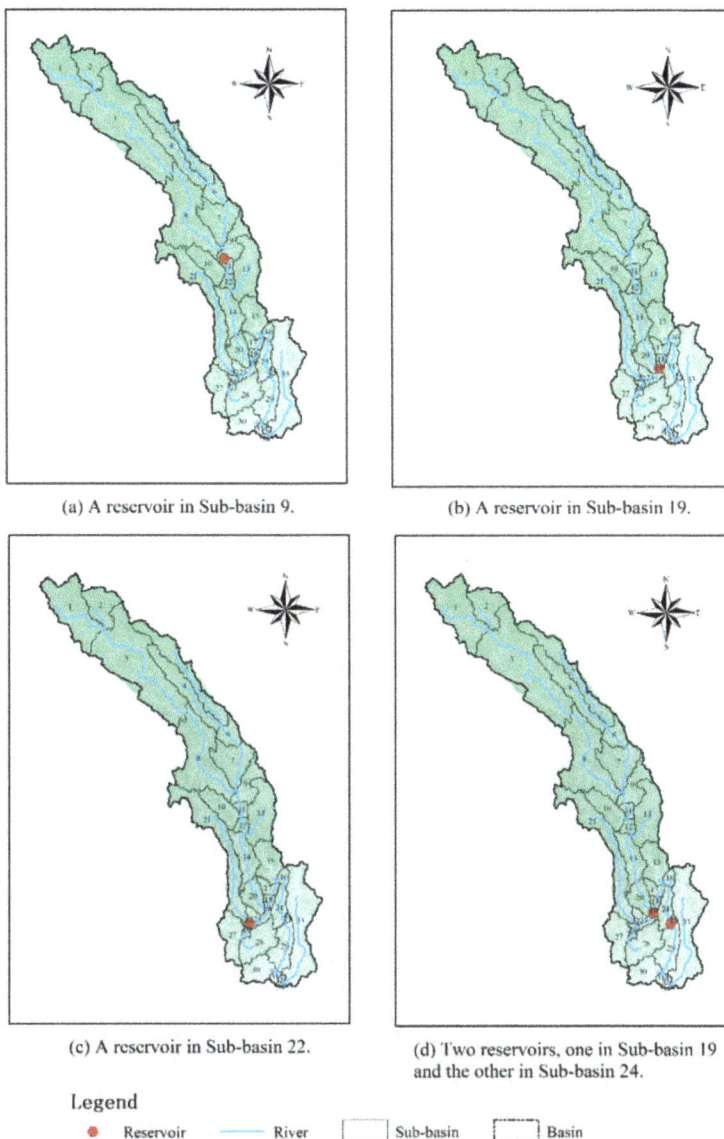

(a) A reservoir in Sub-basin 9.

(b) A reservoir in Sub-basin 19.

(c) A reservoir in Sub-basin 22.

(d) Two reservoirs, one in Sub-basin 19 and the other in Sub-basin 24.

Legend

● Reservoir —— River ☐ Sub-basin ⬚ Basin

Figure 3. Distribution of reservoirs in different scenarios.

Figure 4 shows the relative difference of the annual, seasonal, and daily regulation reservoir on the monthly average flow (2009–2016) at the Ertan control station. The relative difference is calculated as the ratio of the difference between the average monthly flow with a reservoir and without a reservoir, and the monthly average flow without a reservoir. It can be seen from Figure 4 that the average monthly flow with the reservoir is significantly increased compared to the natural monthly average flow in the dry season from October to May, with the most noticeable increase occurring between January and March. Furthermore, a larger increase in dry season runoff can be observed with increasing capacity. The annual regulation reservoir has the greatest increment in the monthly average flow in the dry season. During the flood season, from June to September, the annual regulation reservoir exhibits a strong ability to mitigate floods, especially in the early flood season. The seasonal regulation reservoir and the daily regulation reservoir result in a decrease in runoff from June to July, and a slight increase from August to September. Due to its limited storage capacity, the small reservoir will increase the release rate to ensure its safety in the case of a major flood.

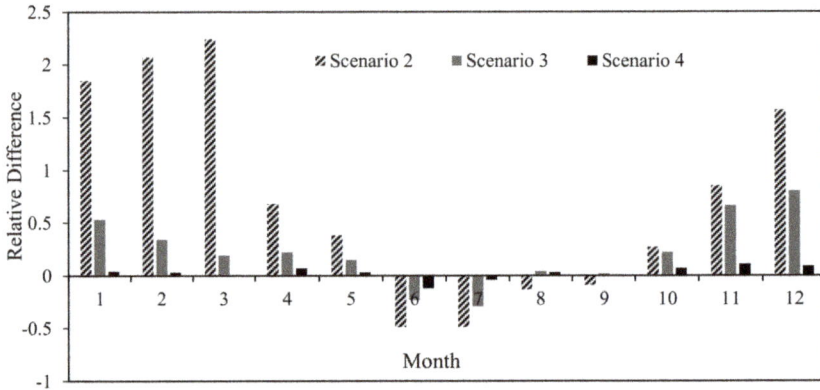

Figure 4. The impact of the reservoir capacity on the monthly average flow (over multiple years) (Scenarios 1–4).

Figure 5 is a simulation result of the daily runoff at Ertan control station under four scenarios. The figure illustrates that the reservoir reduces the flood during the flood season, while it supplies more water during the dry season. The annual regulation reservoir mitigates the flood peak flow the most, and largely weakens the flood peak in the early flood season. Double peak or multi-peak phenomena are most common in the Yalong River Basin. If there is heavy rainfall at the end of the flood season, the reservoir's mitigation of the flood will be decreased due to it having filled earlier in the flood season. The effect of the seasonal regulation reservoir on the peak flow during the flood season is less than that of the annual regulation reservoir, while the daily regulation reservoir has the least impact on the flood. During the dry season, the annual regulation reservoir has the greatest runoff recharge. The seasonal regulation reservoir is slightly inferior, while the daily regulation reservoir is the worst.

In this study, the annual maximum peak flow, peak time, and maximum three-day flood volume were used to evaluate the impact of the reservoir operation on runoff (details in Table 6). As shown in Table 6, the annual regulation reservoir reduces the annual maximum peak flow by 28–48.9% (2011). The maximum reduction of peak flow by the seasonal regulation reservoir is 39.8% (2011), but in the remaining years, the reduction is below 30%. The reduction of the peak flow by the daily regulation reservoir is less than 10%. In the natural scenario, without a reservoir, the annual maximum flow is mainly concentrated between July and August. After regulation by the annual regulation reservoir, the annual maximum flow is basically delayed to September, by more than 30 days (except in 2013 and 2015). The seasonal regulation reservoir has a similar effect to the annual regulation reservoir. The daily regulation reservoir delays the peak flow by less than 8 days, except in 2013, in which it is delayed by 55 days. In the scenarios with reservoirs, the annual maximum three-day flood volume is reduced to varying degrees. Compared with the natural situation, the maximum three-day flood volume in Scenario 2 has been reduced by 24.8–49.8%; in Scenario 3, it is reduced by 14.7–38.5%, and in Scenario 4, it is reduced by less than 11.1%.

Table 6 shows that for scenarios with reservoirs, the reduction of the peak flow and the maximum three-day flood volume was greatest during 2012–2014, and smallest in 2015. This indicates that the reservoir has a stronger mitigation effect on larger flood volumes. The smallest reduction in 2015 may be due to a major flood in 2014, which could still impact the following year, resulting in a decreased ability of the reservoirs to control flooding.

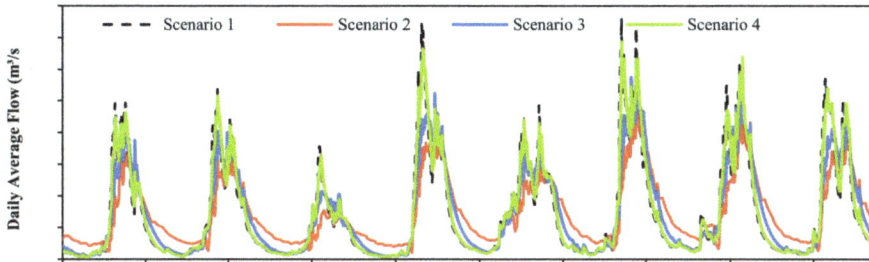

Figure 5. The daily simulation results of runoff (2009–2016) at Ertan station (Scenarios 1–4).

Table 6. Flood characteristics analysis at Ertan station (Scenarios 1–4). (Q_{max} represents annual maximum peak flow, T_{Qmax} represents peak time, and $W_{3,day}$ represents maximum three-day flood volume.).

Index	Scenario	Year	2009	2010	2011	2012	2013	2014	2015	2016
Q_{max} (m^3/s)	1	Value	4929	5353	3565	7419	4832	7616	6200	5701
	2	Relative Difference (%)	−28.9	−29.1	−48.9	−33.8	−30.4	−34.3	−16.8	−31.1
	3	Relative Difference (%)	−18.5	−24	−39.1	−30	−21	−25	−6.6	−25
	4	Relative Difference (%)	−6.2	−4	−8.1	−10.3	−8.8	−9.7	3	−5.2
T_{Qmax}	1	Value	2009/8/17	2010/7/17	2011/7/18	2012/7/16	2013/9/11	2014/7/6	2015/9/5	2016/7/6
	2	Difference (d)	34	45	70	47	0	34	6	76
	3	Difference (d)	−17	45	18	47	0	34	6	76
	4	Difference (d)	0	0	5	5	−55	0	6	8
$W_{3,day}$ $(10^8 m^3)$	1		17.8	19.1	13	26.8	17.1	26.2	21.8	20.7
	2	Relative Difference (%)	−31.5	−31.9	−49.2	−36.9	−33.3	−31.7	−24.8	−31.4
	3	Relative Difference (%)	−18.5	−25.1	−38.5	−32.5	−23.4	−22.1	−14.7	−25.1
	4	Relative Difference (%)	−5.1	−3.7	−10	−10.1	−10.5	−11.1	−3.7	−6.8

4.3. The Impact of Reservoir Location on Runoff

Four scenarios are considered in this section, including Scenarios 1, 2, and 5 (Figure 3b), and Scenario 6 (Figure 3c). The parameters of the reservoirs are shown in Table 2.

Figure 6 shows the relative difference in the monthly average flow (2009–2016) at the Ertan control station depending on whether the annual regulation reservoir is in Sub-basin 19, 9, or 22, or under natural conditions without a reservoir. The relative difference is calculated as above. It can be concluded from Figure 6 that the hydrograph of Scenario 2 becomes flattened compared with the natural situation, and changes even more. On the other hand, the hydrograph experiences only a minor change under Scenario 6.

The average monthly flow under Scenarios 2, 5, and 6 is increased compared with the natural average monthly flow in the dry season (except for the reservoir located in Sub-basin 22 in October). The monthly average flow from December to March changes more and increases month by month. Furthermore, the replenishment effect on runoff in the dry season is enhanced when the reservoir is downstream of the main stream, and is lowest when the reservoir is located in the tributary (Sub-basin 22). Due to small floods in 2013 and 2015, the average monthly flow was larger than usual from April to May. The reservoir located upstream (Sub-basin 19) and in a tributary (Sub-basin 22) had little regulating effect on the runoff. The monthly average flow from April to May therefore differs as follows: Scenario 6 > Scenario 5 > Scenario 2.

During the flood season, Scenario 2 has a greater capacity to impound floods than Scenarios 5 or 6, indicating that a reservoir located downstream of the main stream is best for mitigating floods. In August–September, Scenario 6 is better than Scenarios 2 or 5 in impounding floods. This may be due to the large amount of water stored in the reservoirs during the first flood in Scenario 2 and Scenario 5. When the second flood occurs in the basin, although its peak flow is lower than the first one, the remaining reservoir capacity is smaller. Therefore, the capacity to impound water is limited, and the reduction of the flood peak flow is weakened.

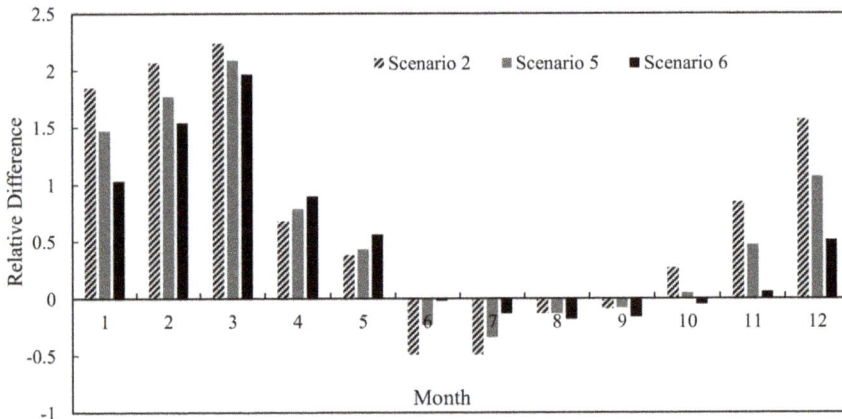

Figure 6. The impact of the reservoir location on the monthly average flow (over multiple years) (Scenarios 1–2 and 5–6).

Figure 7 is a simulation result of the daily runoff at Ertan control station under Scenarios 1, 2, 5, and 6. It is most obvious that a reservoir reduces the flood peak flow, and increases runoff in the dry season, in scenario 2. The effect of Scenario 5 is the second greatest, while the effect of Scenario 6 is the smallest. It illustrates that a reservoir located downstream of the main stream has a better regulation effect on the runoff, when the reservoir capacity is the same.

The annual maximum peak flow, peak time, and maximum three-day flood volume under Scenarios 1, 2, 5, and 6 are shown in Table 7; the annual regulation reservoir in sub-basin 19 reduces the annual maximum peak flow from 16.8% to 48.9%. The absolute value of the annual maximum flow change in Scenario 5 (sub-basin 9) is 0.9–18% lower than that in Scenario 2 (sub-basin 19). When the reservoir is located in the tributary (sub-basin 22), the maximum peak flow is reduced by 7.3–17.9% compared with the natural situation. In this scenario, the change in annual maximum flow is the smallest.

In Scenario 5, the annual maximum peak flow in 2009 and 2011 appeared earlier than in the natural situation. In 2014, the peak flow occurred 19 days later than in Scenario 2. In 2010, 2012, and 2013, it was the same as in Scenario 2. In Scenario 6, the annual maximum peak flow in 2009, 2011, and 2015 appeared earlier than in the natural situation, and in the other years, it was the same as the natural situation. In Scenarios 2, 5, and 6, the annual maximum three-day flood volume was reduced to varying degrees. The absolute change in the annual maximum three-day flood volume under Scenario 5 was 0.8–18.1% smaller than under Scenario 2. The annual maximum three-day volume in Scenario 6 was 8.5–16.9% lower than in the natural situation.

Table 7 shows that a reservoir located in the mainstream (Scenarios 2 and 5) has a stronger decreasing effect on a large flood. When the reservoir is on the tributary (Scenario 6), the flood reduction is better for medium floods.

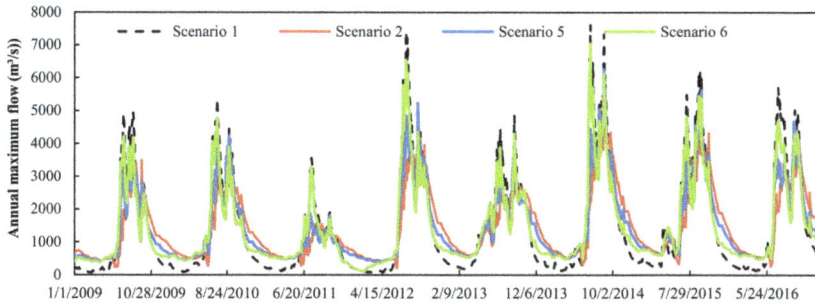

Figure 7. The daily simulation results of runoff (2009–2016) at Ertan station (Scenarios 1–2 & 5–6).

Table 7. Flood characteristics analysis at Ertan station (Scenarios 1–2 & 5–6). (Q_{max} represents annual maximum peak flow, T_{Qmax} represents peak time, and $W_{3,day}$ represents maximum three-day flood volume.).

Index	Scenario	Year	2009	2010	2011	2012	2013	2014	2015	2016
	1	Value	4929	5353	3565	7419	4832	7616	6200	5701
Q_{max} (m^3/s)	2	Relative Difference (%)	−28.9	−29.1	−48.9	−33.8	−30.4	−34.3	−16.8	−31.1
	5	Relative Difference (%)	−22.6	−21.1	−48	−29.7	−12.4	−17.8	−9.4	−18.2
	6	Relative Difference (%)	−14.3	−10.8	−8.1	−11.7	−11.1	−7.3	−12.1	−17.9
	1	Value	2009/8/17	2010/7/17	2011/7/18	2012/7/16	2013/9/11	2014/7/6	2015/9/5	2016/7/6
Time (Qmax)	2	Difference (d)	34	45	70	47	0	34	6	76
	5	Difference (d)	−16	45	−5	47	0	53	6	62
	6	Difference (d)	−37	0	−1	0	0	0	−5	0
	1		17.8	19.1	13	26.8	17.1	26.2	21.8	20.7
$W_{3,day}$ ($10^8 m^3$)	2	Relative Difference (%)	−31.5	−31.9	−49.2	−36.9	−33.3	−31.7	−24.8	−31.4
	5	Relative Difference (%)	−23	−22	−48.5	−31	−15.2	−14.1	−12.8	−19.8
	6	Relative Difference (%)	−14	−11.5	−8.5	−11.6	−11.7	−8.8	−12.8	−16.9

4.4. The Impact of Joint Operation of Dual Reservoirs on Runoff

The four Scenarios considered in this section are Scenarios 1, 2, 7, and 8 (Figure 3d). The parameters of the reservoirs are shown in Table 2.

Figure 8 shows the relative differences of monthly average flow (2009–2016) when Scenarios 2, 7, and 8 are compared with the natural conditions. As can be seen from Figures 10 and 11, the flow curves of Scenario 2 and Scenario 7 are very close. Adjusting the relative location of the two reservoirs results in a small change in runoff, and both Scenarios make the monthly average flow curve flatter. Scenario 8 has a greater impact on flood mitigation during the early flood season (June–July), but a lower impact than Scenario 7 later in the flood season.

In general, when the small reservoir is located upstream of the large reservoir, the reduction of the flood is greater in the flood season, and the increase in runoff is greater in the dry season. When the large reservoir is located upstream of the small reservoir, the runoff regulation of the large reservoir will be affected by the downstream small reservoir, which will weaken the overall performance of the joint operation of the reservoirs. Mainly because of the limited capacity of small reservoirs and the poor performance of runoff regulation, the large reservoir should decreased the release in case of a break of the dams.

When the basin contains only an annual regulation reservoir, the runoff regulation is minimal. However, the difference is small under Scenario 2 and 7. This means that there is little advantage in placing the small reservoir downstream of the large reservoir to improve their overall regulation capacity.

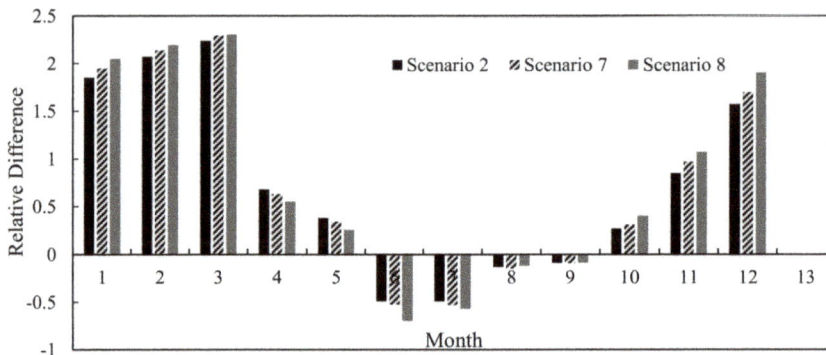

Figure 8. The impact of reservoir location on monthly average flow (over multiple years) (Scenarios 1–2 and 7–8).

Figure 9 shows the daily runoff simulation at Ertan control station under Scenarios 1, 2, 7 & 8. The three Scenarios with reservoirs are essentially the same. However, the runoff in Scenario 2 fluctuates greatly and the flood peak is higher. The increase in runoff during the water supply period differs as follows: Scenario 8 > Scenario 7 > Scenario 2.

The annual maximum peak flow, peak time, and maximum three-day flood volume under Scenarios 1, 7, and 8 are shown in Table 8. In 2011, the relative difference of the annual maximum flow in Scenarios 7 and 8 reached the maximums at the same time, with values of 52.3% and 55%, respectively. In 2015, they reached the minimum values, which were 16.8% and 16.5%, respectively. It can be seen that the change of the annual maximum flow is not much different in the two Scenarios. The differences between the absolute values of the maximum three-day flood volume fluctuate within 4% in the two scenarios. In combination with Table 7, the decrease in the annual maximum flow and maximum three-day flood volume in Scenarios 7 and 8 is higher than that in Scenario 2, except in 2015.

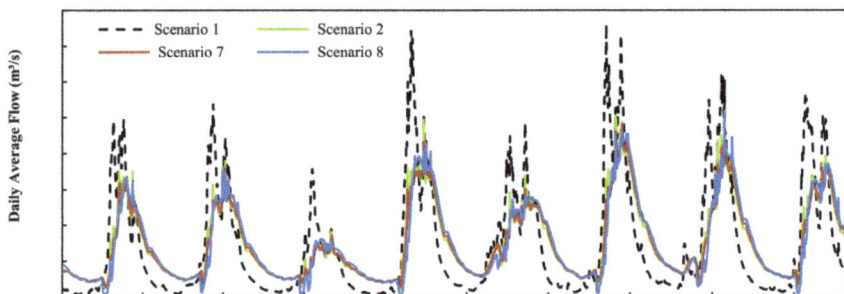

Figure 9. The daily simulation results of runoff (2009–2016) at Ertan station (Scenarios 1–2 & 7–8).

Table 8. Flood characteristics analysis at Ertan station (Scenarios 1 & 7–8). (Q_{max} represents annual maximum peak flow, T_{Qmax} represents peak time, and $W_{3,day}$ represents maximum three-day flood volume.).

Index	Scenario	Year	2009	2010	2011	2012	2013	2014	2015	2016
Q_{max} (m^3/s)	1	Value	4929	5353	3565	7419	4832	7616	6200	5701
	7	Relative Difference (%)	−35.2	−33.3	−52.3	−42.1	−37.6	−37	−16.8	−36.5
	8	Relative Difference (%)	−32.1	−34	−55	−42.4	−41.6	−39.7	−16.5	−35.3
Time (Q_{max})	1		2009/8/17	2010/7/17	2011/7/18	2012/7/16	2013/9/11	2014/7/6	2015/9/5	2016/7/6
	7	Difference (d)	5	45	72	47	0	58	6	74
	8	Difference (d)	12	45	36	57	0	58	6	87
$W_{3,day}$ ($10^8 m^3$)	1		17.8	19.1	13	26.8	17.1	26.2	21.8	20.7
	7	Relative Difference (%)	−34.8	−36.6	−53.1	−43.3	−39.2	−35.5	−25.2	−36.2
	8	Relative Difference (%)	−31.5	−40.3	−54.6	−44.8	−40.9	−37	−24.3	−35.3

In order to clearly show the difference between the seven scenarios with reservoirs, relative to the natural conditions (i.e., the annual maximum flow, peak time, and maximum three-day flood volume distribution and median), a box-line diagram is used to compare the impacts of the reservoir on the runoff processes in all scenarios (Figure 10). For the change in flood peak flow, the medians of Scenarios 7 and 8 are close, and their absolute values are the largest, while the median absolute value of Scenario 4 is the smallest. This shows that the scenarios with two reservoirs have a greater ability to regulate runoff, and the scenario with a daily regulation reservoir has the least ability to regulate runoff, followed by Scenario 6. This means that when the reservoir is located in the tributary, it will greatly weaken its ability to regulate runoff. The median number of days by which the annual maximum flow is delayed under Scenarios 4 and 6 are close, and the range is narrow and concentrated around 0. Other scenarios make little difference in delaying the annual maximum flow. For the reduction of the maximum three-day flood volume, the absolute value of the median becomes smaller as the reservoir capacity decreases.

It can also be seen from Figure 13 that as the reservoir capacity decreases, the absolute value of the median of all flood characteristic values becomes smaller, and the reservoir regulation capacity becomes lower. The medians and ranges under Scenario 2 and Scenario 5 are relatively close, indicating that the regulation capacity of the seasonal regulation reservoir downstream is similar to that of the annual regulation reservoir upstream in the Yalong River Basin. The medians and ranges under Scenario 4 and Scenario 6 are relatively close, indicating that the regulation capacity of the daily regulation reservoir downstream in the Yalong River Basin is similar to that of the annual regulation reservoir upstream in the tributary. The medians and ranges of Scenario 7 and Scenario 8 are close, which indicates that the relative location of the large reservoir and the small reservoir has little effect on their joint operation, but Scenario 8 is better.

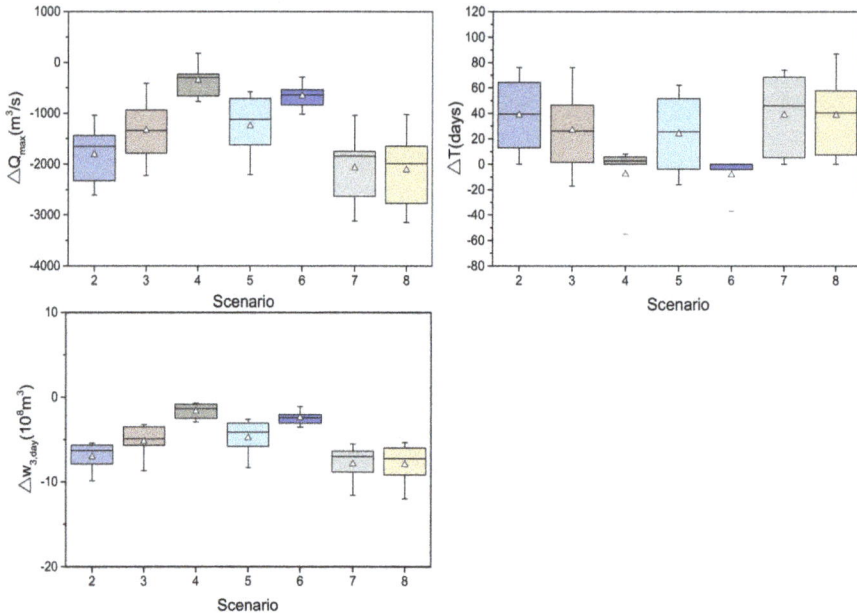

Figure 10. Boxplot of the annual outputs of flood characteristics in 7 Scenarios (2009–2016), where △ represents the mean, and the horizontal lines of the box, from top to bottom, represent the maximum, upper quartile, median, lower quartile, and minimum.

5. Conclusions

Reservoirs are often designed for multiple objectives, such as hydropower generation, flood control, and water supply, and they are operated in such a way that the water can be spatially and temporally redistributed for the benefit of the water users. Obviously, the impact of reservoirs on runoff, especially during floods, is directly correlated to their capacity, location, and joint operation. This study therefore aimes to quantitatively assess the potential impact of different reservoir configurations on the runoff, by changing the capacity and location of a single reservoir, and the relative location of dual reservoirs.

With respect to the influence of the storage capacity on runoff, it was found that the storage capacity is generally in proportion to the effect on runoff regulation, regardless of flood seasons and non-flood seasons. The annual regulation reservoir performs better at reducing the flood peak flow and delaying the flood peak appearance time. Due to a limited storage capacity, the small reservoir may increase its water release in the event of a large flood to ensure dam safety. Therefore, for areas with abundant water resources, like Yalong River Basin, we recommend the construction of large reservoirs if the geographical and economic conditions permit, as this is beneficial to the improvement of flood control in the basin.

As for the influence of the reservoir's location on runoff, it was found that the reservoir in the main stream shows greater regulation capability than the reservoir in the tributary. The regulation capability increases when the reservoir is further downstream, and decreases when the reservoir is further upstream. Therefore, we recommend constructing the reservoir downstream from the mainstream of Yalong River, so that it can play a more vital role in flood control.

An analysis of the influence on runoff of the relative locations of two reservoirs indicates that locating the daily regulation reservoir upstream of the annual regulation reservoir is beneficial to flood control. However, the relative location of two reservoirs makes little difference to the impact on runoff. Runoff regulation capability is generally greater as the number of reservoirs increases.

Therefore, building small reservoirs downstream from large reservoirs in the Yalong River Basin is not recommended, as it is not conducive to flood control.

Overall, reservoir operation can mitigate floods and delay the flood peak appearance time in the flood season, and it can release more water to meet downstream water demands in the dry season. Runoff processes tend to become flattened. Moreover, reservoirs can have a more significant effect on relatively large floods. Therefore, in areas similar to Yalong River Basin, where the annual distribution of runoff is uneven, we recommend building cascade reservoirs to regulate the annual distribution of runoff, and to achieve rational water resources management.

Author Contributions: Conceptualization: M.Y. and X.M. Investigation: X.M. and G.S. Methodology: X.L., M.Y., F.W. and G.S. Formal analysis: X.L. and F.W. Writing: X.L. All the authors have approved the submission of this manuscript.

Funding: This paper was supported by: The National Natural Science Foundation of China (Grant No. 51709271; 41501039); the Young Elite Scientists Sponsorship Program by CAST (2017QNRC001); the project of Power Construction Corporation of China (DJ-ZDZX-2016-02).

Acknowledgments: Special thanks to the editors and the reviewers who helped improve the manuscript a lot.

Conflicts of Interest: The authors declare no conflict of interest.

References

1. Zhang, R.; Zhang, T.; Pei, W. Flood Analysis of Fengshuba Reservoir Considering the Influence of upstream hydraulic engineering. *China R. Water Hydropower* **2008**, *8*, 29–31.
2. Hu, Q.; Yin, T. Impact assessment of climate change and human activities on annual highest water level of Taihu Lake. *Water Sci. Eng.* **2009**, *2*, 1–15.
3. International Commission of Large Dams (ICOLD). *World Register of Dams*; Int. Comm. Large Dams: Paris, France, 2007.
4. World Commission on Dams (WCD). *Dams and Development: A New Framework for Decision-Making*; World Commission on Dams (WCD): London, UK, 2000.
5. Wan, X.; Guan, X.; Zhong, P.; Wang, M.; Mei, J. Impact of large-scale multiple reservoir on basin flood hydrograph. *Adv. Sci. Technol. Water Resources* **2017**, *3*, 66–71.
6. Assani, A.A.; Landry, R.; Daigle, J. Alain, Chalifour. Reservoirs Effects on the Interannual Variability of Winter and Spring Streamflow in the St-Maurice River Watershed (Quebec, Canada). *Water Resources Manag.* **2011**, *25*, 3661–3675. [CrossRef]
7. Vicente-Serrano, S.M.; Zabalza-Martínez, J.; Borràs, G.; López-Moreno, J.I.; Pla, E.; Pascual, D.; Savé, R.; Biel, C.; Funes, I.; Martín-Hernández, N.; et al. Effect of reservoirs on streamflow and river regimes in a heavily regulated river basin of Northeast Spain. *Catena* **2017**, *149*, 727–741. [CrossRef]
8. Moyle, P.B.; Mount, J.F. Homogenous rivers, homogenous faunas. *Proc. Natl. Acad. Sci. USA* **2007**, *104*, 5711–5712. [CrossRef]
9. Döll, P.; Fiedler, K.; Zhang, J. Global-scale analysis of river flow alterations due to water withdrawals and reservoirs. *Hydrol. Earth Syst. Sci.* **2009**, *13*, 2413–2432. [CrossRef]
10. Aus der Beek, T.; Flörke, M.; Lapola, D.M.; Schaldach, R.; Voß, F.; Teichert, E. Modelling historical and current irrigation water demand on the continental scale: Europe. *Adv. Geosci.* **2010**, *27*, 79–85. [CrossRef]
11. Dou, Y.; Yang, W. Effect of Water Resource Projects along Caoe River Basin on Ecological Environments. *Adv. Water Sci.* **1996**, *7*, 260–267.
12. Biemans, H.; Haddeland, I.; Kabat, P.; Ludwig, F.; Hutjes, R.W.A.; Heinke, J.; von Bloh, W.; Gerten, D. Impact of reservoirs on river discharge and irrigation water supply during the 20th century. *Water Resour. Res.* **2011**, *47*, 77–79. [CrossRef]
13. Li, C.; Xue, Z.; Peng, Y.; Zhou, H.; Liu, Y. Impacts of Hydraulic Engineering on Flood. *South-to-North Water Transfers Water Sci. Technol.* **2014**, *12*, 21–25.
14. Zhang, Z.; Zhang, Q.; Deng, X.; Liu, J.; Sun, P. Hydrological Effects of Water Reservoirs on Fluvial Hydrological Process for the East River Basin Using Statistical Modeling Technique. *J. Nat. Resour.* **2015**, *30*, 684–695.

15. Yigzaw, W.; Hossain, F.; Kalyanapu, A. Impact of Artificial Reservoir Size and Land Use/Land Cover Patterns on Probable Maximum Precipitation and Flood: Case of Folsom Dam on the American River. *J. Hydrol. Eng.* **2013**, *18*, 1180–1190. [CrossRef]

16. Chen, T.; Dong, Z.; Jia, B.; Huang, X.; Zhong, D. Analysis of design flood of dam site of Xiaonanhai Hydropower Station considering flood control operation of upstream cascade reservoir group. *J. Hohai Univ. (Nat. Sci.)* **2014**, *42*, 476–480.

17. Yang, D.; Ye, B.; Kane, D.L. Streamflow changes over Siberian Yenisei River Basin. *J. Hydrol. (Amsterdam)* **2004**, *296*, 59–80. [CrossRef]

18. Liu, J.; Shanguan, D.; Liu, S.; Ding, Y. Evaluation and Hydrological Simulation of CMADS and CFSR Reanalysis Datasets in the Qinghai-Tibet Plateau. *Water* **2018**, *10*, 513. [CrossRef]

19. Meng, X.; Wang, H.; Cai, S.; Zhang, X.; Leng, G.; Lei, X.; Shi, C.; Liu, S.; Shang, Y. The China Meteorological Assimilation Driving Datasets for the SWAT Model (CMADS) Application in China: A Case Study in Heihe River Basin. *Preprint* **2016**. [CrossRef]

20. Dong, N.; Yang, M.; Meng, X.; Liu, X.; Wang, Z.; Wang, H.; Yang, C. CMADS-Driven Simulation and Analysis of Reservoir Impacts on the Streamflow with a Simple Statistical Approach. *Water* **2018**, *11*, 178. [CrossRef]

21. Meng, X.; Wang, H.; Lei, X.; Cai, S.; Wu, H. Hydrological Modeling in the Manas River Basin Using Soil and Water Assessment Tool Driven by CMADS. *Tehnički Vjesnik* **2017**, *24*, 525–534.

22. Meng, X.; Dan, L.; Liu, Z. Energy balance-based SWAT model to simulate the mountain snowmelt and runoff – taking the application in Juntanghu watershed (China) as an example. *J. Mt. Sci.* **2015**, *12*, 368–381. [CrossRef]

23. Wang, Y.; Meng, X. Snowmelt runoff analysis under generated climate change scenarios for the Juntanghu River basin in Xinjiang, China. *Tecnología y Ciencias del Agua* **2016**, *7*, 41–54.

24. Meng, X. Simulation and spatiotemporal pattern of air temperature and precipitation in Eastern Central Asia using RegCM. *Sci. Rep.* **2018**, *8*, 3639. [CrossRef]

25. Du, J.; Zhou, G. A Recursive Optimal Algorithm for the Operation of Multireservoir Flood Control System. *Adv. Water Sci.* **1994**, *5*, 134–141.

26. Mei, Y.; Yang, N.; Yan, L. Optimal ecological sound operation of cascade reservoirs in the lower Yalongjiang River. *Adv. Water Sci.* **2009**, *20*, 721–725.

27. Yu, X.; Feng, L.; Yan, D.; Jia, Y.; Yang, S.; Hu, D.; Zhang, M. Development of distributed hydrological model for Yalongjiang River basin. *China Hydrol.* **2008**, *28*, 49–53, (In Chinese with English abstract).

28. Arnold, J.G.; Srinivasan, R.; Muttiah, R.S.; Williams, J.R. Large area hydrologic modeling and assessment part I: Model development. *J. Am. Water Resour. Assoc.* **1998**, *34*, 73–89. [CrossRef]

29. Abbaspour, K.C.; Yang, J.; Maximov, I.; Maximov, I.; Siber, R.; Bogner, K.; Mieleitner, J.; Zobrist, J.; Srinivasan, R. Modelling hydrology and water quality in the pre-alpine/alpine Thur watershed using SWAT. *J. Hydrol.* **2007**, *333*, 413–430. [CrossRef]

30. Chaplot, V. Impact of DEM mesh size and soil map scale on SWAT runoff, sediment, and NO 3 –N loads predictions. *J. Hydrol.* **2005**, *312*, 207–222. [CrossRef]

31. Jayakrishnan, R.; Srinivasan, R.; Santhi, C.; Arnold, J. Advances in the application of the SWAT model for water resources management. *Hydrol. Processes* **2005**, *19*, 749–762. [CrossRef]

32. Li, W.; Chen, X.; He, Y.; Zhang, L. Modification of Reservoir Module in SWAT model and its Application of Runoff Simulation in Highly Regulated Basin. *Trop. Geol.* **2018**, *38*, 226–265.

33. Neitsch, S.L.; Arnold, J.G.; Kiniry, J.R.; Williams, J.R. *Soil and Water Assessment Tool Theoretical Documentation Version 2009*; TWRI Report TR-191; Texas Water Resources Institute: College Station, TX, USA, 2011; pp. 416–422.

34. Meng, X. Spring Flood Forecasting Based on the WRF-TSRM mode. *VJESN* **2018**, *25*, 27–37.

35. Meng, X.; Wang, H.; et al. Investigating spatiotemporal changes of the land-surface processes in Xinjiang using high-resolution CLM3.5 and CLDAS: Soil temperature. *Sci. Rep.* **2017**, *7*, 13286. [CrossRef] [PubMed]

36. Meng, X.; Wang, H. Significance of the China Meteorological Assimilation Driving Datasets for the SWAT Model (CMADS) of East Asia. *Water* **2017**, *9*, 765. [CrossRef]

37. Meng, X.; Wang, H.; Shi, C.; Wu, Y.; Ji, X. Establishment and Evaluation of the China Meteorological Assimilation Driving Datasets for the SWAT Model (CMADS). *Water* **2018**, *10*, 1555. [CrossRef]

38. Fischer, G.; Nachtergaele, F.; Priele, S. *Global Agro-Ecological Zones Assessment for Agriculture (GAEZ 2008)*; IIASA: Laxenburg, Austria; FAO: Rome, Italy, 2008.

39. Zhang, Q.; Zhang, X. Impacts of predictor variables and species models on simulating Tamarix ramosissima distribution in Tarim Basin. Northwestern China. *J. Plant Ecol.* **2012**, *5*, 337–345. [CrossRef]

40. Yang, J.; Reichert, P.; Abbaspour, K.C.; Jun, X.; Hong, Y. Comparing uncertainty analysis techniques for a SWAT application to the Chaohe Basin in China. *J. Hydrol.* **2008**, *358*, 1–23. [CrossRef]

41. Li, Q.; Zhang, J.; Gong, H. Hydrological Simulation and Parameter Uncertainty Analysis Using SWAT Model Based on SUFI-2 Algorithm for Guishuihe River Basin. *Hydrol. China Hydrol.* **2015**, *35*, 43–48.

42. Khoi, D.N.; Thom, V.T. Parameter uncertainty analysis for simulating streamflow in a river catchment of Vietnam. *Glob. Ecol. Conserv.* **2015**, *4*, 538–548. [CrossRef]

43. Zhang, X.; Srinivasan, R.; Bosch, D. Calibration and uncertainty analysis of the SWAT model using Genetic Algorithms and Bayesian Model Averaging. *J. Hydrol.* **2009**, *374*, 307–317. [CrossRef]

44. Nash, J.E.; Sutcliffe, J.V. River flow forecasting through conceptual models: Part I. A discussion of principles. *J. Hydrol.* **1970**, *10*, 282–290. [CrossRef]

45. Zhou, S. Uncertainty Analysis and Dynamic Evaluation of the SWAT Model Parameters. Master's Thesis, Xi'an University of Technology, Xi'an, China, 2008. (In Chinese)

46. Moriasi, D.N.; Arnold, J.G.; van Liew, M.W.; Bingner, R.L.; Harmel, R.D.; Veith, T.L. Model evaluation guidelines for systematic quantification of accuracy in watershed simulations. *Trans. ASABE* **2007**, *50*, 885–900. [CrossRef]

Article

Impact of Climate Variability on Blue and Green Water Flows in the Erhai Lake Basin of Southwest China

Zhe Yuan [1], Jijun Xu [1], Xianyong Meng [2,3,*], Yongqiang Wang [1,*], Bo Yan [1] and Xiaofeng Hong [1]

[1] Changjiang River Scientific Research Institute (CRSRI), Changjiang Water Resources Commission of the Ministry of Water Resources of China, Wuhan 430010, China; yuanzhe_0116@126.com (Z.Y.); xujj07@163.com (J.X.); byanhhu@163.com (B.Y.); hongxiaofeng@mail.crsri.cn (X.H.)

[2] College of Resources and Environmental Science, China Agricultural University (CAU), Beijing 100094, China

[3] Department of Civil Engineering, The University of Hong Kong, Pokfulam 999077, Hong Kong, China

* Correspondence: xymeng@cau.edu.cn or xymeng@hku.hk (X.M.); wangyq@mail.crsri.cn (Y.W.); Tel.: +86-010-60355970 (X.M.); +86-027-82926423 (Y.W.)

Received: 3 November 2018; Accepted: 21 February 2019; Published: 27 February 2019

Abstract: The Erhai Lake Basin is a crucial water resource of the Dali prefecture. This research used the soil and water assessment tool (SWAT) and the China Meteorological Assimilation Driving Datasets for the SWAT model (CMADS) to estimate blue and green water flows. Then the spatial and temporal change of blue and green water flows was investigated. With the hypothetical climate change scenarios, the sensitivity of blue and green water flows to precipitation and temperature has also been analyzed. The results showed that: (1) The CMADS reanalysis dataset can capture the observed probability density functions for daily precipitation and temperature. Furthermore, the CMADS performed well in monthly variables simulation with relative bias and absolute bias less than 7% and 0.5 °C for precipitation and temperature, respectively; (2) blue water flow has increased while green water flow has decreased during 2009 to 2016. The spatial distribution of blue water flow was uneven in the Erhai Lake Basin with the blue water flow increased from low altitudes to mountain areas. While the spatial distribution of green water flow was more homogeneous; (3) a 10% increase in precipitation can bring a 20.8% increase in blue water flow with only a 2.5% increase in green water flow at basin scale. When temperature increases by a 1.0 °C, the blue water flow and green water flow changes by −3% and 1.7%, respectively. Blue and green water flows were more sensitive to precipitation in low altitude regions. In contrast, the water flows were more sensitive to temperature in the mountainous area.

Keywords: blue and green water flows; climate variability; sensitivity analysis; Erhai Lake Basin

1. Introduction

The water resources availability has been affected by climate variability in the past decades, which has caused sustainability concerns in many parts of the world [1,2]. Previous studies have reported that climate variability can alter precipitation, evapotranspiration, soil water, and runoff [3–5] resulting in freshwater resources redistributing in spatial and temporal dimensions [6,7]. With warmer climate conditions, the water-holding capacity of the atmosphere has been increasing, and as a result, the hydrological cycle will be intensified [8,9] posing more challenges to water resource management. Therefore, it is necessary to investigate the impact of climate variability on freshwater resources, which will assist policymakers and administrators to manage water resources in the context of climate change. In general, blue water, namely the surface and groundwater runoff directly generated from precipitation, has been emphasized by water resources assessment and management studies.

While green water, including actual evapotranspiration and soil water, has often been ignored [10]. In fact, green water plays an important role in rain-fed crop production and ecosystem services provision [11,12]. According to the study by Liu et al., more than 80% of the water consumption for global crop production is supported by green water [13]. Natural terrestrial ecosystems, such as grasslands and forests, depend almost entirely on green water [14]. So green water is critical for maintaining the productivity and serviceability of the terrestrial ecosystem. However, in traditional water resources assessment, only the available water resource was taken into consideration. Limitations such as this should be addressed, and temporal–spatial variation should be explored to provide scientific evidence for the construction of water resources management modes and systems.

The Erhai Lake Basin in Southwest China is not only an ecologically fragile area but also a vulnerable area from climate change. As the effects of climate change become more serious, the imbalance between the supply and demand of water resources in the Erhai Lake Basin will be more prominent [15,16]. Thus, it is necessary to assess the impact of climate change on water resources in the Erhai Lake Basin. In the previous studies, the variation of water resources in the Erhai Lake Basin has been analyzed. However, most of the researches focus on the impact of precipitation variations on annual runoff [17,18]. In fact, temperature is also a main factor influencing water resources. The land surface evaporation and water consumption of crops will increase as the temperature rises, leading to a change in water resources [19]. In addition, green water resources should also be considered in this ecologically fragile area. Given the above, the investigation of spatio–temporal distribution characteristics of blue and green water resources is useful for water resources planning and ecological protection in the Erhai Lake Basin.

The concept of blue and green water resources was first proposed by Falkenmark [20]. Since then, numerous methods have been used to assess blue water and green water resources. With the development of distributed hydrological models, the temporal–spatial variations of blue and green water resources can be estimated by methods which have a clearer physical mechanism [11,21–23]. It has been demonstrated that the soil and water assessment tool (SWAT) model can simulate blue and green water resources and detect the impacts of climate variability on hydrological components [24–27]. However, in the basins where the conventional in situ data are not available, the distributed physically-based model cannot estimate the hydrological processes as there are insufficient weather gauges. The satellite-based precipitation datasets, such as Tropical Rainfall Measuring Mission (TRMM) 3B42V7 and the Precipitation Estimation from Remotely Sensed Information using Artificial Neural Networks- Climate Data Record (PERSIANN-CDR), can be used as the forcing data for hydrological models. Nevertheless, the errors would result from measure, resample and retrieval algorithm [28–30]. It has been proved that the reanalysis datasets obtained from observed data and model forecast performance better than satellite-based precipitation [31]. The China Meteorological Assimilation Driving Datasets for the SWAT model (CMADS) developed by Dr. Xianyong Meng from China Agriculture University (CAU) is one of the available reanalysis datasets. This dataset provides multiple meteorological elements with resolutions of $1/3°$, $1/4°$, $1/8°$, and $1/16°$ and can be used to drive various hydrological models [32,33]. Many previous studies have shown that the CMADS reanalysis dataset has a high accuracy for weather element simulation and has been widely used in East Asia, including Heihe River Basin (China), Juntanghu River basin (China), Lijiang River Basin (China), Han River Basin (Korean Peninsula), and so on [34–44]. Based on the above analysis, the CMADS reanalysis dataset and SWAT model can be considered as the important basic data and simulation tool for investigating the impact of climate variability on blue and green water resources in ungauged basins (e.g., Erhai River Basin).

This research selected Erhai River Basin as the study area, and the impact of climate variability on blue and green water flows has been analyzed by the SWAT model and hypothetical climate change scenarios. The remaining sections of this paper are organized as follows: The study area, modeling approach (blue and green water flows simulation based on SWAT model), dataset and hypothetical climate change scenarios are introduced in Section 2; Section 3 shows the evaluation of CMADS reanalysis dataset and SWAT model, spatial and temporal variability of blue and green water flows in the recent eight years, sensitivity of blue and green water flows to climate change; and the discussion and conclusions are summarized in Sections 4 and 5.

2. Materials and Methods

2.1. Study Area

The Erhai Lake Basin (ELB), situated between 99°50′ E and 100°27′ E and between 25°26′ N and 26°26′ N, is the area investigated in this study. The total area is 2496.6 km², accounting for 8.8% of the total area of the Dali prefecture. The elevation of the study area varies from 1958 to 4072 m with an average of 2458 m, dropping off from the edges of the basin to the center (Figure 1). The annual mean precipitation of ELB is about 850 mm with more than 85% falling from May to October. The climate is wetter in the west side, known locally as the famous Eighteen Streams Region, with an annual precipitation of 1072 mm and runoff of more than 200 mm. While on the Eryuan plain, the north side of the ELB, the annual precipitation drops to about 763 mm and the runoff is less than 100 mm. The weather in the basin is mild, with an annual average temperature of 16 °C [45,46].

Impacted by global warming and many other factors, the ELB has witnessed a series of eco–environmental issues, such as reduction in the lake water level, shortage of water resources, and a conflict between water supply and demand. Therefore, an accurate evaluation of water resources in the ELB is essential for water resources planning and management in the Dali prefecture.

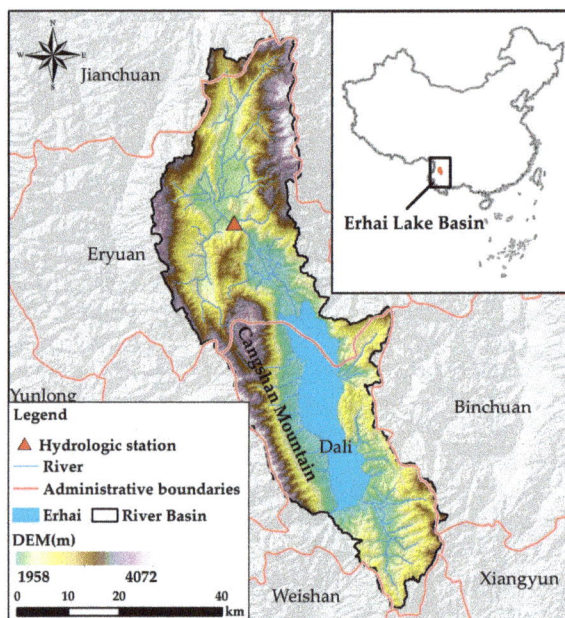

Figure 1. Location of the Erhai Lake Basin (ELB), Southwest China.

2.2. Modeling Approach

The SWAT Model was used to quantify the water flows, including blue water flow (BWF) and green water flow (GWF) in this study. According to the study by Schuol et al. [11], BWF is the river discharge and the deep aquifer recharge, whereas GWF is represented by actual evapotranspiration (Figure 2). All these variables can be simulated by the SWAT Model. In addition, the green water coefficient (GWC) was used to account for the relative importance of BWF and GWF, which can be written as GWC = GWF/(BWF + GWF) [13].

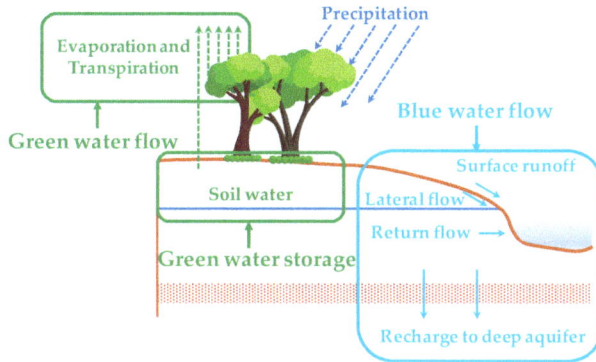

Figure 2. Schematic diagram of simulated components in the soil and water assessment tool (SWAT) model.

The ArcSWAT 2012 is used for the model setup and parameterization. In this study, the ELR was divided into 151 sub-basins with a threshold drainage area of 10 km^2 and further into 722 hydrological response units (HRUs) based on the elevation, land use, and soil type. The monthly hydrological processes were simulated by the SWAT model. The entire simulation period covers 9 years (2008–2016), including a warming up period (2008), calibration period (2009–2014), and validation period (2015–2016). The model's performance of simulating monthly discharge was quantified by the Nash–Sutcliffe values (E_{NS}), determination coefficient (R^2) and relative error (RE) [47,48].

$$E_{NS} = 1 - \frac{\sum_{i=1}^{n} (Qo_i - Qs_i)}{\sum_{i=1}^{n} (Qo_i - \overline{Qo_i})} \tag{1}$$

$$R^2 = \frac{\left[\sum_{i=1}^{n} (Qs_i - \overline{Qs_i}) \sum_{i=1}^{n} (Qo_i - \overline{Qo_i}) \right]^2}{\sum_{i=1}^{n} (Qs_i - \overline{Qs_i})^2 \sum_{i=1}^{n} (Qo_i - \overline{Qo_i})^2} \tag{2}$$

$$RE = \left(\frac{\overline{Qs} - \overline{Qo}}{\overline{Qo}} \right) \times 100\% \tag{3}$$

where, Qo_i and Qs_i are observed discharge and simulated discharge respectively; $\overline{Qo}$ and $\overline{Qs}$ are the average value of observed discharge and simulated discharge respectively; n is the number of observed values. The higher E_{NS} and R^2 and the smaller RE, the better the model performance. According to the suggestion by Kumar and Merwade, the monthly discharge simulations with $E_{NS} > 0.5$ and $RE < \pm15\%$ are acceptable simulations [49].

2.3. Data Sets and Evaluation

The basic data for model setup contains the digital elevation model (DEM), land use, soil, and weather. In this study, the Shuttle Radar Topography Mission (SRTM) 30 m digital elevation data was used for watershed delineation, which was provided by the Advanced Spaceborne Thermal Emission and Reflection Radiometer (Figure 1). The land use map in 2015 for this study was obtained from the Resource and Environment Data Cloud Platform (RESDC, http://www.resdc.cn/) (Figure 3a).

The soil data was obtained from the China Soil Data Set (v1.1), based on the World Soil Database (HSDW) (http://westdc.westgis.ac.cn) (Figure 3b). The above data were used for HRU definition. The daily weather data were collected from the China Meteorological Assimilation Driving Datasets for the SWAT model (CMADS, http://www.cmads.org/) (Figure 3c). The CMADS V1.1 dataset is available from 2008 to 2016 with 0.25° spatial resolution (260 × 400 grid points). This dataset provides the daily max/min-temperatures, 24 h precipitation, solar radiation, air pressure, relative humidity, and wind speed which can be used to initialize SWAT models directly [32]. A total of 17 grid points within and around the ELB were used for the establishment of weather databases in this study. The CMADS V1.1 dataset has been assessed by weather station data collected in Dali station (location is shown as a green dot in Figure 3c). According to the data provided by National Meteorological Information Center (NMIC, http://data.cma.cn/), there is only one meteorological station, Dali, in the study area. Thus, the spatial distribution characteristics of Erhai Lake Basin cannot be fully represented. However, by using the grid data provided by CMADS, this problem can be well solved.

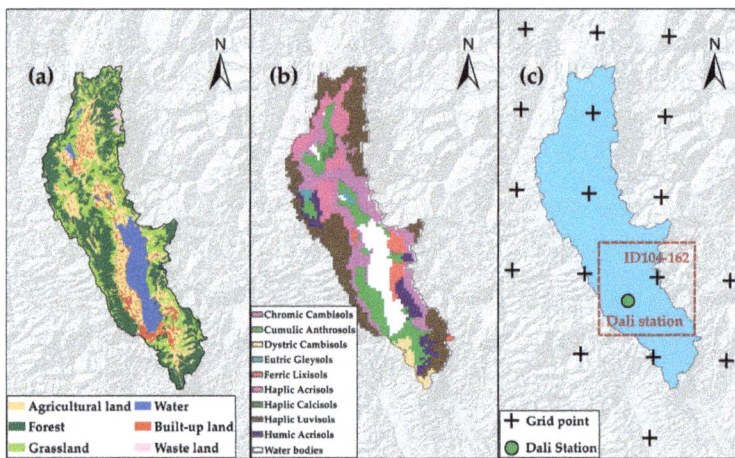

Figure 3. The land use map (**a**), soil type map (**b**) and China Meteorological Assimilation Driving Datasets (CMADS) grid points in the ELB (**c**).

To evaluate the precipitation and temperature provided by the CMADS V1.1, we compared this reanalysis datasets (ID104-162) with the observed precipitation and temperature in the Dali station. This comparison was based on two temporal scales, namely daily scale and monthly scale. In the daily scale, a probability density functions (PDFs)-based assessment was used to illustrate the similarity between observed-PDF and simulated-PDF (Figure 4). An alternative metric-skill score (SS) was defined as Equation (4), which is greater than 0 and smaller than 1. When SS is close to 1, it means that the simulated-PDF fits perfectly with the observed-PDF [50]. At the monthly scale, the bias (B) or absolute bias (B_{abs}) and correlation coefficient (C) were used to evaluate the monthly precipitation and temperature (Equation (5) to Equation (7)) [51].

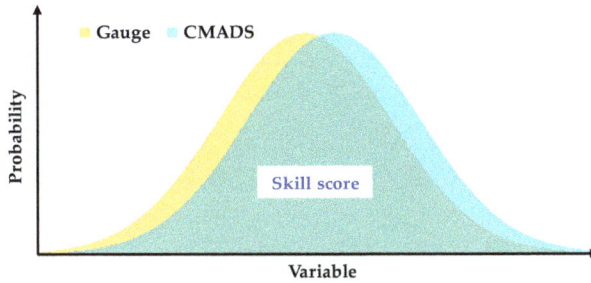

Figure 4. Diagrams of CMADS- probability density functions (PDF) vs Gauge-PDF illustrating the total skill score.

$$SS = \sum_{1}^{n} \min(Fs_n, Fo_n) \tag{4}$$

where, n stands for the number of bins; Fs_n stands for the frequency of values in a given bin from the CMADS; and Fo_n stands for the frequency of values in a given bin from the observed data (Gauge). Summing up the minimum frequency values over all bins and then SS can be obtained.

$$R = \left(\frac{\overline{P_s} - \overline{P_o}}{\overline{P_o}} \right) \times 100\% \tag{5}$$

$$B_{abs} = \overline{T_s} - \overline{T_o} \tag{6}$$

$$C = \frac{Cov(V_s, V_o)}{\sqrt{Var(V_s) Var(V_s)}} \tag{7}$$

where $\overline{P_o}$ and $\overline{P_s}$ are the temporal average of observed precipitation and simulated precipitation, respectively; $\overline{T_o}$ and $\overline{T_s}$ are the temporal average of observed temperature and simulated temperature, respectively; V_o and V_s are the observed value (precipitation or temperature) and simulated value, respectively.

2.4. Climate Change Scenarios and Sensitivity Analysis

The sensitivity of water flows to climate variability can be considered as the proportional change of simulated BWF and GWF comparing with the observed values in the hypothetical climate change scenarios. According to this, the sensitivity can be calculated as follow:

$$\delta(WF, P) = \frac{f(P + \Delta P, T) - f(P, T)}{f(P, T)} \times 100\% \tag{8}$$

$$\delta(WF, T) = \frac{f(P, T + \Delta T) - f(P, T)}{f(P, T)} \times 100\% \tag{9}$$

where $\delta(WF, P)$ and $\delta(WF, T)$ are the response of water flow to precipitation change and temperature change; P and T are observed precipitation and observed temperature; ΔP and ΔT are the change of precipitation and temperature in the hypothetical climate change scenarios. In this study, we assumed that the precipitation in each grid point change by -30% to 30% with an interval of 10% and the temperature in each grid point change by -3 °C to $+3$ °C with an interval of 1 °C.

The Equation (8) and Equation (9) can be used to analyze the basin scale BWF and GWF variation in different precipitation and temperature scenarios. To compare the variation of sensitivity to climate change in different regions, a sensitivity index (SI) is designed in this study to express the change rate between BWF or GWF with precipitation and temperature. The relationship between water flow and precipitation/temperature can be described as:

$$\hat{y} = ax + b. \tag{10}$$

where, $\hat{y}$ is the simulated BWF or GWF; x is the precipitation or temperature; a and b are the coefficients, which can be estimated by the least square method. Then the *SI* can be calculated as:

$$SI = \frac{a}{WF} \tag{11}$$

where, WF is the multi-year average BWF or GWF of each sub-basin in current climate. The *SI* stands for the variation (%) of BWF or GWF as precipitation changes for 1% or temperature changes for 1 °C in each sub-basin. By comparing *SI* in different sub-basins, the difference of BWF and GWF's response to climate change in different regions can be obtained.

3. Results

3.1. Evaluation of CMADS Precipitation and Temperature

Statistical results of CMADS reanalysis data and gauge observations (Dali Station) on daily scale and monthly scale are illustrated in Figures 5 and 6. It can be found that the PDFs of CMADS reanalysis daily temperature are quite tightly clustered around the PDFs of gauge observations. The skill in the CMADS reanalysis Maximum temperature and Minimum temperature were higher than 0.95. CMADS tends to overestimate the amount of drizzle but did quite well for precipitation of more than 4 mm/day. The skill score for daily precipitation approaches 0.8 (Figure 5a). In general, the CMADs showed considerable skill in representing the PDFs of daily gauge observations. The monthly CMADS precipitation and temperature were highly consistent with the monthly gauge observations. In particular, the C values of monthly Maximum temperature and Minimum temperature were nearly 1.0. The relative bias ratio of monthly precipitation was less than 7% and the absolute biases of monthly Maximum temperature and Minimum temperature were less than 0.5 °C (Figure 6). Therefore, the CMADS reanalysis dataset can be used for hydrology process simulation in the ELB.

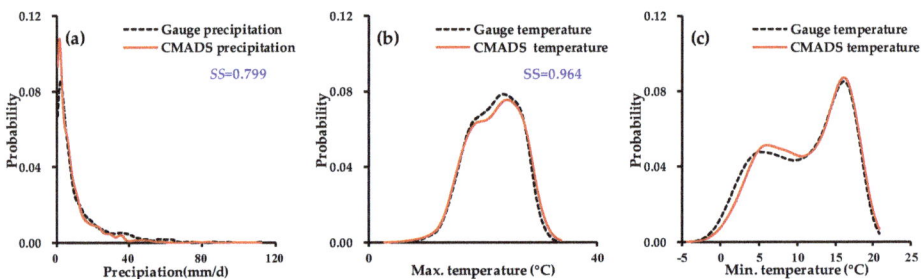

Figure 5. PDFs for the daily CMADS reanalysis data and gauge observations: (**a**) Precipitation; (**b**) Maximum temperature; (**c**) Minimum temperature.

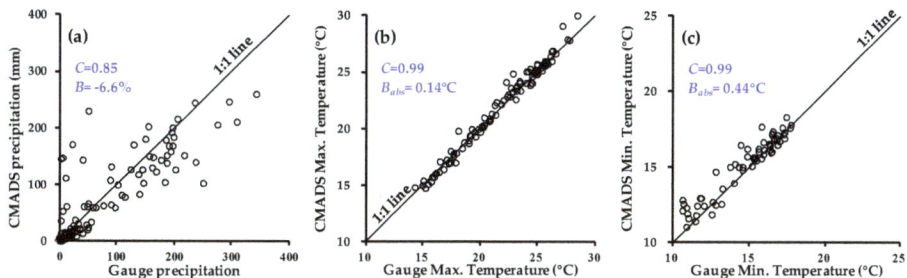

Figure 6. Scatter plots of the monthly CMADS reanalysis data and gauge observations: (**a**) Precipitation; (**b**) Maximum temperature; (**c**) Minimum temperature.

3.2. Evaluation of SWAT Simulation

The observed and SWAT simulated monthly streamflow at Liancheng Station from 2009 to 2016 is illustrated in Figure 7 (the statistical measures are provided in Table 1). It can be found that the simulated streamflow matched well with the observed streamflow except in a few months. E_{NS} and R^2 values were greater than 0.75 and RE value is less than 5% for both the calibration period and validation period. However, the E_{NS} and R^2 decreased for the validation period because the model did not perform well for the wet season in 2015. Generally, a monthly E_{NS} of 0.5 or greater and RE of 15% or less means that the simulation is considered satisfactory. According to these criteria, we can conclude that the SWAT model was a reliable representation of hydrological processes and can be used for the ELB.

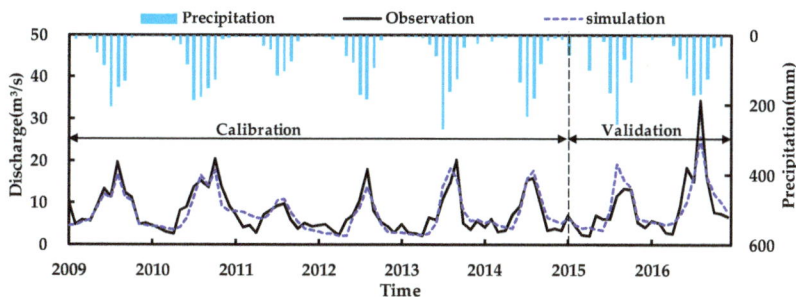

Figure 7. Simulated and observed monthly streamflow at Liancheng station during the calibration period 2009 to 2014 and the validation period 2015 to 2016.

Table 1. The calibration and validation statistics.

Period	E_{NS}	R^2	RE (%)
Calibration (2009 to 2014)	0.802	0.808	−3.7
Validation (2015 to 2016)	0.751	0.754	2.9

The other hydrologic stations in the Erhai Basin were used for water level measurement which cannot use for calibration. According to the study by Huang et al. [18], the average annual discharge into the Erhai Lake was about 683 million m³ during the period 2001 to 2010. This statistic result was based on the data provided by Erhai Administration Bureau of Yunnan Province. In this study, the simulated average annual discharge into the Erhai Lake was about 714 million m³ during the period 2008 to 2016, which is similar to the results counted by Huang et al. In addition, as the variation of water storage in a basin approaches to zero over a long period, the annual average precipitation ($\overline{P}$) should be approximately equal to the sum of annual average blue water flow ($\overline{BWF}$)

and annual average green water flow ($\overline{GWF}$). Based on the SWAT simulated results, the $\overline{P}$, $\overline{BWF}$ and $\overline{GWF}$ during the period 2009 to 2016 are 821.0 mm, 288.9 mm, and 562.3 mm, respectively in the ELB. The difference between $\overline{P}$ and ($\overline{BW}$ + $\overline{GWF}$) is −30.2 mm, accounting for 3.6% of the annual average precipitation. The above analysis proved that the water availability estimated by SWAT model is reasonable.

3.3. Spatial and Temporal Variability of Blue and Green Water Flows in the Erhai Lake Basin

The annual average of BWF and GWF during the period 2009 to 2016 across the ELB were 288.9 mm and 562.3 mm, respectively. The variation coefficients were 0.18 and 0.15 for BWF and GWF, respectively, which means that the change of GWF was relatively more stable than that of BWF. It is mainly because that the GWF is influenced by various factors (e.g., precipitation and temperature) while the precipitation is the major factor affecting the BWF [52–55]. In the ELB, the linear correlation coefficient between precipitation and BWF was high to 0.81. But the relationship between GWF and precipitation/temperature was more complicated.

Both the BWF and GWF increased at the entire basin level in the recent 8 years. As a result, the GWC decreased by 0.01 per year during the study period (Figure 8). The changes in precipitation and temperature in the ELB are illustrated in Figure 9. It could be found that the precipitation has been increasing since 2011 (Figure 9a) which is contrary to the change of temperature (Figures 9b and 9c). In this wetter and colder condition, the runoff (blue water) has increased faster than the evapotranspiration (green water), leading to a lower GWC.

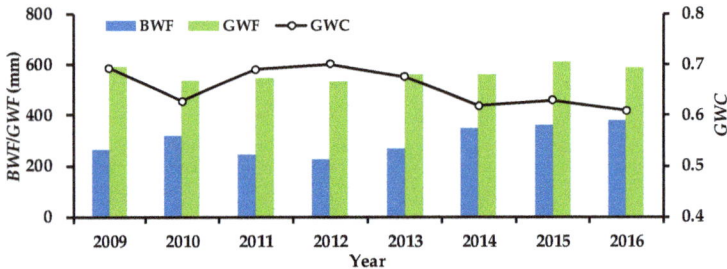

Figure 8. Changes in water flows and green water coefficient from 2009 to 2016: The blue bar represents blue water flow (BWF), and the green bar represents green water flow (GWF). The line with circles represents the green water coefficient (GWC).

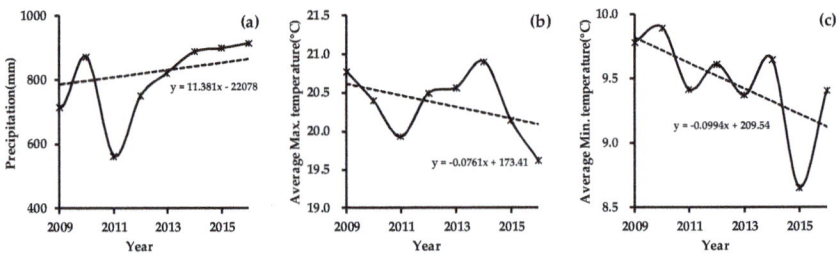

Figure 9. Changes in annual precipitation (**a**), average Maximum temperature (**b**) and average Minimum temperature (**c**) from 2009 to 2016.

The spatial variation of annual average BWF, GWF, and GWC are illustrated in Figure 10. It is obvious that the spatial distribution of BWF was uneven. The BWF shows a higher value in the west of Erhai Lake, where is called Eighteen Streams Region, with a BWF of more than 400 mm/year. The other main area of water-yield is in the mountainous regions located in the north and east of Erhai

Lake, with the BWF ranging from 300 to 400 mm/year (Figure 10a). Compared with BWF, the spatial distribution of GWF is more homogeneous. The GWF in most parts of the ELB changed with a range of 450 to 550 mm. The areas with a high-value of GWF were mainly distributed around Erhai Lake. In the low altitude region, especially in the north and east of the ELB, the GWF is a large percentage of the water flow (Figure 10b), generally the GWC was more than 0.7. The trend of the GWC is downward with altitude (Figure 10c). This is mainly due to higher precipitation at high altitude along with low temperatures and evapotranspiration rates. Consequently water-yield is abundant [56–58].

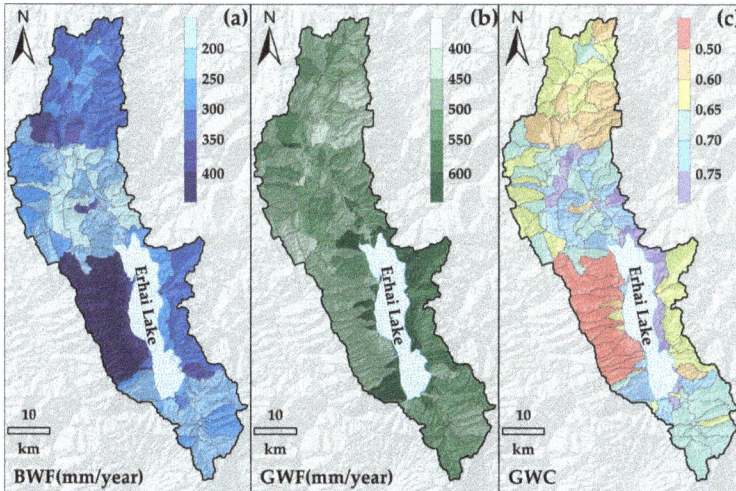

Figure 10. The spatial distribution of annual average BWF (**a**) GWF (**b**), and GWC (**c**) (2009 to 2016).

3.4. Sensitivity of Blue and Green Water Flows to Climate Change

With the parameterized SWAT model, the blue water flow and green water flow can be simulated in the hypothetical climatic scenarios. Then the sensitivity of these water flows to climate change can be estimated.

3.4.1. Sensitivity of Blue and Green Water Flows to Precipitation and Temperature at the Basin Scale

In the hydrographs for precipitation change (−30% to +30%) in Section 2.4, the BWF and GWF would increase with the precipitation. From Figure 11a, we can observe that a 10% increase in precipitation will result in a 20.8% increase in the BWF, but the GWF was less susceptible to precipitation change, showing an increase of 2.5% when precipitation increases by 10%. The GWC would decrease obviously as the precipitation increasing. It varied from 0.80 to 0.54 as precipitation amount changes with −30% and +30%, respectively.

The impact of temperature on the BWF was the opposite to the GWF under the temperature change scenarios (−3 °C to +3 °C). Figure 11b indicates that the BWF would decrease as the temperature increased, owing to a higher evapotranspiration rate in the warmer conditions. It was shown to decrease by 8.8 mm (nearly 3%) with a 1 °C reduction. But both the GWF and temperature would have a similar positive trend when temperature changes between −3.0 and 3.0 °C. The GWF would rise by 10.0 mm (about 1.7%) when temperature increases by 1.0 °C. In the hypothesis for temperature change, GWC would change slightly with an increase of approximately 0.01 for a corresponding 1.0 °C temperature increase.

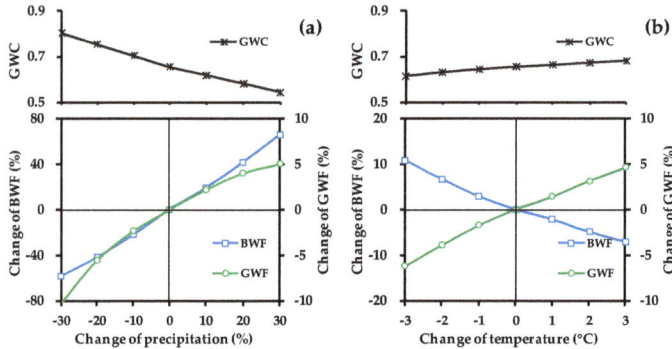

Figure 11. Sensitivity on BWF, GWF, and GWC impacted by precipitation change (**a**) and temperature change (**b**) in the ELB.

3.4.2. Sensitivity of Blue and Green Water Flows to Precipitation and Temperature at the Sub-Basin Scale

With the sensitivity index of water flow to climate change defined in Section 2.4, the spatial difference of sensitivity can be illustrated as Figure 12. In the low altitude regions located on the north and south of Erhai Lake, the BWF was more sensitive to precipitation. These regions are characterized by a lower precipitation–runoff coefficient. Similar results can also be found in Jones et al. [59], Bao et al. [60], and Yuan et al. [61]. The sensitivity of GWF to precipitation has similar spatial distribution characteristics with BWF. The more sensitive areas were also predominately located in the north and south of the Erhai Lake. These spatial distribution characteristics can be scientifically explained according to the Budyko hypothesis [62]. Water availability and energy are major factors that control evapotranspiration (GWF). The lower altitude region usually has a warmer climate. Thus, the GWF was primarily limited by the precipitation and sensitivity to it in the north and south of Erhai Lake. The Budyko hypothesis also explains why the GWF in the mountainous area located on the north and west of the ELB was more sensitive to temperature than that in other sub-basins. The weather condition is usually colder in the mountainous area, and evapotranspiration mainly depends on the energy under the wet condition. Therefore, along with the rising air temperature, the GWF would increase obviously in the mountainous areas, especially in the Cangshan Mountain, where the precipitation is abundant. As a consequence, the BWF would decrease significantly in these regions.

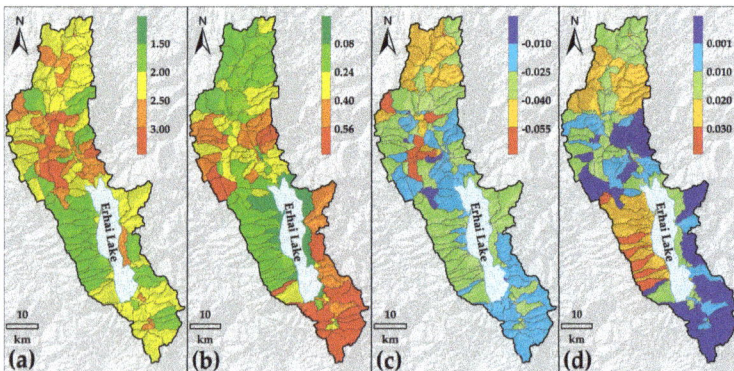

Figure 12. Sensitivity on BWF and GWF impacted by precipitation or temperature change in the sub-basin scale. (**a**) Sensitivity of BWF to precipitation; (**b**) Sensitivity of GWF to precipitation; (**c**) Sensitivity of BWF to temperature; (**d**) Sensitivity of GWF to temperature.

4. Discussion

4.1. Comparison of the Sensitivity of Blue Water Flow and Green Water Flow

The sensitivity of BWF and GWF to precipitation and temperature has been analyzed in this study. However, other climatic factors, such as humidity, radiation, wind speed, have not been taken into consideration. From Figure 11 we can observe that the BWF was more sensitive to precipitation and temperature change compared with GWF. For example, an increase of 20% precipitation will result in an increase of 41.7% and 4.0% in BWF. The BWF was directly formed from precipitation, with correlation coefficient higher than 0.8 (Figure 13a). However, the relationship between precipitation and GWF was less obvious (Figure 13b). It is mainly because precipitation is not the only crucial factor for GWF. The air temperature, solar radiation, relative humidity, and wind speed are also important factors affecting GWF. Previous studies found that the decrease of GWF might be related to solar radiation or wind speed reduction [63–65], while the increase in wind speed or decrease in relative humidity might cause GWF increases [66–68]. Bao et al. have carried similar research in the Haihe River Basin of North China. Their research has also found that the GWF was less sensitivity to precipitation compared with BWF. Taking the Taolinkou catchment in the Haihe River Basin as an example, the BWF and GWF would decrease by 39% and 14% if precipitation decreased by 20% [60]. Besides, the response of BWF and GWF to precipitation and temperature is nonlinear. Thus, the sensitivity of water flows to climate change might be different in different climatic scenarios. But the sensitivity index designed in this study cannot be used to investigate this law. Another sensitivity index is needed to solve the above-mentioned problem in a future study.

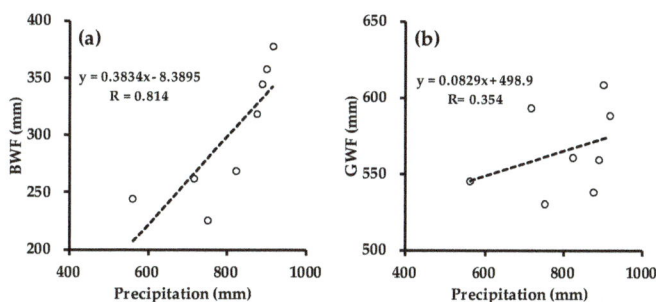

Figure 13. The relationship between precipitation and water flow in the ELB during 2009 to 2016: (**a**) precipitation vs. BWF; (**b**) precipitation vs. GWF.

4.2. Uncertainty Analysis

The major uncertainties in this study come from input data, model parameters, and model structure. To be specific, the gridded data ($0.25° \times 0.25°$) provided by the CMADS was used in this study. Compared with the weather data (only Dali station) provided by National Meteorological Information Center, this dataset can describe the spatial difference of meteorological factors distribution. However, the area of ELB is less than 3000 km^2, and there are only 17 grid points in and around the ELB. Thus, the precipitation data from limited points could not really represent regional precipitation. Higher spatial resolution data might be useful for hydrological simulation in this small watershed. In addition, although there are several hydrologic stations in the study area, only Liancheng Station can provide daily discharge data and be used for parameter calibration, while other stations can only be used to measure water level. Thus, the differences between optimized parameters and real parameters of the ELB cannot be avoided. Therefore, this study will do further research on water flows simulation by SWAT model based on parameters transfer method in the ELB. The different parameters transfer methods, such as spatial proximity, physical similarity, and comprehensive similarity, should be compared. Furthermore, evapotranspiration is an important process for assessing water flows as well as precipitation. The BWF and GWF are associated with the methods used to estimate potential evapotranspiration (ET_0). The SWAT model provides three methods, namely, Penman–Monteith method [69,70], Priestley–Taylor method [71] and Hargreaves method [72] to calculate potential evapotranspiration. In this study, the Penman–Monteith method was selected. Obviously, if the other two methods were used for ET_0 estimation, the sensitivity of water flows to precipitation and temperature would be different from the results in this study. Quantitative uncertainties derived from the model structure in sensitivity analysis should be further analyzed by using different ET_0 estimation methods or even different hydrological models.

4.3. Method for Green Water Flow Estimation

In this study, the GWF was assessed by SWAT model, which is a water balance method. From Section 3.2, we can find that the SWAT model was a reliable representation of streamflow in the ELB according to the observed data. But we can only evaluate the GWF simulation indirectly at long time scales based on the water balance principle. However, the simulation of monthly actual evapotranspiration (ET_a) or GWF has not been verified in this study because of the lack of long-term actual observed data. It can be concluded that the quantitative analysis of BWF's response to climate change has higher reliability than that of GWF.

Beyond water-balance derived ET_a, the GWF can also be estimated by remotely-sensed images. Based on the energy balance Bowen ratio method [73], the relationship between satellite-based vegetation indices and ET_a can be established, and then the GWF can be estimated indirectly [74]. The SWAT-based GWF and satellite-based GWF can be comparatively evaluated by each other. In addition, with the data assimilation methods, e.g., Ensemble Kalman Filter (EnKF) [75], the SWAT-based GWF and satellite-based GWF can also be assimilated. It might be an important work to improve the confidence of for the sensitivity of GWF to climate change and would be carried on in further study.

4.4. Impact of Land Use/Cover Change on Blue and Green Water Flow

Water availability are directly influenced by the climate variability. Since the land use/cover is relatively stable and climate variability affected the water flows more significantly than land use/cover change (LUCC) [76], this research did not analyze the impact of LUCC on blue and green water flow in the ELB. This does not mean that this kind of impact should be ignored. In fact, the components of water availability, such as surface runoff, inter flow, groundwater recharge, evapotranspiration, ect., principally depend on land use/cover [77]. Hence, a change in land use/cover of the ELB can alter the proportions of blue and green water flows. Analysis of multi-year land use data obtained from the RESDC dataset showed that the ELB has witnessed a remarkable expansion of built-up land and rapid shrinkage of agricultural land in the recent 35 years. From 1980 to 2014, the built-up land has significantly increased by 100.8%. The expansion area covered 67.5 km^2, accounting for 2.6% of the study area. In contrast, agricultural land has decreased by 10.7%. This reduced area was as large as 69.8 km^2, representing 2.7% of the ELB (Table 2). Considering these actual situations of land use change, the urban expansion scenarios and ecological restoration scenarios can be further established in the following research. Then the impact of climate and land-use/cover change on water flows can be analyzed comprehensively. Furthermore, the land use/cover in the future (e.g., 2020 year or 2050 year) can be predicted by cellular automata (CA) model [78]. Then the change of blue and green water flows in the future period can be estimated, which would be useful for water resources planning.

Table 2. Land use change from 1980 to 2015.

Land Use	Year 1980		Year 2015		Change	
	Area (km^2)	Percentage (%)	Area (km^2)	Percentage (%)	Area (km^2)	Percentage (%)
Agricultural land	651.8	25.5	582.0	22.8	−69.8	−10.7
Forest	838.8	32.9	851.5	33.4	12.8	1.5
Grassland	703.8	27.6	693.8	27.2	−10.0	−1.4
Water	265.9	10.4	261.0	10.2	−4.9	−1.8
Built-up land	67.0	2.6	134.5	5.3	67.5	100.8
Waste land	25.0	1.0	29.4	1.2	4.4	17.5

5. Conclusions

Using the CMADS reanalysis data, SWAT model, and the hypothetical climatic scenarios, the impact of climate variability on blue and green water flow in Erhai Lake Basin was investigated. According to this research, the following conclusions have been made:

The CMADS performed well in terms of correlation with gauge observations from Dali station. The statistic results showed that the CMADs has a considerable skill in representing the PDFs of daily gauge observations: The skill score was 0.799, 0.964, and 0.957 for daily precipitation, Maximum temperature and Minimum temperature, respectively. At the monthly scale, the CMADS underestimated the precipitation with a bias of −6.6% while ir overestimated the Maximum temperature and Minimum temperature by 0.14 °C and 0.44 °C, respectively. Both precipitation and temperature were highly consistent with the monthly gauge observations. It can be concluded that the CMADS can capture the climate characteristics of the Erhai Lake Basin. In addition, the CMADS reanalysis data can be widely applied in hydrological simulation and water plan and management, especially in basins with no or few data. Moreover, the SWAT model has been proved to be applicable in simulating the hydrologic processes in the Erhai Lake Basin, with E_{NS} and R^2 values greater than 0.75 and RE value less than 5%.

Estimated by the SWAT model, the annual average of blue water flow and green water flow during 2009 to 2016 across the Erhai Lake Basin were 288.9 mm and 562.3 mm, respectively. Blue water flow has increased while green water flow has decreased in the recent 8 years, owing to increasing precipitation and decreasing temperature, leading to a lower GWC. The spatial distribution of blue water flow was uneven in Erhai Lake Basin. It was higher in the mountainous regions with higher precipitation–runoff coefficients, such as Eighteen Streams Region, with blue water flow more than 400 mm/year. However, the spatial distribution of green water flow is more homogeneously, changing with a range of 450 to 550 mm/year in most areas. The trend of the GWC goes downward with the increase of altitude. It is because precipitation is higher while temperatures and evapotranspiration rates are lower at high altitude.

Blue water flow was more sensitive to precipitation and temperature change compared with green water flow. A 10% increase in precipitation can bring about a 20.8% increase in blue water flow while only a 2.5% increase in green water flow at basin scale. When temperature increases by 1.0 °C, blue water flow and green water flow would change by −3% and 1.7%, respectively. Blue water flow and green water flow were more sensitive to precipitation in the low altitude regions located at the north and south of Erhai Lake, which is characterized by a lower precipitation–runoff coefficient and warmer condition. In contrast, blue water flow and green water flow were more sensitive to temperature in the mountainous area, which is characterized by colder and wetter condition.

This study provided insights into blue and green water flows response to climate variability in the Erhai Lake Basin, which will help policymakers and administrators manage water resources in the context of climate change. Spatial variations of sensibility of water flows to climate variability imply that specific adaptation measures in different regions should be taken in the Erhai Lake Basin.

Author Contributions: Z.Y. and Y.W. worked together in forming ideas of this paper; Z.Y., B.Y., J.X. and X.H. worked together in calculating and writing of this manuscript. X.M., J.X. and Y.W. provided supervision during the process.

Funding: This research was funded by National Key Research and Development Project (no. 2016YFC0401306), National Natural Science Foundation of China (no. 41890821; no. 51709008; no. 41601043); National Public Research Institutes for Basic R&D Operating Expenses Special Project (no. CKSF2017029 and no. CKSF2017061/SZ).

Acknowledgments: Special thanks to Guest Editors of this Special Issue and the journal editors for their support.

Conflicts of Interest: The authors declare no conflict of interest. The funding sponsors had no role in the design of the study; in the collection, analyses, or interpretation of data; in the writing of the manuscript, and in the decision to publish the results.

References

1. Zhuang, X.; Li, Y.; Nie, S.; Fan, Y.; Huang, G. Analyzing climate change impacts on water resources under uncertainty using an integrated simulation—Optimization approach. *J. Hydrol.* **2018**, *556*, 523–538. [CrossRef]
2. Miara, A.; Macknick, J.E.; Vörösmarty, C.J.; Tidwell, V.C.; Newmark, R.; Fekete, B. Climate and water resource change impacts and adaptation potential for US power supply. *Nat. Clim. Chang.* **2017**, *7*, 793–798. [CrossRef]
3. Liang, W.; Bai, D.; Wang, F.; Fu, B.; Yan, J.; Wang, S.; Yang, Y.; Long, D.; Feng, M. Quantifying the impacts of climate change and ecological restoration on streamflow changes based on a Budyko hydrological model in China's Loess Plateau. *Water Resour. Res.* **2015**, *51*, 6500–6519. [CrossRef]
4. Chen, Y.; Li, Z.; Fan, Y.; Wang, H.; Deng, H. Progress and prospects of climate change impacts on hydrology in the arid region of northwest China. *Environ. Res.* **2015**, *139*, 11–19. [CrossRef] [PubMed]
5. Mishra, V.; Kumar, R.; Shah, H.L.; Samaniego, L.; Eisner, S.; Yang, T. Multi-model assessment of sensitivity and uncertainty of evapotranspiration and a proxy for available water resources under climate change. *Clim. Chang.* **2017**, *141*, 451–465. [CrossRef]
6. Gohar, A.A.; Cashman, A. A methodology to assess the impact of climate variability and change on water resources, food security and economic welfare. *Agric. Syst.* **2016**, *147*, 51–64. [CrossRef]
7. Simonovic, S.P. Bringing future climatic change into water resources management practice today. *Water Resour. Manag.* **2017**, *31*, 2933–2950. [CrossRef]
8. Arnell, N.W. Climate change and global water resources. *Glob. Environ. Chang.* **1999**, *9*, 31–49. [CrossRef]
9. Huntington, T.G. Evidence for intensification of the global water cycle: Review and synthesis. *J. Hydrol.* **2006**, *319*, 83–95. [CrossRef]
10. Chen, G.; Zhao, W. Green water and its research progresses. *Adv. Earth Sci.* **2006**, *21*, 221–227.
11. Schuol, J.; Abbaspour, K.C.; Yang, H.; Srinivasan, R.; Zehnder, A.J.B. Modeling blue and green water availability in Africa. *Water Resour. Res.* **2008**, *44*, W07406. [CrossRef]
12. Vanham, D. A holistic water balance of Austria-how does the quantitative proportion of urban water requirements relate to other users? *Water Sci. Technol.* **2012**, *66*, 549–555. [CrossRef] [PubMed]
13. Liu, J.; Yang, H. Spatially explicit assessment of global consumptive water uses in cropland: Green and blue water. *J. Hydrol.* **2009**, *384*, 187–197. [CrossRef]
14. Zang, C.; Liu, J. Trend analysis for the flows of green and blue water in the Heihe River basin, northwestern China. *J. Hydrol.* **2013**, *502*, 27–36. [CrossRef]
15. Ding, W. A study on the characteristics of climate change around the Erhai area, China. *Resour. Environ. Yangtze Basin* **2016**, *25*, 599–605. (In Chinese)
16. Huang, H.; Wang, Y.; Li, Q. Climatic characteristics over Erhai Lake basin in the late 50 years and the impact on water resources of Erhai Lake. *Meteorol. Mon.* **2013**, *39*, 436–442. (In Chinese)
17. Li, W.; Yang, C.; Liu, E.; Peng, Z.; Liu, Q. Multiple time scale analysis of water resources in Erhai Lake Basin in recent 59 years. *Chin. J. Agrometeorol.* **2010**, *31*, 10–15. (In Chinese)
18. Li, Y.; Li, B.; Zhang, K.; Zhu, J.; Yang, Q. Study on spatiotemporal distribution characteristics of annual precipitation of Erhai Basin. *J. China Inst. Water Resour. Hydropower Res.* **2017**, *15*, 234–240. (In Chinese)
19. Arias, R.; RodríguezBlanco, M.L.; Taboadacastro, M.M.; Nunes, J.P.; Keizer, J.J.; Taboadacastro, M.T. Water resources response to changes in temperature, rainfall and CO_2 concentration: A first approach in NW Spain. *Water* **2014**, *6*, 3049–3067. [CrossRef]

20. Falkenmark, M. Land-water linkages: A synopsis. Land and water integration and river basin management. *FAO Land Water Bull.* **1995**, *1*, 15–16.

21. Zhang, W.; Zha, X.; Li, J.; Liang, W.; Ma, Y.; Fan, D.; Li, S. Spatiotemporal change of blue water and green water resources in the headwater of Yellow River Basin, China. *Water Resour. Manag.* **2014**, *28*, 4715–4732. [CrossRef]

22. Chen, C.; Hagemann, S.; Liu, J. Assessment of impact of climate change on the blue and green water resources in large river basins in China. *Environ. Earth Sci.* **2015**, *74*, 6381–6394. [CrossRef]

23. Lee, M.H.; Bae, D.H. Climate change impact assessment on green and blue water over Asian monsoon region. *Water Resour. Manag.* **2015**, *29*, 2407–2427. [CrossRef]

24. Glavan, M.; Pintar, M.; Volk, M. Land use change in a 200-year period and its effect on blue and green water flow in two Slovenian Mediterranean catchments-lessons for the future. *Hydrol. Process.* **2013**, *27*, 3964–3980. [CrossRef]

25. Fazeli, F.I.; Farzaneh, M.R.; Besalatpour, A.A.; Salehi, M.H.; Faramarzi, M. Assessment of the impact of climate change on spatiotemporal variability of blue and green water resources under CMIP3 and CMIP5 models in a highly mountainous watershed. *Theor. Appl. Climatol.* **2018**. [CrossRef]

26. Zhou, F.; Xu, Y.; Chen, Y.; Xu, C.; Gao, Y.; Du, J. Hydrological response to urbanization at different spatio-temporal scales simulated by coupling of CLUE-S and the SWAT model in the Yangtze River Delta region. *J. Hydrol.* **2013**, *485*, 113–125. [CrossRef]

27. Fan, M.; Shibata, H. Simulation of watershed hydrology and stream water quality under landuse and climate change scenarios in Teshio River watershed, northern Japan. *Ecol. Indic.* **2015**, *50*, 79–89. [CrossRef]

28. Villarini, G.; Krajewski, W.F.; Smith, J.A. New paradigm for statistical validation of satellite precipitation estimates: Application to a large sample of the TMPA 0.25 3-hourly estimates over Oklahoma. *J. Geophys. Res. Atmos.* **2009**, *114*. [CrossRef]

29. Nijssen, B.; Lettenmaier, D.P. Effect of precipitation sampling error on simulated hydrological fluxes and states: Anticipating the Global Precipitation Measurement satellites. *J. Geophys. Res. Atmos.* **2004**, *109*. [CrossRef]

30. Conti, F.L.; Hsu, K.L.; Noto, L.V.; Sorooshian, S. Evaluation and comparison of satellite precipitation estimates with reference to a local area in the Mediterranean Sea. *Atmos. Res.* **2014**, *138*, 189–204. [CrossRef]

31. Li, C.; Tang, G.; Hong, Y. Cross-evaluation of ground-based, multi-satellite and reanalysis precipitation products: Applicability of the Triple Collocation method across Mainland China. *J. Hydrol.* **2018**, *562*, 71–83. [CrossRef]

32. Meng, X.; Wang, H. Significance of the China meteorological assimilation driving datasets for the SWAT Model (CMADS) of East Asia. *Water* **2017**, *9*, 765. [CrossRef]

33. Meng, X.; Wang, H.; Shi, C.; Wu, Y.; Ji, X. Establishment and Evaluation of the China Meteorological Assimilation Driving Datasets for the SWAT Model (CMADS). *Water* **2018**, *10*, 1555. [CrossRef]

34. Meng, X.; Wang, H.; Cai, S.; Zhang, X.; Leng, G.; Lei, X.; Shi, C.; Liu, S.; Shang, Y. The China Meteorological Assimilation Driving Datasets for the SWAT Model (CMADS) Application in China: A Case Study in Heihe River Basin. *Preprints* **2016**. [CrossRef]

35. Meng, X.; Wang, H.; Wu, Y.; Long, A.; Wang, J.; Shi, C.; Ji, X. Investigating spatiotemporal changes of the land-surface processes in Xinjiang using high-resolution CLM3.5 and CLDAS: Soil temperature. *Sci. Rep.* **2017**, *7*, 13286. [CrossRef] [PubMed]

36. Meng, X.; Dan, L.; Liu, Z. Energy balance-based SWAT model to simulate the mountain snowmelt and runoff—Taking the application in Juntanghu watershed (China) as an example. *J. Mt. Sci.* **2015**, *12*, 368–381. [CrossRef]

37. Meng, X.; Wang, H.; Lei, X.; Cai, S.; Wu, H. Hydrological Modeling in the Manas River Basin Using Soil and Water Assessment Tool Driven by CMADS. *Tehnički Vjesnik* **2017**, *24*, 525–534.

38. Wang, Y.; Meng, X. Snowmelt runoff analysis under generated climate change scenarios for the Juntanghu River basin in Xinjiang, China. *Tecnol. Y Cienc. Agua* **2016**, *7*, 41–54.

39. Meng, X. Simulation and spatiotemporal pattern of air temperature and precipitation in Eastern Central Asia using RegCM. *Sci. Rep.* **2018**, *8*, 3639. [CrossRef] [PubMed]

40. Meng, X. Spring Flood Forecasting Based on the WRF-TSRM mode. *Tehnički Vjesnik* **2018**, *25*, 27–37.

41. Vu, T.; Li, L.; Jun, K. Evaluation of MultiSatellite Precipitation Products for Streamflow Simulations: A Case Study for the Han River Basin in the Korean Peninsula, East Asia. *Water* **2018**, *10*, 642. [CrossRef]

42. Gao, X.; Zhu, Q.; Yang, Z.; Wang, H. Evaluation and hydrological application of CMADS against TRMM 3B42V7, PERSIANN-CDR, NCEP-CFSR, and gauge-based datasets in Xiang River Basin of China. *Water* **2018**, *10*, 1225. [CrossRef]

43. Zhou, S.; Wang, Y.; Chang, J.; Guo, A.; Li, Z. Investigating the dynamic influence of hydrological model parameters on runoff simulation using sequential uncertainty fitting-2-based multilevel-factorial-analysis method. *Water* **2018**, *10*, 1177. [CrossRef]

44. Tian, Y.; Zhang, K.; Xu, Y.-P.; Gao, X.; Wang, J. Evaluation of potential evapotranspiration based on CMADS reanalysis dataset over China. *Water* **2018**, *10*, 1126. [CrossRef]

45. Hu, Y.; Peng, J.; Liu, Y.; Tian, L. Integrating ecosystem services trade-offs with paddy land-to-dry land decisions: A scenario approach in Erhai Lake basin, southwest China. *Sci. Total Environ.* **2018**, *625*, 849–860. [CrossRef] [PubMed]

46. Crook, D.; Elvin, M.; Jones, R.; Ji, S.; Foster, G.; Dearing, J. The History of Irrigation and Water Control in China's Erhai Catchment: Mitigation and Adaptation to Environmental Change. In *Mountains: Sources of Water, Sources of Knowledge*; Springer: Berlin, Germany, 2008.

47. Nash, J.E.; Sutcliffe, J.V. River flow forecasting through conceptual models part I-A discussion of principles. *J. Hydrol.* **1970**, *10*, 282–290. [CrossRef]

48. Gupta, H.; Sorooshian, S.; Yapo, P. Status of automatic calibration for hydrologic models: Comparison with multilevel expert calibration. *J. Hydrol. Eng.* **1999**, *4*, 135–143. [CrossRef]

49. Kumar, S.; Merwade, V. Impact of watershed subdivision and soil data resolution on SWAT model calibration and parameter uncertainty. *J. Am. Water Resour. Assoc.* **2009**, *45*, 1179–1196. [CrossRef]

50. Yin, J.; Yuan, Z.; Yan, D.; Yang, Z.; Wang, Y. Addressing climate change impacts on streamflow in the Jinsha River Basin based on CMIP5 Climate Models. *Water* **2018**, *10*, 910. [CrossRef]

51. Jiang, S.; Ren, L.; Yong, B.; Yuan, F.; Gong, L.; Yang, X. Hydrological evaluation of the TRMM multi-satellite precipitation estimates over the Mishui basin. *Adv. Water Sci.* **2014**, *25*, 641–649. (In Chinese)

52. Zheng, H.; Zhang, L.; Zhu, R.; Liu, C.; Sato, Y.; Fukushima, Y. Responses of streamflow to climate and land surface change in the headwaters of the Yellow River Basin. *Water Resour. Res.* **2009**, *45*, W00A19. [CrossRef]

53. Lan, Y.; Zhao, G.; Zhang, Y.; Wen, J.; Hu, X.; Liu, J.; Gu, M.; Chang, J.; Ma, J. Response of runoff in the headwater region of the Yellow River to climate change and its sensitivity analysis. *J. Geogr. Sci.* **2010**, *20*, 848–860. [CrossRef]

54. Liu, Q.; Mcvicar, T.R. Assessing climate change induced modification of penman potential evaporation and runoff sensitivity in a large water-limited basin. *J. Hydrol.* **2012**, *464*, 352–362. [CrossRef]

55. Xu, C.; Singh, V. Evaluation of three complementary relationship evapotranspiration models by water balance approach to estimate actual regional evapotranspiration in different climatic regions. *J. Hydrol.* **2005**, *308*, 105–121. [CrossRef]

56. Wang, C.; Zhou, X. Effect of the recent climate change on water resource in Heihe river basin. *J. Arid Land Resour. Environ.* **2010**, *24*, 60–65. (In Chinese)

57. Guo, Q.; Yang, Y.; Chen, X.; Chen, Z. Annual Variation of Heihe River Runoff during 1957–2008. *Prog. Geogr.* **2011**, *30*, 550–556. (In Chinese)

58. Zang, C.; Liu, J.; Velde, M.; Kraxner, F. Assessment of spatial and temporal patterns of green and blue water flows under natural conditions in inland river basins in Northwest China. *Hydrol. Earth Syst. Sci.* **2012**, *16*, 2859–2870. [CrossRef]

59. Jones, R.N.; Chiew, F.H.S.; Boughton, W.C.; Zhang, L. Estimating the sensitivity of mean annual runoff to climate change using selected hydrological models. *Adv. Water Resour.* **2006**, *29*, 1419–1429. [CrossRef]

60. Bao, Z.; Zhang, J.; Liu, J.; Wang, G.; Yan, X.; Wang, X.; Zhang, L. Sensitivity of hydrological variables to climate change in the Haihe river basin, China. *Hydrol. Process.* **2012**, *26*, 2294–2306. [CrossRef]

61. Yuan, Z.; Yan, D.; Yang, Z.; Yin, J.; Zhang, C.; Yuan, Y. Projection of surface water resources in the context of climate change in typical regions of China. *Hydrol. Sci. J.* **2017**, *62*, 283–293. [CrossRef]

62. Budyko, M.I. Climatic factors of the external physical-geographical processes. *Gl Geofiz Obs.* **1950**, *19*, 25–40. (In Russian)

63. Brutsaert, W.; Parlange, M.B. Hydrologic cycle explains the evaporation paradox. *Nature* **1998**, *396*, 30. [CrossRef]

64. Golubev, V.S.; Lawrimore, J.H.; Groisman, P.Y.; Speranskaya, N.A.; Zhuravin, S.A.; Menne, M.J.; Peterson, T.C.; Thomas, C.; Malone, R.W. Evaporation changes over the contiguous united states and the former USSR: A reassessment. *Geophys. Res. Lett.* **2001**, *28*, 2665–2668. [CrossRef]

65. Xu, C.; Gong, L.; Jiang, T.; Chen, D.; Singh, V.P. Analysis of spatial distribution and temporal trend of reference evapotranspiration and pan evaporation in Changjiang (Yangtze river) catchment. *J. Hydrol.* **2006**, *327*, 81–93. [CrossRef]

66. Yu, P.; Yang, T.; Chou, C. Effects of climate change on evapotranspiration from paddy fields in southern Taiwan. *Clim. Chang.* **2002**, *54*, 165–179. [CrossRef]

67. Burn, D.H.; Hesch, N.M. Trends in evaporation for the Canadian prairies. *J. Hydrol.* **2007**, *336*, 61–73. [CrossRef]

68. Dinpashoh, Y.; Jhajharia, D.; Fakheri-Fard, A.; Singh, V.P.; Kahya, E. Trends in reference crop evapotranspiration over Iran. *J. Hydrol.* **2011**, *399*, 422–433. [CrossRef]

69. Monteith, J.L. Evaporation and environment. In Proceedings of the 19th Symposium of the Society for Experimental Biology, New York, NY, USA, 1 January 1965; Cambridge University Press: Cambridge, UK, 1965; pp. 205–233.

70. Allen, R.G.; Jensen, M.E.; Wright, J.L.; Burman, R.D. Operational estimates of reference evapotranspiration. *Agron. J.* **1989**, *81*, 650–662. [CrossRef]

71. Priestley, C.H.B.; Taylor, R.J. On the assessment of surface heat flux and evaporation using large-scale parameters. *Mon. Weather Rev.* **1972**, *100*, 81–92. [CrossRef]

72. Hargreaves, G.L.; Hargreaves, G.H.; Riley, J.P. Agricultural benefits for Senegal River Basin. *J. Irrig. Drain. Eng.* **1985**, *111*, 113–124. [CrossRef]

73. Allen, R.G.; Pereira, L.S.; Howell, T.A.; Jensen, M.E. Evapotranspiration information reporting: I. Factors governing measurement accuracy. *Agric. Water Manag.* **2011**, *98*, 899–920. [CrossRef]

74. Nagler, P.L.; Glenn, E.P.; Kim, H.; Emmerich, W.; Scott, R.L.; Huxman, T.E.; Huete, A.R. Relationship between evapotranspiration and precipitation pulses in a semiarid rangeland estimated by moisture flux towers and MODIS vegetation indices. *J. Arid. Environ.* **2007**, *70*, 443–462. [CrossRef]

75. Sun, L.; Seidou, O.; Nistor, I.; Liu, K. Review of the Kalman type hydrological data assimilation. *Hydrol. Sci. J.* **2016**, *61*, 2348–2366. [CrossRef]

76. Zhao, A.; Zhu, X.; Liu, X.; Pan, Y.; Zuo, D. Impacts of land use change and climate variability on green and blue water resources in the Weihe river basin of northwest china. *Catena* **2016**, *137*, 318–327. [CrossRef]

77. Sajikumar, N.; Remya, R.S. Impact of land cover and land use change on runoff characteristics. *J. Environ. Manag.* **2015**, *161*, 460–468. [CrossRef] [PubMed]

78. Deng, Z.; Zhang, X.; Li, D.; Pan, G. Simulation of land use/land cover change and its effects on the hydrological characteristics of the upper reaches of the Hanjiang Basin. *Environ. Earth Sci.* **2015**, *73*, 1119–1132. [CrossRef]

water

MDPI

Article

The Impacts of Climate Variability and Land Use Change on Streamflow in the Hailiutu River Basin

Guangwen Shao, Yiqing Guan, Danrong Zhang *, Baikui Yu and Jie Zhu

College of Hydrology and Water Resources, Hohai University, Nanjing 210098, China;
guangwenshao@hhu.edu.cn (G.S.); yiqingguan@hhu.edu.cn (Y.G.); yubaikui@126.com (B.Y.);
zhujie58603586@163.com (J.Z.)
* Correspondence: danrong_zhang@hhu.edu.cn; Tel.: +86-177-6172-4730

Received: 6 March 2018; Accepted: 17 June 2018; Published: 20 June 2018

Abstract: The Hailiutu River basin is a typical semi-arid wind sandy grass shoal watershed in northwest China. Climate and land use have changed significantly during the period 1970–2014. These changes are expected to impact hydrological processes in the basin. The Mann–Kendall (MK) test and sequential *t*-test analysis of the regime shift method were used to detect the trend and shifts of the hydrometeorological time series. Based on the analyzed results, seven scenarios were developed by combining different land use and/or climate situations. The Soil Water Assessment Tool (SWAT) model was applied to analyze the impacts of climate variability and land use change on the values of the hydrological components. The China Meteorological Assimilation Driving Datasets for the SWAT model (CMADS) was applied to enhance the spatial expressiveness of precipitation data in the study area during the period 2008–2014. Rather than solely using observed precipitation or CMADS precipitation, the precipitation values of CMADS and the observed precipitation values were combined to drive the SWAT model for better simulation results. From the trend analysis, the annual streamflow and wind speed showed a significant downward trend. No significant trend was found for the annual precipitation series; however, the temperature series showed upward trends. With the change point analysis, the whole study period was divided into three sub-periods (1970–1985, 1986–2000, and 2001–2014). The annual precipitation, mean wind speed, and average temperature values were 316 mm, 2.62 m/s, and 7.9 °C, respectively, for the sub-period 1970–1985, 272 mm, 2.58 m/s, and 8.4 °C, respectively, for the sub-period 1986–2000, and 391 mm, 2.2 m/s, and 9.35 °C, respectively, for the sub-period 2001–2014. The simulated mean annual streamflow was 35.09 mm/year during the period 1970–1985. Considering the impact of the climate variability, the simulated mean annual streamflow values were 32.94 mm/year (1986–2000) and 36.78 mm/year (2001–2014). Compared to the period 1970–1985, the simulated mean annual streamflow reduced by 2.15 mm/year for the period 1986–2000 and increased by 1.69 mm/year for the period 2001–2014. The main variations of land use from 1970 to 2014 were the increased area of shrub and grass land and decreased area of sandy land. In the simulation it was shown that these changes caused the mean annual streamflow to decrease by 0.23 mm/year and 0.68 mm/year during the periods 1986–2000 and 2001–2014, respectively. Thus, the impact of climate variability on the streamflow was more profound than that of land use change. Under the impact of coupled climate variability and land use change, the mean annual streamflow decreased by 2.45 mm/year during the period 1986–2000, and the contribution of this variation to the decrease in observed streamflow was 27.8%. For the period 2001–2014, the combined climate variability and land use change resulted in an increase of 0.84 mm/year in annual streamflow. The results obtained in this study could provide guidance for water resource management and planning in the Erdos plateau.

Keywords: statistical analysis; SWAT; CMADS; climate variability; land use change; streamflow

1. Introduction

Climate variability and land use change are recognized as two important factors influencing hydrological processes. Studies showed that the increased temperature could result in corresponding changes in the timing and volume of spring flood and evapotranspiration [1,2]. The variability of precipitation, especially the changes in frequency and intensity of extreme precipitation events, could lead to a variation in stream flow and peak flow. Land use changes are directly linked to changes in the hydrological processes, such as evapotranspiration, interception, and infiltration, etc., and then to the impact of groundwater recharge, baseflow, and streamflow, etc. [3–6].

In general, the methods of assessing the impacts of climate variability/land use change on hydrological processes can be mainly divided into three groups, which are time series analysis, paired catchments experiments, and hydrological modeling [7]. Time series analysis is easy to implement as it is a statistical method [8–10]. However, it can only be applied to simple analysis of the hydrological effects of climate variability/land use change, and lacks physical mechanisms. Paired catchments experiments and hydrological modeling can be applied to research into the interaction between hydrological and climate variability/land use change [11]. The advantage of paired catchments experiment is that it can remove climate variability/land use change through the comparison of two catchments under the similar climatic conditions/land use [12]. Nevertheless, this approach is difficult to apply to larger basins and is very time consuming [13]. In recent years, hydrological models have often been applied to study the relationship between climate variability/land use change and hydrological processes, for the inhomogeneity of climate and land use can be introduced to these models [3,14–16].

Among the widely used hydrological models, the Soil Water Assessment Tool (SWAT) is a utility model, as it has the capability for simulating many processes such as the hydrologic cycle, sediment transportation, and soil erosion. In the SWAT model, with GIS (Geographic Information System) and other interface tools, information on topographic, land use, and soil data can be conveniently written to input parameters of the model. Because SWAT is an open-source model, researchers can conveniently improve the simulation performance in specific study areas. Meanwhile, the SWAT model can be easily calibrated and incorporate changes in land use [17]. As a result, it is a widely used tool for the study of the hydrological effects of environmental change [18–20]. Yin [21] applied the SWAT model to quantify the impact of climate variability/land use change on the streamflow over a long historical time period. Kim [22] used SWAT model to evaluate the separated and combined impacts of future changes and land use changes on the streamflow. Commonly, researchers analyze the trend and change points of historical hydro-meteorology data before building the SWAT model [13,23]. The former can help to understand the effects of climate variability on streamflow and the latter can help to select a period with no significant human activity in order to build the SWAT model [23].

Precipitation is an important input variable in the SWAT model. The accuracy of hydrological simulation could be directly affected by the quantity and quality of precipitation data. Regular and sufficient rain gauges can aid in revealing the exact spatial distribution of precipitation. However, in some areas there are fewer rain gauges and it may difficult to exactly reflect the spatial distribution of precipitation [24,25]. Combining satellite data and observed data to estimate rainfall is an effective method to solve the problem. Based on this method, several products have been developed and widely applied in hydrological modeling [26,27]. The China Meteorological Assimilation Driving Datasets for the SWAT model (CMADS), which was developed by Dr. Xianyong Meng from the China Institute of Water Resources and Hydropower Research (IWHR), is one such product and has been widely applied in numerous studies [28–31]. For example, Liu [28] found that CMADS had unique advantages in hydrological simulations compared with observed data and Climate Forecast System Reanalysis (CFSR). Zhao [30], based on CMADS, applied three methods to analyze the parameter uncertainty in a mountain-loess transitional watershed. Cao [31] applied CMADS to estimate hydrological elements in the Lijiang River basin.

The Hailiutu River basin, which belongs to the middle Yellow River basin, is a typical sandy grass shoal watershed in the Erdos plateau in northwest China. This basin has undergone climate variability and land use change in recent decades [32]. Studies showed that in the central-south Yellow River basin the intensity of precipitation extremes increased between the years 1986 and 2011, and may have resulted in increased surface runoff [33,34]. The annual average air temperature showed a significant increasing trend in recent decades which may induce the occurrence of increased evapotranspiration [13,35]. Studies also showed that streamflow change is more sensitive to precipitation than temperature in the Yellow River basin, especially in the middle section [36,37]. Because of the implementation of the desert greening policy, the land use of study area has changed, i.e., increasing the area of shrub and grass land and decreasing the area of sand land [32]. The increased shrub and grass land could result in decreased baseflow and increased surface runoff and evapotranspiration [6,15]. Meanwhile, the water demand has increased in recent decades, especially after the Erdos Plateau was targeted as a priority area of the western development strategy for China in the 21st century. Accordingly, groundwater extraction, construction of reservoirs, and diversion of dams have increased [38–40]. The streamflow in the Hailiutu River basin is mainly recharged by groundwater, especially in the dry season. More extraction of groundwater would result in decreased streamflow [41,42]. Engineering measures such as reservoir construction and diversion of dams may affect high flow by reducing overland flow and peak streamflow [43]. The annual streamflow of Huangfuchuan basin, which is located in the same climate zone of Hailiutu River basin, has decreased significantly because the increased of construction of check dams [10,44,45]. Therefore, it is necessary to study the impacts of climate variability/land use change and other factors such as the increase of ground water extraction, and construction of reservoirs and diversion dams on the streamflow in the Hailiutu River basin.

The overall objective of this study is to assess the impacts of climate variability and land use change on hydrological components in the Hailiutu River basin in the Erdos plateau. The specific objectives are: (1) to analyze the temporal variation of hydrometeor-ology with trend analysis and change points testing; (2) to evaluate the performance of SWAT model for simulation of streamflow in the Hailiutu River basin; (3) to compare the impact of climate variability on hydrological components with land use change; and (4) to evaluate the effects of climate variability/land use change and other factors, such as the increase of ground water extraction and construction of reservoirs and diversion dams, on streamflow. The results obtained in this study could provide guidance for water resource management and planning in the Erdos plateau.

2. Materials and Methods

2.1. Study Area

The Hailiutu River is located in the Erdos Plateau in northwest China, and is a tributary of the Wuding River of the Yellow River basin. The Hailiutu River basin (38°02′–38°50′, 108°37′–109°14′) is near the Maowusu desert and covers an area of 2473 km^2 (Figure 1). It has a relatively flat topography and the main landscape is of undulating sand dunes [19]. The elevation of the Hailiutu River basin ranges from 1471 m in the north west region to 1016 m at the watershed outlet. The geological formation in the Hailiutu River basin can be divided into four strata: (1) the Holocene Aeolian sand with a thickness of 0 to 50 m; (2) the upper Pleistocene Shalawusu sandstone formation of the Quaternary age with a thickness of 5 to 90 m; (3) the Luohe sandstone of Cretaceous age with a thickness of 180 to 330 m, overlain with the Shalawusu formation; and (4) the bedrock, which consists of impermeable Jurassic sediments [39,46–48]. The study area has a semi-arid climate with unevenly distributed precipitation. The mean annual precipitation is 330 mm, with about 80% of the annual rainfall occurring during the rainy season (June to September). Yang et al. [39,49] found that hydrological processes were mainly affected by direct infiltration and evapotranspiration in the study area. The Hailiutu River has a relatively steady discharge even in the dry season (October to May) with groundwater being the primary source of streamflow [32].

2.2. Data Collection

The geographic data (topography, soil and land use) and hydrometeorological data are used in this study (Table 1). A 30 × 30 m resolution of digital elevation model (DEM) was accessed from the Geospatial Data Cloud of China to calculate the slope (Figure 2a). The soil data, with 1 × 1 km resolution, was obtained from the Harmonized Word Soil Database (HWSD) supplied by the Environmental and Ecological Science Data Center for West China. The spatial distribution of soil types is shown in the Figure 2b. Cambic Arenosols represent the most widely distributed soil type, consisting of 89% sand, 5% clay, and 6% silt, and account for 43% of all soil types in the Hailiutu River basin. Data for land use types with spatial and temporal variation in 1986, 1995, and 2010 were provided by the Data Center for Resources and Environmental Sciences of the Chinese Academy of Sciences (RESDC) (Figure 3). There are six main land use types, including agricultural land, shrub and grass land, forest land, water body, urban land, and sandy land in the Hailiutu River basin. The daily meteorological data from 1970 to 2014, which include wind speed, minimum and maximum temperature, and relative humidity, were acquired from the China Meteorological Sharing Service System. The observed daily rainfall and streamflow data of Hailiutu River basin were provided by the Yellow River Conservancy Commission.

Table 1. Summarized information on the data used for this study. CMADS: China Meteorological Assimilation Driving Datasets for the Soil Water Assessment Tool model.

Data Layer	Description of Data Layer	Data Sources
Topographic	30 × 30 m resolution digital elevation model (DEM) applied to calculate slopes and slope lengths	Geospatial Data Cloud of China
Soil map/layer	1 × 1 km resolution map; soil layer attributes for each soil layer	Environmental and Ecological Science Data Center for West China
Land use	30 × 30 m resolution map of 1986, 1995, and 2010	Data Center for Resources and Environmental Sciences, Chinese Academy of Sciences
Daily meteorological data	Daily wind speed, minimum and maximum temperature and relative humidity from 1970 to 2014	China Meteorological Sharing Service System.
Daily rainfall and monthly streamflow	Daily precipitation in/around watershed; monthly streamflow for the outlet of watershed	Yellow River Conservancy Commission
CMADS V1.0	Assimilation driving datasets applied to better reflect the spatial distribution of precipitation and meteorology	http://www.cmads.org

There are four rainfall stations and two meteorology stations in the Hailiutu River basin, of which only three rainfall stations are located in the study area. In order to better reflect the spatial distribution of precipitation, the China Meteorological Assimilation Driving Datasets for the SWAT model version 1.0 (CMADS V1.0) was introduced in this study [50]. The CMADS was developed by Dr. Xianyong Meng from the China Institute of Water Resources and Hydropower Research (IWHR) and has received worldwide attention [28–30]. In the CMADS, the integration of air temperature, air pressure, humidity, and wind velocity data were mainly achieved through the LAPS (Local Analysis and Prediction System)/STMAS (Space-Time Multiscale Analysis System). The data sources for the CMADS series include nearly 40,000 regional automatic stations and China's 2421 national automatic and business assessment centers [51–54]. This ensures that the CMADS has wide applicability within the country, and that data accuracy is vastly improved. The CMADS has been widely applied in hydrological simulations and produced better results in China [53,55]. Climate data from the CMADS can be downloaded from the official CMADS website (http://www.cmads.org).

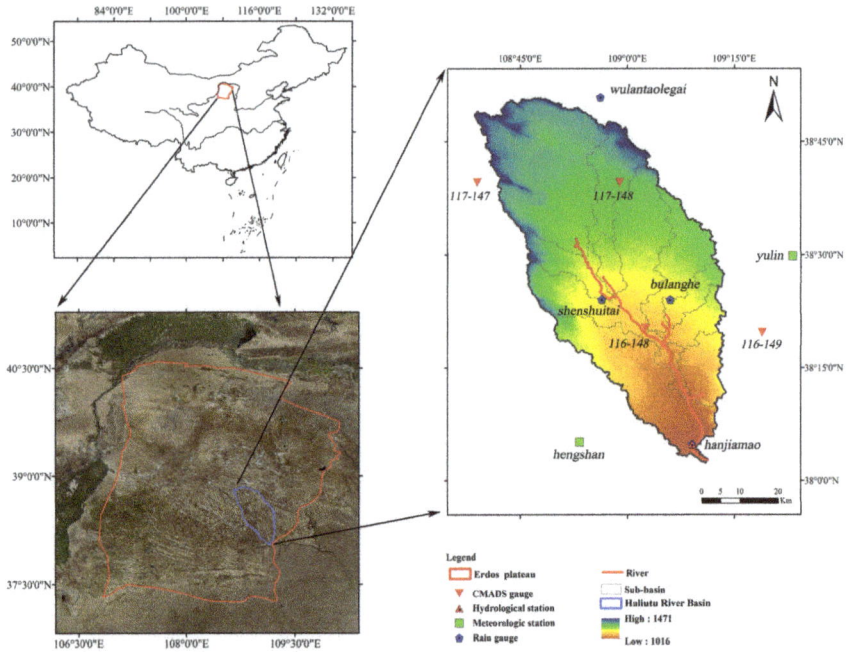

Figure 1. The location of the Hailiutu River basin and its digital elevation model with hydrometeorological stations.

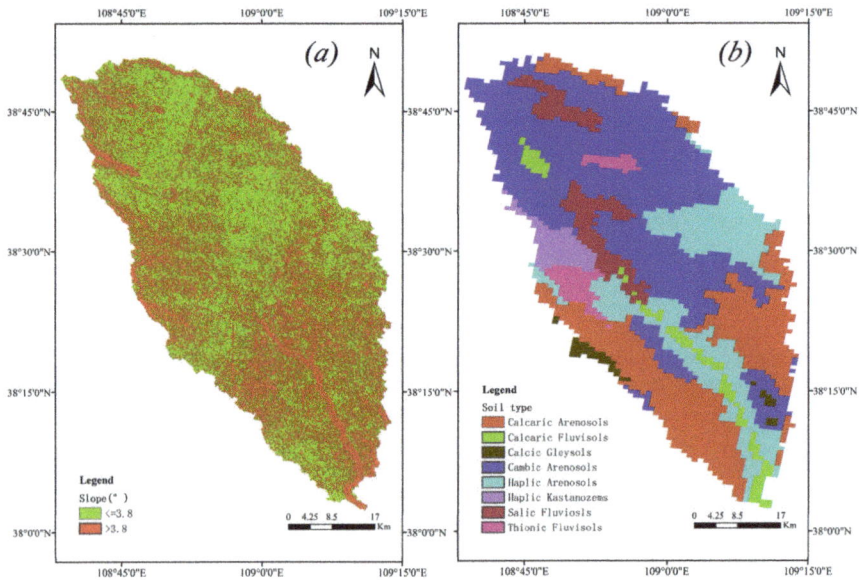

Figure 2. The (**a**) slope classes and (**b**) soil types of the Hailiutu River basin.

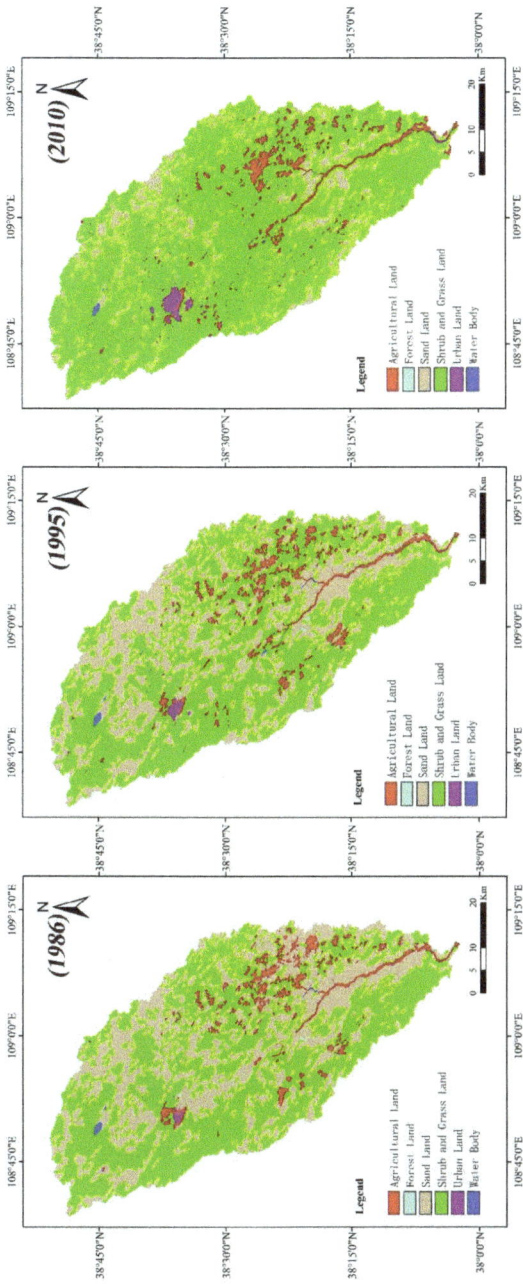

Figure 3. The land use patterns of the Hailiutu River basin for the years 1986, 1995, and 2010.

2.3. Methods

2.3.1. Time Series Analysis

1. Trend Analysis

The non-parameter Mann–Kendall (MK) test has been widely applied to identify the monotonic tendency and significance in hydro-meteorological time series such as those of precipitation, streamflow, and temperature [45,56,57]. The MK test statistic S is calculated as:

$$S = \sum_{i=1}^{n-1} \sum_{k=i+1}^{n} \text{sgn}(x_k - x_i) \tag{1}$$

where, x_k and x_i are the sequential data values, n is the length of time series, and the function $\text{sgn}(\theta)$ is defined as:

$$\text{sgn}(\theta) = \begin{cases} 1 & if \ \theta > 0 \\ 0 & if \ \theta = 0 \\ -1 & if \ \theta < 0 \end{cases} \tag{2}$$

For an independent data sample without tied values, the test statistic S is approximate normal distribution, and the variance of S can be calculated by Equation (3).

$$Var(S) = \frac{[n(n-1)(2n+5)]}{18} \tag{3}$$

Then, the MK test statistic Z can be estimated by Equation (4), where n is larger than 10.

$$Z = \begin{cases} \frac{S-1}{\sqrt{Var(S)}} & if \quad S > 0 \\ 0 & if \quad S = 0 \\ \frac{S+1}{\sqrt{Var(S)}} & if \quad S < 0 \end{cases} \tag{4}$$

In the two-sided test, the level of significance was assumed as α, and the critical value was defined as $Z_{\alpha/2}$. Therefore, the null hypothesis H_0 should be accepted if $|Z| \leq Z_{\alpha/2}$. The time series has an upward tendency for $Z > 0$ and there is a downward tendency for $Z < 0$.

2. Change Point Detection Method

There are numerous methods to identify the change points of time series, such as Pettitt test, the Mann–Kendall test, sliding t detection, etc. These methods commonly adopt confirmatory statistical techniques with a priori hypothesis to detect the change points of time series. That means substantial amounts of data (at least 10 or more time-series points) should be accumulated to apply a formal statistical test. Therefore, the change points are usually detected long after they actually occurred [58]. In contrast, the sequential t-test analysis of regime shift (STARS) method proposed by Rodionov belongs to the category of exploratory or data-driven analysis and does not need an a priori hypothesis that abrupt change has occurred at a certain time [59]. Consequently, the STARS method can detect the change point relatively early when a regime shift occurs [44,60]. In this study, the STARS method is used to detect the regime shift of hydrometeorological data.

2.3.2. Assessing the Impacts of Climate Variability and Land Use Change

In this study, based on the change point analysis, the hydrometeorology data were divided into three sub-periods (see Section 3.1). Three land use patterns were collected as described in the data collection section during each sub-period. For simplicity, two sub-periods (the prior phase and the latter phase) are displayed in this section (Figure 4). Firstly, the land use and climate condition of the prior phase, as the original scenario, was used to calibrate and validate the SWAT model. Subsequently, the changed climate/land use (latter phase) were applied to develop scenarios. Then, based on the calibrated model, the simulation results of each scenario were applied to evaluate the impact of climate variability/land use change on hydrological components. For instance, the impact of land use change on the hydrological components can be evaluated solely by replacing the prior land use pattern with the latter land use pattern. The combined impact of climate variability and land use change on the streamflow also can be assessed by using climate and land use data of the latter phase to drive the calibrated model.

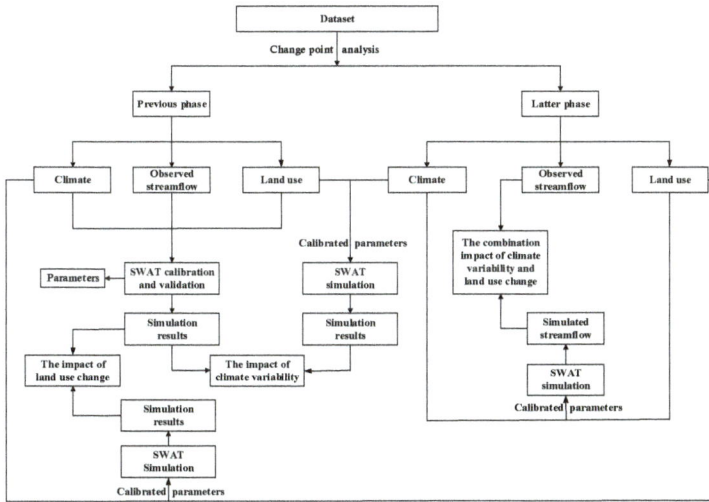

Figure 4. The flowchart for assessing the impacts of climate variability and land use change (refer to Yin et al. [21]). SWAT: Soil Water Assessment Tool.

2.3.3. SWAT Model and Setup

The Soil Water Assessment Tool (SWAT) is a comprehensive semi-distributed ecohydrological model that was developed by the United States Department of Agricultural Research Service to simulate the water cycle, transportation of agricultural chemicals, and sediment. In the SWAT model, Hydrological Response Units (HRUs) represent the minimum unit of calculation. Based on the daily components of the hydrological cycle, a daily water budget in each HRU is calculated [61].The main hydrological components of SWAT include surface flow, lateral flow, groundwater flow, percolation, evapotranspiration and transmission losses [6]. In this study, the surface flow is estimated by the modification of the SCS (Soil Conservation Service) curve number procedure method. The lateral flow is estimated by percolation using a kinematic storage model. The groundwater flow is estimated by the linear-reservoir model method. The Penman–Monteith method is applied to calculate potential evapotranspiration (PET) and then the actual evapotranspiration is estimated by PET. Further descriptions of the SWAT model can be found in theoretical documents [61].

Despite its widespread application, the performance of SWAT model varies significantly. Therefore, some researchers have modified the SWAT model to meet their specific requirements. Panagopoulos et al. [62] applied a greatly refined sub watershed structure based on 12-digit hydrological units to improve simulating performance for large-scale watershed. Qi [63] modified the SWAT with an energy balance module to improve the simulation accuracy for snowmelt. In the SWAT model, two forms of groundwater storage (shallow and deep) are applied, of which only the shallow storage is active to simulate the groundwater flow. Researchers found that SWAT has poor performance in dry seasons or in areas where groundwater system is complicated and groundwater is the main source of streamflow [64–66]. Therefore, using more active groundwater storage to simulate the groundwater flow could result in a more realistic representation of hydrological cycle in these special cases. According to geological characteristics of the Hailiutu River basin, three active types of groundwater storage (upper, medium, and lower groundwater storage) and one inactive type of groundwater storage (i.e., deep groundwater storage) are introduced in the original SWAT model code of the groundwater module to simulate groundwater flow in the SWAT model (hereafter SWAT means modified SWAT).

In the SWAT model, there are several steps to set up the model. First, based on the DEM, the watershed is delineated and divided into several sub-basins. Then, the sub-basins were further subdivided into HRUs of homogeneous characteristics by the overlaying of slope (calculated from the DEM), land use, and soil layers. Lastly, the information on the DEM, land use, and soil was written into the parameters of the SWAT model. In the watershed delineation and subdivision processes, the actual river system is applied to force the generated streams of the SWAT model to follow the actual stream. In this study, the whole study area was divided into 10 sub-basins based on the actual river system and DEM data (Figure 1). Based on the spatial distribution of land use type (1986), soil type, and slope, the sub-basins were further subdivided into 146 HRUs.

2.3.4. SWAT Model Calibration and Evaluation

The sequential uncertainty fitting (SUFI-2) method was applied to calibrate the parameters of SWAT model [67]. The SUFI-2 method combines optimization and uncertainty analysis to adopt a global search procedure. Through Latin hypercube sampling, this method can deal with numerous parameters. In this paper, the Nash–Sutcliffe efficiency (NSE), the determination coefficient (R^2), and percent bias (PBIAS) were employed to evaluate model performance.

For model evaluation, the dataset of the period 1970–1985 was divided into the calibration period (1970–1980) and the validation period (1981–1985). A warm up period from 1970–1974 was chosen to achieve a steady state for modeling. The SWAT model was calibrated by the SUFI-2 method; the calibrated parameters, the best simulated streamflow, and the performance of the SWAT model are presented in Table 2, Figure 5, and Table 3, respectively. For model prediction, it could be found that the 95PPU, i.e., 95% prediction uncertainty, contained 93% and 80% of the observations during the calibration and validation periods, respectively. For monthly streamflow simulation, the model performance can be identified as satisfactory if $0.70 < NSE < 0.80$, $0.75 < R^2 < 0.85$, and $\pm5 < PBIAS < \pm10$ [68]. Based on 95PPU and criterion values of best simulation, the SWAT model can satisfactorily simulate the monthly streamflow in the Hailiutu River basin.

Table 2. The initial and calibrated parameters selected for the SWAT model for the period 1970–1980.

Parameter	Description	Initial Range	Calibrated Range	Best Value
v_alpha_bf_u	Recession factor for upper aquifer	0.3 to 0.8	0.406 to 0.412	0.408
v_gw_delay_u	Delay factor for upper aquifer	1 to 10	1.075 to 1.084	1.081
v_alpha_bf_m	Recession factor for middle aquifer	0.005 to 0.1	0.069 to 0.076	0.07
v_gw_delay_m	Delay factor for middle aquifer	30 to 350	248 to 295	283
v_alpha_bf_l	Recession factor for lower aquifer	0.001 to 0.05	0.033 to 0.036	0.035
v_gw_delay_l	Delay factor for lower aquifer	250 to 500	452 to 458	452
v_rchdp_mld	Percolation factor from upper aquifer	0.92 to 1	0.967 to 0.971	0.969
v_rchrg_ld	Percolation factor from middle aquifer	0.6 to 0.95	0.935 to 0.946	0.937
v_rchrg_d	Percolation factor from lower aquifer	0.1 to 0.4	0.296 to 0.368	0.347
v_gw_revap	Revap	0.02 to 0.2	0.063 to 0.066	0.064
r_CN2	SCS curve number for soil condition II	−0.5 to 0.5	−0.135 to −0.12	−0.129
v_esco	Soil evaporation fraction	0.1 to 0.8	0.55 to 0.58	0.564
a_awc	Available water capacity of soil layer	−0.04 to 0.1	−0.04 to −0.036	−0.038
r_sol_k	Saturated hydraulic conductivity	−0.9 to 0.1	−0.677 to −0.664	−0.676

Note: r means to multiply by original value, a means to add or subtract original value and v means to replace original value.

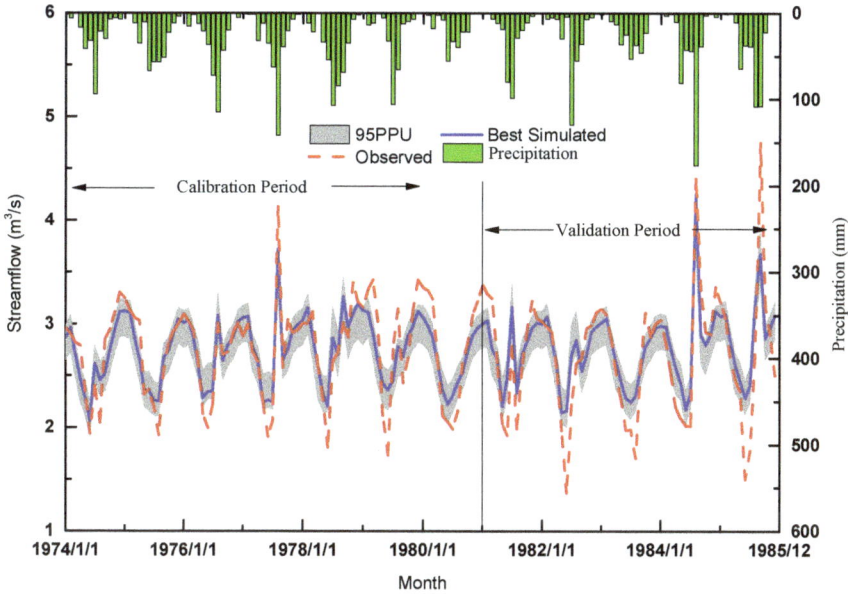

Figure 5. Observed and simulated monthly streamflow of the Hailiutu River basin.

Table 3. The performance of the SWAT model during the calibration and validation periods.

Periods \ Criterion	p-Factor	r-Factor	NSE	R^2	PBIAS
Calibration period	0.93	0.76	0.78	0.80	−0.43
Validation period	0.8	0.59	0.70	0.79	−0.30

3. Results and Discussion

3.1. The Hydrometeorology Analysis for Annual Time Series

The MK method was applied to analyze trend of hydrometeorology which included annual streamflow, annual precipitation, annual win speed, annual maximum temperature, annual minimum temperature, and annual average temperature in the Hailiutu River basin (Table 4). The test statistic Z value of annual streamflow is −3.69 and its absolute value is larger than the 2.58. This means the annual streamflow showed a significant downward trend at the 0.01 level of significance during the period 1970–2014. The test statistic Z values of annual precipitation is 1.41, which is lower than the critical value of 1.64 at the 0.1 level of significance. Therefore, the annual precipitation presented an insignificant upward trend during the study period. The test statistic Z value of the wind speed is −5.57 and its absolute value is larger than 2.58. This means the annual wind speed showed a significant downward trend at the 0.01 level of significance during the period 1970–2014. The test statistic Z values of annual maximum temperature, annual minimum temperature and annual average temperature are positive and larger than 1.96. Therefore, all temperature series show a significantly upward trend at the 0.05 level of significance. The variation of annual minimum temperature (Z = 5.13) is the most pronounced of the three temperature variables.

Table 4. Results of temporal trends in annual streamflow, precipitation, wind speed, maximum temperature, minimum temperature and mean temperature of the Hailiutu River basin.

	Test Statistic Z	Threshold of Different Confidence Levels			Tendency	Significant
		0.01	0.05	0.1		
Annual streamflow	−3.69	2.58	1.96	1.64	Downward	***
Annual precipitation	1.41	2.58	1.96	1.64	Upward	-
Annual wind speed	−5.57	2.58	1.96	1.64	Downward	***
Annual maximum temperature	2.52	2.58	1.96	1.64	Upward	**
Annual minimum temperature	5.13	2.58	1.96	1.64	Upward	***
Annual average temperature	4.50	2.58	1.96	1.64	Upward	***

Note: ** and *** mean through confidence test under 0.05 and 0.01 level of significance respectively.

The STARS method was used to detect the change points for hydrometeorology during the period 1970–2014 and the results are presented in the Figure 6. As shown in Figure 6, 1986 and 2001 were detected as change points for the annual streamflow. The mean annual streamflow from 1970 to 1985 is 2.73 m^3/s, a value that then sharply decreases to 2.03 m^3/s during the period 1986–2000, and finally increases to 2.41 m^3/s during the 21st century. No significant change point was detected for the annual precipitation, the mean of which was 326 mm for the whole study period. The years 1979 and 1998 were detected as change points for the annual wind speed. Before the year 1979, mean annual wind speed was 2.71 m/s, which then decreased to 2.59 m/s and 2.2 m/s during the periods 1980–1998 and 1999–2014, respectively. For three temperature series, two change points (1999 and 2005) were detected in the annual maximum temperature and only one change point (1999) was detected in the annual average temperature. Before the year 1999, mean annual maximum temperature was 15.48 °C, which then rose to 16.97 °C during the period 1999–2004, and finally fell to 15.78 °C from 2005 to 2014. Figure 6e,f show that annual minimum temperature and average temperature increased abruptly after the year 1999. The mean annual minimum temperature increased from 1.76 °C (before 1999) to 3.64 °C (after 1999) while the mean annual average temperature increased from 8.15 °C (before 1999) to 9.46 °C (after 1999).

According to the above change points detection results, the data were divided into three sub-periods (1970–1985(C0), 1986–2000 (C1), and 2001–2014 (C2)) by two change points (1986 and 2001). The land use patterns of 1986, 1995, and 2010 were chosen to represent the land use variation for the three sub-periods. In order to analyze the impacts of climate variability and land use change in the Hailiutu River basin, seven scenarios were developed based on the combination of different sub-periods and land use patterns (Table 5). For example, the scenario S0 reflected the climate condition during period C0 and the land use pattern of the year 1986. The scenarios S1 and S2, based on the climate condition during period C1and C2, respectively, were developed to reflect the climate variability with the same land use pattern of S0. The scenarios S0, S5, and S6 can be considered as the actual conditions during the periods C0, C1, and C2. S0 was applied to the SWAT model as a baseline scenario. The comparison among the simulated results of S0, S1, and S2 revealed the impacts of climate variability on the values of hydrological components. The comparison among the simulated results of S0, S3, and S4 revealed the impacts of land use change on the values of hydrological components. The comparison among the simulated results of S0, S5, and S6 revealed the impacts of the combination of climate variability and land use change on the streamflow.

Table 5. Different scenarios for SWAT modeling.

Land Use Patterns \ Sub-Periods	1970–1985 (C0)	1986–2000 (C1)	2001–2014 (C2)
1986	S0	S1	S2
1995	S3	S5	
2010	S4		S6

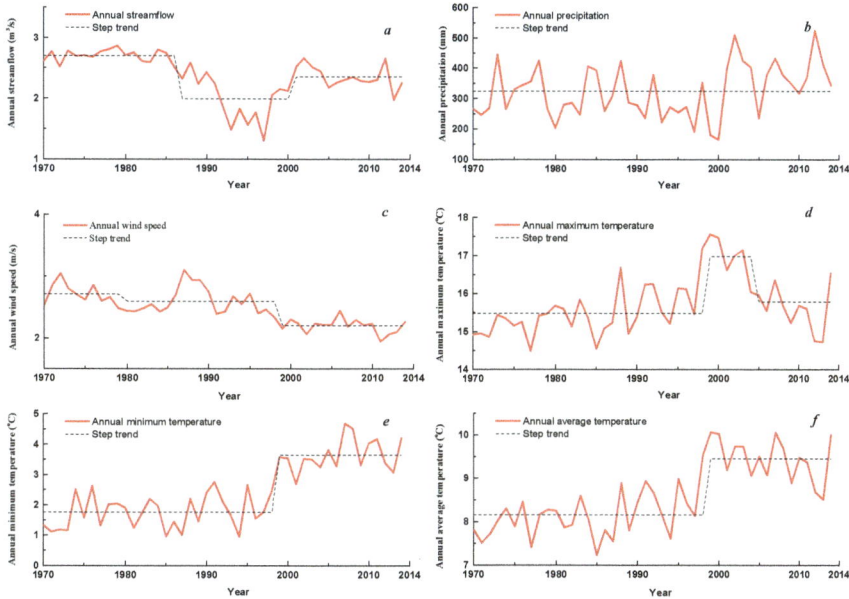

Figure 6. The (**a**) temporal variation of annual streamflow; (**b**) precipitation; (**c**) wind speed; (**d**) maximum temperature; (**e**) minimum temperature; and (**f**) average temperature of the Hailiutu River basin. The dashed lines are the step trends.

3.2. The Hydrometeorology Analysis for Montly Time Series

3.2.1. The Variation of Climate

The variation of mean monthly streamflow and climate was analyzed during three periods (C0, C1, and C2) and the results are shown in Figure 7. The streamflow during periods C1 and C2 decreased as compared with C0, in accordance with Figure 6. Compared with C0, the streamflow during the period C1 decreased in all months, especially in September and October with a streamflow value of 1 m^3/s. The mean monthly streamflow during period C2 has almost no change compared with C1 in rainy season, i.e., May to September. However, it increased in other seasons, especially from December to February, raising to similar values as C0. Compared with C0, the precipitation decreased substantially during the period C1. There was no apparent variation in precipitation between C0 and C2. Compared with C0, the mean wind speed during period C1 showed a decreasing trend from November to May, while it decreased in all months during period C2. In recent decades, the maximum temperature presents upward trend from February to June, especially with respect to the maximum temperature during period C2 which rose about 3 °C as compared with C0. However, in September and December, the maximum temperature during period C2 decreased compared with C1, although it had a higher value compared to C0. The minimum temperature values increased in all months during C1 and C2 compared with C0, and the variation in February was more obvious.

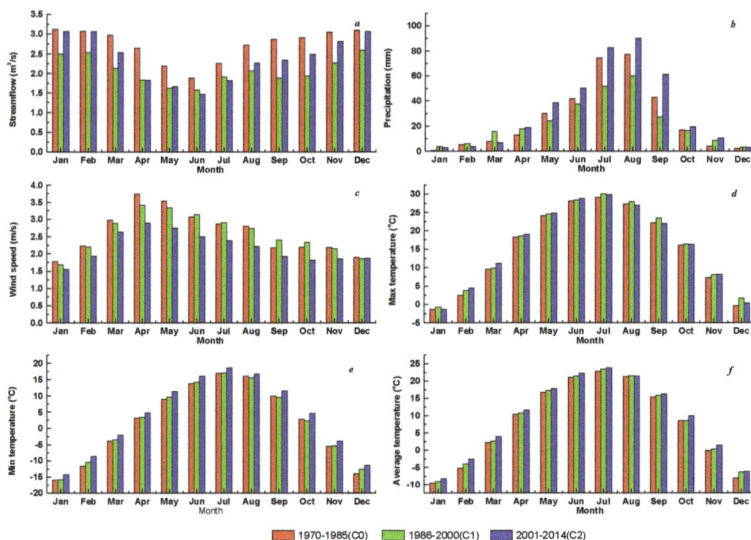

Figure 7. Change variation of (**a**) mean monthly streamflow; (**b**) precipitation; (**c**) wind speed; (**d**) maximum temperature; (**e**) minimum temperature; and (**f**) average temperature of the Hailiutu River basin.

3.2.2. The Variation of Land Use

Figure 8 displays the land use change of the study area from 1986 to 2010. As shown in the Figure 8, the main land use types of the year 1986 were: sandy land (43.31%), shrub and grass land (51.10%), and agricultural land (4.79%) in the Hailiutu River basin. The predominant three land use types vastly changed during the whole period. The area of shrub and grass land presented a prominent upward trend. This area covered 1797 km² in 2010 and increased 533.5 km² or 42.23% compared with 1986. On the contrary, the sandy land area shows a significant downward trend. This area measured 481.6 km² in 2010, a decrease of 589.2 km² or 55.02% compared with 1986. The amount of agricultural land area slightly increased in the 1995, then fell to a similar value as in 1986 in 2010. The area of forest land and urban land account for less than 3% of the Hailiutu River basin, and both of them showed an apparent upward trend over the past 3 decades. Compared with 1986, the increment in forest land area and urban land area was of 39.8 km² and 14.4 km², respectively. At the beginning of the 1980s, an afforestation project named "Three North Forest Shelterbelts" was started by the Chinese government. At the beginning of the 2000s, the local government implemented the policy of "Closing Sandy Land and Forbidding Herding" [32]. The increment of shrub and grass land and the decrement of sandy land can be explained by the implementation of the project and policy.

The transition matrix is commonly applied to analyze the variation in each land use category compared to the other categories. The transition matrix of land use changes in the Hailiutu River basin from 1986 to 2010 is displayed in the Table 6. The columns represent the land use categories in 1986 and the rows represent the categories in 2010. In the third row of Table 6, the values are 55, 0.5, 41.6, 0.1, 0.7, 22.3, and 120.2. That means 55 km² (marked in bold) of agricultural land was maintained as agricultural land in 2010; 0.5 km² of forest land changed to agricultural land in 2010; 41.6 km² of shrub and grass land changed to agricultural land in 2010; 0.1 km² of water changed to agricultural land in 2010; 0.7 km² of urban land changed to agricultural land in 2010; 22.3 km² of sandy land changed to agricultural land in 2010; and the area of agricultural land was 120.2 km² in 2010. In the third column of Table 6, the values are 55, 3.6, 53.4, 0.1, 4.7, 1.7, and 118.5. That means 55 km² of agricultural land was maintained as agricultural land in 2010; 3.6 km² of agricultural land changed to forest land in 2010;

53.4 km^2 of agricultural land changed to shrub and grass land in 2010; 0.1 km^2 of agricultural land changed to water bodies; 4.7 km^2 of agricultural land changed to urban land; 1.7 km^2 of agricultural land changed to sandy land; and the area of agricultural land was 118.75 km^2 in 1986.

In the Hailiutu River basin, 53.4 km^2 of agricultural land converted into shrub and grass land, while 41.6 km^2 of shrub and grass land converted to agricultural land. In total, 644.6 km^2 of sandy land converted to shrub and grass land, while only 95.1 km^2 of shrub and grass land converted to sandy land. This shows that the project and policy of greening the desert achieved remarkable results. Overall, the apparent transformation of land use types is related to sandy land and shrub and grass land, which indicates that the phenomenon of land degradation and desertification still exists in the Hailiutu River basin.

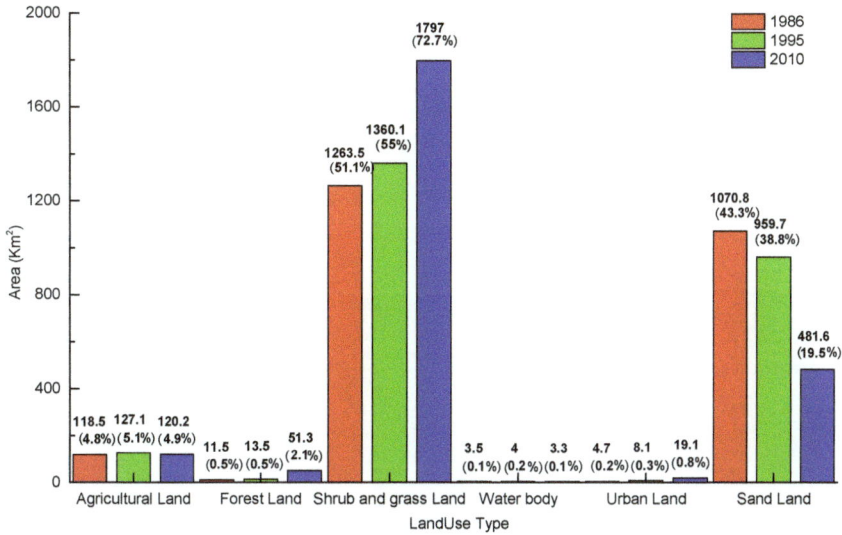

Figure 8. The variation of land used in the three eras. The values in the brackets are the percentages for each type of land use.

Table 6. Transition matrix of land use changes in Hailiutu River basin over the different periods.

		Agricultural Land	Forest Land	Shrub and Grass Land	Water Body	Urban Land	Sand Land	Total
				1986				
	Agricultural land	55	0.5	41.6	0.1	0.7	22.3	120.2
	Forest land	3.6	4	26.2	0.1	0	17.4	51.3
	Shrub and grass land	53.4	6.5	1091.2	0.9	0.4	644.6	1797
2010	Water body	0.1	0	0.2	2.2	0	0.8	3.3
	Urban land	4.7	0	9.2	0	3.6	1.6	19.1
	Sand land	1.7	0.5	95.1	0.2	0	384.1	481.6
	Total	118.5	11.5	1263.5	3.5	4.7	1070.8	

Note: The unchanged area of each land use category is marked in bold.

3.3. The Impacts of Climate Variability on the Value of Hydrological Components

Table 7 lists the mean annual hydrological components simulated by the SWAT model under different climate variability scenarios. The hydrological components include surface flow, lateral flow, baseflow, and evapotranspiration. All simulated hydrological components under S1 decreased as compared with S0. From the perspective of quantity, the largest decrement occurs in the evapotranspiration (−12.3 mm/year), followed by baseflow (−2.13 mm/year). However, from the perspective of proportion, an apparent variation occurs in the lateral flow (−34.5%), followed by surface flow (−25.00%). For S2, all hydrological components increased as compared with S0 except for the baseflow. The simulated surface flow under S2 from the perspective of proportion substantially increased. Also, the simulated evapotranspiration in S2 had a marginal increase as compared with S0 (246.83 vs. 292.36 mm/year).

Under the impact of climate variability, the simulated mean annual streamflow values during the periods C1 and C2 are 29.73 mm/year and 38.76 mm/year, respectively (Table 8). Compared with the period 1970–1985, the variation in mean annual streamflow for the two periods is of −2.15 mm/year and 1.69 mm/year, respectively. That means the climate variability resulted in a decrease of mean annual streamflow by 6.13% and an increase of mean annual streamflow by 4.82% during the periods C1 and C2, respectively.

Table 7. Changes in annual mean hydrological components under different climate variability scenarios (mm/year).

	S0	S1	S2	S1–S0	S2–S0
Surface flow	0.07	0.05	0.52	−0.02(−25.00%)	0.45(642.85%)
Lateral flow	1.37	1.07	1.71	−0.3(−34.50%)	0.34(24.82%)
Baseflow	38.16	36.03	38.08	−2.13(−12.42%)	−0.087(−0.21%)
Evapotranspiration	246.83	234.53	292.36	−12.3(−19.89%)	45.53 (18.45%)

Note: The values in the brackets are the percentage of variation.

Table 8. The observed streamflow and simulated streamflow under different scenarios.

Scenarios	Climate	Land Use	Streamflow (mm/Year)		Observed Change		Simulated Change	
			Observed	Simulated	mm/Year	%	mm/Year	%
S0	1970–1985	1986	34.83	35.09	-	-	-	
S1	1986–2000	1986	-	32.94	-	-	−2.15	−6.13
S2	2001–2014	1986	-	36.78	-	-	1.69	4.82
S3	1970–1985	1995	-	34.86	-	-	−0.23	−0.66
S4	1970–1985	2010	-	34.41	-	-	−0.68	−1.94
S5	1986–2000	1995	25.90	32.61	−2.48	−25.63	−2.45	−7.07
S6	2001–2014	2010	30.79	35.93	−0.84	−11.60	0.84	2.39

3.4. The Impacts of Land Use Change on the Value of Hydrological Components

The mean annual hydrological components of surface flow, lateral flow, baseflow and evapotranspiration under different land use scenarios are listed in the Table 9. Compared with S0, the increments in mean annual surface flow are of 0.154 mm/year and 0.363 mm/year under S1 and S2, respectively. The mean annual lateral flow decreased from 1.365 mm/year under S0 to 1.353 mm/year under S2 (a decrease of 0.88%). Meanwhile, the mean annual baseflow decreased from 38.17 mm/year under S0 to 36.83 mm/year under S2 (a decrease of 3.51%). Mean annual evapotranspiration increased from 246.825 mm/year under S0 to 267.334 mm/year under S2.

The forest land is identified as an "other" land use type when generating HRUS, because the forest land is scattered under S0 and S1 in the Hailiutu River basin. Therefore, the contribution of forest land to hydrological processes under S0 and S1 can be ignored. The water yield is mainly controlled by shrub and grass land and sandy land. It is worth mentioning that although the area of urban land accounts for small proportion of the entire basin, it contributes a large proportion of surface flow.

Under the impact of land use change, the simulated mean annual streamflow values during periods C1 and C2 are 34.86 mm/year and 34.41 mm/year, respectively (Table 8). Compared with the period 1970–1985, the variation in simulated mean annual streamflow during the two periods is of −0.23 mm/year and −0.68 mm/year, respectively. Considering the effect of land use change, the mean annual streamflow decreased by 0.66% and 1.94% during periods C1 and C2, respectively.

Table 9. The contribution of different land use types to the value of hydrological components under different land use scenarios (mm/year).

Scenarios		Agricultural Land	Forest Land	Shrub and Grass Land	Urban Land	Sandy Land	Total
S0	Surface flow	0.035	-	0.02	0.011	0.000	0.066
	Lateral flow	0.184	-	0.606	0.002	0.573	1.365
	Baseflow	1.778	-	19.433	0.005	16.955	38.171
	Evapotranspiration	21.913	-	122.558	0.042	102.312	246.825
S3	Surface flow	0.04	-	0.03	0.15	0.000	0.22
	Lateral flow	0.192	-	0.659	0.002	0.508	1.361
	Baseflow	1.863	-	20.397	0.025	15.283	37.568
	Evapotranspiration	23.688	-	132.637	0.712	90.945	247.982
S4	Surface flow	0.039	0.001	0.121	0.268	0.000	0.429
	Lateral flow	0.171	0.023	0.879	0.003	0.277	1.353
	Baseflow	1.688	0.799	26.522	0.129	7.693	36.831
	Evapotranspiration	21.596	5.982	172.306	1.519	45.931	247.334

3.5. Combined Impacts of Climate Variability and Land Use Change on Streamflow

The simulated streamflow and observed streamflow under combined scenarios (S5 and S6) are listed in the Table 8. Under the combined impact of climate variability and land use change, the simulated mean annual streamflow values during periods C1 and C2 are 32.61 mm/year and 35.93 mm/year, respectively. The simulated mean annual stream flow decreased by 2.45 mm/year (7.07%) during the period C1 and the contribution of these variations to the decrease in observed streamflow was 27.8%. For period C2, the combined climate variability and land use change induced an increment of 0.84 mm/year (2.39%) in the annual streamflow.

3.6. Discussion

In this study, observed precipitation data from 1970 to 2014 were applied to drive SWAT model. The CMADS was introduced to enhance the spatial expressiveness of precipitation data in the Hailiutu River basin during the period 2008–2014. To evaluate the performance of CMADS in the study area, the spatially averaged precipitation of CMADS is compared with observed data as shown in Figure 9. Based on the linear-regression analysis, it is shown that the precipitation from CMADS has a high correlation with the observed data, with a coefficient of determination value of 0.81 (Figure 9a). It can be seen from Figure 9b that the precipitation duration curves obtained from CMADS and observed data are quite close to each other, especially for daily areal precipitation greater than 15 mm. For daily areal precipitation of less than 15 mm, the duration curve from CMADS is slightly lower than that of the observed data.

In order to further display the role of CMADS in the study area, three precipitation datasets, i.e., of observed precipitation (OBS), CMADS precipitation, and a combination of both (OBS+CMADS), were compared in the temporal and spatial scale. Figure 10 shows the comparison of areal precipitation of the whole basin obtained using different precipitation datasets. It was found that CMADS could better reflect the precipitation variation in the monthly scale. As mentioned earlier, CMADS trends to underestimate the precipitation when its value is less than 15 mm. This would result in CMADS providing a lower precipitation value than OBS in the rainy season (May–September). For the other seasons, the areal precipitation obtained by CMADS was significantly lower than observed. In November 2011 in particular, the observed value was 58.2 mm while the value from CMADS was only 10.9 mm. As shown in the Figure 1, only four rain gauges can participate to calculate areal precipitation in the study area. It is difficult to reproduce the actual areal precipitation as there is lack of adequate rain gauges. In the northwest of the study area in particular there is only one rain gauge which is located in the outside of the basin. It can be seen that introducing CMADS better reflects the spatial distribution of precipitation than OBS (Figure 11). Therefore, combined observed precipitation and precipitation of CMADS better reflected the temporal and spatial variation of the study area.

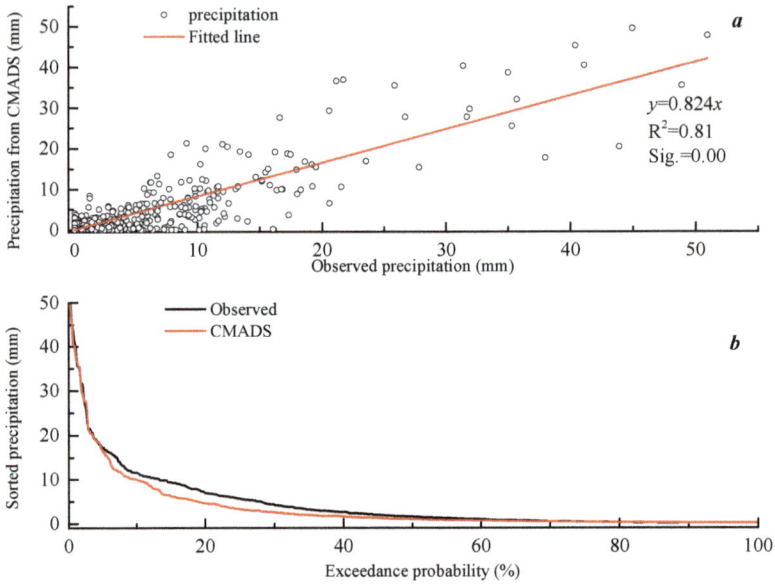

Figure 9. The evaluation of precipitation from CMADS. (**a**) A scattered plot of observed precipitation and CMADS precipitation; (**b**) the duration curve of observed precipitation and CMADS precipitation.

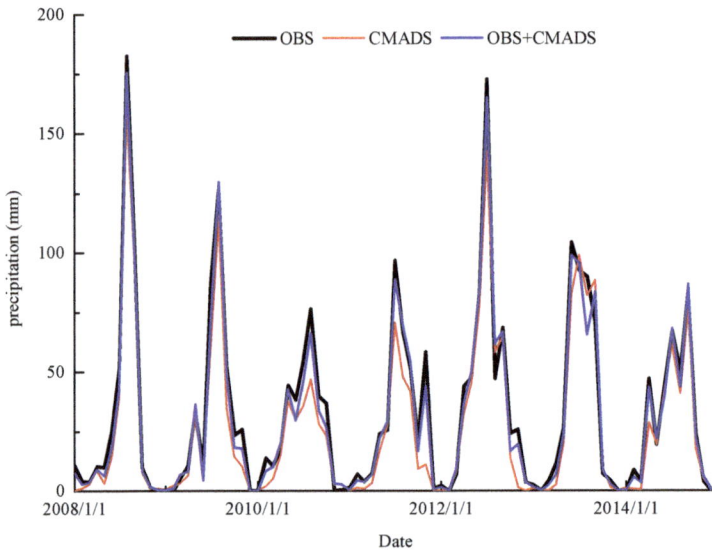

Figure 10. The comparison of monthly precipitation obtained by different precipitation datasets.

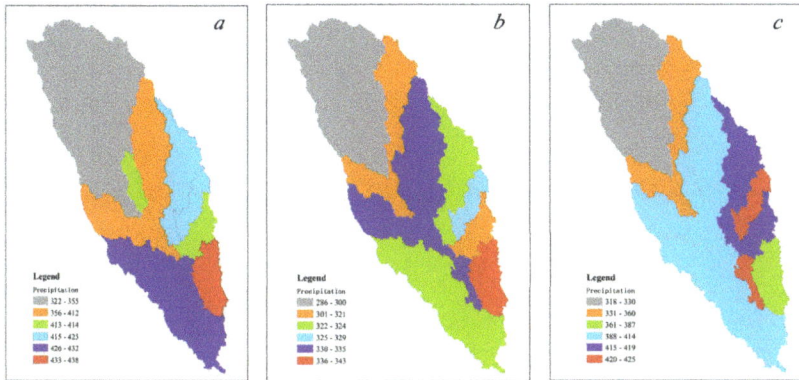

Figure 11. The spatial distribution of areal precipitation obtained by (**a**) OBS; (**b**) CMADS; and (**c**) OBS+CMADS.

The SWAT model was calibrated by observed precipitation (OBS) during the period 2008–2011 and values of calibrated parameters are listed in the Table 10. Then, the calibrated SWAT model was applied to evaluate the performance of CMADS and combination of both datasets (OBS+CMADS). Figure 12 shows that the criterion value ranges for the simulated streamflow of SWAT model with different driving data during the calibration and validation periods. For NSE values, OBS+CMADS generates the highest NSE with an average of 0.72 (calibration period) and 0.42 (validation period). The variation of the NSE value of OBS+CMADS is also less than for the other precipitation datasets. During the validation period, the CMADS has a lightly higher NSE than OBS, with an average of 0.12, while the OBS mean is 0.10. For the R^2 values, OBS+CMADS and OBS have similar performance during the calibration period, while OBS+CMADS has a higher R^2 (0.68) than OBS (0.63) during the validation period. For the PBIAS values, during the calibration and validation periods, OBS and OBS+CMADS overestimate the streamflow, while the CMADS underestimates the streamflow. The simulation results from OBS+CMADS are relatively close to those of the observed data. To summarize, among all the precipitation data inputs for the simulation, the SWAT model with OBS+CMADS datasets was found to be able to give the best simulation results. Meanwhile, the performance of the SWAT model using OBS and OBS+CMADS was found to be better than only using CMADS data.

Table 10. The initial and calibrated parameters selected for the SWAT model during the period 2008–2011.

Parameter	Initial Range	Calibrated Range	Best Value
v_alpha_bf_u	0.3 to 0.8	0.584 to 0.652	0.619
v_gw_delay_u	1 to 10	1.525 to 2.088	1.651
v_alpha_bf_m	0.005 to 0.1	0.032 to 0.053	0.041
v_gw_delay_m	30 to 350	212 to 259	245
v_alpha_bf_l	0.001 to 0.05	0.023 to 0.032	0.027
v_gw_delay_l	250 to 500	405 to 500	492
v_rchdp_mld	0.92 to 1	0.988 to 0.992	0.989
v_rchrg_ld	0.6 to 0.95	0.915 to 0.956	0.937
v_rchrg_d	0.1 to 0.4	0.146 to 0.310	0.212
v_gw_revap	0.02 to 0.2	0.108 to 0.131	0.109
r_CN2	−0.5 to 0.5	−0.237 to −0.191	−0.211
v_esco	0.1 to 0.8	0.530 to 0.691	0.589
a_awc	−0.04 to 0.1	−0.024 to −0.016	−0.022
r_sol_k	−0.9 to 0.1	−0.876 to −0.739	−0.744

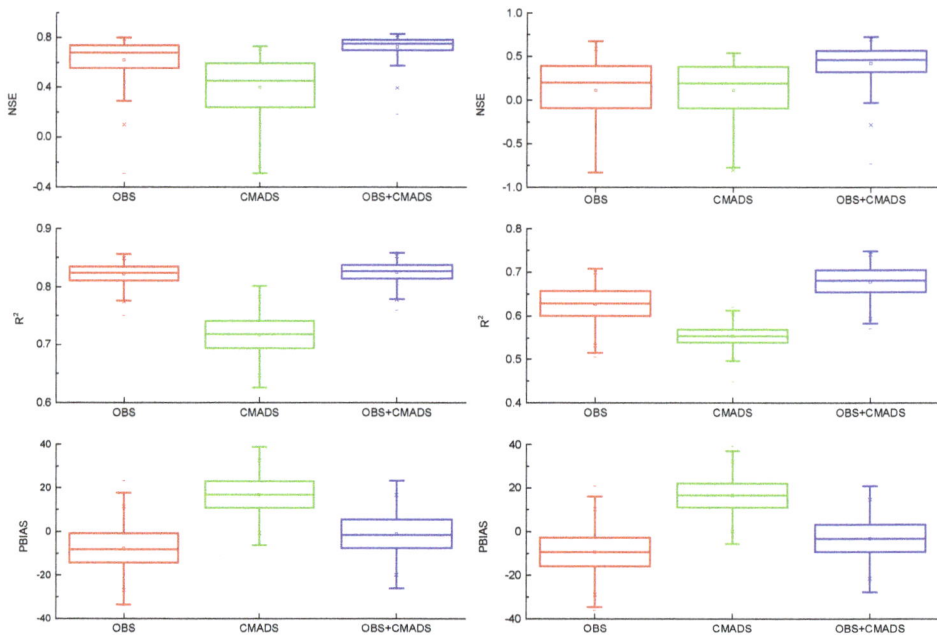

Figure 12. The box plots for the criteria of NSE (**top**), R^2 (**medium**) and PBIAS (**bottom**) during calibration period (**left**) and validation period (**right**). The square symbol and middle line in the box represent the mean value and median value, respectively. Each box ranges from the lower (25th) to upper quartile (75th). PBIAS: percent bias.

Figure 13 shows that the monthly streamflow simulated by using different precipitation datasets as inputs. Based on the Figure 13, OBS+CMADS outperforms the other precipitation datasets. For CMADS, although the CMADS trends to underestimate the precipitation, the streamflow simulated by CMADS is relatively close to observed streamflow. During the period 2011 to 2012, the simulated streamflow was significantly lower than the observed streamflow. This phenomenon could be explained by lower precipitation than OBS. Table 11 lists the p-factor, r-factor, and criterion value of best simulated streamflow for different precipitation input datasets. It was found that OBS+CMADS had the highest p-factor value in both the calibration and validation periods. That means the streamflow simulated with OBS+CMADS could include more observation data than OBS and CMADS. During the validation period, CMADS has the lowest r-factor value and that means the streamflow simulated with CMADS more closely represents the observed streamflow than OBS and OBS+CMADS. For calibration period, OBS and OBS+CMADS have similar performance and OBS+CMADS outperforms than OBS. During the validation period, compared with OBS, the NSE value of OBS+CMADS was significantly improved (from 0.45 to 0.55).

Table 11. The criterion value for the simulated streamflow of SWAT model with different driving data during the calibration and validation periods.

Criterion	Calibration Period			Validation Period		
	OBS	CMADS	OBS+CMADS	OBS	CMADS	OBS+CMADS
p-factor	0.9	0.75	0.92	0.86	0.58	0.89
r-factor	0.87	0.51	0.94	1.43	1.14	1.34
NSE	0.80	0.73	0.83	0.45	0.46	0.55
R^2	0.83	0.74	0.83	0.63	0.50	0.65
PBIAS	1.02	2.66	−1.72	−0.24	2.88	−2.78

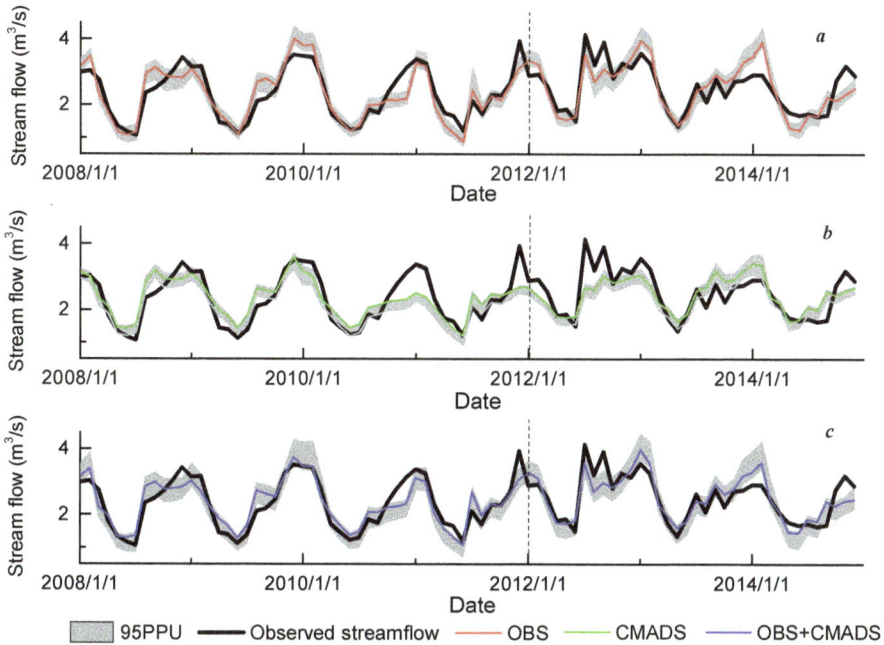

Figure 13. Simulation results of SWAT with (**a**) OBS; (**b**) CMADS; (**c**) OBS+CMADS in the Hailiutu River basin during the period 2008–2014.

To set up a hydrological model, for example the SWAT model, we usually use the measured precipitation data from rainfall gauge in the past. In some basins, there are more rain gauges [13,69,70]. For example, the Huangfuchuan basin is located near the Hailiutu River basin and covers an area of 3246 km^2. This basin has a similar area to the Hailiutu River basin (2473 Km^2) and it has 12 rain gauges [13]. However, the Hailiutu River basin only has four rain gauges, of which only three rainfall stations are located in the watershed. Now, CMADS is available for the SWAT model users. It is free and very convenient to use. From the above results, it can be found that introducing the precipitation from CMADS can improve the simulation accuracy. This improvement can be attributed to introducing the CMADS to better reflect the precipitation spatial distribution. Thus, CMADS will be a very useful data source for hydrological modeling in China in the future.

In this study, as mentioned earlier, the whole study period was divided into three sub-periods. The calibrated model was expected to reflect the real hydrological process of the first sub-period. Nevertheless, there were some uncertain factors, e.g., the coarse spatial resolution of precipitation and the deficiency of model that affected the accuracy of the simulation. Those factors led to a difference between modeled and measured streamflow. However, streamflow/baseflow comparisons were made between different scenarios. Comparisons were not made with the measured streamflow/baseflow. The difference in the results could be attributed to the applied scenario changes only. Hessel et al. [71] also stated that all scenarios for one watershed are subjected to the same uncertain factors.

Based on the results of scenarios, the impact of climate variability and land use change on hydrological components can be compared. It is also observed that the impact of climate variability on the variation of streamflow is more profound compared to land use change. Compared with S0, in addition to land use change, the mean annual precipitation during C1 decreased. The variation of land use change and precipitation induced a decrease of 2.45 mm/year in the mean annual streamflow during the period C1 (Table 8). An increment of 0.84 mm/year in mean annual streamflow was observed during period C2 because of the effect of the climate variability. During the periods of C1 and C2, the observed mean annual streamflow recorded relatively lesser flows than the simulated streamflow, and the observed streamflow decreased while the simulated streamflow increased in the period of C2. After the 1980s, many diversion dams were constructed in the Hailiutu River basin for local water supply [72]. In recent years, water demand has increased rapidly with the development of energy and agriculture. This has resulted in more exploitation of groundwater resources and streamflow reduction [40,73]. Therefore, the impact of these factors on streamflow reduction in Hailiutu River basin should be considered in the hydrological modeling in the future.

4. Conclusions

In this paper, the statistics analysis methods and hydrological modeling were applied to assess the impacts of climate variability and land use change on the hydrological components in the Hailiutu River basin of the Chinese Erdos Plateau, and the combined impacts of climate and land use change on streamflow reduction were also analyzed. The results are listed as follows:

The CMADS was introduced to improve the spatial expressiveness of precipitation data in the study area over the period 2008–2014. There is a good correlation between CMADS and the observed data during this period. The SWAT model was calibrated by observed data during the period 2008–2011. Compared with only using observed precipitation data, the NSE value increased from 0.80 to 0.83 when using combined data during 2008–2011 and increased from 0.45 to 0.55 when using combined data during 2012–2014. The R^2 value increased from 0.63 to 0.65 when using combined data during 2012–2014. This proves that the performance of SWAT model when using combined observed and CMADS precipitation values was significantly improved. It suggests that combining the CMADS with traditional hydrological measurements might be very helpful for improving hydrological modeling.

The Mann–Kendall test of annual hydrometeorology series indicated that streamflow and wind speed showed a significantly downward trend during the period 1970–2014. A slight upward trend was detected for precipitation while a significantly upward trend was detected for temperature. From the analysis of the STARS test, the years 1986 and 2001 were detected as change points for annual streamflow. The years 1979 and 1998 were detected as change points for annual wind speed. The year 1999 was detected as a change point for all temperature series, and 2005 as the change point for annual maximum temperature series. Based on the change points of annual streamflow, the whole study period was divided into three sub-periods, i.e., 1970–1985, 1986–2000, and 2001–2014.

Compared with 1970–1985, the mean monthly streamflow during the period 1986–2000 decreased, especially in September and October. The streamflow increased during the period 2001–2014, with almost no variation compared with the period 1986–2000 from May to July. The precipitation during the period 1986–2000 substantially decreased as compared with the period 1970–1985. During the period 2001–2014, the precipitation rose to similar values as the period 1970–1985. From the comparison among monthly temperature in three periods, the annual maximum temperature showed an upward trend during January–June, and the annual minimum and average temperature presented upward trends in all the months. In the study area, the main variations of land use were increased area of shrub and grass land (from 51.10% of 1986 to 72.68% of 2010) and decreased area of sandy land (from 43.31% of 1986 to 19.48% of 2010). These variations can be explained by the implementation of the project of "Three North Forest Shelterbelts" and policy of "Closing Sandy Land and Forbidding Herding".

As compared to the period 1970–1985, the climate variability led to a significant decrease in streamflow during the period 1986–2000, and induced a moderate increase in streamflow during the period 2000–2014. The land use changes caused increases in surface runoff, evapotranspiration, and decreases in baseflow, lateral flow, and streamflow. These changes are mainly attributed to the increase of shrub and grass land area and a decrease of sandy land area. In general, the impacts of climate variability on the value of hydrological components were more profound than land use change in the Hailiutu River basin. Therefore, the importance of increasing adaptation to climate variability should be considered when planning and managing water resources.

Compared to the period 1970–1985, the observed mean annual streamflow decreased during the periods 1986–2000 and 2001–2014. During the period 1986–2000, the climate variability and land use change resulted in a decrease of mean annual streamflow by 6.13% and 0.66%, respectively. The combined climate variability and land use change induced a decrease in mean annual streamflow by 7.07%, and was responsible for 27.87% of the decrease in observed mean annual streamflow. For the period 2001–2014, land use change induced a 1.94% decrease in mean annual streamflow. There was a positive impact of climate variability on the streamflow, with an increment of 4.82%. Under the impact of combined climate variability and land use change, it was observed that mean annual streamflow increased by 2.39%. The discrepancy between observed and simulated streamflow during the period 2001–2014 implies that the impact of other factors such as local water supply and exploitation of groundwater on the streamflow in Hailiutu River basin may not be ignored in hydrological modeling.

Author Contributions: Y.G. and D.Z. were primarily accountable for data collection and design and coordination of the study. G.S. were responsible for data analysis and writing of the paper. B.Y. and J.Z. were responsible for results presentation.

Funding: This research was funded by the National Natural Science Foundation of China (NSFC51579067 and NSFC51209064), the Qing Lan Project, and the Postgraduate Research & Practice Innovation Program of Jiangsu Province (KYCX17_0416).

Acknowledgments: Great thanks to Victoria Adjei from Hohai University (China) for the editing of the English language and style of this manuscript. The authors also thank the anonymous reviewers for their very valuable comments.

Conflicts of Interest: The authors declare no conflict of interest.

References

1. Oki, T.; Kanae, S. Global hydrological cycles and world water resources. *Science* **2006**, *313*, 1068–1072. [CrossRef] [PubMed]
2. Stocker, T. *Climate Change 2013: The Physical Science Basis: Working Group I Contribution to the Fifth Assessment Report of the Intergovernmental Panel on Climate Change*; Cambridge University Press: Cambridge, UK, 2014.
3. Chen, Y.; Xu, Y.; Yin, Y. Impacts of land use change scenarios on storm-runoff generation in Xitiaoxi basin, China. *Quat. Int.* **2009**, *208*, 121–128. [CrossRef]
4. Sajikumar, N.; Remya, R.S. Impact of land cover and land use change on runoff characteristics. *J. Environ. Manag.* **2015**, *161*, 460–468. [CrossRef] [PubMed]
5. Tadesse, W.; Whitaker, S.; Crosson, W.; Wilson, C. Assessing the Impact of Land-Use Land-Cover Change on Stream Water and Sediment Yields at a Watershed Level Using SWAT. *Open J. Mod. Hydrol.* **2015**, *5*, 68. [CrossRef]
6. Woldesenbet, T.A.; Elagib, N.A.; Ribbe, L.; Heinrich, J. Hydrological responses to land use/cover changes in the source region of the Upper Blue Nile Basin, Ethiopia. *Sci. Total Environ.* **2017**, *575*, 724–741. [CrossRef] [PubMed]
7. Li, Z.; Liu, W.; Zhang, X.; Zheng, F. Impacts of land use change and climate variability on hydrology in an agricultural catchment on the Loess Plateau of China. *J. Hydrol.* **2009**, *377*, 35–42. [CrossRef]
8. Liu, Z.; Yao, Z.; Huang, H.; Wu, S.; Liu, G. Land Use and Climate Changes and Their Impacts on Runoff in the Yarlung Zangbo River Basin, China. *Land Degrad. Dev.* **2014**, *25*, 203–215. [CrossRef]

9. López-Moreno, J.I.; Vicente-Serrano, S.M.; Moran-Tejeda, E.; Zabalza, J.; Lorenzo-Lacruz, J.; García-Ruiz, J.M. Impact of climate evolution and land use changes on water yield in the Ebro basin. *Hydrol. Earth Syst. Sci.* **2011**, *15*, 311–322. [CrossRef]

10. Li, L.J.; Zhang, L.; Wang, H.; Wang, J.; Yang, J.W.; Jiang, D.J.; Li, J.Y.; Qin, D.Y. Assessing the impact of climate variability and human activities on streamflow from the Wuding River basin in China. *Hydrol. Process.* **2007**, *21*, 3485–3491. [CrossRef]

11. Du, J.; Rui, H.; Zuo, T.; Li, Q.; Zheng, D.; Chen, A.; Xu, Y.; Xu, C.-Y. Hydrological Simulation by SWAT Model with Fixed and Varied Parameterization Approaches Under Land Use Change. *Water Resour. Manag.* **2013**, *27*, 2823–2838. [CrossRef]

12. Brown, A.E.; Zhang, L.; McMahon, T.A.; Western, A.W.; Vertessy, R.A. A review of paired catchment studies for determining changes in water yield resulting from alterations in vegetation. *J. Hydrol.* **2005**, *310*, 28–61. [CrossRef]

13. Zuo, D.; Xu, Z.; Yao, W.; Jin, S.; Xiao, P.; Ran, D. Assessing the effects of changes in land use and climate on runoff and sediment yields from a watershed in the Loess Plateau of China. *Sci. Total Environ.* **2016**, *544*, 238–250. [CrossRef] [PubMed]

14. Park, J.-Y.; Park, M.-J.; Joh, H.-K.; Shin, H.-J.; Kwon, H.-J.; Srinivasan, R.; Kim, S.-J. Assessment of MIROC3.2 HiRes Climate and CLUE-s Land Use Change Impacts on Watershed Hydrology Using SWAT. *Trans. ASABE* **2011**, *54*, 1713–1724. [CrossRef]

15. Khoi, D.N.; Suetsugi, T. The responses of hydrological processes and sediment yield to land-use and climate change in the Be River Catchment, Vietnam. *Hydrol. Process.* **2014**, *28*, 640–652. [CrossRef]

16. Tu, J. Combined impact of climate and land use changes on streamflow and water quality in eastern Massachusetts, USA. *J. Hydrol.* **2009**, *379*, 268–283. [CrossRef]

17. Ashagre, B.B. SWAT to Identify Watershed Management Options: Anjeni Watershed, Blue Nile Basin, Ethiopia. Master's Thesis, Cornell University, New York, NY, USA, 2009.

18. Baker, T.J.; Miller, S.N. Using the Soil and Water Assessment Tool (SWAT) to assess land use impact on water resources in an East African watershed. *J. Hydrol.* **2013**, *486*, 100–111. [CrossRef]

19. Bieger, K.; Hörmann, G.; Fohrer, N. The impact of land use change in the Xiangxi Catchment (China) on water balance and sediment transport. *Reg. Environ. Chang.* **2015**, *15*, 485–498. [CrossRef]

20. Rahman, K.; da Silva, A.G.; Tejeda, E.M.; Gobiet, A.; Beniston, M.; Lehmann, A. An independent and combined effect analysis of land use and climate change in the upper Rhone River watershed, Switzerland. *Appl. Geogr.* **2015**, *63*, 264–272. [CrossRef]

21. Yin, Z.; Feng, Q.; Yang, L.; Wen, X.; Si, J.; Zou, S. Long Term Quantification of Climate and Land Cover Change Impacts on Streamflow in an Alpine River Catchment, Northwestern China. *Sustainability* **2017**, *9*, 1278. [CrossRef]

22. Kim, J.; Choi, J.; Choi, C.; Park, S. Impacts of changes in climate and land use/land cover under IPCC RCP scenarios on streamflow in the Hoeya River Basin, Korea. *Sci. Total Environ.* **2013**, *452–453*, 181–195. [CrossRef] [PubMed]

23. Zhang, A.; Zhang, C.; Fu, G.; Wang, B.; Bao, Z.; Zheng, H. Assessments of Impacts of Climate Change and Human Activities on Runoff with SWAT for the Huifa River Basin, Northeast China. *Water Resour. Manag.* **2012**, *26*, 2199–2217. [CrossRef]

24. Chen, Z.; Chen, Y.; Li, B. Quantifying the effects of climate variability and human activities on runoff for Kaidu River Basin in arid region of northwest China. *Theor. Appl. Climatol.* **2013**, *111*, 537–545. [CrossRef]

25. Miao, C.; Ashouri, H.; Hsu, K.-L.; Sorooshian, S.; Duan, Q. Evaluation of the PERSIANN-CDR Daily Rainfall Estimates in Capturing the Behavior of Extreme Precipitation Events over China. *J. Hydrometeorol.* **2015**, *16*, 1387–1396. [CrossRef]

26. Stisen, S.; Sandholt, I. Evaluation of remote-sensing-based rainfall products through predictive capability in hydrological runoff modelling. *Hydrol. Process.* **2010**, *24*, 879–891. [CrossRef]

27. Su, F.; Hong, Y.; Lettenmaier, D.P. Evaluation of TRMM Multisatellite Precipitation Analysis (TMPA) and Its Utility in Hydrologic Prediction in the La Plata Basin. *J. Hydrometeorol.* **2008**, *9*, 622–640. [CrossRef]

28. Liu, J.; Shanguan, D.; Liu, S.; Ding, Y. Evaluation and Hydrological Simulation of CMADS and CFSR Reanalysis Datasets in the Qinghai-Tibet Plateau. *Water* **2018**, *10*, 513. [CrossRef]

29. Vu, T.T.; Li, L.; Jun, K.S. Evaluation of Multi-Satellite Precipitation Products for Streamflow Simulations: A Case Study for the Han River Basin in the Korean Peninsula, East Asia. *Water* **2018**, *10*, 642. [CrossRef]

30. Zhao, F.; Wu, Y.; Qiu, L.; Sun, Y.; Sun, L.; Li, Q.; Niu, J.; Wang, G. Parameter Uncertainty Analysis of the SWAT Model in a Mountain-Loess Transitional Watershed on the Chinese Loess Plateau. *Water* **2018**, *10*, 690. [CrossRef]

31. Cao, Y.; Zhang, J.; Yang, M.; Lei, X.; Guo, B.; Yang, L.; Zeng, Z.; Qu, J. Application of SWAT Model with CMADS Data to Estimate Hydrological Elements and Parameter Uncertainty Based on SUFI-2 Algorithm in the Lijiang River Basin, China. *Water* **2018**, *10*, 742. [CrossRef]

32. Zhou, Y.; Wenninger, J.; Yang, Z.; Yin, L.; Huang, J.; Hou, L.; Wang, X.; Zhang, D.; Uhlenbrook, S. Groundwater–surface water interactions, vegetation dependencies and implications for water resources management in the semi-arid Hailiutu River catchment, China—A synthesis. *Hydrol. Earth Syst. Sci.* **2013**, *17*, 2435–2447. [CrossRef]

33. Gao, T.; Wang, H. Trends in precipitation extremes over the Yellow River basin in North China: Changing properties and causes. *Hydrol. Process.* **2017**, *31*, 2412–2428. [CrossRef]

34. Jiang, P.; Yu, Z.; Gautam, M.R.; Acharya, K. The Spatiotemporal Characteristics of Extreme Precipitation Events in the Western United States. *Water Resour. Manag.* **2016**, *30*, 4807–4821. [CrossRef]

35. Li, H.; Zhang, Q.; Singh, V.P.; Shi, P.; Sun, P. Hydrological effects of cropland and climatic changes in arid and semi-arid river basins: A case study from the Yellow River basin, China. *J. Hydrol.* **2017**, *549*, 547–557. [CrossRef]

36. Li, B.; Li, C.; Liu, J.; Zhang, Q.; Duan, L. Decreased Streamflow in the Yellow River Basin, China: Climate Change or Human-Induced? *Water* **2017**, *9*, 116. [CrossRef]

37. Wang, S.; Yan, Y.; Yan, M.; Zhao, X. Quantitative estimation of the impact of precipitation and human activities on runoff change of the Huangfuchuan River Basin. *J. Geogr. Sci.* **2012**, *22*, 906–918. [CrossRef]

38. Hou, L.; Zhou, Y.; Bao, H.; Wenninger, J. Simulation of maize (*Zea mays* L.) water use with the HYDRUS-1D model in the semi-arid Hailiutu River catchment, Northwest China. *Hydrol. Sci. J.* **2017**, *62*, 93–103. [CrossRef]

39. Yang, Z.; Zhou, Y.; Wenninger, J.; Uhlenbrook, S. A multi-method approach to quantify groundwater/surface water-interactions in the semi-arid Hailiutu River basin, northwest China. *Hydrogeol. J.* **2014**, *22*, 527–541. [CrossRef]

40. Yang, Z.; Zhou, Y.; Wenninger, J.; Uhlenbrook, S.; Wang, X.; Wan, L. Groundwater and surface-water interactions and impacts of human activities in the Hailiutu catchment, northwest China. *Hydrogeol. J.* **2017**, *25*, 1341–1355. [CrossRef]

41. Xu, Z.X.; Takeuchi, K.; Ishidaira, H.; Zhang, X.W. Sustainability Analysis for Yellow River Water Resources Using the System Dynamics Approach. *Water Resour. Manag.* **2002**, *16*, 239–261. [CrossRef]

42. Huo, Z.; Feng, S.; Kang, S.; Li, W.; Chen, S. Effect of climate changes and water-related human activities on annual stream flows of the Shiyang river basin in arid north-west China. *Hydrol. Process.* **2008**, *22*, 3155–3167. [CrossRef]

43. Zhang, X.; Zhang, L.; Zhao, J.; Rustomji, P.; Hairsine, P. Responses of streamflow to changes in climate and land use/cover in the Loess Plateau, China. *Water Resour. Res.* **2008**, *44*. [CrossRef]

44. Zhou, Y.; Yang, Z.; Zhang, D.; Jin, X.; Zhang, J. Inter-catchment comparison of flow regime between the Hailiutu and Huangfuchuan rivers in the semi-arid Erdos Plateau, Northwest China. *Hydrol. Sci. J.* **2015**, *60*, 688–705. [CrossRef]

45. Chen, Y.; Guan, Y.; Shao, G.; Zhang, D. Investigating Trends in Streamflow and Precipitation in Huangfuchuan Basin with Wavelet Analysis and the Mann-Kendall Test. *Water* **2016**, *8*, 77. [CrossRef]

46. Huang, J.; Hou, G.; Li, H.; Yin, L.; Lu, H.; Zhang, J.; Dong, J. Estimating subdaily evapotranspiration rates using the corrected diurnal water-table fluctuations in a shallow groundwater table area. In Proceedings of the 2011 International Symposium on Water Resource and Environmental Protection, Xi'an, China, 20–22 May 2011; Volume 4, pp. 3093–3099.

47. Wang, X.-S.; Zhou, Y. Shift of annual water balance in the Budyko space for a catchment with groundwater dependent evapotranspiration. *Hydrol. Earth Syst. Sci. Discuss.* **2015**, *12*, 11613–11650. [CrossRef]

48. Yin, L.; Zhou, Y.; Huang, J.; Wenninger, J.; Hou, G.; Zhang, E.; Wang, X.; Dong, J.; Zhang, J.; Uhlenbrook, S. Dynamics of willow tree (*Salix matsudana*) water use and its response to environmental factors in the semi-arid Hailiutu River catchment, Northwest China. *Environ. Earth Sci.* **2014**, *71*, 4997–5006. [CrossRef]

49. Yang, Z.; Zhou, Y.; Wenninger, J.; Uhlenbrook, S.; Wan, L. Simulation of Groundwater-Surface Water Interactions under Different Land Use Scenarios in the Bulang Catchment, Northwest China. *Water* **2015**, *7*, 5959–5985. [CrossRef]

50. Meng, X.Y. *China Meteorological Assimilation Driving Datasets for the SWAT Model*, version 1.0.; Cold and Arid Regions Science Data Center at Lanzhou: Lanzhou, China, 2016. [CrossRef]

51. Meng, X.; Wang, H. Significance of the China Meteorological Assimilation Driving Datasets for the SWAT Model (CMADS) of East Asia. *Water* **2017**, *9*, 765. [CrossRef]

52. Meng, X.; Long, A.; Wu, Y.; Yin, G.; Wang, H.; Ji, X. Simulation and spatiotemporal pattern of air temperature and precipitation in Eastern Central Asia using RegCM. *Sci. Rep.* **2018**, *8*, 3639. [CrossRef] [PubMed]

53. Meng, X.; Wang, H.; Lei, X.; Cai, S.; Wu, H.; Ji, X.; Wang, J. Hydrological modeling in the Manas River Basin using soil and water assessment tool driven by CMADS. *Teh. Vjesn.* **2017**, *24*, 525–534.

54. Meng, X.; Wang, H.; Wu, Y.; Long, A.; Wang, J.; Shi, C.; Ji, X. Investigating spatiotemporal changes of the land-surface processes in Xinjiang using high-resolution CLM3. 5 and CLDAS: Soil temperature. *Sci. Rep.* **2017**, *7*, 13286. [CrossRef] [PubMed]

55. Meng, X.; Wang, H.; Cai, S.; Zhang, X.; Leng, G.; Lei, X.; Shi, C.; Liu, S.; Shang, Y. The China Meteorological Assimilation Driving Datasets for the SWAT Model (CMADS) Application in China: A Case Study in Heihe River Basin. *Preprints* **2017**. [CrossRef]

56. Yue, S.; Wang, C. The Mann-Kendall Test Modified by Effective Sample Size to Detect Trend in Serially Correlated Hydrological Series. *Water Resour. Manag.* **2004**, *18*, 201–218. [CrossRef]

57. Sang, Y.-F.; Wang, Z.; Liu, C. Comparison of the MK test and EMD method for trend identification in hydrological time series. *J. Hydrol.* **2014**, *510*, 293–298. [CrossRef]

58. Belete, M.D. The impact of sedimentation and climate variability on the hydrological status of Lake Hawassa, South Ethiopia. Ph.D. Thesis, Universitäts- und Landesbibliothek Bonn, Bonn, Germany, 2013.

59. Rodionov, S.N. A sequential algorithm for testing climate regime shifts. *Geophys. Res. Lett.* **2004**, *31*, L09204. [CrossRef]

60. Gong, X.; Liu, Z.; Gao, H.; Yang, H.; Wang, H. Regime Shifts in the Huanghe Freshwater Discharge During the Past Fifty Years. In Proceedings of the 2009 3rd International Conference on Bioinformatics and Biomedical Engineering, Beijing, China, 11–13 June 2009; pp. 1–3.

61. Neitsch, S.L.; Arnold, J.G.; Kiniry, J.R.; Williams, J.R. *Soil and Water Assessment Tool Theoretical Documentation*, version 2009; Texas Water Resources Institute: College Station, TX, USA, 2011.

62. Panagopoulos, Y.; Gassman, P.W.; Jha, M.K.; Kling, C.L.; Campbell, T.; Srinivasan, R.; White, M.; Arnold, J.G. A refined regional modeling approach for the Corn Belt–Experiences and recommendations for large-scale integrated modeling. *J. Hydrol.* **2015**, *524*, 348–366. [CrossRef]

63. Qi, J.; Li, S.; Jamieson, R.; Hebb, D.; Xing, Z.; Meng, F.-R. Modifying SWAT with an energy balance module to simulate snowmelt for maritime regions. *Environ. Model. Softw.* **2017**, *93*, 146–160. [CrossRef]

64. Shirmohammadi, A.; Chu, T.W. Evaluation of the swat model's hydrology component in the piedmont physiographic region of Maryland. *Trans. ASAE* **2004**, *47*, 1057–1073.

65. Amatya, D.M.; Jha, M.; Edwards, A.E.; Williams, T.M.; Hitchcock, D.R. SWAT-based streamflow and embayment modeling of karst-affected Chapel branch watershed, South Carolina. *Trans. ASABE* **2011**, *54*, 1311–1323. [CrossRef]

66. Cheng, L.; Xu, Z.X.; Luo, R. SWAT application in arid and semi-arid region: A case study in the Kuye River Basin. *Geogr. Res.* **2009**, *28*, 65–73.

67. Abbaspour, K.C.; Johnson, C.A.; van Genuchten, M.T. Estimating Uncertain Flow and Transport Parameters Using a Sequential Uncertainty Fitting Procedure. *Vadose Zone J.* **2004**, *3*, 1340. [CrossRef]

68. Moriasi, D.N.; Gitau, M.W.; Pai, N.; Daggupati, P. Hydrologic and water quality models: Performance measures and evaluation criteria. *Trans ASABE* **2015**, *58*, 1763–1785.

69. Zhang, S.; Fan, W.; Li, Y.; Yi, Y. The influence of changes in land use and landscape patterns on soil erosion in a watershed. *Sci. Total Environ.* **2017**, *574*, 34–45. [CrossRef] [PubMed]

70. Tian, Y.; Wang, S.; Bai, X.; Luo, G.; Xu, Y. Trade-offs among ecosystem services in a typical Karst watershed, SW China. *Sci. Total Environ.* **2016**, *566–567*, 1297–1308. [CrossRef] [PubMed]

71. Hessel, R.; Messing, I.; Liding, C.; Ritsema, C.; Stolte, J. Soil erosion simulations of land use scenarios for a small Loess Plateau catchment. *CATENA* **2003**, *54*, 289–302. [CrossRef]

72. Yang, Z.; Zhou, Y.; Wenninger, J.; Uhlenbrook, S. The causes of flow regime shifts in the semi-arid Hailiutu River, Northwest China. *Hydrol. Earth Syst. Sci.* **2011**, *8*, 5999–6030. [CrossRef]

73. Yin, L.; Hu, G.; Huang, J.; Wen, D.; Dong, J.; Wang, X.; Li, H. Groundwater-recharge estimation in the Ordos Plateau, China: Comparison of methods. *Hydrogeol. J.* **2011**, *19*, 1563–1575. [CrossRef]

Article

Simulated Runoff and Sediment Yield Responses to Land-Use Change Using the SWAT Model in Northeast China

Limin Zhang [1,2], Xianyong Meng [3,4,*], Hao Wang [2,*] and Mingxiang Yang [2,*]

1   College of Architecture and Civil Engineering, Beijing University of Technology (BJUT), Beijing 100124, China; zhanglm152@163.com
2   State Key Laboratory of Simulation and Regulation of Water Cycle in River Basin & China Institute of Water Resources and Hydropower Research (IWHR), Beijing 100038, China
3   College of Resources and Environmental Science, China Agricultural University (CAU), Beijing 100094, China
4   Department of Civil Engineering, The University of Hong Kong (HKU), Pokfulam 999077, Hong Kong, China
*   Correspondence: xymeng@cau.edu.cn or xymeng@hku.hk (X.M.); wanghao@iwhr.com (H.W.); yangmx@iwhr.com (M.Y.)

Received: 31 January 2019; Accepted: 27 April 2019; Published: 1 May 2019

Abstract: Land-use change is one key factor influencing the hydrological process. In this study, the Hun River Basin (HRB) (7919 km^2), a typical alpine region with only four gauge meteorological stations, was selected as the study area. The China Meteorological Assimilation Driving Datasets for the SWAT model (CMADS), widely adopted in East Asia, was used with the Soil and Water Assessment Tool (SWAT) model to simulate runoff and sediment yield responses to land-use change and to examine the accuracy of CMADS in the HRB. The criteria values for daily/monthly runoff and monthly sediment yield simulations were satisfactory; however, the validation of daily sediment yield was poor. Forestland decreased sediment yield throughout the year, increased water percolation, and reduced runoff during the wet season, while it decreased water percolation and increased runoff during the dry season. The responses of grassland and forestland to runoff and sediment yield were similar, but the former was weaker than the latter in terms of soil and water conservation. Cropland (urban land) generally increased (increased) runoff and increased (decreased) sediment yield; however, a higher sediment yield could occur in urban land than that in cropland when precipitation was light.

Keywords: runoff; sediment yield; land-use change; SWAT; CMADS

1. Introduction

Water quantity and quality have become serious concerns in many countries around the world, negatively affecting the human survival environment and sustainable economic and social development [1–3]. Land-use and land-cover change (LUCC), one direct means for human activities (e.g., afforestation, deforestation, urbanisation, and reservoir construction) to alter the landscape, plays an important role in influencing the hydrologic response of watersheds in multiple ways [4,5]. LUCC is considered directly linked to changes in the hydrologic components in a watershed, such as evapotranspiration, surface runoff, groundwater, and stream flow, and hence can change flood frequency and severity, base flow, and annual mean discharge [6–9]. Following a conversion in land use, soil erosion will be altered in turn affecting the sediment yield [10–12]. Many previous studies throughout the world have shown that the changes in runoff and sediment yield are influenced by LUCC at different spatial and temporal scales [6–16]. For example, Buendia et al. [13] found that increased forest area was the major driver of reduced stream flows and peak magnitudes as well as prevention of increases in sediment load in an upland Mediterranean catchment. Lin et al. [14] applied the Soil and Water Assessment Tool (SWAT) model to investigate runoff responses at annual, monthly, and daily time scales using the same meteorological input but two different land-use scenarios (1985 and 2006, with reduced forest and increased cropland and urbanized area); the results showed a varying change in runoff among the three time scales and three catchments in the Jinjiang catchment. Mueller et al. [15] investigated the impact of LUCC on sediment yield for a meso-scale catchment in the Southern Pyrenees and showed that LUCC had greater impacts on sediment yield than climate change. Most investigations have focused on the whole changes in runoff/sediment yield influenced by different historical land-use scenarios at annual scale; however, few studies regarding the impacts of individual land-use types on runoff and sediment at annual and seasonal scales have been conducted.

To evaluate the hydrologic and sediment effects of environmental change, several methods have been developed and are mainly divided into three categories: paired catchment approach, time series analysis (statistical method), and hydrological modelling [7,17]. The paired catchment approach is often considered to be the best method for the compensation of climatic variability in small experimental catchment during the observation period. However, it is typically carried out in small catchments because it is difficult to identify two medium- or large-sized catchments areas that are similar in area, shape, geology, climate, and vegetation [18,19]. The time series analysis method is a simple mathematical statistical model that cannot show the physical mechanism of a hydrological response. At the same time, owing to the complexity of factors affecting hydrological changes in the basin, it is easy to misjudge if other factors are excluded from the time series changes in the characteristic variables [20]. Among these approaches, the hydrological modelling method, particularly a physically based distributed hydrological model, is the most suitable for use in scenario studies and in assisting in understanding the mechanisms of influence resulting from land-use impacts [6,14]. Compared with other hydrological models, the SWAT model [21] is among the appropriate models for simulating runoff and sediment yield under different land-use scenarios and it has been widely used throughout the world (see SWAT Literature Database [22]). The SWAT model was used in this study because of its availability, convenience, friendly interface, and simple operation; it can be downloaded from the official website [23].

Spatial variabilities in precipitation and temperature are key factors in impacting the water balance of large watersheds, particularly in mountainous areas [21,24,25]. However, sediment yield is closely related to precipitation and runoff. Therefore, meteorological input data with high precision and quality aids in reducing the uncertainty in the model input and improves model simulation reliability. In this study, the SWAT model was driven by the China Meteorological Assimilation Driving Datasets for the SWAT model (CMADS) established by Dr. Xianyong Meng from China Agricultural University (CAU) [26]. The atmospheric re-analysis dataset CMADS was established by using Space-Time Multiscale Analysis System (STMAS) assimilation and big data technology that corrects the European Centre for Medium Range Weather Forecasts (ECMWF) with nearly 40,000 regional automatic stations in China such that it can better reflect the real surface meteorological conditions [27,28]. CMADS has formats suitable for SWAT, consisting of max/min temperature, precipitation, relative humidity, solar radiation, and wind speed and can drive the SWAT model without changing the data type. This is very convenient for users. Many researchers have shown the relatively higher accuracy of CMADS, which will provide important basic data for hydrological research in East Asia, especially in areas with scarce data [29–36]. For example, Liu et al. [29] applied CMADS, Climate Forecast System Reanalysis (CFSR), and observed meteorological data to the SWAT model and found that CMADS + SWAT obtained a better result than that of the other modes on the Qinghai–Tibet Plateau, northwest China, where meteorological stations are scarce. Compared with observed meteorological data (OBS) or CMADS, the SWAT model driven by OBS + CMADS provided the best simulation result for runoff in the Hailiutu River Basin [31]. Thom et al. [32] used rainfall data from the Tropical Rainfall Measuring Mission (TRMM) and CMADS to simulate runoff of the Han River Basin on the Korean Peninsula and the accuracy was acceptable.

The Hun River basin (HRB) has a large population and is among the areas subject to serious water shortages in China. During recent years, the Chinese government has made a major strategic decision of 'Revitalization of Old Industrial Bases in Northeast China'. Meanwhile, the Liaoning province government has proposed construction of the Shenfu Connection Area (Shenfu New Town) with an area of 605 km^2 to connect Shenyang and Fushun. Moreover, many studies have demonstrated that the government policies can have a considerable impact on land use, such as "Household Responsibility System" and "Grain to Green Program" [37,38]. In addition, in cold areas, the climate is characterized by a longer freezing period and winter every year. Therefore, it is essential to analyze the runoff and sediment yield under individual land-use types and it needs to be considered at a local scale.

The objectives of this study were to (1) calibrate and validate the SWAT model in terms of runoff and sediment yield based on CMADS in the HRB, (2) evaluate the impacts of individual land-use types on annual and seasonal runoff and sediment yield at a sub-basin scale; and (3) analyze the differences in runoff and sediment yield under individual land-use types during the wet and dry seasons.

2. Materials and Methods

2.1. Study Area

The total area of the HRB is 11,481 km^2 and the average annual runoff is approximately 3.05 billion m^3. The study region (7919 km^2) is upstream of Shenyang station (41°29′–42°16′ N, 123°22′–125°17′ E) within the HRB in Liaoning Province (Figure 1). The Dahuofang Reservoir is 18 km from Fushun and 68 km from Shenyang. The upstream of the reservoir is mountainous with elevations ranging from 400 to 800 m. The downstream of the reservoir is hills with the elevations of 100–200 m. The study region is affected by temperate semi-humid and semi-arid continental winds with an annual mean temperature of 4–8 °C. The area has a long and cold winter and a hot and rainy summer. Precipitation is mainly occurs during the period of June–September or May–August, accounting for 70%–80% of the total annual precipitation overall [39]. Grassland, cropland, and forestland are the dominant land-use types in the catchment.

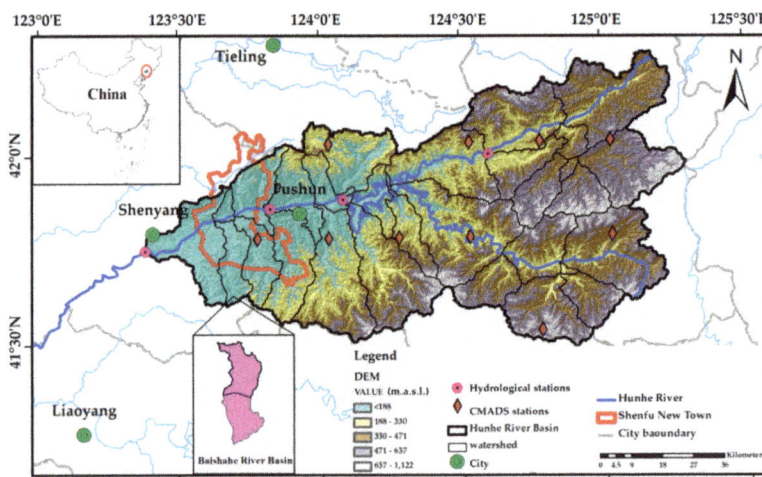

Figure 1. The range and location of the study region in China.

The Baishahe River is an important tributary in the Hun River downstream, and the main river flows to Shenfu New Town (SNT) [40]. The Baishahe River Basin (BRB) covers an area of 162.41 km^2; the overlap area of BRB and SNT is 85.85 km^2. Analyzing the changes in land use in SNT from 1997 to 2011, we found that the conversion of land use was mainly among forestland, cropland, grassland, and residential land, such as returning cropland to forestland and/or residential land. SNT is a developing city, the land-use conversion of which could be a representative of urban expansion in China.

2.2. Data Collection

The spatial data used in this study included digital elevation model (DEM), soil type, and land-use data. The 90 m × 90 m resolution DEM was derived from CGIAR—Consortium for Spatial Information (CGIAR-CSI) [41]. The soil data set was provided by the Environmental and Ecological Science Data Center for West China, National Natural Science Foundation of China [42]. The Chinese soil dataset (v1.1) was based on the Harmonized World Soil Database (HWSD) at a scale of 1:1,000,000 and a resolution of 1 km [43]. The percentages of the total area of the seven soil types in the basin are as follows: Calcaric Cambisols: 0.13%, Haplic Phaeozems: 17.59%, Cumulic Anthrosols: 4.67%, Haplic Luvisols: 56.61%, Gleyic Luvisols: 18.30%, Urban: 1.67%, and water bodies: 0.96%. Soil water characteristics for each soil type were obtained by using the SPAW (Soil-Plant-Air-Water) computer model developed by the U.S. Department of Agriculture (USDA) [44]. The land-cover map was drawn using the global land cover for the year 2000 (GLC 2000) project coordinated by the Global Vegetation Monitoring Unit of the European Commission Joint Research Centre with a resolution of 1 km, as loaded from http://westdc.westgis.ac.cn [45]. The land-use types were classified into five categories according to the SWAT code. The land-use types and their distribution in the HRB are shown in Figure 2. The percentages of the total area of different land-use types in the basin are as follows: Agricultural Land-Generic (AGRL): 16.59%, Forest-Mixed (FRST): 79.17%, Pasture (PAST): 3.76%, Residential-High Density (URHD): 0.39%, and Water (WATR): 0.09%. The BRB, whose sub-basin number is 22 in this study, was selected to investigate the impacts of individual land-use types on runoff and sediment yield. Its main land use types are forest land (42.0 km^2, 31.1%) and cropland (120.41 km^2, 68.9%) with the land-use data of GLC2000 and its main soil types are Cumulic Anthrosols (18.82%), Haplic Luvisols (47.99%), and Haplic Phaeozems (33.54%).

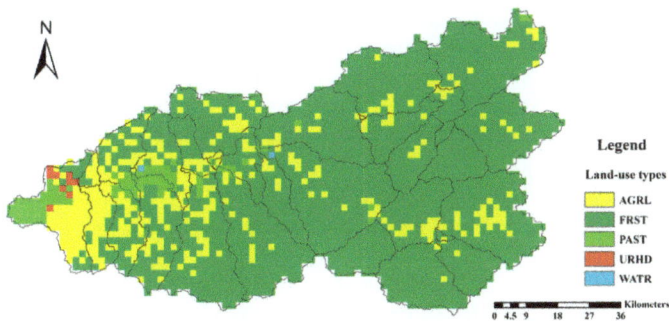

Figure 2. The land use types and their distribution in the HRB. Remarks: FRST—Forestland; PAST—Grassland; AGRL—Cropland; URHD—Urban land; and WATR—Water.

Meteorological data including daily precipitation, maximum and minimum air temperature, relative humidity, wind speed, and solar radiation were obtained from the CMADS official website (http://www.cmads.org/) [26]. The spatial resolution of CMADS1.1 is 1/4^0 and the period is 2008–2016. The observed runoff (unit: m^3/s) and sediment data (unit: kg/s) at Beikouqian, Dahuofang Reservoir (no sediment yield data), Fushun, and Shenyang gauge stations between 2008 and 2014 were derived from the Annual Hydrological Report of the People's Republic of China published by the Ministry of Water Resources of the People's Republic of China.

2.3. Methods

2.3.1. Hydrological Modelling

The SWAT model developed by Jeffrey G. Arnold is a continuous-time, semi-distributed, and process-based river basin model [46]. It is a physically based model operating on a daily time step and was initially developed to predict the impact of management on water, sediment, and agriculture chemical yields [47]. SWAT is a deterministic model, which means that each successive model run using the same inputs will result in the same outputs. Therefore, this type of model is suitable to separate and evaluate the effects of a single variable, and it is easy to compare the relative effects from one to another [48]. In applying the SWAT model, the study area is first divided into sub-basins based on the DEM, and then these are further divided into one or more hydrologic response units (HRUs). Each HRU, consisting of unique land use, management, topographical, and soil characteristics, is an independent unit of the SWAT model and does not interact with the other HRUs. Simulation of the watershed hydrology is separated into a land phase, which controls the amount of water, sediment, nutrient, and pesticide loadings to the main channel in each sub-basin, and an in-stream or routing phase, which is the movement of water, sediment, etc., through the channel network of the watershed to the outlet [46]. The SWAT model is based on the water balance in the soil profile, and the processes simulated include infiltration, surface runoff, evapotranspiration (ET), lateral flow, and percolation. All the water balance processes in the same HRU are regarded as being consistent.

The study area was divided into 34 sub-basins and 317 HRUs. The Soil Conservation Service (SCS) curve method [49] was used to calculate the surface runoff generated by each independent HRU resulting from the daily input precipitation; the confluence was eventually obtained in the exit section. The Penman–Monteith method [50] was used to calculate the potential evapotranspiration in the study area and the variable storage routing method developed by Williams (1969) [51] was selected to calculate the water evolution in the main channel. Soil erosion was computed using the Modified Universal Soil Loss Equation (MUSLE) [52]. Sediment loadings from each HRU in the landscape were then summed at the sub-basin level and the resulting loads were routed by runoff and distributed to the watershed outlet [53]. Sediment transport in the channel network was simultaneously controlled by deposition and degradation processes, which depended on the sediment loads originating from upland areas and the channel transport capacity [53]. A more detailed description of the mechanisms and structure of the SWAT model can be found in the theoretical documentation [23] and in Arnold et al. [21].

2.3.2. SWAT Calibration, Validation and Performance Evaluation

The SWAT Calibration Uncertainty Program (SWAT-CUP) was selected for the sensitivity analysis, calibration, and validation. SUFI-2 (Sequential Uncertainty Fitting) is a comprehensive optimization and gradient search method able to simultaneously calibrate multiple parameters and with a global search function. It also considers the uncertainty of the input data, model parameters, and model structure [54,55].

The calibration and validation of runoff and sediment were completed at monthly and daily scales at the gauge stations in the HRB. The calibration period was 2008–2012 encompassing average, wet, and dry years and the validation period was 2013–2014. Model performance, defined as the goodness of fit between observed and predicted runoff/sediment yield, was quantitatively evaluated using the Nash–Sutcliffe efficiency coefficient (NSE), the coefficient of determination (R^2), and percent bias (PBIAS) [56]. R^2 ranges from 0 to 1; and typically values greater than 0.5 are considered acceptable [57,58]. When NSE and PBIAS are consistent with the given criteria proposed by Moriasi et al. [56] (NSE > 0.5; $\pm 15\% \leq$ PBIAS $< \pm 25\%$ for runoff; and $\pm 30\% \leq$ PBIAS $< \pm 55\%$ for sediment yield) for a monthly time step, the model is considered to be applicable to the catchment and the impact analyses.

2.3.3. Data Simulation

In this study, data simulation was based on four land-use and three precipitation scenarios. We assumed four land-use scenarios (forestland, grassland, cropland, and urban land, respectively) in extreme cases throughout the whole SNT region and as a part of land use within the HRB in each model run. Calibrated parameters were used and the simulation period was set as 2008 to 2012 every time using the same climatic forcing data because of the serious debris flow that occurred during 2013. The land-use condition of GLC2000 in SNT was set as the baseline scenario. Then, we assessed the contribution of changes in individual land-use types on runoff and sediment in the selected sub-basin (BRB) at annual and seasonal scales. Under Scenario 1 (S1), SNT was covered only by forestland, referring to the conversion of cropland to forestland in the BRB. Under Scenario 2 (S2), SNT was covered only by grassland, transforming forestland and cropland into grassland in the BRB. Under Scenario 3 (S3), SNT was covered only by cropland, assuming a change from forestland into cropland in the BRB. Scenario 4 (S4), SNT was covered only by urban land, converting forestland and cropland into urban land in the BRB. The information of land-use scenarios were shown Figure 3 and Table 1. Under the different scenarios, the land-use change in the BRB (as shown in Figure 3) only occurred in the overlap area of the BRB and SNT, and the land use of other areas and the meteorological inputs remained constant; therefore, it was easy to distinguish the contribution of changes in individual land-use types on runoff and sediment yield.

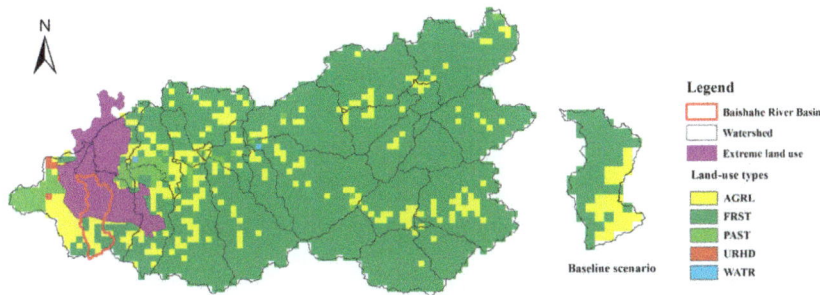

Figure 3. Land-use change scenarios in the BRB in the HRB. "Extreme land use" means the land-use types are FRST, PAST, AGRL, and URHD under S1, S2, S3, and S4, respectively. Remarks: FRST—Forestland; PAST—Grassland; AGRL—Cropland; URHD—Urban land; and WATR—Water.

Table 1. Land-use types under different scenarios in the BRB.

Scenarios	LUCC Description in BRB	Forestland		Grassland		Cropland		Urban	
		Area (km²)	%	Area (km²)	%	Area (km²)	%	Area (km²)	%
Baseline	GLC2000 (forestland and cropland)	42	25.86	—	—	120.41	74.14	—	—
S1	Cropland changed into forestland	118.1	72.72	—	—	44.31	27.28	—	—
S2	Forestland and cropland changed into grassland	32.25	19.86	85.85	52.86	44.31	27.28	—	—
S3	Forestland changed into cropland	32.25	19.86	—	—	130.16	80.14	—	—
S4	Forestland and cropland changed into urban land	32.25	19.86	—	—	44.31	27.28	85.85	52.86

To further research the responses of runoff and sediment yield to land-use change, three main time scales (annual and wet and dry seasons) were adopted in this study. The wet season was further divided into wet season 1 (W1) and wet season 2 (W2) and the dry season was divided into dry season 1 (D1) and dry season 2 (D2). W1 was from May to September, and correspondingly, D1 was October to April. W2 ranged from June to September while D2 was October to May.

3. Results

3.1. Model Calibration and Validation

The calibrated parameters with their initial values and descriptions are presented in Table 2. The model results for calibration and validation based on performance criteria in the present study are listed in Table 3. The runoff and sediment yield for the calibration and validation at the gauge stations between simulated and observed data in the HRB are shown in Figures 4–7 and Table 3.

Table 2. Parameters used to calibrate runoff and sediment yield in the HRB.

Parameter	Definition	Min. Value	Max. Value
	Parameters used to calibrate runoff		
ALPHA_BNK	Baseflow alpha factor for bank storage	0	1
CN2	SCS runoff curve number for moisture condition II	35	98
SOL_K	Soil conductivity	0	2000
CH_K2	Effective hydraulic conductivity in the main channel	0	150
SOL_BD	Soil bulk density	1.1	1.9
GWQMN	Threshold depth of water in the shallow aquifer required for return flow to occur	0	500
GW_REVAP	Groundwater "revap" coefficient	0.02	0.2
ESCO	Soil evaporation compensation factor	0.01	1
CH_N2	Manning's n value for main channel	0	0.3
GW_DELAY	Groundwater delay time	0	500
SOL_AWC	Soil available water storage capacity	0	1
ALPHA_BF	Baseflow alpha factor	0	1
SFTMP	Snowfall temperature	−5	5
SMTMP	Snow melt base temperature	−5	5
	Parameters used to calibrate sediment yield		
SPCON	Linear parameters for calculating the channel sediment rooting	0.001	0.01
SPEXP	Exponent parameter for calculating the channel sediment routing	1	1.5
CH_COV	Channel cover factor	0	1
CH_EROD	Channel erodibility factor	0	0.6
PRF	Peak rate adjustment factor for sediment routing in the main channel	0	2
USLE_K	USLE equation soil erodibility (K) factor	0	0.65

Table 3. Criteria for examining model calibration and validation accuracy.

Time Scales	C/V	Criteria	Runoff				Sediment		
			Beikouqian	Dahuofang Reservoir	Fushun	Shenyang	Beikouqian	Fushun	Shenyang
Monthly	Calibration (2008–2012)	NSE	0.92	0.78	0.82	0.83	0.56	0.93	0.95
		R²	0.94	0.8	0.82	0.84	0.57	0.96	0.96
		PBIAS (%)	19.56	21.62	13.02	15.12	27.54	7.92	23.84
	Validation (2013–2014)	NSE	0.89	0.67	0.7	0.75	0.002	0.41	0.84
		R²	0.94	0.69	0.74	0.7	0.99	0.96	0.9
		PBIAS (%)	16.75	18.11	22.81	23.72	97.4	68.9	41.1
	Validation (2014)	NSE					0.62	0.65	0.72
		R²					0.68	0.71	0.8
		PBIAS (%)					11.86	22.16	32.9
Daily	Calibration (2008–2012)	NSE	0.79	0.53	0.64	0.65	0.29	0.4	0.53
		R²	0.8	0.55	0.65	0.67	0.5	0.47	0.57
		PBIAS (%)	11.14	24.52	25.3	24.84	60.68	25.77	16.98
	Validation (2013–2014)	NSE	0.85	0.56	0.57	0.67	0.12	0.36	0.44
		R²	0.86	0.63	0.66	0.69	0.94	0.85	0.67
		PBIAS (%)	1.44	20.72	11.6	11.5	91.96	41.52	5.56

Note: "C/V" means "Calibration/Validation".

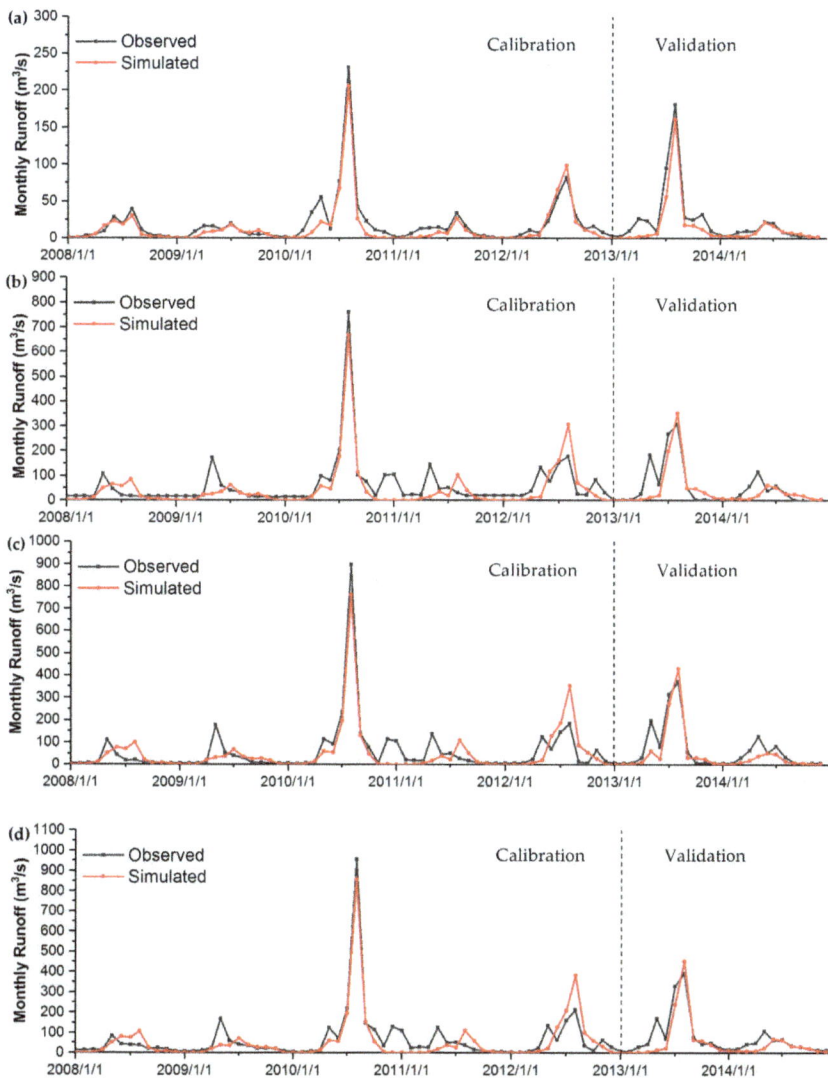

Figure 4. Comparison of observed and simulated monthly runoff during the calibration (2008–2012) and validation periods (2013–2014): (**a**) Beikouqian, (**b**) Dahuofang Reservoir, (**c**) Fushun, and (**d**) Shenyang.

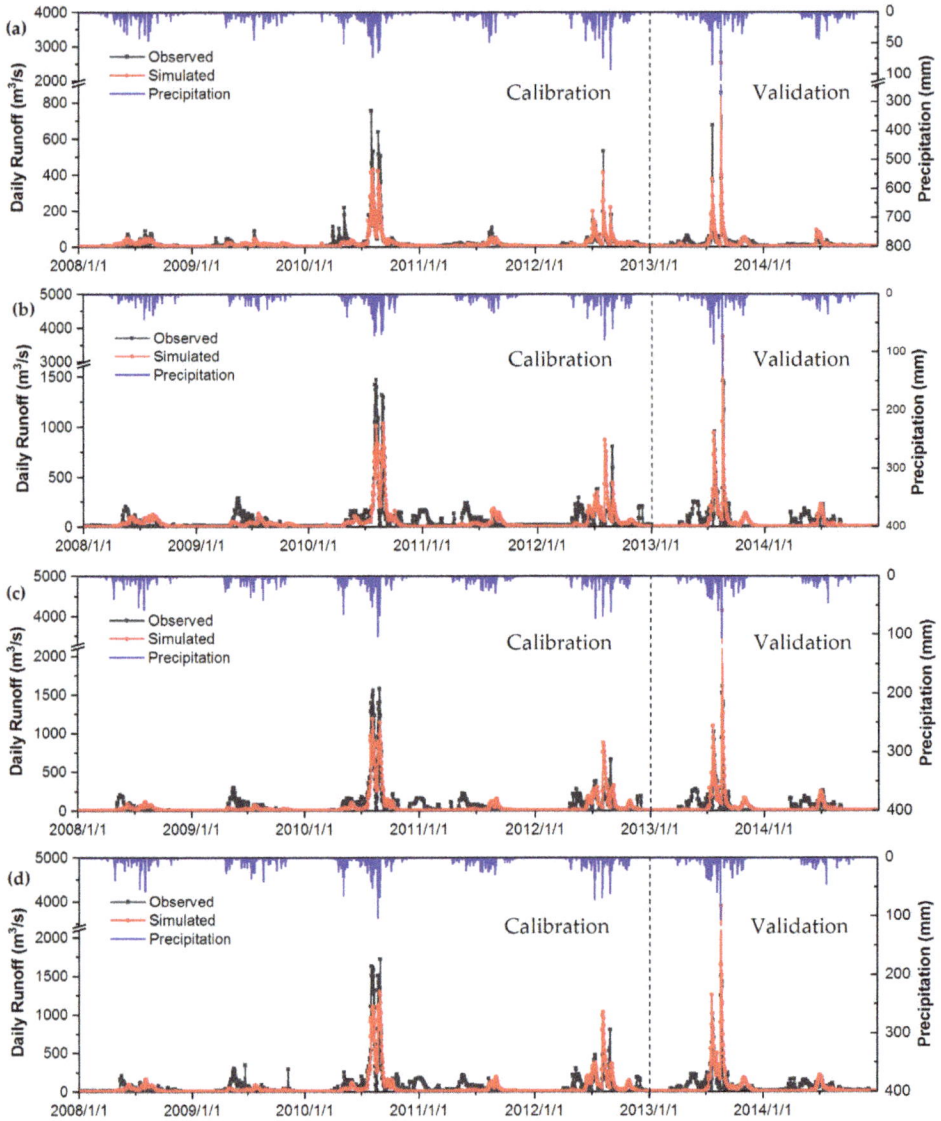

Figure 5. Comparison of observed and simulated daily runoff during the calibration (2008–2012) and validation periods (2013–2014): (**a**) Beikouqian, (**b**) Dahuofang Reservoir, (**c**) Fushun, and (**d**) Shenyang.

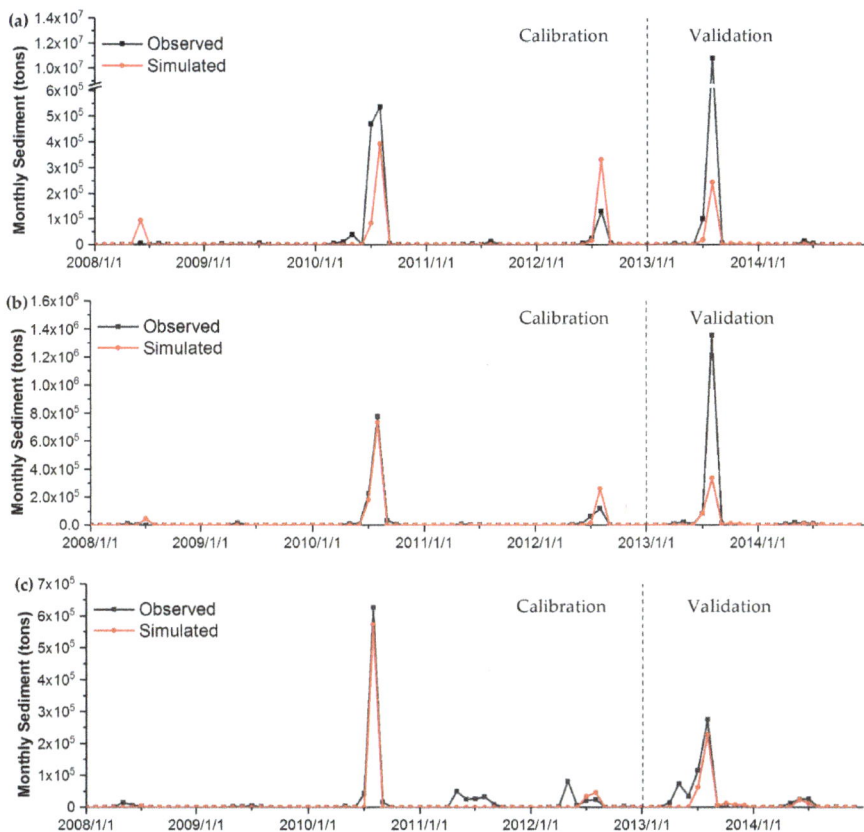

Figure 6. Comparison of observed and simulated monthly sediment yield during the calibration (2008–2012) and validation periods (2013–2014): (**a**) Beikouqian, (**b**) Fushun, and (**c**) Shenyang.

For monthly and daily runoff, all the NSE and R^2 values were greater than 0.5 and the PBIAS values were within a range of ±25%, suggesting satisfactory simulation was achieved according to Moriasi et al. [52].

For monthly sediment, the calibration result was satisfied; however, the validation result was poor for the Beikouqian and Fushun stations, particularly for the upstream Beikouqian station (the NSE and R2 were 0.002 and 0.99, respectively). The total observed sediment yield during August 2013 was approximately 1.08 million tons; however, the simulated was only 0.24 million tons, 0.84 million tons lower than the observed (Figure 6). The maximum daily average sediment transport rate was 94 tons per second, i.e., 8,121,600 tons per day, observed at Beikouqian station on 17 August 2013. According to the 2013 Bulletin of Flood and Drought Disaster in China [59] published on the internet by the Ministry of Water Resources of the People's Republic of China, a heavy rainstorm occurred north of Liaoning Province from 15–17 August 2013, and the maximum point of rainfall accumulation (456 mm) was observed at Hongtoushan Station, Fushun, Liaoning Province. The upstream Hun River was subject to an extraordinary 50-year flood leading to a large-scale debris-flow-dominated geohydrological process [60]. This debris flow disaster might have resulted in a considerable increase in sediment and resulting in the poor evaluation during the validation period (2013–2014). At the same time, a similar reason or more complex factors led to a disappointing result (most NSE values were lower than 0.5) for daily sediment during the calibration and validation periods, particularly for the upstream and middle stream hydrological stations.

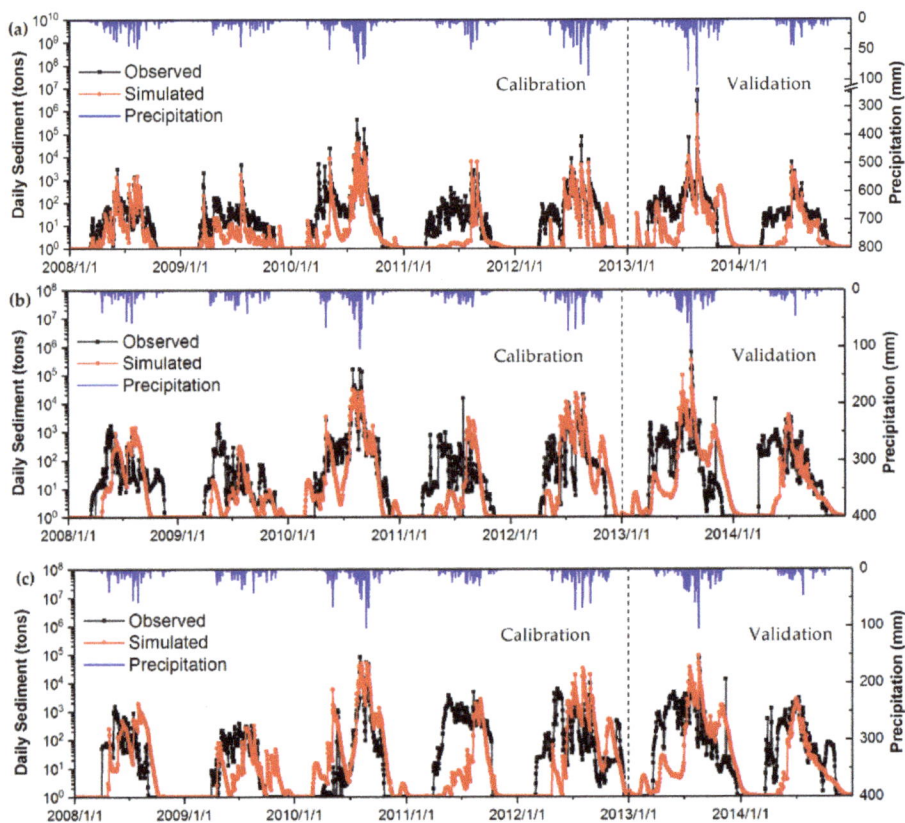

Figure 7. Comparison of observed and simulated daily sediment yield during the calibration (2008–2012) and validation periods (2013–2014): (**a**) Beikouqian, (**b**) Fushun, and (**c**) Shenyang.

In practice, the model performance was good during both calibration and validation for runoff, particularly for the upstream region of the HRB. The cause of the difference in consistency of the simulation hydrographs among the gauge stations may be related with the human activities (e.g., irrigation, reservoir operation, and water diversion), which can greatly affect hydrological processes [48,61]. In this study the upstream is a reservoir water source protection region with a larger area of forestland and fewer external factors; in contrast, two big cities (Fushun and Shenyang) and considerable cropland occur in the downstream of the HRB. This is a probable reason why the simulated flood peak was higher than the observed at Dahuofang Reservoir, Fushun, and Shenyang stations on 17 August 2013 (as shown in Figure 5b–d).

If the validation for the monthly sediment yield were to be changed to 2014, all the indicator values met the requirements (Table 3). Therefore, the SWAT model was thought to be a reliable representation of hydrologic processes and sediment export and can be used to simulate responses for the studied catchment at a monthly scale.

3.2. Water Balance Components Under Different Land-Use Scenarios

To further analyze the impacts of LUCC changes on the hydrological cycle, the hydrological components of individual land-use scenarios were assessed based on model simulation. The analysis results showed that the water balance varied among different land-use scenarios at different time scales in the BRB (Figure 8).

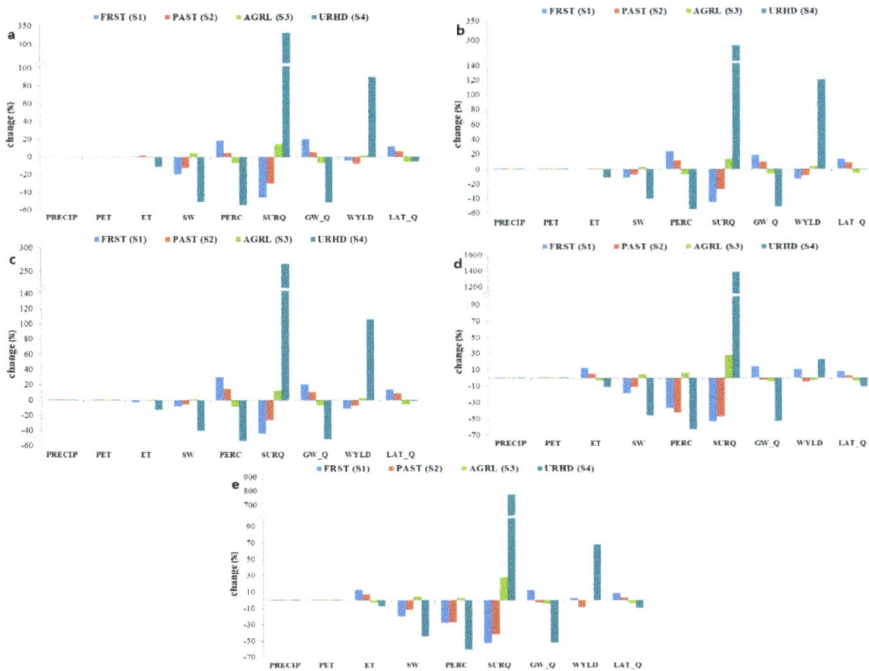

Figure 8. Impact of land-use change on water balance components at all statistical time scales: (**a**) refers to "annual" scale; (**b**) refers to "W1"; (**c**) refers to "W2"; (**d**) refers to "D1"; and (**e**) refers to "D2". FRST—Forestland; PAST—Grassland; AGRL—Cropland; URHD—Urban land.

The impact of land-use change on the annual water balance components is shown in Figure 8a. Under the four hypothetical land-use change scenarios applied in our study sub-basin (the BRB), when the baseline land use (consisting of cropland and forestland) was converted to cropland/urban land, or to grassland/forestland, this would lead to the same change trend of the water balance components, excluding soil water content. For example, when changing the baseline scenario to the urban land scenario, the amount of water percolation (PERC) decreased 54.22% (2.66 mm), groundwater discharge (GW_Q) decreased by 51.32% (2.14 mm), soil water content (SW) decreased by 50.15% (17.02 mm), surface runoff (SURQ) increased by 332.27% (8.76 mm), and water yield (WYLD) increased 90.31% (6.61 mm). However, when by changing the baseline scenario to the forestland scenario, PERC and GW_Q increased by 18.77% and 20.03% (1.18 mm and 0.76 mm), respectively, and SW, SURQ, and WYLD decreased by 19.72%, 45.83%, and 4.29% (5.05 mm, 1.33 mm, and 0.55 mm), respectively. Other water balance components (e.g., precipitation (PRECIP), potential evapotranspiration (PET), actual evapotranspiration (ET)) insignificantly changed, except for a 10.75% reduction in ET under urban land scenario.

During the wet season (W1 and W2), the change trend of the water balance components (Figure 8b,c) was similar to that at annual scale.

During the dry season (D1 and D2), the changes in the water balance components influenced by land-use change scenarios were consistent (Figure 8d,e). For D1, the forestland scenario (S1) increased ET, GW_Q, and WYLD by 12.45%, 14.00%, and 10.32% (1.45 mm, 0.34 mm, and 0.27 mm), respectively, and decreased SW, PERC, and SURQ by 18.79%, 37.40%, and 53.62% (6.50 mm, 0.32 mm, and 0.07 mm), respectively. The urban land scenario (S4) decreased ET, SW, PERC, and GW_Q by 11.51%, 45.94%, 63.20%, and 53.00% (1.34 mm, 15.89 mm, 0.54 mm, and 1.29 mm), respectively, and increased SURQ and WYLD by 1385.94% and 23.04% (1.90 mm and 0.61 mm), respectively. There were few water balance components that changed more than 5% under the grassland (S2) and cropland scenarios (S3). The grassland scenario (S2) resulted in a decrease in SW, PERC, and SURQ of 11.17%, 42.19%, and 47.38% (3.86 mm, 0.36 mm, and 0.07 mm), respectively. Cropland scenario (S3) increased PERC and SURQ by 6.21% and 27.76% (0.05 mm and 0.04 mm), respectively. Therefore, during the dry season, forestland and urban land have a great effect on the water balance components.

3.3. Contribution of Changes in Individual Land-Use Scenarios on Runoff

Simulated runoff changes resulting from the land-use change scenarios at different time scales under the same meteorological inputs (2008–2012) are summarized and compared in Table 4.

(1) Annual scale

The impact of land-use change on annual runoff is shown in Figure 9 and Table 4. The urban land scenario (S4) dramatically increased runoff by 90.23% (0.62 m^3/s), followed by the cropland scenario (S3) with a minor increase in runoff (1.08%). However, the scenarios of forestland (S1) and grassland (S2), respectively, reduced runoff by 4.05% and 7.27% (0.03 m^3/s and 0.05 m^3/s). Meanwhile, when compared with forestland, grassland plays an important role in reducing runoff at annual scale. The contribution of changes in individual land-use scenarios on runoff at annual scale was urban land > cropland > baseline > forestland > grassland.

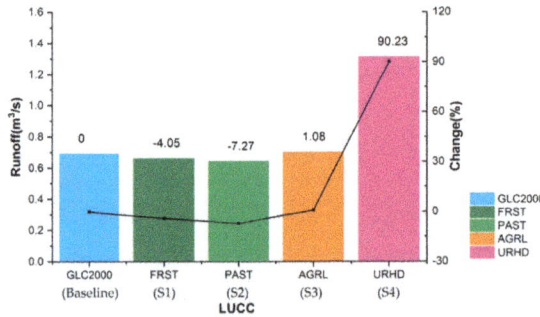

Figure 9. Responses of different land-use scenarios to annual runoff. Remarks: FRST—Forestland; PAST—Grassland; AGRL—Cropland; URHD—Urban land.

Moreover, at annual scale, the coefficients of determination (n = 12) at the sub-basin scale under the scenarios of urban land, cropland, baseline, grassland, and forestland, were 0.87, 0.41, 0.36, 0.25, and 0.15, respectively. The high R^2 for the rainfall–runoff relationship under the urban land scenario suggests that changes in runoff in the sub-basin under this scenario (covered by a large area of impervious surface) were actually mainly a function of the changes in precipitation, while the weakest rainfall–runoff relationship occurred under the forestland scenario (Figure 10).

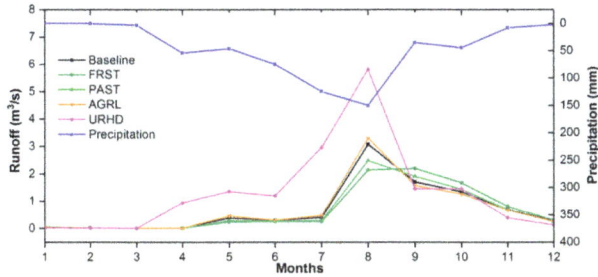

Figure 10. Responses of different land-use scenarios to average monthly runoff. Remarks: FRST—Forestland; PAST—Grassland; AGRL—Cropland; URHD—Urban land.

(2) Wet season and dry season

The contribution of changes on runoff during the wet seasons (W1 and W2) under different scenarios was similar to that at annual scale (Figures 9 and 11). In addition, the scenarios of urban land (S4) and cropland (S3) increased runoff (from 90.23% to 117.5% and from 1.08% to 3.87%, respectively) while forestland (S1) and grassland (S2) reduced runoff (from −4.05% to −13.83% and from −7.27% to −11.27%, respectively), compared with the values at annual scale (Table 4). For example, the urban land scenario (S4) increased runoff by 117.5% (1.382 m^3/s) and the forestland scenario (S1) decreased runoff by 13.83% (0.163 m^3/s) during W1. Reviewing Figure 11 and Table 4, we found that the range of the change rate of runoff was influenced by individual land-use scenarios during W2 less than that during W1. The reason might be that the differences among the effects of land-use change on runoff decrease if the denominator runoff (or precipitation) increases during wet season (Figure 11a,b). The contribution of changes in individual land-use scenarios on runoff during the wet season was urban land > cropland > baseline > grassland > forestland.

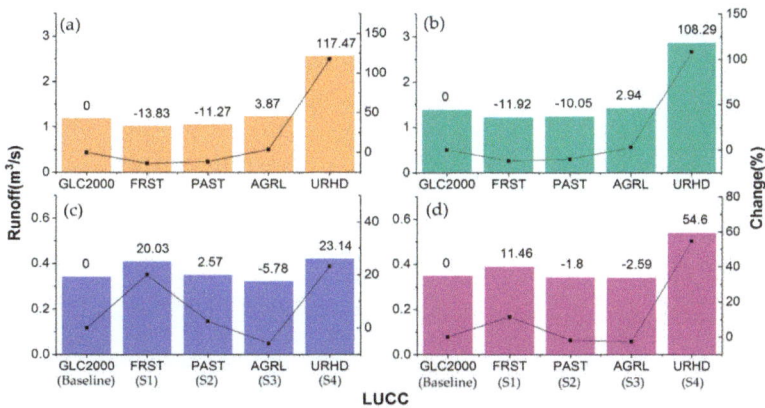

Figure 11. Runoff responses of different land-use scenarios during the wet season (W1 and W2) and dry season (D1 and D2): (**a**) refers to "W1"; (**b**) refers to "W2"; (**c**) refers to "D1"; and (**d**) refers to "D2". FRST—Forestland; PAST—Grassland; AGRL—Cropland; URHD—Urban land.

Table 4. Simulated runoff changes resulting from the land-use change scenarios at different time scales.

Time Scales	Average Monthly Precipitation (mm)	GLC2000 (Baseline)		FRST (S1)		PAST (S2)		AGRL (S3)		URHD (S4)	
		Q (m³/s)	R (%)	Q (m³/s)	R (%)	Q (m³/s)	R (%)	Q (m³/s)	R (%)	Q (m³/s)	R (%)
D1	16.2	0.341	—	0.409	20.03	0.35	2.57	0.321	−5.78	0.42	23.14
D2	19.9	0.348	—	0.388	11.46	0.341	−1.8	0.339	−2.59	0.538	54.6
Annual	45.38	0.69	—	0.66	−4.05	0.64	−7.27	0.7	1.08	1.31	90.23
W1	86.3	1.176	—	1.013	−13.83	1.044	−11.27	1.222	3.87	2.558	117.5
W2	96.4	1.372	—	1.208	−11.92	1.234	−10.05	1.412	2.94	2.857	108.3
D1–D2	3.7	0.007	—	−0.021	−8.57	−0.009	−4.37	0.018	3.19	0.118	31.46
W1–W2	10.1	0.196	—	0.195	1.91	0.19	1.22	0.19	−0.93	0.299	−9.18

Note: "W1–W2" means variable change from W1 to W2; and similarly for "D1–D2". FRST—Forestland; PAST—Grassland; AGRL—Cropland; URHD—Urban land.

Table 4 and Figure 11c,d show that in contrast to the responses of runoff to land use change at annual and wet seasonal scales, the forestland scenario (S1) increased runoff by 20.03% and 11.46% (0.068 m³/s and 0.04 m³/s) during D1 and D2, respectively, while the cropland scenario (S3) decreased runoff by 5.78% and 2.59% (0.02 m³/s and 0.01 m³/s) during D1 and D2, respectively. Moreover, during D1 and D2, the urban land scenario (S4) increased runoff by a relatively moderate amount (23.14% (0.079 m³/s) and 54.60% (0.19 m³/s), respectively) compared with the sharp contrast at annual and wet seasonal scales (all were greater than 90% (0.62 m³/s)). Nevertheless, when the baseline land use (consisting of cropland and forestland) was converted to grassland, the effects of land-use change on runoff were opposite in the D1 and D2 (Table 4). The grassland scenario (S2) increased 2.57% (0.009 m³/s) of runoff during D1 but reduced runoff by 1.80% (0.007 m³/s) during D2. The contribution of changes in individual land-use scenarios on runoff during D1 and D2 were urban land > forestland > grassland > baseline > cropland; and urban land > forestland > baseline > grassland > cropland, respectively.

3.4. Contribution of Changes in Individual Land-Use Scenarios on Sediment Yield

Simulated sediment yield changes resulting from the land-use change scenarios at different time scales were summarized and compared as listed in Table 5.

Table 5. Simulated sediment yield changes resulting from land-use change scenarios at different time scales.

Time Scales	Average Monthly Precipitation (mm)	GLC2000 (Baseline)		FRST (S1)		PAST (S2)		AGRL (S3)		URHD (S4)	
		S (tons)	R (%)	S (tons)	R (%)	S (tons)	R (%)	S (tons)	R (%)	S (tons)	R (%)
D1	16.2	4.73	—	1.47	−68.99	1.47	−68.83	7.25	53.29	11.31	139.32
D2	19.9	25.34	—	7.69	−69.65	7.75	−69.42	38.82	53.22	17.82	−29.68
Annual	45.38	122.8	—	33.23	−72.94	33.64	−72.61	184.16	49.97	43.11	−64.89
W1	86.3	288.1	—	77.69	−73.03	78.68	−72.69	431.83	49.89	87.63	−69.58
W2	96.4	317.73	—	84.3	−73.47	85.43	−73.11	474.83	49.45	93.7	−70.51
D1–D2	3.7	20.61	—	6.23	−0.65	6.28	−0.59	31.57	−0.07	6.51	−169
W1–W2	10.1	29.62	—	6.61	−0.44	6.75	−0.42	43	−0.44	6.07	−0.93

Notes: "W1–W2" means variable change from W1 to W2, and similarly for "D1–D2". FRST—Forestland; PAST—Grassland; AGRL—Cropland; URHD—Urban land.

(1) Annual scale

The impacts of land-use change on annual sediment yield are shown in Figure 12 and Table 5. It is shown that when the baseline land use (consisting of cropland and forestland) was converted to cropland, the sediment yield increased 50.40% (61.36 tons). The scenarios of forestland (S1), grassland (S2) and urban land (S4) decreased the sediment yield by 72.96%, 72.60%, and 65.41% (89.57 tons, 89.16

tons, and 79.69 tons), respectively. The contribution of changes in individual land-use scenarios on sediment yield at annual scale was cropland > baseline > urban land > grassland > forestland.

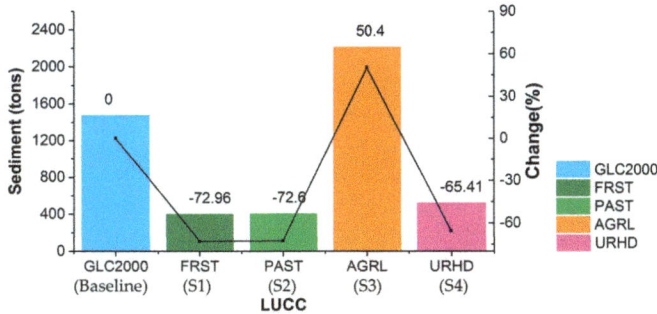

Figure 12. Responses of annual sediment yield to different land-use scenarios. Remarks: FRST—Forestland; PAST—Grassland; AGRL—Cropland; URHD—Urban land.

(2) Wet season and dry season

The effects of land-use change on sediment yield during the wet season (W1 and W2) were consistent to those at annual scale (Figure 13a,b). The change rates of sediment yield induced by different land-use scenarios during the W1 and W2 were nearly the same (Table 5). The urban land scenario (S4) during other statistical periods in this study typically decreased the sediment yield, while it produced substantial sediment (an increase of 139.32% (6.58 tons)) during D1, even more than cropland scenario (an increase of 53.29% (2.52 tons)) (Figure 13c and Table 5). The average reduction changes in sediment yield caused by forestland/grassland were approximately 73% and 69% during the wet and dry seasons, respectively. The contribution of changes in individual land-use scenarios on sediment yield during the wet season (W1 and W2) and D2 was cropland > baseline > urban land > grassland > forestland; however, it was urban land > cropland > baseline > grassland > forestland during D1.

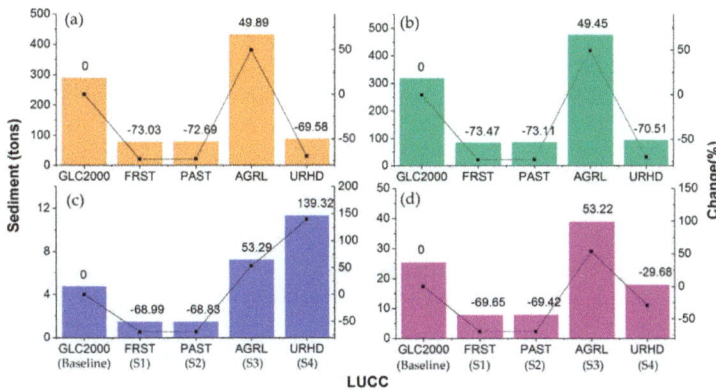

Figure 13. Sediment yield responses to different land-use scenarios during the wet season (W1 and W2) and dry season (D1 and D2): (**a**) refers to "W1"; (**b**) refers to "W2"; (**c**) refers to "D1"; and (**d**) refers to "D2". FRST—Forestland; PAST—Grassland; AGRL—Cropland; URHD—Urban land.

4. Discussion

In this study, the CMADS + SWAT mode was used to calibrate and validate the runoff and sediment yield at monthly and daily scales. In general, satisfactory results have been achieved for monthly runoff, daily runoff, and monthly sediment yield. In the HRB, some studies in the upstream catchment area of the Dahuofang Reservoir using the SWAT model and driven by OBS have been conducted [39,62]. Compared with Han [39] and Yang [62], the CMADS + SWAT mode achieved a better result for monthly runoff and a worse but satisfactory result for monthly sediment yield. However, both of the researchers did not simulate runoff and sediment yield on the daily scale. In general, soil erosion is controlled by a variety of factors [6,37]. Regarding the sediment yield simulation, it tended to be underestimated during most summers. A similar phenomenon was shown in the Miyun Reservoir catchment [63]. In relatively extreme cases, high-speed surface runoff may cause strong erosion to the surface, and even landslides under the action of hydrodynamic force, resulting in a large amount of sediment transport in the river. However, there may be some uncertainties in the observed sediment data [64] for the model calibration and validation, although these data were obtained from authoritative reports issued by the Ministry of Water Resources of the P. R. China. Extraction techniques and sediment sampling intervals were not shown in the reports, and there were few monitoring points [64]. Thus, it is impossible to check whether sediment yield was underestimated or overestimated. Another possible reason for underestimating the sediment yield could be the limitations of the SCS-CN method. The duration and intensity of rainfall are not considered; instead, the average daily rainfall is used as a SWAT input [65]. In fact, high-intensity and even short-duration rainfall could generate more sediment than actualized in the model based on daily rainfall [63]. Overall, the CMADS can well reflect the climatic characteristics of the study area. The CMADS + SWAT mode is reliable and CMADS can be used for other studies in the HRB.

The runoff responses under each land-use scenario at the dry seasonal scale were different from that at annual and wet seasonal scales (Table 4); while the sediment yield responses under each land-use scenario were similar at annual and seasonal scales (Table 5). At annual and wet seasonal scales, the scenarios of forestland (S1) and grassland (S2) increased percolation loss to depth, decrease WYLD, and lead to decrease in runoff. However, it was the opposite during the dry season. Deforestation and urbanization usually decrease ET and percolation, increase WYLD and SURQ, and lead to increase in runoff [14,37]. Guo et al. [66] reported that the increase in forest cover, due to the conversion of agricultural lands to forest, reduced runoff in the wet season but increased it in the dry season in the Poyang Lake basin. Fu et al. [67] found that in Northeastern China, the conversion of cropland, forestland, and grassland into forestland/grassland resulted in the reduction of the monthly runoff from March to August, while the runoff for other months increased; however, the situation was the opposite when the cropland, forestland, and grassland were converted into croplands. The change ratios for the forestland scenario (from −13.83% to 20.03%) and grassland scenario (from −11.27% to 2.57%) (Table 4) showed that forestland had a greater advantage than grassland in reducing flood potentials during the wet season and in moderating drought severity during the dry season. A similar pattern of runoff responses was noted by Lin et al. [14] and Guo et al. [66]. In contrast, Huang et al. [68] found that the forestland was no better than grassland in conserving water when the canopy cover of forestland was less than grassland; we also noticed that some studies found the forest changes has limited [69] or no influence [70] on watershed flows of rivers. In addition, forestland was expected to greatly reduce flood for small storms but less significance for the largest storms [71]. As shown in Table 4, when the rainfall was large, the hydrological changes could be weakened due to land-use change. Because of the retention effect of forests on floods, the remaining water flow may overlap with the latter flood peaks, thus causing greater disasters than that of nonforestland, if the previous flood outflow process was not over yet and a new storm occurred [72]. In this study, the precipitation was 246.27 mm, but runoff yield was 264.62 mm in the Beikouqian region (almost covered by forestland) during August 2013. It might be the overlapping of flood peaks that resulted in an extraordinary flood disaster. The urban land, covered by many impervious layers, usually significantly decreased ET,

SW, PERC, and GW_Q, and increased SURQ and WYLD, leading to a relative significant increase in runoff. Similar conclusions have been drawn by many studies [34,65,73]. Chang et al. [74] indicated that only the urban watershed exhibited a significant increase in both wet and dry season runoff ratios. In fact, because the runoff generated by urban land is closely related to precipitation, it was less than that of forestland in the early dry season (September to November) (Figure 10). In all the statistical periods, both forestland (S1) and grassland (S2) scenarios reduced sediment yield by approximately 70%, while sediment yield increased by about 50% under cropland scenario (S3). Forestland and grassland can protect soil from erosion, while cropland generated more sediment [6]. Yan et al. [37] reported that cropland (forestland) had a greater positive (negative) effect on sediment yield, with negative regression coefficients between grassland and urban land and the sediment yield in the Upper Du Watershed. Serpa et al. [53] found the cropland depicted the highest erosion rates, and forests typically had lower erosion rates than grasslands. Urban land scenario (S4) decreased sediment yield by more than 60% at annual and wet seasonal scales. However, sediment yield reduced by 29.68% in D1 while it increased by 139.32% (the change rate even more than cropland scenario) in D2. The primary reason for this might be that the urban land scenario (S4) is prone to produce more runoff than cropland scenario (S3) in D1, leading to higher sediment transport to the rivers (Figures 11c and 13c). However, when the precipitation is no longer a factor restricting the runoff caused by cropland, more sediment will be produced by the cropland (Table 5). Moreover, heavy rainfall can weaken the hydrological processes such as interception and infiltration of the underlying surface (Table 4) [14], as well as the influence of underlying surface conditions on the relationship between rainfall and sediment yield (Table 5) [75].

In this study, we only investigated the effects of land-use change (LUCC only occurred in the downstream of the BRB) on runoff and sediment yield at annual and seasonal (W1, W2, D1, and D2) scales. However, we noticed many studies have demonstrated that runoff and sediment transport processes can be significantly influenced by the spatial distribution patterns of land use at different temporal and spatial scales. Yin et al. [76] reported that the impacts of LUCC changes on flow regimes were greater after the "Grain for Green Program" in the downstream areas of the Jinghe River Basin than in the upstream. Lin et al. [14] analyzed the land-use change impacts on catchment runoff using different time indicators and the results showed a varying change in runoff among three time scales and three catchments. Khoi et al. [6] compared their results to several studies in Vietnam and found that a smaller change because of LUCC occurred in streamflow and sediment yield, which could be explained by the approximately 50% change in forest land that occurred in the downstream. Interestingly, in our study, although the LUCC only occurred in the downstream of the BRB, significant change ratios were predicted for runoff and sediment yield under individual land-use scenarios at annual and seasonal scales. Although the area of BRB is small, the effects of individual land-use types on runoff and sediment yield at annual and seasonal scales were relatively clearly shown. Due to the heterogeneity in landscape, climate, and geology in the watersheds, it is essential to quantify the effects of spatial distribution patterns of land use for a local scale. Our further work will analyze the impact of land-use pattern changes on runoff and sediment in the HRB from the perspective of large-scale and multi-time.

Undeniably speaking, the SWAT model is useful and effective to investigate the responses of runoff and sediment yield to land-use change, but some limitations as well as uncertainties exist in both the data and the model. The runoff and sediment yield may be influenced by the resolution of the DEM, soil, and land use and the watershed and HRU delineation [77]. Precipitation is a key factor affecting runoff events and sediment exports [78,79], whereas precipitation is adopted in each sub-basin in a basin according to the nearest rain gauge to the sub-basin centroid [80]. Therefore, having a good number and distribution of rain gauges in a basin is very beneficial for simulations of hydrological processes. In this study, ten CMADS stations selected by SWAT were relatively uniform distribution in the HRB (Figure 1). Acceptable and relatively stable model performance was achieved when the rain gauge density ranged between 1.0 and 1.4 per 1000 km^2 [81], and the station density was 1.26 per 1000 km^2 in the HRB. Moreover, a number of empirical and quasiphysical equations in SWAT model were developed based on climatic conditions in the U.S., which may not be appropriate for the climate in China, such as the SCS-CN method and MUSLE method [82]. The MUSLE method does not include the processes occurring in the watershed for both erosions cause by landslides and the "second-storm effect" effecting the mobilization of particulates from soil surface [83]. This may lead to great effects on the results for sediment yield simulations, especially at daily scale. The databases in SWAT were also mostly developed to reflect North American conditions, while the transferability of that to Chinese conditions is doubted by Ongley et al. [84]. It is suggested that some parameters in the empirical equations and the SWAT databases should be modified to suit Chinese conditions in order to improve the simulation results. In addition, calibrated parameters are conditioned on the choice of objective function, the type and length of measured data, and the procedure used for calibration, etc. [55]. Of course, the accuracy of the data is necessary for model simulation. Despite these limitations and uncertainties, the simulation results for monthly/daily runoff and monthly sediment yield were satisfactory according to Moriasi et al. [56].

5. Conclusions

In this research, the SWAT model driven by the CMADS was successfully applied to the HRB, specifically to determine the monthly monthly and daily runoffs. Because many factors can affect the calibration and verification of sediment yield, the simulations of sediment yield in this study were not perfect, but the production of monthly sediment was simulated to meet the requirements (2014 selected as the validation period). The results demonstrated that CMADS was a reliable meteorological data source for the HRB, and the model application was a valid tool for investigating the impacts of land-use changes on runoff and sediment yield. CMADS has been successfully applied in some areas of China and other East Asian countries (e.g., South Korea), and can reflect the climatic characteristics of these areas well, perhaps providing valuable meteorological data for the regions with scarce gauge stations. Moreover, CMADS can be further improved its accuracy of data in other East Asian countries through revising by local measured data. CMADS covered only nine years (2008—2016); if the time series could be extended for decades, it may help researchers to conduct more extensive research in East Asia.

At the sub-basin scale (BRB), the responses of individual land-use types to annual and seasonal runoff and sediment yield using scenario assumptions were determined. Our research indicated that the effects of land-use type on water balance components, runoff, and sediment yield might be altered when different time scales were considered. The changes of water balance components, runoff, and sediment yield under individual land-use scenarios were similar between the annual and wet seasonal scale. Compared with grassland, forestland played a more active role in water redistribution from the wet season to the dry season by promoting infiltration-recharging-discharging processes, and in reducing the soil erosion. However, during the wet season, forestland may lead to greater flood disaster when the subsurface water flow overlaps with the latter flood peaks, because more water is stored in soil and bedrock in forestland. Deforestation and urbanization usually reduce evapotranspiration and water penetration, and increase SURQ, thereby leading to an increase in runoff. Relatively lower runoff occurred under the cropland and urban land scenarios in the early dry season, because the water stored by grassland and forestland during the wet season recharged the river flow, particularly for forestland. Cropland, due to high soil erosion, increased the amount of sediment to be transported into the river along with runoff. On the other hand, under light precipitation (or runoff) conditions, urban land might increase sediment export more, resulting from a relatively larger runoff. In addition, heavy rainfall can weaken the hydrological processes such as interception and infiltration of the underlying surface, as well as the influence of underlying surface conditions on the relationship between rainfall and sediment yield. The limitations as well as uncertainties in SWAT may influence the results in terms of runoff and sediment yield simulations. In general, this study provides useful information for policy makers about the influences of each land-use type on soil and water conservation at annual and seasonal scales in the HRB during urbanization. The influence of land-use pattern change on runoff and sediment yield in river basins needs to be further studied.

Author Contributions: The modeling and writing of this work by L.Z.; X.M. designed the experiments; Observed data was provided by X.M. and M.Y.; the improvement of manuscript writing by X.M. and H.W.

Funding: This research was financially joint supported by the National Science Foundation of China (41701076); the National key Technology R & D Program of China (2017YFC0404305); Young Elite Scientists Sponsorship Program by CAST (2017QNRC001); National Science Foundation for Young Scientists of China (Grant No.51709271).

Conflicts of Interest: The authors declare no conflict of interest.

References

1. Armanini, D.G.; Horrigan, N.; Monk, W.A.; Peters, D.L.; Baird, D.J. Development of a benthic macroinvertebrate flow sensitivity index for Canadian rivers. *River Res. Appl.* **2011**, *27*, 723–737. [CrossRef]
2. Piao, S.L.; Ciais, P.; Huang, Y.; Shen, Z.H.; Peng, S.S.; Li, J.S.; Zhou, L.P.; Liu, H.Y.; Ma, Y.C.; Ding, Y.H.; et al. The impacts of climate change on water resources and agriculture in China. *Nature* **2010**, *467*, 43–51. [CrossRef] [PubMed]
3. Notter, B.; Hurni, H.; Wiesmann, U.; Abbaspour, K.C. Modelling water provision as an ecosystem service in a large East African river basin. *Hydrol. Earth Syst. Sci.* **2012**, *16*, 69–86. [CrossRef]
4. Wolka, K.; Mengistu, T.; Taddese, H.; Tolera, A. Impact of Land Cover Change on Water Quality and Stream Flow in Lake Hawassa Watershed of Ethiopia. *Agric. Sci.* **2014**, *5*, 647–659. [CrossRef]
5. Zhang, M.F.; Liu, N.; Harper, R.; Li, Q.; Liu, K.; Wei, X.; Ning, D.Y.; Hou, Y.P.; Liu, S.R. A global review on hydrological responses to forest change across multiple spatial scales: Importance of scale, climate, forest type and hydrological regime. *J. Hydrol.* **2017**, *546*, 44–59. [CrossRef]
6. Khoi, D.N.; Suetsugi, T. The responses of hydrological processes and sediment yield to land-use and climate change in the Be River Catchment, Vietnam. *Hydrol. Process.* **2014**, *28*, 640–652. [CrossRef]
7. Li, Z.; Liu, W.Z.; Zhang, X.C.; Zheng, F.L. Impacts of land use change and climate variability on hydrology in an agricultural catchment on the Loess Plateau of China. *J. Hydrol.* **2009**, *377*, 35–42. [CrossRef]
8. Nunes, J.P.; Naranjo Quintanilla, P.; Santos, J.M.; Serpa, D.; Carvalho-Santos, C.; Rocha, J.; Keizer, J.J.; Keesstra, S.D. Afforestation, subsequent forest fires and provision of hydrological services: A model-based analysis for a Mediterranean mountainous catchment. *Land Degrad. Dev.* **2018**, *29*, 776–788. [CrossRef]
9. Ghaffari, G.; Keesstra, S.; Ghodousi, J.; Ahmadi, H. SWAT-simulated hydrological impact of land-use change in the Zanjanrood basin, Northwest Iran. *Hydrol. Process.* **2010**, *24*, 892–903. [CrossRef]
10. Cerdan, O.; Govers, G.; Le Bissonnais, Y.; Van Oost, K.; Poesen, J.; Saby, N.; Gobin, A.; Vacca, A.; Quinton, J.; Auerswald, K.; et al. Rates and spatial variations of soil erosion in Europe: A study based on erosion plot data. *Geomorphology* 2010, *122*, 167–177. [CrossRef]
11. García-Ruiz, J.M. The effects of land uses on soil erosion in Spain: A review. *Catena* 2010, *81*, 1–11. [CrossRef]
12. Keesstra, S.D. Impact of natural reforestation on floodplain sedimentation in the Dragonja basin, SW Slovenia. *Earth Surf. Process. Landf.* **2007**, *32*, 49–65. [CrossRef]
13. Buendia, C.; Bussi, G.; Tuset, J.; Vericat, D.; Sabater, S.; Palaub, A.; Batalla, R.J. Effects of afforestation on runoff and sediment load in an upland Mediterranean catchment. *Sci. Total Environ.* **2016**, *540*, 144–157. [CrossRef]
14. Lin, B.Q.; Chen, X.W.; Yao, H.X.; Chen, Y.; Liu, M.B.; Gao, L.; James, A. Analyses of landuse change impacts on catchment runoff using different time indicators based on SWAT model. *Ecol. Indic.* **2015**, *58*, 55–63. [CrossRef]
15. Mueller, E.N.; Francke, T.; Batalla, R.J.; Bronstert, A. Modelling the effects of land-use change on runoff and sediment yield for a meso-scale catchment in the Southern Pyrenees. *Catena* 2009, *79*, 288–296. [CrossRef]
16. Zhan, C.S.; Jiang, S.S.; Sun, F.B.; Jia, Y.W.; Yue, W.F.; Niu, C.W. Quantitative contribution of climate change and human activities to runoff changes in the Wei River basin, China. *Hydrol. Earth Syst. Sci.* **2014**, *18*, 3069–3077. [CrossRef]
17. Li, H.Y.; Zhang, Y.Q.; Vaze, J.; Wang, B.D. Separating effects of vegetation change and climate variability using hydrological modeling and sensitivity-based approaches. *J. Hydrol.* **2012**, *420*, 403–418. [CrossRef]

18. Lørup, J.K.; Refsgaard, J.C.; Mazvimavi, D. Assessing the effect of land use change on catchment runoff by combined use of statistical tests and hydrological modelling: Case studies from Zimbabwe. *J. Hydrol.* **1998**, *205*, 147–163. [CrossRef]

19. Wei, X.H.; Li, W.H.; Zhou, G.Y.; Liu, S.R.; Sun, G. Forests and Streamflow—Consistence and Complexity. *J. Nat. Resour.* **2005**, *20*, 761–770.

20. Yao, Y.L.; Lu, X.G.; Wang, L. A Review on Study Methods of Effect of Land Use and Cover Change on Watershed Hydrology. *Wetl. Sci.* **2009**, *7*, 83–88.

21. Arnold, J.G.; Srinivasan, R.; Muttiah, R.S.; Williams, J.R. Large area hydrologic modeling and assessment. Part I: Model development. *J. Am. Water Resour. Assoc.* **1998**, *34*, 73–89. [CrossRef]

22. SWAT Literature Database. Available online: https://www.card.iastate.edu/swat_articles/ (accessed on 15 July 2018).

23. SWAT Official Website. Available online: https://swat.tamu.edu/ (accessed on 21 May 2018).

24. Gassman, P.W.; Sadeghi, A.M.; Srinivasan, R. Applications of the SWAT Model Special Section: Overview and Insights. *J. Environ. Qual.* **2014**, *43*, 1–8. [CrossRef]

25. Abbaspour, K.C.; Yang, J.; Maximov, I.; Siber, R.; Bogner, K.; Mieleitner, J.; Zobrist, J.; Srinivasan, R. Modelling hydrology and water quality in the pre-alpine/alpine Thur watershed using SWAT. *J. Hydrol.* **2007**, *333*, 413–430. [CrossRef]

26. Meng, X.; Wang, H.; Shi, C.; Wu, Y.; Ji, X. Establishment and Evaluation of the China Meteorological Assimilation Driving Datasets for the SWAT Model (CMADS). *Water* **2018**, *10*, 1555. [CrossRef]

27. Meng, X.; Wang, H.; Wu, Y.P.; Long, A.H.; Wang, J.H.; Shi, C.X.; Ji, X.N. Investigating spatiotemporal changes of the land-surface processes in Xinjiang using high-resolution CLM3.5 and CLDAS: Soil temperature. *Sci. Rep.* **2017**, *7*, 13286. [CrossRef]

28. Meng, X.; Wang, H.; Lei, X.H.; Cai, S.Y.; Wu, H.J.; Ji, X.N.; Wang, J.H. Hydrological Modeling in the Manas River Basin Using Soil and Water Assessment Tool Driven by CMADS. *Teh. Vjesn.* **2017**, *24*, 525–534. [CrossRef]

29. Liu, J.; Shangguan, D.; Liu, S.; Ding, Y. Evaluation and Hydrological Simulation of CMADS and CFSR Reanalysis Datasets in the Qinghai-Tibet Plateau. *Water* **2018**, *10*, 513. [CrossRef]

30. Meng, X.; Wang, H. Significance of the China Meteorological Assimilation Driving Datasets for the SWAT Model (CMADS) of East Asia. *Water* **2017**, *9*, 765. [CrossRef]

31. Shao, G.; Guan, Y.; Zhang, D.; Yu, B.; Zhu, J. The Impacts of Climate Variability and Land Use Change on Streamflow in the Hailiutu River Basin. *Water* **2018**, *10*, 814. [CrossRef]

32. Thom, V.; Li, L.; Jun, K.S. Evaluation of Multi-Satellite Precipitation Products for Streamflow Simulations: A Case Study for the Han River Basin in the Korean Peninsula, East Asia. *Water* **2018**, *10*, 642. [CrossRef]

33. Cao, Y.; Zhang, J.; Yang, M.; Lei, X.; Guo, B.; Yang, L.; Zeng, Z.; Qu, J. Application of SWAT Model with CMADS Data to Estimate Hydrological Elements and Parameter Uncertainty Based on SUFI-2 Algorithm in the Lijiang River Basin, China. *Water* **2018**, *10*, 742. [CrossRef]

34. Dong, N.P.; Yang, M.X.; Meng, X.Y.; Liu, X.; Wang, Z.; Wang, H.; Yang, C. CMADS-Driven Simulation and Analysis of Reservoir Impacts on the Streamflow with a Simple Statistical Approach. *Water* **2018**, *11*, 178. [CrossRef]

35. Yuan, Z.; Xu, J.; Meng, X.; Wang, Y.; Yan, B.; Hong, X. Impact of Climate Variability on Blue and Green Water Flows in the Erhai Lake Basin of Southwest China. *Water* **2019**, *11*, 424. [CrossRef]

36. Meng, X.; Wang, H.; Chen, J. Profound Impacts of the China Meteorological Assimilation Driving Datasets for the SWAT Model (CMADS). *Water* **2019**, *11*, 832. [CrossRef]

37. Yan, B.; Fang, N.F.; Zhang, P.C.; Shi, Z.H. Impacts of land use change on watershed streamflow and sediment yield: An assessment using hydrologic modelling and partial least squares regression. *J. Hydrol.* **2013**, *484*, 26–37. [CrossRef]

38. Qiu, L.; Wu, Y.; Wang, L.; Lei, X.; Liao, W.; Hui, Y.; Meng, X. Spatiotemporal response of the water cycle to land use conversions in a typical hilly-gully basin on the Loess Plateau, China. *Hydrol. Earth Syst. Sci.* **2017**, *21*, 6485–6499. [CrossRef]

39. Han, C.W. *Environmental Behavior, Simulation and Prediction of Nonpoint Source Nitrogen and Phosphorus in Cold Regions*; Dalian University of Technology: Dalian, China, 2011.

40. Liu, Y. *Research on Water Pollution Control Planning in Shenfu New Town*; Shenyang Jianzhu University: Shenyang, China, 2013.

41. DEM Derived from CGIAR-CSI. Available online: http://srtm.csi.cgiar.org/SELECTION/inputCoord.asp (accessed on 16 June 2018).

42. China Soil Map Based Harmonized World Soil Database (v1.1). Available online: http://westdc.westgis.ac.cn (accessed on 20 June 2018).

43. Fischer, G.; Nachtergaele, F.; Prieler, S.; Velthuizen, H.T.; Verelst, L.; Wiberg, D. *Global Agro-ecological Zones Assessment for Agriculture (GAEZ 2008)*; IIASA: Laxenburg, Austria; FAO: Rome, Italy, 2008.

44. Saxton, K.E.; Rawls, W.J. Soil water characteristic estimates by texture and organic matter for hydrologic solutions. *Soil Sci. Soc. Am. J.* **2006**, *70*, 1569–1578. [CrossRef]

45. Ran, Y.H.; Li, X.; Lu, L. Evaluation of four remote sensing based land cover products over China. *Int. J. Remote Sens.* **2010**, *31*, 391–401. [CrossRef]

46. Arnold, J.G.; Moriasi, D.N.; Gassman, P.W.; Abbaspour, K.C.; White, M.J.; Srinivasan, R.; Santhi, C.; Harmel, D.; van Griensven, A.; Van Liew, M.W.; et al. SWAT: Model use, calibration, and validation. *Trans. ASABE* **2012**, *55*, 1491–1508. [CrossRef]

47. Arnold, J.G.; Allen, P.M. Estimating hydrologic budgets for three Illinois watersheds. *J. Hydrol.* **1996**, *176*, 57–77. [CrossRef]

48. Luo, K.S.; Tao, F.L.; Moiwo, J.P.; Xiao, D.P. Attribution of hydrological change in Heihe River Basin to climate and land use change in the past three decades. *Sci. Rep.* **2016**, *6*, 33704. [CrossRef]

49. USDA Soil Conservation Service. *National Engineering Handbook Section 4 Hydrology*; USDA-Soil Conservation Service: Washington, DC, USA, 1972; Chapters 4–10.

50. Monteith, J.L. Evaporation and the Environment in the State and Movement of Water in Living Organisms. In *Proceedings of the 19th Symposia of the Society for Experimental Biology*; Cambridge University Press: London, UK, 1965; pp. 205–234.

51. Williams, J.R. Flood routing with variable travel time or variable storage coefficients. *Trans. ASAE* **1969**, *12*, 100–103. [CrossRef]

52. Williams, J.R.; Berndt, H.D. Sediment Yield Prediction Based on Watershed Hydrology. *Trans. ASAE* **1977**, *20*, 1100–1104. [CrossRef]

53. Serpa, D.; Nunes, J.P.; Santos, E.S.; Jacinto, R.; Veiga, S.; Lima, J.C.; Moreira, M.; Corte-Real, J.; Keizer, J.J.; Abrantes, N. Impacts of climate and land use changes on the hydrological and erosion processes of two contrasting Mediterranean catchments. *Sci. Total Environ.* **2015**, *538*, 64–77. [CrossRef]

54. Faramarzi, M.; Abbaspour, K.C.; Schulin, R.; Yang, H. Modelling blue and green water resources availability in Iran. *Hydrol. Process.* **2009**, *23*, 486–501. [CrossRef]

55. Yang, J.; Reicher, P.; Abbaspour, K.C.; Xia, J.; Yang, H. Comparing uncertainty analysis techniques for a SWAT Application to the Chaohe Basin in China. *J. Hydrol.* **2008**, *58*, 1–23. [CrossRef]

56. Moriasi, D.N.; Arnold, J.G.; Van Liew, M.W.; Bingner, R.L.; Harmel, R.D.; Veith, T.L. Model evaluation guidelines for systematic quantification of accuracy in watershed simulations. *Am. Soc. Agric. Biol. Eng.* **2007**, *50*, 885–900. [CrossRef]

57. Santhi, C.; Arnold, J.G.; Williams, J.R.; Dugas, W.A.; Srinivasan, R.; Hauck, L.M. Validation of the SWAT model on a large river basin with point and nonpoint sources. *J. Am. Water Resour. Assoc.* **2001**, *37*, 1169–1188. [CrossRef]

58. Garbrecht, J.D.; Van Liew, M.W.; Arnold, J.G. Hydrologic simulation on agricultural watersheds: Choosing between two models. *Trans. ASAE* **2003**, *46*, 1539–1551. [CrossRef]

59. 2013 Bulletin of Flood and Drought Disaster in China. Available online: http://www.mwr.gov.cn/sj/tjgb/zgshzhgb/201612/t20161222_776091.html (accessed on 2 July 2018).

60. Zhang, P. *Study on Risk Assessment and Prevention of Huangdai Gully Debris Flow in Changsha Village, Qingyuan Country*; Jilin University: Changchun, China, 2017.

61. Bieger, K.; Hormann, G.; Fohrer, N. Simulation of Streamflow and Sediment with the Soil and Water Assessment Tool in a Data Scarce Catchment in the Three Gorges Region, China. *J. Environ. Qual.* **2014**, *43*, 37–45. [CrossRef]

62. Yang, W. *Smimulation on Agricultural Non-Point Sources Pollution and Risk Assessment in Dahuofang Reservoir Catchment of Liaoning Province*; Jilin University: Changchun, China, 2012.

63. Xu, Z.X.; Pang, J.P.; Liu, C.M.; Li, J.Y. Assessment of runoff and sediment yield in the Miyun Reservoir catchment by using SWAT model. *Hydrol. Process.* **2009**, *23*, 3619–3630. [CrossRef]

64. Wang, S.; Luan, J.; Liu, Q. Analysis of Sediment Characteristics in Upper Hunhe River Basin (Fushun Section). *Water Resour. Hydr. Northeast. Chin.* **2008**, *26*, 28–29.

65. Nie, W.M.; Yuan, Y.P.; Kepner, W.; Nash, M.S.; Jackson, M.; Erickson, C. Assessing impacts of landuse changes on hydrology for the upper San Pedro watershed. *J. Hydrol.* **2011**, *407*, 105–114. [CrossRef]

66. Guo, H.; Hu, Q.; Jiang, T. Annual and seasonal streamflow responses to climate and land-cover changes in the Poyang Lake basin, China. *J. Hydrol.* **2008**, *355*, 106–122. [CrossRef]

67. Fu, Q.; Shi, R.; Li, T.; Sun, Y.; Liu, D.; Cui, S.; Hou, R. Effects of land-use change and climate variability on streamflow in the Woken River basin in Northeast China. *River Res. Appl.* **2019**, *35*, 121–132. [CrossRef]

68. Huang, M.B.; Kang, S.Z.; Li, Y.S. A comparison of hydrological behavior of forest and grassland watersheds in gully region of the Loess Plateau. *J. Nat. Resour.* **1999**, *14*, 226–231.

69. Buttle, J.M.; Metcalfe, R.A. Boreal forest disturbance and streamflow response, northeastern OntarioCan. *J. Fish. Aquat. Sci.* **2000**, *57*, 5–18. [CrossRef]

70. Ceballos-Barbancho, A.; Morán-Tejeda, E.; Luengo-Ugidos, M.A.; Llorente-Pinto, J.M. Water resources and environmental change in a Mediterranean environment: The south-west sector of the Duero river basin (Spain). *J. Hydrol.* **2008**, *351*, 126–138. [CrossRef]

71. Calder. *Land Use Impacts on Water Resources. Land-Water Linkages in Rural Watersheds Electronic Workshop*; FAO: Rome, Italy, 2000.

72. Sun, H. Progress of the research on the role of the forest during the past 20 years. *J. Nat. Resour.* **2001**, *16*, 407–412.

73. Wang, Y.; Qi, S.; Sun, G.; Steve, G. Impacts of climate and land-use change on water resources in a watershed: A case study on the Trent River basin in North Carolina, USA. *Adv. Water Sci.* **2011**, *22*, 51–58.

74. Chang, H. Comparative streamflow characteristics in urbanizing basins in the Portland Metropolitan Area, Oregon, USA. *Hydrol. Process.* **2007**, *21*, 211–222. [CrossRef]

75. Hao, F.; Chen, L.; Liu, C.; Dai, D. Impact of Land Use Change on Runoff and Sediment Yield. *J. Soil Water Conserv.* **2004**, *18*, 5–8.

76. Yin, J.; He, F.; Xiong, Y.J.; Qiu, G.Y. Effects of land use/land cover and climate changes on surface runoff in a semi-humid and semi-arid transition zone in northwest China. *Hydrol. Earth Syst. Sci.* **2017**, *21*, 183–196. [CrossRef]

77. Gassman, P.W.; Reyes, M.R.; Green, C.H.; Arnold, J.G. The Soil and Water Assessment Tool: Historical development, applications, and future research directions. *Trans. ASABE* **2007**, *50*, 1211–1240. [CrossRef]

78. Keesstra, S.D.; Davis, J.; Masselink, R.H.; Casalí, J.; Peeters, E.T.; Dijksma, R. Coupling hysteresis analysis with sediment and hydrological connectivity in three agricultural catchments in Navarre, Spain. *J. Soil. Sediment.* **2019**, *19*, 1598–1612. [CrossRef]

79. Masselink, R.H.; Temme, A.J.A.M.; Giménez Díaz, R.; Casalí Sarasíbar, J.; Keesstra, S.D. Assessing hillslope-channel connectivity in an agricultural catchment using rare-earth oxide tracers and random forests models. *Cuad. Investig. Geogr.* **2017**, *43*, 19–39. [CrossRef]

80. Douglas-Mankin, K.R.; Srinivasan, R.; Arnold, A.J. Soil and Water Assessment Tool (SWAT) Model: Current Developments and Applications. *Trans. ASABE* **2010**, *53*, 1423–1431. [CrossRef]

81. Xu, H.; Xu, C.Y.; Chen, H.; Zhang, Z.; Li, L. Assessing the influence of rain gauge density and distribution on hydrological model performance in a humid region of China. *J. Hydrol.* **2013**, *505*, 1–12. [CrossRef]

82. Neitsch, S.L.; Arnold, J.G.; Kiniry, J.R.; Williams, J.R. *Soil and Water Assessment Tool Theoretical Documentation Version 2009*; Texas Water Resources Institute: College Station, TX, USA, 2011.

83. Abbaspour, K.C. *SWAT-CUP: SWAT Calibration and Uncertainty Programs—A User Manual (Version 2012)*; Swiss Federal Institute of Aquatic Science and Technology: Dübendorf, Switzerland, 2011.

84. Ongley, E.D.; Zhang, X.; Yu, T. Current status of agricultural and rural non-point source pollution assessment in China. *Environ. Pollut.* **2010**, *158*, 1159–1168. [CrossRef]

water

MDPI

Article

Investigating Spatial and Temporal Variation of Hydrological Processes in Western China Driven by CMADS

Yun Li [1], Yuejian Wang [2,*], Jianghua Zheng [1,*] and Mingxiang Yang [3,*]

1   College of Resources and Environmental Sciences, Xinjiang University (XJU),
    Urumqi 830046, China; 0603liyun@163.com
2   Department of Geography, Shihezi University, Shihezi 832000, Xinjiang, China
3   China Institute of Water Resource and Hydropower Research (IWHR), Beijing 100038, China
*   Correspondence: wangyuejian0808@163.com (Y.W.); zheng_jianghua@126.com (J.Z.);
    yangmx@iwhr.com (M.Y.); Tel.: +86-180-9993-9983 (Y.W.); +86-135-7988-0590 (J.Z.); +86-180-4655-5306 (M.Y.)

Received: 28 December 2018; Accepted: 18 February 2019; Published: 28 February 2019

Abstract: The performance of hydrological models in western China has been restricted due to the scarcity of meteorological observation stations in the region. In addition to improving the quality of atmospheric input data, the use hydrological models to analyze Hydrological Processes on a large scale in western China could prove to be of key importance. The Jing and Bortala River Basin (JBR) was selected as the study area in this research. The China Meteorological Assimilation Driving Datasets for the SWAT model (CMADS) is used to drive SWAT model, in order to greatly improve the accuracy of SWAT model input data. The SUFI-2 algorithm is also used to optimize 26 sensitive parameters within the SWAT-CUP. After the verification of two runoff observation and control stations (located at Jing and Hot Spring) in the study area, the temporal and spatial distribution of soil moisture, snowmelt, evaporation and precipitation were analyzed in detail. The results show that the CMADS can greatly improve the performance of SWAT model in western China, and minimize the uncertainty of the model. The NSE efficiency coefficients of calibration and validation are controlled between 0.659–0.942 on a monthly scale and between 0.526–0.815 on a daily scale. Soil moisture will reach its first peak level in March and April of each year in the JBR due to the snow melting process in spring in the basin. With the end of the snowmelt process, precipitation and air temperature increased sharply in the later period, which causes the soil moisture content to fluctuate up and down. In October, there was a large amount of precipitation in the basin due to the transit of cold air (mainly snowfall), causing soil moisture to remain constant and increase again until snowmelt in early spring the following year. This study effectively verifies the applicability of CMADS in western China and provides important scientific and technological support for the spatio-temporal variation of soil moisture and its driving factor analysis in western China.

Keywords: CMADS; SWAT; JBR; soil moisture; hydrological processes; spatio-temporal

---

1. Introduction

In recent years, there have been a number of serious ecological and water crises in the world, with changes to the land surface process in arid areas having a significant influence on the whole inland river water cycle and the ecological environment of vast areas. These changes include arid inland water, the water quality and the continuous degradation, all of which make a frequent contribution to adverse water events. Due to the unique structure of arid oases, their water and energy cycles have their own laws, and local climate conditions are inextricably linked to them. Therefore, it is necessary to systematically analyze the temporal and spatial variations of the surface hydrological components in the arid areas of Xinjiang, which may provide important technical support for the ecological and hydrological restoration and sustainable development of arid areas.

The JBR are located in Xinjiang, western China, and their basins enclose the total area of the surface's hydrological components' spatial variation in the arid alpine area of Xinjiang; the middle of this area is a valley, whilst the east is a basin. Furthermore, its geomorphology and terrain can be connected with the Junggar Basin and divided into three major types of landforms: mountain, valley, and basin. Because the watershed interchanges are located in the sinking area of the Alashan Mountain, wind is the largest meteorological hazard in the Bo River Basin. Due to the fragile ecological structure of the basin, the impact of both salinization and desertification is serious. In recent years, with climate change and large-scale human reclamation activities, the amount of water flowing into the Ebinur Lake in the JBR has decreased and the ecological environment has continued to degenerate. According to statistics, between the 1950s and 1970s, the Ebinur Lake in the JBR shrunk in size by nearly 678 square kilometers, and lost nearly 2.3 billion cubic meters of water. Due to the sharp decline in water resources, the lake's salinity has greatly increased, the wetland area has become smaller, and the lake's role in climate regulation has reduced, which is devastating to both the ecosystem and local residents. The impact of this makes the already very fragile ecological environment of the JBR further deteriorate, resulting in a biological chain chasm, reduced biological diversity, intensified desertification and other ecological problems. As the JBR contributes greatly to the ecological balance and social economy in Xinjiang, it is vital to develop the surface related parameters (such as the ecological hydrological parameters) of the JBR and provide the locality with feasible and sustainable development based on the simulation of the land surface process. However, due to the scarcity of traditional observation sites in the region, large differences in the underlying surface, and the increased influence of climate change and human activities, it is difficult to simulate the temporal and spatial changes of the surface components in Xinjiang. In other words, the uncertainty of the atmospheric-related data will lead to a significant increase in the uncertainty of the region's analog output [1].

Numerous analyses have shown that if the atmospheric dataset contains more observation data, the simulation results can be improved significantly [2–16]. The NCEP and NCAR are cooperating in a project denoted "reanalysis" to produce a 40-year record of global analyses of atmospheric fields in support of the needs of the research and climate monitoring communities. ERA is a re-analysis of meteorological observations from September 1957 to August 2002 produced by the European Centre for Medium-Range Weather Forecasts (ECMWF) in collaboration with many institutions. There are currently many kinds of atmospheric datasets in use in China and overseas, such as the NCAR/DOE of NCEP [17,18], ERA-15, ERA-40, ERA-Interim reanalysis data [19], JRA-25 reanalysis data [20] and the Princeton dataset, all of which can enhance the model's performance. As traditional atmospheric observing stations do not cover the whole world, the above datasets provide an important basis for data analysis [17]. However, despite the continuous emergence of various types of atmospheric reanalysis datasets or atmospheric-driven fields, these datasets need to be further validated for a study of large and medium-sized areas. For example, Jeremy et al. used the regional climate model (RegCM) to simulate and evaluate the monthly variation of precipitation in the winter and summer seasons in the East Asian monsoon region, and found that the RegCM model made large errors when forecasting precipitation. It is believed that this phenomenon is particularly evident in winter. Furthermore, numerous studies have shown that NCEP, ERA and JRA reanalysis datasets can leads to obvious seasonal and regional differences over China [21–28]. For example, Shi et al. [29] used various technical means to evaluate the usability of NCAR-driven data (e.g., air temperature, wind speed) in China, with their results showing that the wind field anomalies were negatively correlated with altitude. Ning et al. [30] used Noah LSM-HMS driven by model (CMADS) with a very good performance, indicating that streamflow is increased in dry seasons and decreased in wet seasons, and large and small reservoirs can have equally large effects on the streamflow.

The above studies have proved that although all kinds of reanalysis products and datasets can reflect large-scale meteorological elements; however, for China, especially the western region, the regional surface differences are large and observation stations are relatively scarce. Moreover, most of the reanalysis datasets used in previous research have not been assimilated with data from the regional automatic stations within China; therefore, they are unable to reflect the strength and frequency of the various driving factors effectively [31]. In summary, it is very important to simulate and analyze the relevant surface components by using a meteorological background field which has been corrected by a regional automatic station so as to develop a more credible hydrological simulations [32,33].

The China Meteorological Assimilation Datasets for the SWAT model (CMADS) was completed over the 9-year period of 1 January 2008 through 31 December 2016, and has been used in many watersheds throughout East Asia [34–43]. Furthermore, Researchers from China also used CMADS data and Penman-Monteith method to calculate Potential evapotran-spiration (PET) across China with good performance [44–51]. However, although CMADS has many applications in China, there are few applications in western China, where traditional weather stations are scarce. Especially in the western part of China, the surface space-time differentiation and complexity, coupled with the region's economic and other objective factors, limit the local establishment of a large number of meteorological observation systems [50–52]. Therefore, the use of limited traditional observatories by local scientists does not for the basis for good scientific and effective study of the local underlay surface. Since analyses of surface processes (such as soil moisture, temperature, melting snow, etc.) based on the CMADS+ SWAT mode are not being used very well until now, the purpose of this paper is try to use high-precision meteorological data (CMADS) to drive the SWAT model. When the SWAT model is localized, we will effectively analyze and examine other surface processes (snowmelt, soil moisture, etc.).

2. Study Area

The JBR are located on the northern slope of the western Tianshan Mountains (Figure 1), (81°46′–83°51′ E, 44°02′–45°10′ N), with a total area of 11,300 km^2. The precipitation is mostly from the Arctic Ocean and the Atlantic water vapor, and shows an overall pattern of (1) more in the mountains than the plains, (2) more in the west than the east, and (3) more along the shady slope than the sunny slope. The basin has a distribution of nearly 460 glaciers, with a total of 15.4 km^2 of glacier reserves. Among them, the glacier areas in the JBR regions reach 96.2 km^2 and 110.3 km^2 respectively. The amount of glacier replenishment in the Jinghe is about 96 million m^3, whilst in the Bo River it is about 105 million m^3, accounting for 20.6% and 21.4% of their total river runoffs respectively. The combined effects of temperature, precipitation and topography have led to the runoff replenishment of the region depending mainly on snow, ice, rainfall and groundwater, making the area a typical arid basin. However, the mineralization of Ebinur Lake has gradually increased in recent years, and the competition between ecological and domestic water usage has become increasingly intensive. In addition, the land degradation within the basin is also very serious, and 1500 km^2 of the underwater region of the Ebinur Lake has been degraded into a salt desert, with a salinization area of 71 km^2.

In recent years, researchers have only used various types of reanalysis data (or regional climate models) and single point observation data to study the JBR and the whole Xinjiang region. As the Xinjiang regional meteorological stations are relatively scarce, the region has not yet undergone a more thorough and reliable simulation and analysis of the hydrological process. However, it is very important that the temporal and spatial evolution of the hydrological correlation components are simulated using the SWAT model, which has a high resolution and reliable driving field.

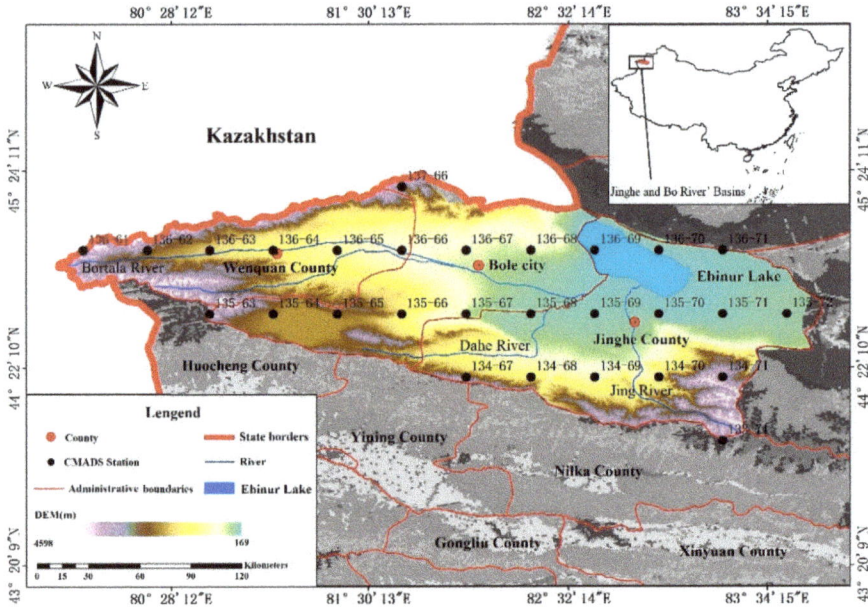

Figure 1. Location of Jinghe and Bortala River Basins (JBR).

3. Material and Methods

3.1. Material

The China Meteorological Assimilation Driving Datasets for the SWAT model (CMADS) is a public datasets developed by Xianyong Meng from China agriculture university (http://www.cmads. org/).The input datasets for the SWAT model mainly includes land use, the river network, digital elevation model (DEM) and soil distribution, as well as other data (Figure 2). Among them, the SRTM90mdigital elevation is obtained from the CGIAR-CSI SRTM database: (http://srtm.csi.cgiar. org/SELECTION/inputCoord.asp).

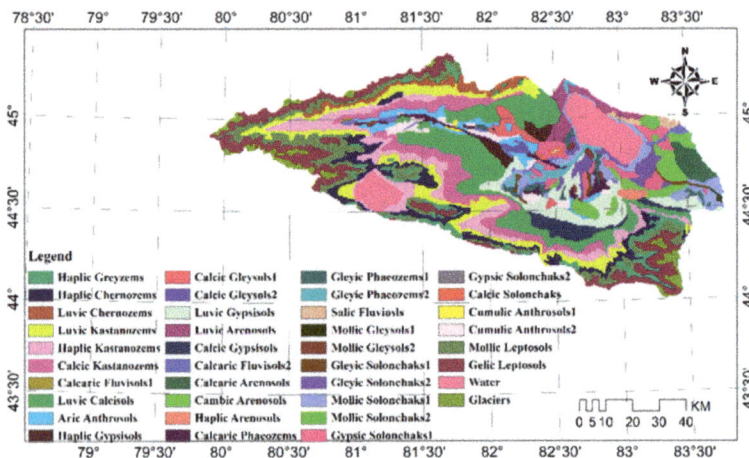

(a)

Figure 2. *Cont.*

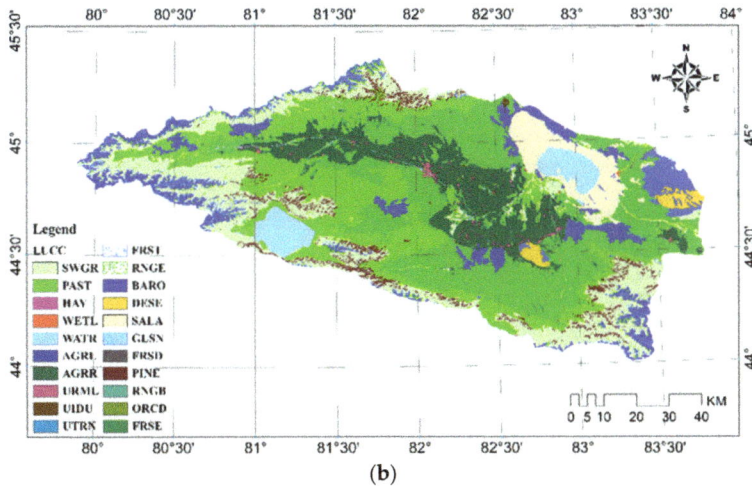

Figure 2. (**a**) Soil classification and (**b**) Land usage distribution.Remarks: SWGR: Slender Wheat grass; PAST: Pasture; HAY: Hay; WETL: Wetlands-Mixed; WATR: Water; AGRL: Agricultural Land-Generic; AGRR: Agricultural Land-Row Crops; URML: Residential-Med/Low Density; UIDU: Industrial; UTRN: Transportation; FRST: Forest-Mixed; RNGE: Range-Grasses; BARO: Bare rock; DESE: Desert; SALA: Saline land; GLSN: Glacier and snow; FRSD: Forest-Deciduous; PINE: Pine; RNGB: Range-Brush; ORCD: Orchard; FRSE: Forest-Evergreen.

The physical properties of soil determines the characteristics of the production and confluence of the different hydrological units in the SWAT model, and also provide a reference for the definition of the hydrological response unit (HRU). The soil input data selected in this study is the Harmonized World Soil Database version 1.1 (HWSD 1.1), which has been derived from the World Soil Database (HWSD). The land use comes from the Management Office of the JBR. The distribution of the original land use data in the study area is analyzed in the JBR (see Figure 3B). The land use data is superimposed on the Second Glacier Inventory Dataset of China (Version 1.0) [25].

In order to ensure the consistency of the resolution within the SWAT model, the spatial resolution of DEM and soil distribution, land use data is unified into 1 km, and the projection coordinates are set uniformly to wgs_1984_utm_zone_44n projection.

3.1.1. The Atmosphere-Driven Data

CMADS incorporated technologies of LAPS/STMAS and was constructed using multiple technologies and scientific methods, including loop nesting of data, projection of resampling models, and bilinear interpolation. The above process is carried out under conditions of strict quality control [26,27]. Some of the specific information and site distribution of the CMADS V1.0 used in this study are shown in Table 1.

This study uses the CMADS V1.0 version as the SWAT atmospheric driving dataset, with a spatial resolution of 1/3 degrees, a day-to-day time resolution, and a data scale of 2008–2013. The SWAT model has read the driving elements (i.e., temperature, humidity, wind, precipitation and radiation data) of the 23 sites of the CMADS V1.0 in the JBR.

Table 1. The CMADS information.

Dataset	CMADS Serial Dataset
Elements provided	Daily maximum/minimum temperature, relative humidity, wind speed, precipitation, solar radiation
Spatial scale of dataset	0–65° N, 60°–160° E
Temporal scope of dataset	37.5°–39.2° N, 98.5°–101.2° E
Data temporal scale	1 January 2008–31 December 2013
Data original resolution	0.333°, 0.25°, 0.125°, 0.0625°
Spatial resolution	0.25°
Number of CMADS sites in the study	28

3.1.2. Hydrological Data

In this study, observed daily runoff datasets of two hydrological stations in JBR are selected for calibration and validation, detailed information relating to these stations is shown in Table 2.

Table 2. Information relating to the hydrological stations in the JBR.

Station Name	Latitude	Longitude	Altitude(m)	Data Time (Year)
Hot Spring	44°59′	81°02′	1310	2009-2013
Jinghe Mountain	44°22′	82°55′	620	2009-2013

During the period of 2009–2013, the maximum daily discharge of Jinghe Mountain Hydrological Station appeared in June, July and August, with a maximum value of 85.70 m^3/s, occurring on 19 June 2010; the minimum daily discharge appeared on February and March, with a minimum value of 0.62 m^3/s, occurring on 12 March 2013; and the maximum daily discharge of hot spring hydrological station appeared on June and July, with a maximum value of 77.20 m^3/s, occurring on July 2010. The minimum daily flow occurred in May and August, with a minimum value of 3.24 m^3/s, occurring on 19 May 2010. The daily maximum and minimum flow of Jinghe Mountain Station tends to decrease, while the daily maximum and minimum flow of hot spring station tends to increase first and then decrease.

3.2. SWAT Hydrological Model

In this study, the semi-distributed hydrological model, SWAT, is used to simulate the hydrological correlation components. Unlike the fully distributed model, this model treats the homogeneous land cover/utilization, soil distribution and management unit as a hydrological response unit (HRU). The model considers that all the water balance processes within the HRU are consistent. The SWAT model has been updated to SWAT 2012 since its release (https://swat.tamu.edu/). The simulation process of SWAT is divided into two steps: the first is the runoff stage, which can import pesticides, sediment and various nutrients in each natural sub-basin into the main watercourse; the second, the convergence process, mainly refers to the migration process of sediments and water flow to the major water outlet in the river basin. In the SWAT model, water balance plays an important role [6]. Furthermore, the SWAT model uses the Soil Conservation Service's (SCS) precipitation runoff curve to calculate the daily runoff process. This method supposes that the water flow along the slope's surface is the surface runoff when the surface soil moisture is low and the infiltration rate is larger, which will decrease with the increase of the soil moisture. At this point, if the infiltration rate is less than the rainfall intensity, then the phenomenon of the low-lying land fill arises. After filling the land, the surface runoff quickly forms and merges into the river watercourse. The SWAT model provides three methods for calculating the potential evaporation: the Penman–Monteith (P-M) method [6–8], the Priestley–Taylor method [9] and the Hargreaves method. In this study, the P-M method was chosen as the one for evaporation simulation because the CMADS can provide all the input elements (the traditional weather station could not provide solar radiation data). In addition, the SWAT model also assumes that the evaporation of the canopy-intercepted rainfall is calculated, and the evaporation and sublimation components are estimated using the Ritchie method, with the corresponding results being obtained. In this process, if there is snow in the HRU, the snow sublimation will be calculated first and then the soil evaporation process will be calculated. At present, the SWAT snow melting module uses the method of degree day factor, which suggests that the snowmelt process is influenced by the air and snow cover temperatures, the snowmelt rate and snow cover area.

3.3. The SWAT Model's Scheme Settings

The SWAT model extracts the river network information based on the DEM to divide the basin area into smaller sub-basins. The area of the study area is 11275 km^2, which is divided into 39 sub-basins and 1648 HRUs. Due to the advantages that the CMADS has in providing additional solar radiation elements, the SWAT model is used to calculate the potential evaporation by the P-M method, which requires the input of solar radiation, temperature, wind speed and relative humidity. As the precipitation data are derived daily, the surface runoff simulation is calculated using the SCS curve. The runoff of the surface will be simulated in different HRUs and eventually converge to the main river basin. Finally, the SWAT model will select the river watercourse accumulation method, based on a continuous equation, to calculate the water evolution of the main watercourse. In order to reduce the error (especially in the high altitude area) of the spatial interpolation, the SWAT model is used to interpolate the single-point meteorological data space in the basin using the Centroid method. In this paper, in order to identify the different elevation zones' precipitation distribution accurately, the basins are divided into different elevation zones.

Due to the limitations of the CMADS drive field time scale (2008–2014) and runoff observation data, and in order to make all the hydrological processes in the initial stage of the simulation transform from the initial state to one of equilibrium, this study will set the warm-up period to 1 year (i.e., 2008), the calibration period to 2009–2010, and the verification period to 2011–2013.

3.3.1. Sensitivity Analysis

In this study, SWAT-CUP is used to calibrate the SWAT model that is driven by the CMADS. SWAT-CUP is an automated calibration and uncertainty analysis program developed by the EWAGE Institute for the SWAT model [14]. A sensitivity analysis analyzes the sensitivity of the model's parameters to the simulation results. In this study, a total of 26 parameters were analyzed for the sensitivity of the runoff-related parameters in order to obtain the ranking of the pattern sensitivity parameters (see Table 3).

Table 3. The final values of the SWAT model's parameters.

Parameter Names	Parameter Definitions	Parameters Final Values
CN2.mgt	SCS runoff curve values	34
ALPHA_BF.gw	Baseflow α factor	0.453
GW_DELAY.gw	Aquiclude replenish delay time (days)	42
GWQMN.gw	Water level threshold for shallow aquifers when groundwater is imported into the main river course	8.5
GW_REVAP.gw	Groundwater re-evaporation coefficient	0.03
ESCO.hru	Soil evaporation replenish coefficient	1.008
CH_L(1).rte	Manning value of the main water course	94.174
CH_K1.rte	River course effective infiltration coefficient	60.302
ALPHA_BNK.rte	Baseflow water recession coefficient	0.45
SFTMP	Average temperature on snowmelt day (°C)	4.8
PLAPS	Lapse rate of precipitation day (mm/km)	44.5
SMFMN	Snowmelt factor on 21 December (mm/day-°C)	9.407
SMFMX	Snowmelt factor on June 21	0.1302
TLAPS	Temperature lapse rate (°C/km)	−4.3039

3.3.2. Calibration scheme of SWAT

The first 14 sensitivity parameters are used as those for the later period's calibration and the model's parameters are calibrated in the model's calibration period (2009–2010). In this study, SWAT-CUP is used to calibrate the annual observation data of two stations in the JBR. The calibration process takes into account the annual average evaporation and precipitation of the basins. Then the calibration is extended to a monthly period. After the monthly data calibration has been completed, a daily observation of the data, parameters calibration and fine-tuning are conducted. In the calibration process, the relationship between the annual evaporation and runoff is considered first, and then it is confirmed that the simulation results are in accordance with the total annual evaporation and precipitation, and thus, that the runoff appears feasible. It was found that the precipitation reduction rate (PLAPS) in the variables was 44.5 mm/km. By analyzing the final values of the SWAT model's parameters (Table 3), it was also found that the temperature gradient model considers −4.3039 °C /km as its optimal value, and the above two parameters' calibration values agree with the actual average values over multiple years in the basin.

3.3.3. The Model's Evaluation

The Nash-Sutcliffe Efficiency (NSE) coefficient and R^2 deterministic coefficient are used as model evaluation indicators, as both have been widely used to evaluate the performance of the model [15,16]. Out of these, the NSE coefficient is a normal statistical equation, which reflects the observed value and the corresponding analog value of the degree of fit. The NSE can be calculated using Equation (1):

$$NSE = 1 - \frac{\sum_i (Q_m - Q_s)_i^2}{\sum_i (Q_{m,i} - \overline{Q_m})^2} \tag{1}$$

where, Q is the runoff value variable, Q_m and Q_s represent the observed and model values respectively, and $\overline{Q_m}$ indicates the mean observed runoff. The NSE is in the range of $-\infty$ to 1: when the NSE calculation result is 1, the observed value and the modal value can be regarded as consistent; when the value is between 0.5–1, the model's results are acceptable; and when the NSE is less than 0, the model's results can be considered as poor. The deterministic coefficient R^2 determines the degree of correlation between the variables (see Equation (2)).

$$R^2 = \frac{\left[\sum_i (Q_{m,i} - \overline{Q_m})(Q_{s,i} - \overline{Q_s}) \right]^2}{\sum_i (Q_{m,i} - \overline{Q_m})^2 \sum_i (Q_{s,i} - \overline{Q_s})^2} \tag{2}$$

where Q_m and Q_s are the same as in equation 6, and i indicates the *i*th observation or simulation value. Many researchers have used $R^2 > 0.5$ and NSE > 0.5 as a satisfactory criterion for the SWAT model [52], whilst others also believe that NSE > 0.4 can also be used as a criterion for a satisfactory model indicator. This study uses the criteria set by Moriasi et al [53]; that is, when the model is in its calibration stage, if the monthly scale simulation results are NSE ≥ 0.65, or if the daily scale simulation results are NSE ≥ 0.5, the model simulation results are acceptable [54].

4. Results and Discussion

4.1. Simulated runoff by CMADS+SWAT

In this study, the CMADS+SWAT model was used to export the monthly runoff of the two hydrological stations (Jinghe mountain station and Hot Spring station), and the parameters were calibrated (see Figures 3–6).

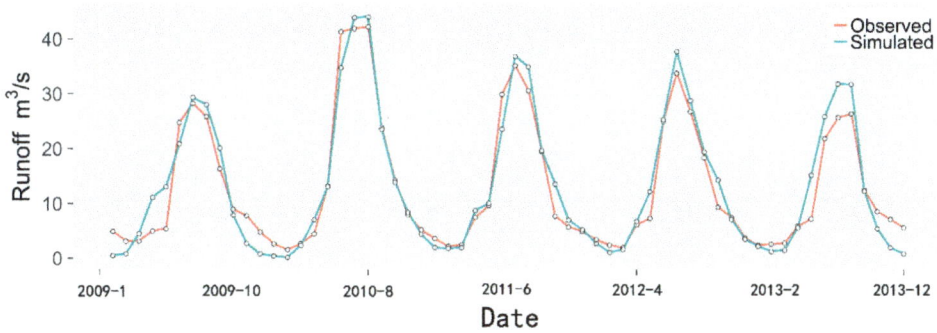

Figure 3. Monthly runoff (2009–2013) of Jinghe mountain station derived from the SWAT model driven by the CMADS.

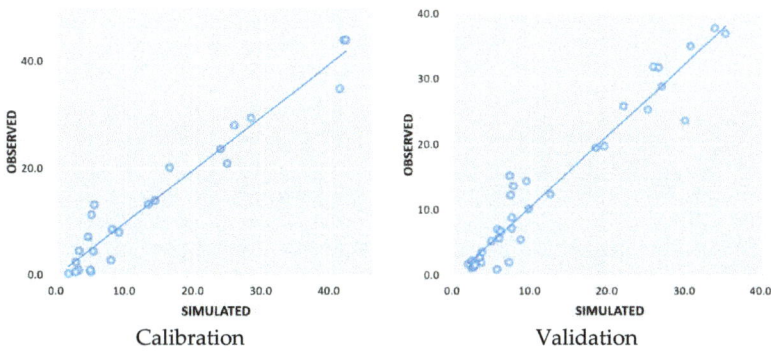

Calibration   Validation

Figure 4. Deterministic coefficients of the monthly observed and simulated runoff of Jinghe mountain station.

It was found that on a monthly scale, the SWAT model that is driven by CMADS has reached the satisfactory index (see Table 4) in the two control stations of the JBR. Moreover, on a monthly scale, the simulation results obtained by CMADS achieved satisfactory results (NSE = 0.939, R^2 = 0.942) at the Jinghe Mountain Station (calibration period).

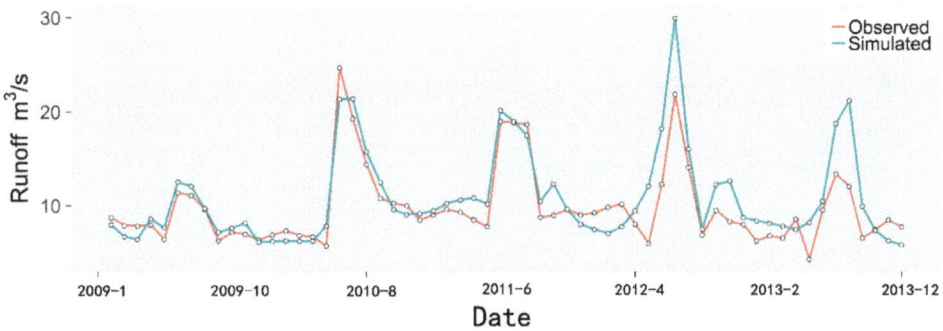

Figure 5. Monthly runoff (2009–2013) of Hot Spring station derived from the SWAT model driven by the CMADS.

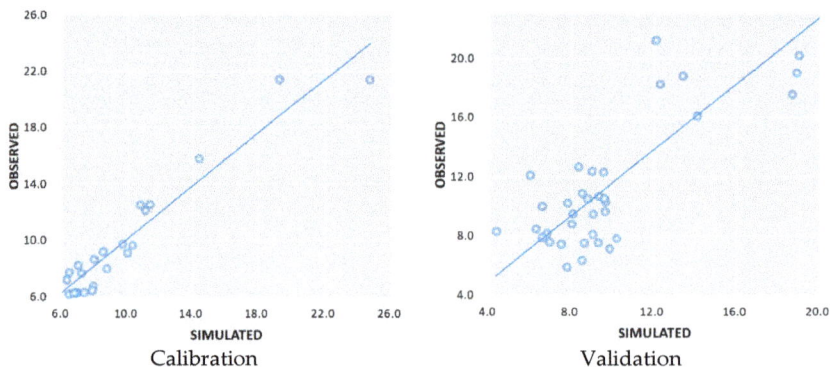

Calibration   Validation

Figure 6. Deterministic coefficients of the monthly observed and simulated runoff of the Hot Spring station.

Table 4. The assessment under a monthly and daily scale simulated by the CMADS+SWAT model.

	Jinghe Mountain Station				Hot Spring Station		
NSE	CP R²	VP NSE	VP R²	CP NSE	CP R²	VP NSE	VP R²
Month	0.939	0.942	0.904	0.934	0.917	0.914	0.659
Day	0.801	0.815	0.802	0.851	0.796	0.791	0.526

Note: CP represents Calibration Period, VP represents Verification Period.

During the validation period, although the NSE efficiency coefficient and the R^2 deterministic coefficient were slightly lower than those of the calibration period, they were overall satisfactory (NSE = 0.904, R^2 = 0.934). Compared with the Jinghe mountain station, the simulation accuracy of the Hot Spring station in the calibration and the verification periods were slightly lower. The authors believe that the glacier in the upper reaches of the Hot Spring has a great influence on the simulation results of the Hot Spring station. Furthermore, in the SWAT model, the degree-day factor only considers the snowmelt factor carefully, which leads to the simulation accuracy of the river basin in the Hot Spring station to be lower than that of the Jinghe Control Station (where the glacier replenishment rate is smaller than that of the former).

After completing the monthly scale calibration and verification, this study inputs the monthly scale optimal parameter value into the SWAT model for daily fine tuning and calibration. The results show that the SWAT model driven by the CMADS achieves acceptable results on a daily scale for the two control stations (see Figures 7–10, and Table 4). The runoff simulation results of the SWAT model driven by the CMADS demonstrate good consistency in the daily hydrological graphs of the two sub-watersheds.

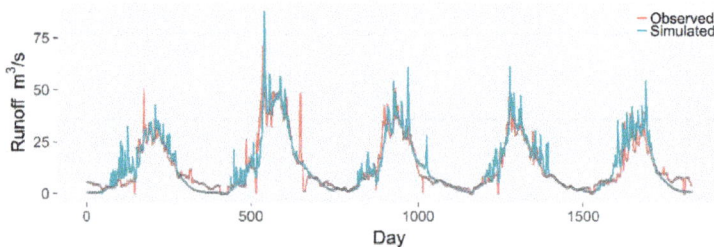

Figure 7. Daily runoff (2009–2013) of Jinghe mountain station derived from the SWAT model driven by the CMADS.

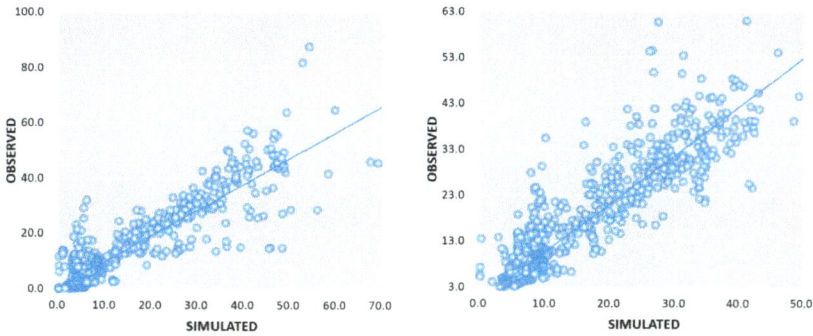

Figure 8. Deterministic coefficients of the daily observed and simulated runoff of Jinghe mountain station.

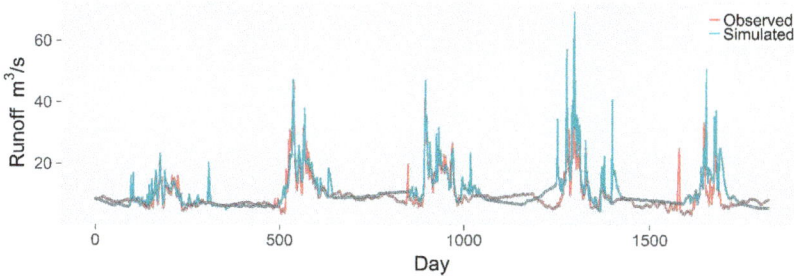

Figure 9. Daily runoff (2009–2013) of Hot Spring station derived from the SWAT model driven by the CMADS.

In the calibration period, the results for the Jinghe Mountain Station (NSE = 0.801, R^2 = 0.815) were similar to those of the Hot Spring Station (NSE = 0.0.796, R^2 = 0.791), and the results were simulated on the basis of the daily simulation driven by the CMADS. During the model's validation period, although the SWAT model was acceptable at both sites, the simulation results (NSE = 0.851, R^2 = 0.796) from the model in the Jinghe mountain Station were superior to those of the Hot Spring Station (NSE = 0.526, R^2 = 0.592).

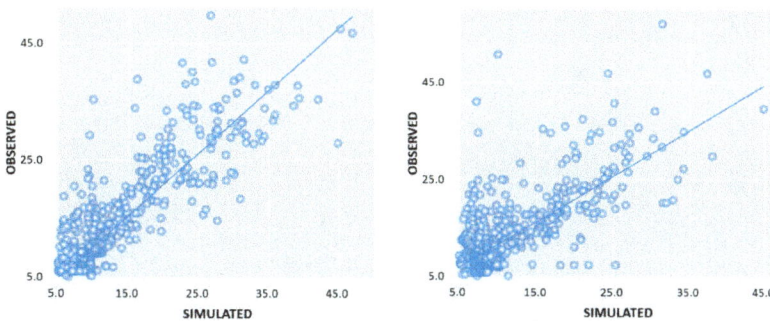

Figure 10. Deterministic coefficients of the daily observed and simulated runoff of Hot Spring station.

4.2. Spatial and Temporal Distribution Variation of Hydrological Processes

Since the SWAT model have been localized to the JBR and obtained satisfactory results, in order to analyze the ability of the CMADS+SWAT model to simulate the spatial and temporal evolution of the soil moisture and snowmelt variables from time and space perspectives, and to quantitatively analyze the response between the components, this study uses the Jinghe sub-basin as the main object of analysis object, and extracts the relationship between the multiple elements.

4.2.1. Response of Snowmelt Process and Soil Moisture

In order to study the effect of snowmelt on the soil moisture, this part of the study extracted the spatial variation of the soil moisture and the corresponding snowmelt (Figure 11) of the whole river basin on 8 April 2012. The left figure displays the soil moisture distribution, whilst the right one shows the corresponding time of the snowmelt's spatial distribution. In order to analyze the relationship between the various surface components in the basin quantitatively, this study also extracted the time series of various surface components (including the soil moisture, potential evapotranspiration, precipitation and snowmelt) in the river basin (see Figure 12).

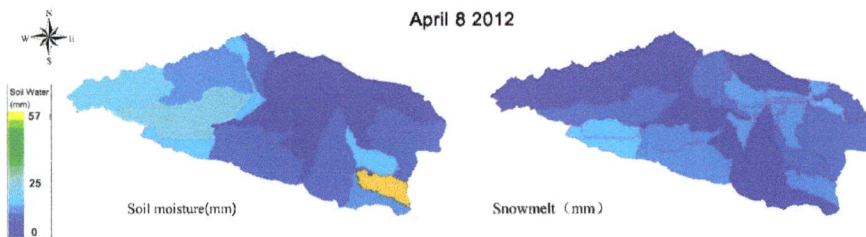

Figure 11. Spatial distribution of soil moisture and snowmelt rateof the JBR.

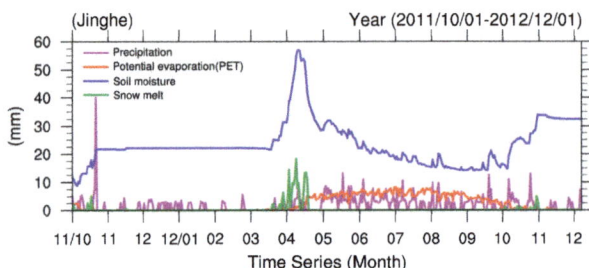

Figure 12. Time series of the parameters of the JBR (take the Jinghe mountain station as an example).

Through the distribution of the soil moisture throughout the JBR on 8 April 2012, it was found that the soil moisture in the entire basin was in a wet state on that day (see Figure 14A). Furthermore, where the Hot Spring station controls the sub-watershed soil, its moisture was between 18 mm and 20 mm, whereas the soil moisture of the Jinghe Control Station and its nearby sub-basin reached nearly 57 mm. It was also found that on 8 April 2012, by analyzing the spatial distribution of the corresponding snowmelt in the basin (Figure 14A1), there was a large snowmelt in the JBR. Moreover, the main area where snowmelt occurred was just the northern slope of the western Tianshan Mountains, which experiences a huge snow cover for 3-4 months in a year. In order to analyze the magnitude of snowmelt and soil moisture in the JBR quantitatively, and to analyze the direct response relationship, this study carried out a time series analysis of the soil moisture and snowmelt in a single sub-basin that is under the control of the Jinghe Hydrological Station (Jinghe Mountain Station).

Figure 12 shows the various types of surface or near-surface components (i.e., potential evapotranspiration, soil moisture, precipitation and amount of snowmelt) in the river basin. It was found that the snow melting phenomenon began to occur in the middle of March 2012, and the snowmelt phenomenon appeared on 8 April 2012. The snowfall of the Jinghe was 18 mm/day and the soil moisture in the natural sub-basin of the Jinghe Control Station also reached a high level (about 56 mm) during this period. The analysis also found that the most important contributor of the soil moisture's increase when the snow fell was the snow itself, with only a small amount coming from precipitation. In addition, the soil moisture was in inverse proportion to the potential evaporation values.

4.2.2. Response of Precipitation and Soil Moisture

This paper has analyzed the Jinghe sub-basin in the JBR, and focused on the influence of snowmelt on the soil moisture. However, in addition to the snow melt phenomenon, the impact of precipitation on the soil moisture cannot be ignored. This section analyzes the effects of precipitation on the soil moisture from summer to autumn, since the precipitation of snowfall in early spring has been analyzed. As the soil moisture in the study area will fluctuate during the long period of time between the snow melting period and September, and the sudden rise of the soil during the middle and late period of October will remain constant, the authors will compare the above two stages of precipitation and soil moisture using a response analysis of both time and space extraction; this will also serve to verify the changes of other variables (such as permafrost and snow depth) during the change of soil moisture in the Jinghe Mountain Station. This study also extracted the observation data of ice and snow (between 2010 and 2011) (Figure 13A,B,A1,B1).

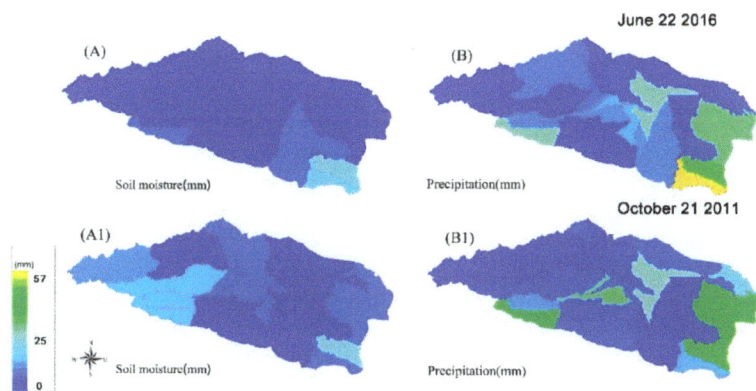

Figure 13. Correlation analysis of the soil moisture and precipitation in the JBR.

Figure 13A,B1 shows the spatial distribution of the soil moisture and precipitation in the JBR during the days of 22 June 2010 and 21 October 2011. Figure 13A shows the distribution of the soil moisture in the Bo River Basin on 22 June, 10 June, and Figure 13A1 shows the distribution of precipitation in the corresponding watershed. The results show that the soil moisture in the control basin was 24.8 mm, and the precipitation in the control area of the Jinghe Mountain Station was 30.1mm. The results also show that the increase of precipitation during the day led to the rapid increase of soil moisture in a short period. A similar situation also occurred in 2011, as shown in Figure 13A1, for the spatial distribution of soil moisture on 21 October. Figure 13B1 shows the distribution of precipitation over the entire basin in the corresponding period. It was found that the soil moisture in the sub-basin of the Jinghe mountain station and Hot Spring station is substantial, with 22 mm and 18mm, repectively. The analysis also found that, while the soil moisture in the basin reached a high level in the same period, the precipitation occurred in the whole sub-basin of the Jinghe and Bortala Rivers. In order to accurately verify the response relationship between the soil moisture and precipitation in order of magnitude, this study extracted other surface components (i.e., potential evapotranspiration, soil moisture, precipitation and snow melt) of the Jinghe mountain station (2011–2012) that corresponded with the other results (Figure 14AB).

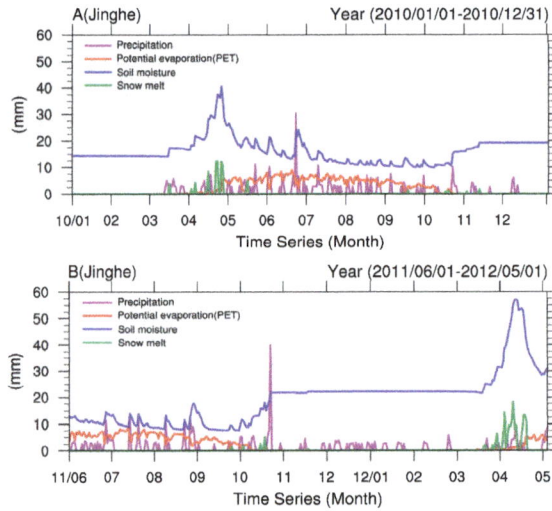

Figure 14. Time series of the soil moisture and precipitation in the JBR (Take the Jinghe Mountain station as an example).

The results show that the precipitation in the basin was close to 30 mm on 22 June 2010, and the soil moisture also reached 24.8 mm (Figure 14A), corresponding to the above spatial distribution. On 21 October 2011, the precipitation in the sub-basin of the Jinghe Yamaguchi Control Station reached nearly 40 mm. At the same time, the soil moisture value in the sub-basin also climbed to about 22 mm and remained constant thereafter (Figure 14B). The rapid increase of the soil moisture at the above two streams of the Jinghe Mountain station indicates that the precipitation in the late autumn in the Jinghe Mountain sub-basin plays an important role in the later changes of the soil moisture.

The overall change of the soil moisture from the annual circulation's perspective was analyzed. In the study area, the snow depth analyzed in the basin is from the third to fourth month of the research. The snow depth in the basin has experienced the rising and falling processes, and is close to 0 at the end of the snow melting. This is consistent with the snow depth observations extracted from the Jinghe Station (Figure 15A,C). In the later period, the climatic temperature of the study area rapidly increased, and as the air temperature is higher in this period, the frozen soil melts (Figure 15B,D). The soil evaporation also increased and accelerated the evaporation of soil water, which influenced the soil moisture's trend at this time. Ten months after the snow arrived, due to the cold and humid air flow over a wide range of precipitation (snow), the soil water content in the river basin increased significantly, which, as well as the cold air, is also caused by a significant reduction in evaporation. At this point, the soil liquid water content freezes (from November to December each year) and frozen soil is produced. Additionally, until the melting season occurs in the following year, when the frozen soil melting phenomenon occurs once again, part of the permafrost will be converted into soil liquid water.

In summary, the soil moisture in the basin will reach its first high level from March to April each year, mainly due to the snowmelt in the basin. After the end of the snowmelt period, the increase in precipitation and air temperature, as well as other phenomena, will lead to a trend in the soil temperature fluctuation. In mid-October the cold air arrives, leading to a large amount of precipitation (snow), and eventually turning the water in the soil into a frozen state. During the snow melt period in the following year, the soil liquid water increases again until the end of the snowmelt period.

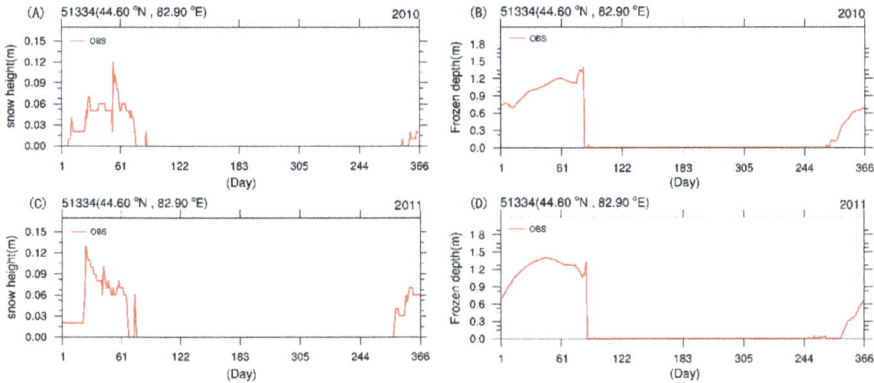

Figure 15. Snow depth and frozen depth values in 2010 and 2011 from the Jinghe Control Station.

Most researchers in China used traditional meteorological observatories with relatively scarce distribution when they studied the Western basin in China in the past. Similar studies have shown that scarce meteorological observatories do not represent the real underlying surface [48], resulting in greater uncertainty in meteorological inputs and the uncertainty of model output is affected [33]. For example, some researchers were able to obtain only one weather station on the north slope of the Tianshan mountains for surface analysis of Juntanghu watershed [29]. In addition, the mode output of climate change scenario prediction will be very unreliable when traditional weather stations are unable to effectively calibrate the SWAT model [1,32,34]; some scholars have realized that the western region of China needs to use the re-analysis data set to make up for the lack of data to drive SWAT. However, these researchers did not achieve good results by directly using reanalysis-data-driven SWAT without local correction. This is because most meteorological reanalysis data did not use automatic stations in China for correction, and cannot accurately reflect the intensity and frequency of the underlying surface in real weather [16]. Taking central Asia as an example, some scholars believe that an improper setting of climate simulation may lead to errors in simulation of ground climate field in regions with complex terrain and obvious spatial differences of meteorological elements [28]. For example, when the CFSR is used to drive the SWAT model, it can be found that the CFSR precipitation is overestimated in summer, resulting in overestimated runoff [44–46].

Different from other studies, CMADS driving SWAT model was selected in this study, and important surface process components (such as soil temperature, soil moisture, snowmelt, etc.) were obtained. Under the condition of SWAT calibration, we believed that these surface processes could be trusted. In the future, in order to reduce the output uncertainty of the SWAT model and obtain the best simulation results, we will focus on adjusting the parameterized scheme based on CMADS+SWAT mode. At the same time, the real-time data (CMADS-WRF) can be further used to drive the SWAT model, providing early warning for the western region of China to minimize local environmental pollution and property loss.

5. Conclusions

Based on the typical analysis and verification area of the JBR in Xinjiang, the China Meteorological Assimilation Driving Datasets for the SWAT model (CMADS) was selected to drive the SWAT model. The spatiotemporal distribution of soil moisture, snowmelt, evaporation and precipitation was analyzed within this area to verify the availability of the CMADS in western China.

The study found that:

(1)   The SWAT model was well calibrated by the CMADS. On the monthly scale, the performance of the SWAT model reached a satisfactory level in two control stations (Hot Spring and Jinghe

Mountain) within the JBR (see Table 4). During the validation period, the simulation result (NSE = 0.851, R^2 = 0.796) of the model at Jinghe Station was better than that in Hot Spring station (NSE = 0.526, R^2 = 0.592). After localization the SWAT model, the daily simulated runoff (year 2012) was extracted in the study area, and we found that the SWAT model reproduced the runoff perfectly in year 2012.

(2) From the perspective of time and space: During the snowmelt period, the main source of the soil moisture increase in the JBR is snowmelt, while only a small amount comes from precipitation; soil moisture and potential evaporation value show an inverse phenomenon; precipitation formed at the end of autumn in the sub-basin controlled by the Jinghe Mountain has played an important role in the later change of soil moisture. Soil moisture in the basin reaches its first high level in March-April each year, which is mainly caused by snow melting in the basin. After the snowmelt period, the soil temperature fluctuates upward and downward due to the increase of precipitation and the warming of air temperature. Until mid-October, the cold air transits and produces heavy precipitation (snow) and eventually converts the soil water into frozen soil. The next year, when the snowmelt period arrives, the liquid water of the soil increases again until the end of the snowmelt period.

(3) The overall research shows that the CMADS can localization the SWAT model perfectly and and can effectively calculate other surface processes (Soil moisture, snowmelt, etc.), We believe that CMADS will provide an important data base for the lack of sites in western China, and provide more raw material for scientific discovery.

Author Contributions: Y.W. were primarily accountable for data collection and design and coordination of the study. Y.L. were responsible for data analysis and writing of the paper. J.Z. and M.Y. were responsible for results presentation.

Funding: The research was founded by the National Science Foundation of China (41701076, 51709271, 41661040), the Young Elite Scientists Sponsorship Program by CAST (2017QNRC001) and the Xinjiang Youth Science and Technology Innovation Talents Training Project (2015), (2017).

Conflicts of Interest: The authors declare no conflict of interest.

References

1. Meng, X.; Wang, H.; Lei, X.; Cai, S.; Wu, H.; Ji, X.; Wang, J. Hydrological Modeling in the Manas River Basin Using Soil and Water Assessment Tool Driven by CMADS. *Teh. Vjesn.* **2017**, *24*, 525–534.
2. Berg, A.A.; Famiglietti, J.S.; Walker, J.P.; Houser, P.R. Impact of bias correction to reanalysis products on simulations of North American soil moisture and hydrological fluxes. *J. Geophys. Res. Atmos.* **2003**, *108*, 4490. [CrossRef]
3. Maurer, E.P.; Wood, A.W.; Adam, J.C.; Lettenmaier, D.P. A long-term hydrologically based dataset of LAND SURFACE FLUX and states for the conterminous United States. *J. Clim.* **2002**, *15*, 3237–3251. [CrossRef]
4. Fekete, B.M.; Vörösmarty, C.J.; Roads, J.O.; Willmott, C.J. Uncertainties in precipitation and their impacts on runoff estimates. *J. Clim.* **2004**, *17*, 294–304. [CrossRef]
5. Sheffield, J.; Ziegler, A.D.; Wood, E.F.; Chen, Y. Correction of the high-latitude rain day anomaly in the NCEP-NCAR reanalysis for land surface hydrological modeling. *J. Clim.* **2004**, *17*, 294–304. [CrossRef]
6. Monteith, J.L. Evaporation and the environment. In *The State and Movement of Water in Living Organisms, 19th Symposium of the Society for Experimental Biology*; Cambridge University Press: Swansea, UK, 1965; pp. 205–234.
7. Allen, R.G. A Penman for all seasons. *J. Irrig. Drain. Eng.* **1986**, *112*, 348–368. [CrossRef]
8. Allen, R.G.; Jensen, M.E.; Wright, J.L.; Burman, R.D. Operational estimates of reference evapotranspiration. *Agron. J.* **1989**, *81*, 650–662. [CrossRef]
9. Priestley, C.H.B.; Taylor, R.D. On the assessment of surface heat flux and evaporation using large-scale parameters. *Mon. Weather Rev.* **1972**, *100*, 81–92. [CrossRef]
10. Hargreaves, G.H. Moisture availability and crop production. *Trans. ASAE* **1975**, *18*, 980–984. [CrossRef]
11. Hargreaves, G.H.; Samani, Z.A. Reference crop evapotranspiration from temperature. *Appl. Eng. Agric.* **1985**, *1*, 96–99. [CrossRef]

12. Hargreaves, G.H.; Samani, Z.A. Estimating potential evapotranspiration. *J. Irrig. Drain. Eng.* **1982**, *108*, 225–230.
13. Leo Hargreaves, G.; Hargreaves, G.H.; Paul Riley, J. Agricultural benefits for Senegal River Basin. *J. Irrig. Drain. Eng.* **1985**, *111*, 113–124. [CrossRef]
14. Abbaspour, K.C.; Vejdani, M.; Haghighat, S. SWAT-CUP calibration and uncertainty programs for SWAT. In Proceedings of the MODSIM 2007 International Congress on Modelling and Simulation, Christenchurch, New Zealand, 10–13 December 2007; pp. 1596–1602.
15. Nash, J.E.; Sutcliffe, J.V. River Flow Forecasting through conceptual models part 1—A discussion of principles. *J. Hydrol.* **1970**, *10*, 282–290. [CrossRef]
16. Meng, X.; Wang, H.; Cai, S.; Zhang, X.; Leng, G.; Lei, X.; Shi, C.; Liu, S.; Shang, Y. The China Meteorological Assimilation Driving Datasets for the SWAT Model (CMADS) Application in China: A Case Study in Heihe River Basin. *Preprints.* **2016**. [CrossRef]
17. Kalnay, E.; Kanamitsu, M.; Kistler, R.; Collins, W.; Deaven, D.; Gandin, L.; Iredell, M.; Saha, S.; White, G.; Woollen, J.; et al. The NCEP/NCAR 40-year reanalysis project. *Bull. Amer. Meteorol. Soc.* **1996**, *77*, 437–472. [CrossRef]
18. Kanamitsu, M.; Ebisuzaki, W.; Woollen, J.; Yang, S.-K.; Hnilo, J.J.; Fiorino, M.; Potter, G.L. NCEP-DOE AMIP-II Reanalysis (R-2). *Bull. Am. Meteorol. Soc.* **2002**, *83*, 1631–1644. [CrossRef]
19. Uppala, S.M.; Kållberg, P.M.; Simmons, A.J.; Andrae, U.; Da Costa Bechtold, V.; Fiorino, M.; Gibson, J.K.; Haseler, J.; Hernandez, A.; Kelly, G.A.; et al. The ERA-40 re-analysis. *Q. J. R. Meteorol. Soc.* **2005**, *131*, 2961–3012. [CrossRef]
20. Onogi, K.; Tsutsui, J.; Koide, H.; Sakamoto, M.; Kobayashi, S.; Hatsushika, H.; Matsumoto, T.; Yamazaki, N.; Kamahori, H.; Takahashi, K.; et al. The JRA-25 Reanalysis. *J. Meteorol. Soc. Jpn. Ser. II* **2007**, *85*, 369–432. [CrossRef]
21. Zhao, T.; Ailikun; Feng, J. An intercomparison between NCEP reanalysis and observed data over China. *Clim. Environ. Res.* **2004**, *9*, 278–294.
22. Zhao, T.; Fu, C. Applicability evaluation of surface air temperature form several reanalysis dataset in China. *Plateau Meteorol.* **2009**, *28*, 595–607.
23. Zhao, T.; Fu, C. Preliminary comparison and analysis between ERA-40, NCEP-2 reanalysis and observations over Chnia. *Clim. Environ. Res.* **2006**, *11*, 15–33.
24. Shi, X.H.; Xu, X.D.; Xie, L.A. Reliability analyses of anomalies of NCEP/NCAR reanalysis wind speed and surface temperature in climate change research in China. *Acta Meteorol. Sin.* **2006**, *64*, 709–722.
25. Liu, S.; Yao, X.; Guo, W.; Xu, J.; Shangguan, D.; Wei, J.; Bao, W.; Wu, L. The contemporary glaciers in China based on the Second Chinese Glacier Inventory. *Acta Geogr. Sin.* **2015**, *70*, 3–16.
26. Meng, X.; Wang, H.; Shi, C.; Wu, Y.; Ji, X. Establishment and Evaluation of the China Meteorological Assimilation Driving Datasets for the SWAT Model (CMADS). *Water* **2018**, *10*, 1555. [CrossRef]
27. Meng, X.; Wang, H.; Wu, Y.; Long, A.; Wang, J.; Shi, C.; Ji, X. Investigating spatiotemporal changes of the land-surface processes in Xinjiang using high-resolution CLM3.5 and CLDAS: Soil temperature. *Sci. Rep.* **2017**, *7*, 13286. [CrossRef] [PubMed]
28. Meng, X.; Wang, H. Significance of the China Meteorological Assimilation Driving Datasets for the SWAT Model (CMADS) of East Asia. *Water* **2017**, *9*, 765. [CrossRef]
29. Shi, C.X.; Xie, Z.H.; Qian, H.; Liang, M.L.; Yang, X.C. China land soil moisture EnKF data assimilation based on satellite remote sensing data. *Sci. China Earth Sci.* **2011**, *54*, 1430–1440. [CrossRef]
30. Dong, N.; Yang, M.; Meng, X.; Liu, X.; Wang, Z.; Wang, H.; Yang, C. CMADS-Driven Simulation and Analysis of Reservoir Impacts on the Streamflow with a Simple Statistical Approach. *Water* **2018**, *11*, 178. [CrossRef]
31. Stamnes, K.; Tsay, S.C.; Wiscombe, W.; Jayaweera, K. Numerically stable algorithm for discrete-ordinate-method radiative transfer in multiple scattering and emitting layered media. *Appl. Opt.* **1988**, *27*, 2502–2509. [CrossRef] [PubMed]
32. Meng, X.-Y.; Yu, D.-L.; Liu, Z.-H. Energy balance-based SWAT model to simulate the mountain snowmelt and runoff—Taking the application in Juntanghu watershed (China) as an Example. *J. Mt. Sci.* **2015**, *12*, 368–381. [CrossRef]
33. Wang, Y.J.; Meng, X.Y.; Liu, Z.H. Snowmelt Runoff Analysis under Generated Climate Change Scenarios for the Juntanghu River Basin, in Xinjiang, China. *Tecnol. Cienc. Agua* **2017**, *7*, 41–54.
34. Meng, X.; Long, A.; Wu, Y.; Yin, G.; Wang, H.; Ji, X. Simulation and spatiotemporal pattern of air temperature and precipitation in Eastern Central Asia using RegCM. *Sci. Rep.* **2018**, *8*, 3639. [CrossRef] [PubMed]
35. Meng, X.; Sun, Z.; Zhao, H.; Ji, X.; Wang, H.; Xue, L.; Wu, H.; Zhu, Y. Spring Flood Forecasting Based on the WRF-TSRM Mode. *Teh. Vjesn.* **2018**, *1*, 141–151.

36. Zhao, F.; Wu, Y.; Qiu, L.; Sun, Y.; Sun, L. Parameter Uncertainty Analysis of the SWAT Model in a Mountain-Loess Transitional Watershed on the Chinese Loess Plateau. *Water* **2018**, *10*, 690. [CrossRef]
37. Vu, T.T.; Li, L.; Jun, K.S. Evaluation of Multi-Satellite Precipitation Products for Streamflow Simulations: A Case Study for the Han River Basin in the Korean Peninsula, East Asia. *Water* **2018**, *10*, 642. [CrossRef]
38. Liu, J.; Shanguan, D.; Liu, S.; Ding, Y. Evaluation and Hydrological Simulation of CMADS and CFSR Reanalysis Datasets in the Qinghai-Tibet Plateau. *Water* **2018**, *10*, 513. [CrossRef]
39. Cao, Y.; Zhang, J.; Yang, M.; Lei, X. Application of SWAT Model with CMADS Data to Estimate Hydrological Elements and Parameter Uncertainty Based on SUFI-2 Algorithm in the Lijiang River Basin, China. *Water* **2018**, *10*, 742. [CrossRef]
40. Shao, G.; Guan, Y.; Zhang, D.; Yu, B.; Zhu, J. The Impacts of Climate Variability and Land Use Change on Streamflow in the Hailiutu River Basin. *Water* **2018**, *10*, 814. [CrossRef]
41. Zhou, S.; Wang, Y.; Chang, J.; Guo, A.; Li, Z. Investigating the Dynamic Influence of Hydrological Model Parameters on Runoff Simulation Using Sequential Uncertainty Fitting-2-Based Multilevel-Factorial-Analysis Method. *Water* **2018**, *10*, 1177. [CrossRef]
42. Gao, X.; Zhu, Q.; Yang, Z.; Wang, H. Evaluation and Hydrological Application of CMADS against TRMM 3B42V7, PERSIANN-CDR, NCEP-CFSR, and Gauge-Based Datasets in Xiang River Basin of China. *Water* **2018**, *10*, 1225. [CrossRef]
43. Tian, Y.; Zhang, K.; Xu, Y.-P.; Gao, X.; Wang, J. Evaluation of Potential Evapotranspiration Based on CMADS Reanalysis Dataset over China. *Water* **2018**, *10*, 1126. [CrossRef]
44. Qin, G.; Liu, J.; Wang, T.; Xu, S.; Su, G. An Integrated Methodology to Analyze the Total Nitrogen Accumulation in a DrinkingWater Reservoir Based on the SWAT Model Driven by CMADS: A Case Study of the Biliuhe Reservoir in Northeast China. *Water.* **2018**, *10*, 1535. [CrossRef]
45. Ebita, A.; Kobayashi, S.; Ota, Y.; Moriya, M.; Kumabe, R.; Onogi, K.; Harada, Y.; Yasui, S.; Miyaoka, K.; Takahashi, K.; et al. The Japanese 55-Year Reanalysis "JRA-55": An interim report. *SOLA* **2011**, *7*, 149–152. [CrossRef]
46. Dee, D.P.; Uppala, S.M.; Simmons, A.J.; Berrisford, P.; Poli, P.; Kobayashi, S.; Andrae, U.; Balmaseda, M.A.; Balsamo, G.; Bauer, P.; et al. The ERA—Interim reanalysis: Configuration and performance of the data assimilation system. *Q. J. R. Meteor. Soc.* **2011**, *137*, 553–597. [CrossRef]
47. Saha, S.; Moorthi, S.; Pan, H.L.; Wu, X.; Wang, J.; Nadiga, S.; Tripp, P.; Kistler, R.; Woollen, J.; Behringer, D.; et al. The NCEP climate forecast system reanalysis. *Bull. Am. Meteorol. Soc.* **2010**, *91*, 1015–1057. [CrossRef]
48. Rienecker, M.M.; Suarez, M.J.; Gelaro, R.; Todling, R.; Bacmeister, J.; Liu, E.; Bosilovich, M.G.; Schubert, S.D.; Takacs, L.; Kim, G.; et al. MERRA: NASA's modern-era retrospective analysis for research and applications. *J. Clim.* **2011**, *24*, 3624–3648. [CrossRef]
49. Li, L.; Bai, L.; Yao, Y.; Yang, Q.; Zhao, X. Patterns of climate change in Xinjiang projected by IPCC SRES. *J. Resour. Ecol.* **2013**, *4*, 27–36.
50. Lioubimtseva, E.; Cole, R. Uncertainties of climate change in arid environments of Central Asia. *Rev. Fish. Sci.* **2006**, *14*, 29–49. [CrossRef]
51. Xu, Y.; Gao, X.; Shen, Y.; Xu, C.; Shi, Y.; Giorgi, F. A daily temperature dataset over China and its application in validating a RCM simulation. *Adv. Atmos. Sci.* **2009**, *26*, 763–772. [CrossRef]
52. Santhi, C.; Arnold, J.G.; Williams, J.R.; Hauck, L.M.; Dugas, W.A. Application of a watershed model to evaluate management efforts on point and nonpoint source pollution. *Trans. ASAE* **2011**, *44*, 1559–1570.
53. Nafees Ahmad, H.M.; Sinclair, A.; Jamieson, R.; Madani, A.; Hebb, D.; Havard, P.; Yiridoe, E.K. Modeling sediment and nitrogen export from a rural watershed in Eastern Canada using the soil and water assessment tool. *J. Environ. Qual.* **2011**, *40*, 1182–1194. [CrossRef] [PubMed]
54. Moriasi, D.N.; Arnold, J.G.; Van Liew, M.W.; Bingner, R.L.; Harmel, R.D.; Veith, T.L. Model Evaluation Guidelines for Systematic Quantification of Accuracy in Watershed Simulations. *Trans. ASABE* **2007**, *50*, 885–900. [CrossRef]

water

MDPI

Article

Moisture Distribution in Sloping Black Soil Farmland during the Freeze–Thaw Period in Northeastern China

Xianbo Zhao [1,2], Shiguo Xu [1,*], Tiejun Liu [3], Pengpeng Qiu [2] and Guoshuai Qin [1]

[1]   The Institution of Water and Environment Research, Dalian University of Technology, Dalian 116024, China; xianbozhao2004@126.com (X.Z.); qgs1991@mail.dlut.edu.cn (G.Q.)
[2]   Institute of Soil and Water Conservation, Heilongjiang Province Hydraulic Research Institute, Harbin 150080, China; qiupengpeng_01@163.com
[3]   Institute of Water Resources for Pastoral Area, Huhhot 010021, China; mksltj@126.com
*   Correspondence: sgxu@dlut.edu.cn; Tel.: +86-0411-8470-7680

Received: 30 January 2019; Accepted: 6 March 2019; Published: 14 March 2019

Abstract: This paper outlines dynamics of near-surface hydrothermal processes and analyzes the characteristics of moisture distribution during the freeze–thaw period in a typical black soil zone around Harbin, Northeastern China, a region with a moderate depth of seasonally frozen ground and one of the most important granaries in China. At Field Site 1, we analyzed the soil temperature and soil moisture content data from November 2011 to April 2012 from soil depths of 1, 5, 10, and 15 cm in sunny slope, and from depths of 1, 5, and 10 cm in shady slope black soil farmland. At Field Site 2, soil samples were collected from a 168 m long sloping black soil field at locations 10, 50, 100, and 150 m from the bottom of the slope at different depths of 0–1 cm, 1–5 cm, and 5–10 cm at the same location. Analysis of the monitored Site 1 soil temperature and soil moisture content data showed that the soil moisture content and soil temperature fit line is consistent with a Gaussian distribution rather than a linear distribution during the freeze–thaw period. The soil moisture content and time with temperature fit line is in accordance with a Gaussian distribution during the freeze–thaw period. Site 2 soil samples were analyzed, and the soil moisture contents of the sloping black soil farmland were obtained during six different freeze–thaw periods. It was verified that the soil moisture content and time with temperature fit line is in accordance with a Gaussian distribution during the six different freeze–thaw periods. The maximum surface soil moisture content was reached during the early freeze–thaw period, which is consistent with the natural phenomenon of early spring peak soil moisture content under temperature rise and snow melt. The soil moisture contents gradually increased from the top to the bottom in sloping black soil farmland during the freeze–thaw period. Since the soil moisture content is related to soil temperature during the freeze–thaw cycle, we validated the correlation between soil temperature spatiotemporal China Meteorological Assimilation Driving Datasets for the Soil and Water Assessment Tool (SWAT) model–Soil Temperature (CMADS-ST) data and monitored data. The practicality of CMADS-ST in black soil slope farmland in the seasonal frozen ground zone of the study area is very good. This research has important significance for decision-making for protecting water and soil environments in black soil slope farmland.

Keywords: sloping black soil farmland; soil moisture content; freeze–thaw period; soil temperature; CMADS-ST

1. Introduction

In this paper, the distribution of soil moisture in sloping black soil farmland during freeze–thaw cycles in Northeastern China is discussed. Soil moisture has great impacts on food security, human health, and ecosystem function; however, it is difficult to quantify the regional distribution and dynamics of soil moisture [1]. There are many factors that can influence soil moisture distribution. Previous research has shown that the soil moisture content and freezing/thawing processes have strong correlations with soil temperature [2]. Research in the Loess Plateau of China indicated that soil clay content and topography were the most important factors affecting soil water content at soil depths of 0–500 cm in the gully [3,4]. Studies in Northwestern China showed that land use and vegetation also have impacts on the soil moisture distribution [5–8]. Besides physicochemical factors, rainfall-runoff processes could also have great impacts on the soil moisture dynamics [9]. Furthermore, the antecedent soil moisture conditions also play an important role in rainfall-triggered shallow landslide events, which has been debated by Lazzari et al [10]. Large seasonal frozen soil regions exist all over the world. The freeze–thaw period is an important process influencing the soil moisture dynamic process. Snowmelt runoff is an important part of spring runoff and has an important impact on soil moisture distribution in frozen soil regions [11,12]. Studies of the soil moisture contents during freeze–thaw periods have been conducted. For instance, researchers have quantitatively studied the heat transfer of soil moisture during the freezing period and have provided water–heat coupling equations for freezing soil [13–16]. The first such equation was the moisture migration model, published in 1973 [17], which is based on the theory that the migration of unfrozen water in partially frozen ground is similar to the migration of moisture in unsaturated soil. Quantitative studies have produced numerical solutions for modeling water-heat coupling in freeze–thawing soils [18]. The application of these models to soils at mid and high latitudes requires the inclusion of freeze–thaw processes within the soil and the accumulation and ablation of snow cover [19]. Simulated soil temperature and soil liquid water contents are comparable to measured values, which show the isolating effect of snow cover. Under freeze–thaw conditions, plots have been studied to determine the effects of soil moisture on freeze–thaw processes [20–24]. The processes of soil freeze–thawing and the dynamic variations of moisture–heat transfer, including the soil water content, temperature, frost depth, soil evaporation, and water flux in the seasonal freeze–thawing period, were used in a one-dimensional simulation of soil water and heat dynamics during the winter period [25]. The importance of the heterogeneity of the infiltration process that results from the effects of the ground freezing, snow melting (including the contact between the melting snow cover and the soil), and unsaturated flow, is emphasized [26].

The black soil zone in Northeastern China is located in a region with a moderate depth of seasonally frozen ground (-2 ± 1 m), and which is one of the most important granaries in China. Plain-and-hill sloping black soil farmland ($<10°$) accounts for 60% of the total cultivated land (9.5×10^6 ha in Heilongjiang Province) in the black soil zone [27–30]. However, few studies have been conducted to determine the distribution of soil moisture with depth over time in sloping black soil farmland in Northeastern China during freeze–thaw cycles. Under the relevant boundary conditions of sloping farmlands in black soil plains and hills, exploring the changing process of freeze–thawing soils with soil moisture, and correlations between soil moisture and temperature, is important for protecting soil moisture in sloping black soil farmland. Besides this, soil temperature plays a key role in the land surface processes, since this parameter affects a series of physical, chemical, and biological processes in the soil, such as water and heat fluxes [31]. In the study of moisture distribution during the freeze–thaw period, it is very important to obtain soil temperature changes during soil freezing and thawing. However, soil temperature datasets are often difficult to obtain, requiring a large amount of equipment and time. Some researchers have used reanalysis datasets to investigate the relationship between soil temperature and soil moisture and obtained satisfactory results [32–36]. The China meteorological assimilation driving datasets for the soil and water assessment tool (SWAT) model–soil temperature (CMADS-ST) are the soil temperature reanalysis datasets of CMADS series datasets developed by Prof. Dr. Xianyong Meng from China and have attracted great attention [37]. CMADS incorporated technologies of local analysis and prediction system/space time multiscale analysis system (LAPS/STMAS) and was constructed using multiple technologies and scientific methods, including loop nesting of data, the projection of resampling models, and bilinear interpolation. The

CMADS has been used successfully in different basins, such as the Heihe River Basin, Juntanghu Basin, Manas River Basin, and Han River Basin, indicating good applicability of CMADS in East Asia [38–50]. However, the relative studies mainly focused on the surface hydrological process and meteorological data, whereas the application of the CMADS-ST to soil temperature and soil moisture distribution has been rarely studied, especially in the black soil zone [51–54].

In this background, it is necessary to compare and verify the applicability of the reanalysis datasets of CMADS-ST. Therefore, the aims of this paper are:

(1) To analyze the relationship between the reanalysis datasets of CMADS-ST and the observed values;
(2) to study the characteristics of moisture distribution during the freeze–thaw period; and
(3) to identify the effects of soil temperature on the moisture distribution in the black soil zone.

2. Materials and Methods

2.1. Experimental Site

Site 1: The Water Conservancy Comprehensive Experimental Research Center of Heilongjiang Province, China, is located on the outskirts of Harbin. Here, a typical moderately deep seasonally frozen ground experimental observation field was established from November 2011 to April 2012. Site 1 for field observations of natural freezing and thawing is located at 45°38′16.48″ N 126°22′51.31″ W at an elevation of 138 m a.s.l. and has a 5° slope. The site contains sunny slope and shady slope black soil farmland. The lowest monthly average temperature occurs in January, which was −24.6 °C in 2012 and −21.3 °C in 2014.

Site 2: An area of sloping black soil farmland in Longquan Town, Bayan County, Heilongjiang Province, China, was chosen as the experimental observation area, as shown in Figure 1. Site 2 for field observations of natural freezing and thawing is located at 46°12′08″ N 127°28′34″ W at an elevation of 158 m a.s.l. and has a 1.3° slope with a length of 168 m.

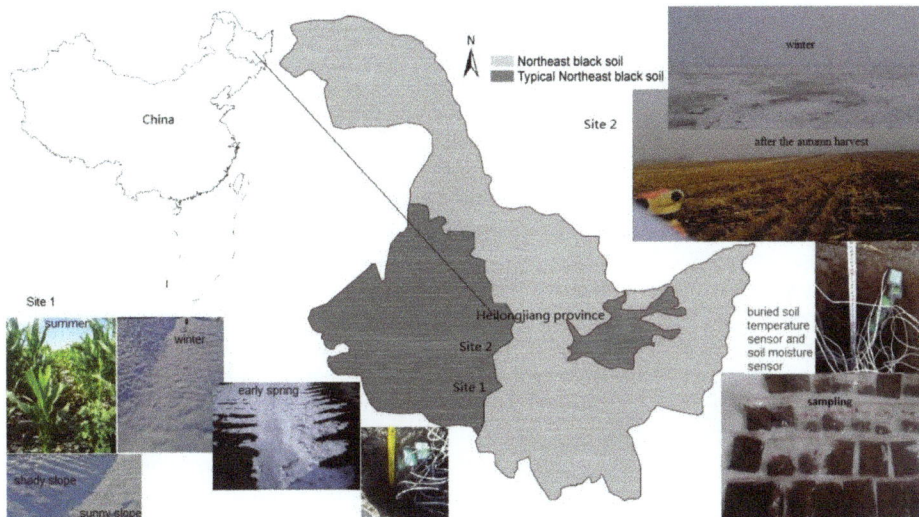

Figure 1. Map of the observation point locations and site case diagram, Heilongjiang Province, Northeastern China. Light gray indicates the northeast black soil zone and dark gray indicates the typical northeast black soil zone. Site 1—corn is grown in summer, snow cover in winter, shady slope and sunny slope obvious, early spring snowmelt. Site 2—using the level gauge to measure the height of black soil slope farmland after autumn harvest, buried soil temperature sensor and soil moisture sensor, snow in winter, black soil sampling.

The climate of the research zone is continental monsoon and cold temperate. The multi-year mean annual air temperature is 5.6 °C, the lowest monthly average temperature is −15.8 °C, and the mean annual precipitation is 423 mm. In the region of seasonally frozen ground, precipitation mainly occurs from May to September each year, with less precipitation occurring during late autumn, winter, and early spring. Four meteorological seasons occur during the year: spring, summer, autumn, and winter. The summer season is hot and rainy, and the winter season is cold and long. Planting occurs during the spring season in the middle of May, and crops are harvested during the autumn in early October. Crops are planted each year in monoculture using ridge tillage, and the main crop grown in the region is corn. The area consists of sloping black soil farmlands in the region of moderately deep seasonal frozen ground of Northeastern China. The average depth of the seasonal frozen ground between the years 1949 and 2010 was 1.95 ± 0.25 m at the site.

2.2. Experiment and Data Collection

Site 1: After the autumn harvest in 2011, soil temperature and moisture sensors were buried at depths of 1, 5, 10, and 15 cm in sunny slope black soil farmland, and at depths of 1, 5, and 10 cm in shady slope black soil farmland during the freeze–thaw periods. After the autumn harvest in 2013, the level was used to measure the height, the distance of the measuring tape is the slope distance, and then the slope of the slope farmland was calculated. Site 2: The soil temperature and moisture sensors were buried at depths of 1, 5, 10, and 15 cm in slope black soil farmland at a distance of 100 m from the bottom of the slope during the freeze–thaw periods. We observed the soil moisture content and temperature at various depths during the freeze–thaw period for the plots containing black soils in the region of seasonally frozen ground at sites 1 and 2. Precautions were taken during the installation of sensors used in the field, especially for the problem of soil temperature and the direction in which the soil moisture sensor was buried. For sites 1 and 2, firstly, pits were dug to a depth of 60 cm and a diameter of about 40 cm in the black soil slope farmland after the autumn harvest. When installing the soil temperature and soil moisture sensors, special attention was paid to the direction of burial. The direction of the sensor probe was toward the top of the slope. The soil temperature and soil moisture sensor were aligned parallel to the slope. The depth of the sloping farmland from the surface of the sloping farmland was measured with a steel tape measure. Finally, the soil from the pit was backfilled into the pit of the embedded instrument, and the compaction was made to be as close as possible to the original surface. The soil moisture and soil temperature sensor probes were oriented in the direction of the top of the slope to prevent disturbance of the excavation and to prevent the soil moisture from collecting, as the disturbance would affect the monitoring of soil moisture content.

Site 1: we collected data on the soil temperature and volumetric soil moisture content from November 2011 to April 2012 at depths of 1, 5, 10, and 15 cm in sunny slope black soil farmland, and at depths of 1, 5, and 10 cm in shady slope black soil farmland.

Site 2: Soil samples were collected from the 168 m long sloping black soil field at locations 10, 50, 100, and 150 m from the bottom of the slope for testing and analysis. During the freeze–thaw periods, including the pre-freezing period, freezing period, early freeze–thaw period, mid freeze–thaw period, late freeze–thaw period, and post-thawing period, surface-soil samples were taken from the sloping black soil farmland at depths of 0–1 cm, 1–5 cm, and 5–10 cm at the same location. The soil samples were tested to obtain mass soil moisture content (different from the volumetric soil moisture content obtained from Site 1), soil ammonia nitrogen content, and soil available phosphorus content. This paper only studies the distribution of soil moisture content; the distribution of soil ammonia nitrogen and soil available phosphorus are studied separately. At Site 2, soil layering temperature data were collected at hourly intervals on the day of sampling. The soil temperature layering depths were 1, 5, 10, and 15 cm, respectively. Site 2 sampling points, in the pre-freezing period, were at distances of 10, 50, 100, and 150 m from the bottom of the slope, and sampling from the surface of the black soil slope farmland was performed every 5 cm to 50 cm depth. An aluminum box was used to collect the marked soil. The soil samples were weighed and dried in the laboratory to obtain the soil bulk density.

We used high-resolution time series data from the CMADS-ST to conduct spatial- and temporal-scale analyses of meteorological data. We investigated the collection of the CMADS soil temperature and precipitation data with the test monitoring location close to the coordinates of the points, and the time synchronization of data. The moisture in the plow layer of black soils is significantly different at different depths during the freeze–thaw process depending on precipitation and soil temperature.

3. Results

3.1. Observed Soil Moisture and Soil Temperature Values

Figure 2 shows the soil moisture contents and soil temperature scatter diagrams that were observed at various depths during the 2011–2012 freeze–thaw period, including at the sunny slope depths of 1, 5, 10, and 15 cm and the shady slope depths of 1, 5, and 10 cm. The observation dates were from 20 November 2011 to18 April 2012. The daily volumetric soil moisture contents and soil temperatures were measured at experimental Site 1.

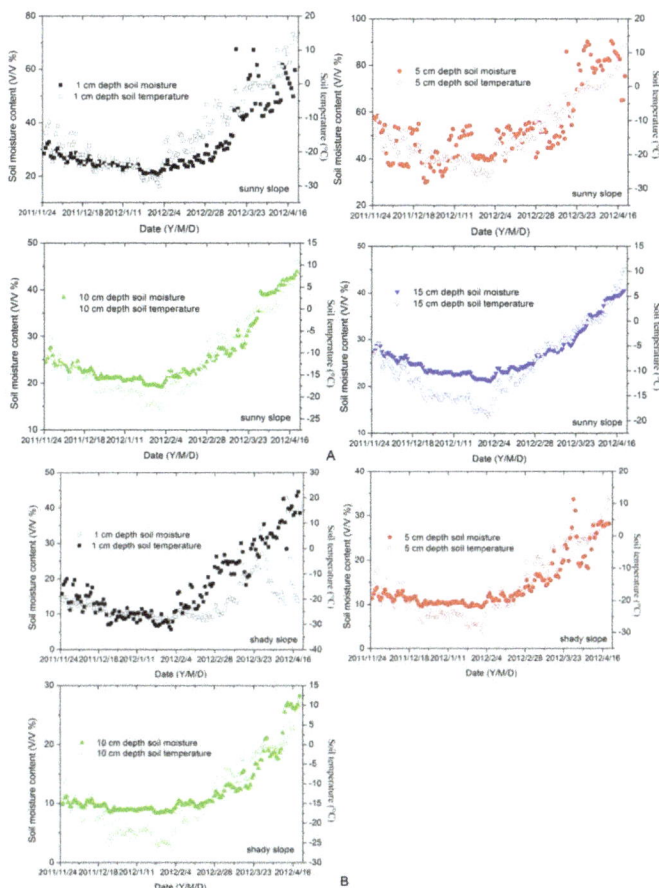

Figure 2. Moisture contents and soil temperature distribution at different depths during the freeze–thaw period. (**A**) Sunny slope black soil farmland; (**B**) shady slope black soil farmland. Solid symbols (squares, circle, up triangle, down triangle) represent soil moisture and open symbols represent soil temperature. Different colors represent soil moisture content at different depths of soil temperature. The abscissa is always the same and the ordinate is different each time.

Figure 2A shows that, for the sunny slope, the surface black soil temperature at a depth of 1 cm in the study area ranged from −31.9to 15 °C. The surface of the black soil at a depth of 1 cm was seasonally covered by ice and snow and experienced approximately 39 freeze–thaw cycles in the spring season of 2012. The shady slope experienced 47 freeze–thaw cycles in the spring season of 2012, as shown in Figure 3.

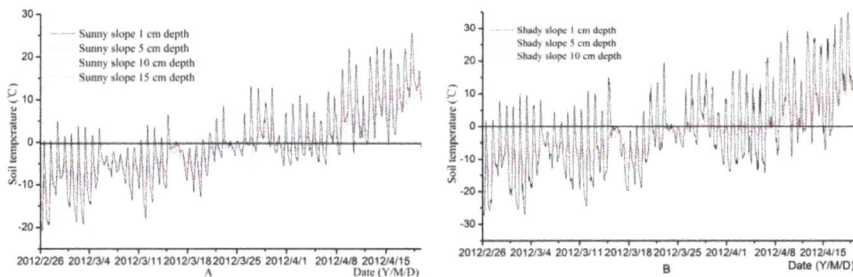

Figure 3. Variations of soil temperature in the black soil plow layer during the soil thawing period at different underground depths. (**A**) is sunny slope black soil farmland, soil temperature recorded hourly, 39 freeze–thaw cycles in the freeze–thaw period; (**B**) is shady slope black soil farmland, 47 freeze–thaw cycles in the freeze–thaw period. Data are from the period 26 February–20 April 2012.

The distribution of soil temperature and soil moisture was measured within the black soil plow layer during the freeze–thaw process. The soil temperature and moisture data were obtained from typical moderately deep seasonally frozen ground. This was an experimental observation field setup shown in Figure 2. Soil temperature changes affect the variations of soil moisture contents in seasonally frozen ground regions with black topsoil during soil freezing and thawing period. This shows that the soils under the plow layer in moderately deep seasonally frozen regions of black soil are usually subjected to a freeze–thaw period of approximately five months.

Figure 2 shows the distribution of the soil moisture contents and soil temperature at depths of 1, 5, 10, and 15 cm in the sunny slope black soil farmland and at depths of 1, 5, and 10 cm in the shady slope black soil farmland during the freeze–thaw period. The abscissa is always the same time of the freeze–thaw period from 20 November 2011 to18 April 2012.The left side of the ordinate Y-axis represents the soil temperature, and the temperature scale range is different at different depths. The right side of the ordinate Y-axis represents the soil moisture content, and the soil moisture content scale range is also different at different depths. Solid black squares, solid red circles, solid green up triangles, and solid blue down triangles represent soil moisture at depths of 1, 5, 10, and 15 cm, respectively. Open symbols represent soil temperature, and different symbol colors represent soil moisture content at different depths of soil temperature. It was found that the scatter patterns of soil moisture content and soil temperature are similar in the same freeze–thaw cycle period, at different depths.

Figure 4 shows changes in the soil moisture contents in the plow layer of black soils with temperature during a period of freeze–thaw cycles in an area of seasonal frozen ground. This figure shows that the moisture in the plow layer of black soils is significantly different at different depths during the freeze–thaw process depending on precipitation and soil temperature.

Figure 4 shows the relationships between precipitation and soil moisture contents at different depths during the freeze–thaw period. Precipitation data were obtained from the CMADS data-sharing website (http://www.cmads.org). The surface temperatures of the black soils at different depths (1, 5, and 10 cm) were measured at the experimental observation Site 1.

During the winter, precipitation occurs in the form of snowfall. The black sagging histogram in Figure 4 shows the precipitation in the study area. During the freeze–thaw period from20 November 2011 to18 April 2012, the daily maximum snowfall was 12 mm snow water equivalent. The total precipitation during the freeze–thaw period accounted for 17.3% of the annual amount (10 April 2011–10 April 2012).

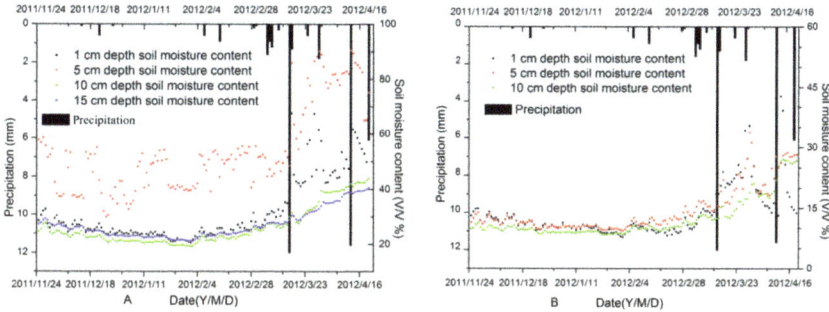

Figure 4. Moisture content and precipitation during the freeze–thaw periods. (**A**) is sunny slope black soil farmland; (**B**) is shady slope black soil farmland.

3.2. The Relationship between Soil Moisture and Soil Temperature

The relationship between soil temperature and soil moisture content in sloping black soil farmland during the freeze–thaw period is shown in Figure 5.

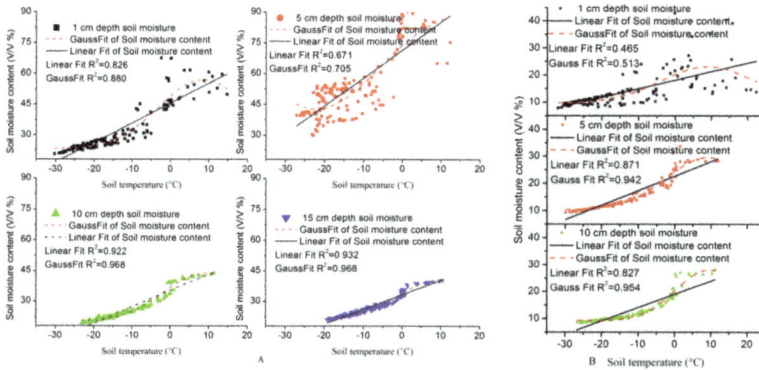

Figure 5. Comparison of Gaussian and linear fits of soil moisture content and temperature during the freeze–thaw periods. (**A**) is in sunny slope black soil farmland; (**B**) is in shady slope black soil farmland. The black solid line represents linear fit and the red dashed line represents Gaussian fit.

Figure 5A shows abscissa is according to the temperature with time as the axis of freeze–thawing periods from 20 November 2011 to 18 April 2012. Ordinate represent values of volumetric soil moisture contents. Small black solid squares, red solid circles, green solid up triangles, and blue solid down triangles respectively represent soil moisture contents at depths of 1, 5, 10, and 15 cm. The fitted curves plot of the solid lines correspond to a linear distribution of soil moisture contents and soil temperature. The fitted curves plot of the red dashed lines correspond to a Gaussian distribution of soil moisture and soil temperature. Figure 5A,B shows a comparison of Gaussian and linear fits of soil moisture content and temperature during the freeze–thaw periods in sun slope black soil farmland and shady slope black soil farmland, respectively.

As shown in Figure 5A, the linear fit for soil moisture content and temperature at depths of 1, 5, 10, and 15 cm in sunny slope black soil farmland during the freeze–thaw periods has coefficients of determination (R^2) of 0.826, 0.671, 0.922, and 0.932, respectively. The Gaussian fit for soil moisture content and temperature at depths of 1, 5, 10, and 15 cm have R^2 values of 0.880, 0.705, 0.968, and 0.968, respectively.

In Figure 5B, the linear fit for soil moisture content and temperature at depths of 1, 5, and 10 cm in shady slope black soil farmland during the freeze–thaw periods has R^2 values of 0.465, 0.871, and 0.827, respectively. The Gaussian fit for soil moisture content and temperature at depths of 1, 5, and 10 cm has R^2 values of 0.513, 0.942, and 0.954, respectively.

It is therefore found that the soil moisture content and soil temperature fit line is consistent with a Gaussian distribution rather than a linear distribution during the freeze–thaw period.

3.3. Soil Moisture Distribution in Freeze–Thaw Period

The relationship between soil moisture content and time during the freeze–thaw period in sloping black soil farmland is considered. The distribution of soil moisture content in the sunny and shady slope black soil farmland during the freeze–thaw periods is shown in Figure 6.

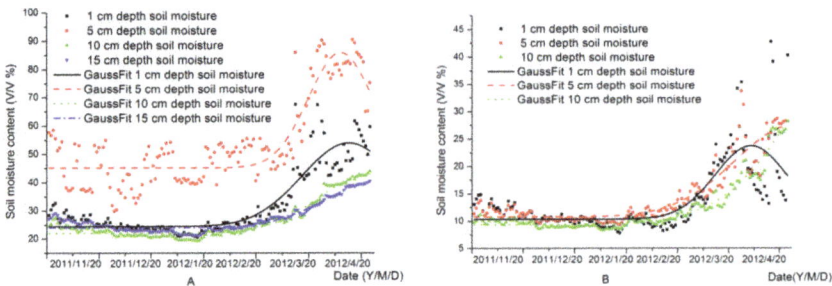

Figure 6. Distribution of soil moisture content during the freeze–thaw periods. (**A**) is in sunny slope black soil farmland; (**B**) is in shady slope black soil farmland.

Figure 6A shows abscissa is according to the time as the axis of freeze–thawing periods from 20 November 2011 to 18 April 2012. Ordinate represent the values of soil volumetric moisture contents. Small black solid squares, red solid circles, green solid up triangles, and blue solid down triangles respectively represent soil moisture contents at depths of 1, 5, 10, and 15 cm. The fitted curves plot of the black solid lines, red dashed lines, green dotted lines, and blue dashed and dotted lines respectively represent depths of 1, 5, 10, and 15 cm, and correspond to a Gaussian distribution of soil moisture and time. Figure 6A shows a Gaussian fit of soil moisture content and time in sunny slope black soil farmland during the freeze–thaw periods. Figure 6B shows the same as Figure 6A but for the shady slope.

A comparison of curve-fitting parameters of the distributions of soil moisture content is shown in Table 1.

Table 1 show that the soil moisture content and time obeys a Gaussian fit at depths of 1, 5, 10, and 15 cm in sunny slope black soil farmland during the freeze–thaw periods, with R^2 values of 0.835, 0.788, 0.948, and 0.910, respectively. The soil moisture content and time obeys a Gaussian fit at depths of 1, 5, and 10 cm in shady slope black soil farmland during the freeze–thaw periods, with R^2 values of 0.558, 0.867, and 0.954, respectively.

Therefore, the soil moisture content and time with temperature are in accordance with a Gaussian distribution during the freeze–thaw period.

Table 1. Curve-fitting parameters of the distribution of soil moisture content, 20 November 2011–18 April 2012.

Model				Gaussian				
Equation				$y = y_0 + \dfrac{A}{w\sqrt{\pi/2}} e^{-2(\frac{x-x_c}{w})^2}$				
	Depth	1 cm		5 cm		10 cm		15 cm
	Reduced Chi-Sqr	22.52		52.48		2.5		2.4
	R^2	0.835		0.788		0.948		0.91
Sunny slope	Prob > F	0.0001		0.0001		0.0001		0.0001
	Parameter	Value	Standard Error	Value	Standard Error	Value	Standard Error	Value	Standard Error
	y_0	24.46	0.53	45.22	0.74	21.93	0.18	23.93	0.17
	x_c	138.29	2.22	134.59	1.07	148.18	2.57	152.63	4.79
	w	42.78	3.98	29.31	2.41	46.92	3.28	49.92	5.37
	A	1581.8	159.9	1496.6	114.2	1240.6	124	1039.1	182.8
	Depth	1 cm		5 cm		10 cm		
	Reduced Chi-Sqr	18		4.4		1.14		
	R^2	0.558		0.867		0.954		
Shady slope	Prob > F	0.0001		0.0001		0.0001		
	Parameters	Value	Standard Error	Value	Standard Error	Value	Standard Error	
	y_0	10.38	0.46	10.8	0.25	9.31	0.13	
	x_c	130.54	1.86	151.28	6.02	204.25	26.85	
	w	33.8	4.37	56.95	7.18	82.07	15.14	
	A	565.5	69.2	1193	233.8	5023.9	3807.4	

Note: Where y_0, A, w, and x_c are parameters. $A > 0$; offset: y_0; center: x_c; width: w; area: A. Prob > F: Probability is greater than F test (Also called Homogeneity test of variance).

4. Analysis and Discussion

The soil moisture content and soil temperature or time fit line is in accordance with a Gaussian distribution during the freeze–thaw period. In this verification, soil samples were collected from Site 2, which is a 168 m long sloping black soil field, at distances of 10, 50, 100, and 150 m from the bottom of the slope for testing and analysis. During the freeze–thaw periods, surface-soil samples were taken from the sloping black soil farmland at depths of 0–1 cm, 1–5 cm, and 5–10 cm at the same location. At the Site 2 sampling point, in the pre-freezing period, at depths of 10, 50, 100, and 150 m from the bottom of the slope, sampling from the surface of the black soil slope farmland was conducted every 5 to 50 cm depth, and an aluminum box was used to collect the marked soil. In the laboratory, the soil samples were weighed and dried to obtain a soil bulk density measurement chart, as shown in Figure 7.

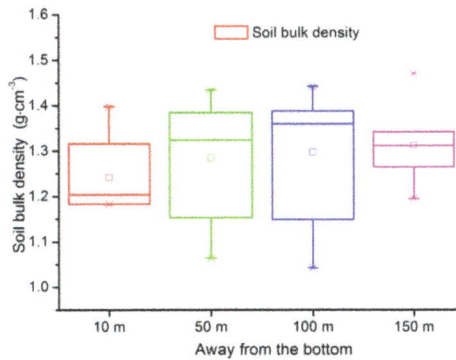

Figure 7. Soilbulk density measurement chart.

4.1. Freeze–Thaw Periods Divided into Six Periods

The freeze–thaw periods are abbreviated by F–T. The natural freeze–thaw cycle from late autumn to winter and early spring was divided into six periods: The pre-freezing period (F–T_p), freezing period (F–T_f), early freeze–thaw period (F–T_e), mid freeze–thaw period (F–T_m), late freeze–thaw period (F–T_l), and post-thawing period (F–T_{po}); which is a function of temperature related ($f(t)$).

These observations are presented in Table 2 temperature for n freeze–thaw periods (F–T_n), $n = 1,2,3,4,5,6$, for the observational study. Taking November 2013–April 2014 as an example, the freeze–thaw period is divided into six periods of different air temperatures and different depths of soil temperature. A schematic is shown in Figure 8.

Table 2. Freeze–thaw periods divided into six periods.

n	Periods	$F - T_n = f(t)$	Season	t_a	t_s
1	Pre-freezing period	F–$T_1 = F$–$T_P = f(t)$	After the autumn harvest to winter	>0 °C	>0 °C (all day)
2	Freezing period	F–$T_2 = F$–$T_f = f(t)$	Winter	<0 °C	<0 °C
3	Early freeze–thaw period	F–$T_3 = F$–$T_e = f(t)$	Early spring	>0 °C at daytime <0 °C at night	1 cm depth >0 °C at noon 1 cm depth <0 °C in the morning and at night
4	Mid freeze–thaw period	F–$T_5 = F$–$T_1 = f(t)$	Early-mid spring	>0 °C at daytime <0 °C at night	>0 °C at daytime <0 °C at night
5	Late freeze–thaw period	F–$T_5 = F$–$T_1 = f(t)$	Mid spring	>0 °C at daytime >0 °C during night (most of the time, and occasionally below 0 °C)	>0 °C at daytime ≈0 °C ↓↑ at night 1 cm depth >0 °C 5 cm depth <0 °C at night
6	Post-thawing period	F–$T_6 = F$–$T_{Po} = f(t)$	Late spring	>0 °C	>0 °C at depths of 1 cm, 5 cm, to 10 cm and 15 cm

Notes: The freeze–thaw periods are abbreviated by F–T, which is a function of temperature related ($f(t)$); F–$T_n = f(t)$, $n = 1,2,3,4,5,6$; > is more than, < is less than, ≈ is about equal, ↓↑ is up and down fluctuating; t is short for temperature, including air temperature and soil temperature, t_a is short for air temperature, and t_s is short for soil temperature.

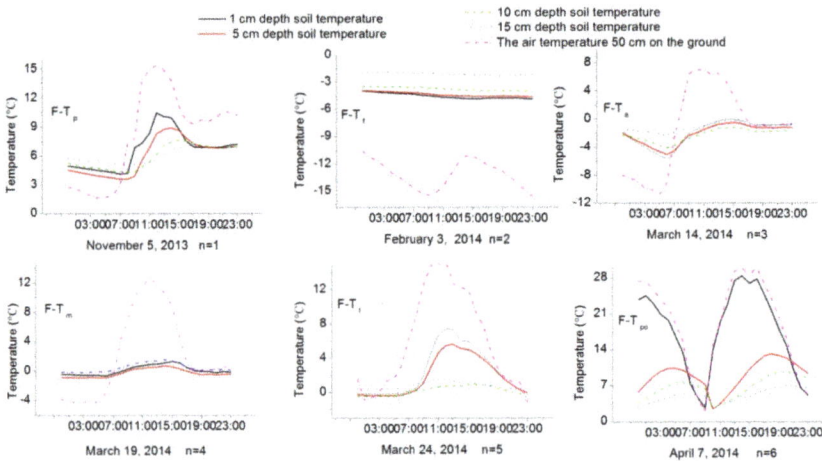

Figure 8. Temperatures on the day of sampling at different depths. Freeze–thaw periods are divided into six periods, including the pre-freezing period (F–T_p), freezing period (F–T_f), early freeze–thaw period (F–T_e), mid freeze–thaw period (F–T_m), late freeze–thaw period (F–T_l), and post-thawing period (F–T_{po}), according to the soil temperature.

Figure 8 shows the temperatures on the day of sampling at different depths. Time is the x-coordinate, from 00:00 to 23:00 per hour. Temperature is the y-coordinate, including the soil temperature at depths of 1, 5, 10, and 15 cm and the air temperature 50 cm above the ground. The soil moisture contents at the same locations during different periods were tested at depths of 1, 5, and 10 cm, respectively. Sampling was taken at different depths in the different freeze–thaw periods including F–T$_p$, F–T$_f$, F–T$_e$, F–T$_m$, F–T$_l$, and F–T$_{po}$, according to the soil temperature. Take the freeze–thaw period from October 2013 to April 2014 as an example. F–T$_p$ is from 20 October 2013 to 6 November 2013; F–T$_f$ is from 7 November 2013 to 9 March 2014; F–T$_e$ is from 10 March 2014 to 16 March 2014; F–T$_m$ is from 17 March 2014 to 22 March2014; F–T$_l$ is from 23 March 2014 to 25 March 2014; and F–T$_{po}$ is after 26 March 2014 to the end of April 2014; which is a function of temperature related. The specific sampling time selection is the F–T$_p$ on 5 November 2013, the F–T$_f$ on 3 February 2014, the F–T$_e$ on 14 March 2014, the F–T$_m$ on 19 March 2014, the F–T$_l$ on 24 March2014, and the F–T$_{po}$ on 7 April 2014.

4.2. Changes of Soil Moisture Content in Freeze–Thawing Period

Soil samples were collected at depths of 0–1 cm, 1–5 cm, and 5–10 cm, and at distances of 10 m, 50 m, 100 m, and 150 m from the bottom of the sloping black soil farmland in six different periods in time according to the temperature, F–T$_n$, n = 1,2,3,4,5,6.The soil moisture contents at the same locations during different periods were tested at depths of 0–1 cm, 1–5 cm, and 5–10 cm, respectively. Sampling was performed at different depths in the different freeze–thaw periods including F–T$_p$, F–T$_f$, F–T$_e$, F–T$_m$, F–T$_l$, and F–T$_{po}$, according to the soil temperature.

4.2.1. Plots of Soil Moisture Contents at the Same Depth

Soil moisture contents at the same depth (0–1 cm, 1–5 cm, 5–10 cm) during different freeze–thaw periods are shown in Figure 9.

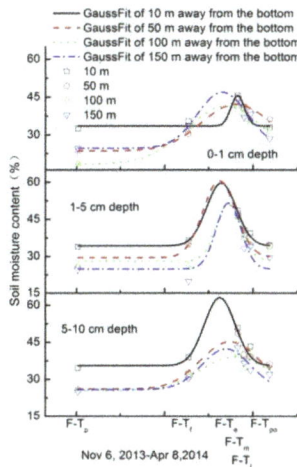

Figure 9. Distribution of soil moisture content at the same depth during the freeze–thaw periods.

Figure 9 shows abscissa is according to the temperature with time as the axis of freeze–thaw periods from F–T$_p$, F–T$_f$, F–T$_e$, F–T$_m$, F–T$_l$, and F–T$_{po}$; which is from 6 November 2013 to 8 April 2014. Ordinate represent depths of 0–1 cm, 1–5 cm, and 5–10 cm of soil moisture contents, respectively. Small black open squares represent a distance of 10 m away from the bottom to the base of the 168 m long sloping black soil land, in soil moisture contents at the F–T$_p$, F–T$_f$, F–T$_e$, F–T$_m$, F–T$_l$, and F–T$_{po}$ periods. Open red circles, open green up triangles, and open blue down triangles respectively represent distances of 50, 100, and 150 m away from the bottom in soil moisture contents in the six freeze–thaw periods.

Figure 9 shows soil moisture contents at the same depths during different freeze–thaw periods. The figure shows fitted curves plot of the black solid, red dashed, green dotted, and blue dashed and dotted lines, which respectively correspond to soil moisture contents at distances of 10, 50, 100, and 150 m away from the bottom to the base of the 168 m long sloping black soil land sampling point.

The soil moisture content is influenced by the initial soil moisture content before freezing and snowmelt [11,12,20,23,30,42] in the freeze–thaw period. The soil moisture content decreased slightly after the soil from a depth of 0–1 cm was thawed and increased at depths of 1–5 cm and 5–10 cm compared with the soil moisture content during $F-T_p$. The maximum soil moisture content at depths of 0–1 cm and 1–5 cm was reached during $F-T_e$ ($t_a > 0$ °C at daytime, 1 cm depth $t_s > 0$ °C at noon; $t_a < 0$ °C at night, 1 cm depth $t_s < 0$ °C in the morning and at night) due to snowmelt and decreased as the snow melted more with time and temperature. Additionally, the soil moisture content at depths of 0–1 cm, 1–5 cm, and 5–10 cm during $F-T_{po}$ was less than that during $F-T_l$ but greater than that during $F-T_p$. Due to the melting of snow in $F-T_e$, the surface soil moisture content is significantly increased in the $F-T_e$ compared to $F-T_f$, so the soil moisture content fit line is in accordance with Gaussian distribution during the six freeze–thaw periods. The fitted soil moisture content curves conform to Gaussian distributions at the same depths of 0–1 cm, 1–5 cm, and 5–10 cm at different locations in the six freeze–thaw periods.

4.2.2. Soil Moisture Contents Measured at the Same Location

Figure 10 shows plots of soil moisture content during the six different periods according to the temperature. Abscissa is according to the temperature with time as the x-axis showing the freeze–thawing periods $F-T_p$, $F-T_f$, $F-T_e$, $F-T_m$, $F-T_l$, and $F-T_{po}$; which is from 6 November 2013 to 8 April 2014. Ordinate represent distances of 10, 50, 100, and 150 m away from the bottom to the base of the 168 m long sloping black soil land sampling point of soil moisture contents, respectively. The solid black squares, solid red circles, and blue up triangles correspond to soil moisture contents at depths of 1, 5, and 10 cm, respectively.

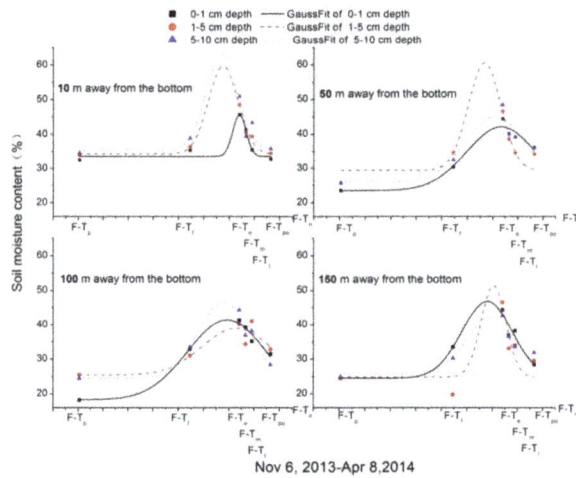

Figure 10. Plots of the soil moisture contents at the same location during the freeze–thaw periods.

Figure 10 shows fitted curves plot of soil moisture contents at the same location during different freeze–thaw periods according to the temperature. The black solid, red dashed, and blue dotted lines correspond to soil moisture contents at depths of 0–1 cm, 1–5 cm, and 5–10 cm, respectively.

At the same location 10 m away from the bottom of the 168 m long sloping black soil land sampling point, and at depths of 0–1 cm, 1–5 cm, and 5–10 cm, the soil moisture content, increased slightly for a while, then increased and finally decreased during the six different periods according to the temperature.

The fitted soil moisture content curves conform to Gaussian distributions at the same location at different depths in the six freeze–thaw periods. From 150 m away from the bottom of the 168 m long sloping black soil land sampling point, 100 m away from the bottom, 50 m away from the bottom, to 10 m away from the bottom, the soil moisture contents gradually increased at depths of 0–1 cm, 1–5 cm, and 5–10 cm during the six freeze–thaw periods. That is to say, in the sloping black soil farmland during the freeze–thaw periods, the soil moisture contents gradually increased from the top to the bottom.

4.3. Verification Analysis of Soil Moisture Contents Distribution

The fitted soil moisture content curves are shown in Figures 9 and 10, whether it is at the same depth different locations in the six freeze–thaw periods or at the same location different depths in the six freeze–thaw periods. The soil moisture content increased slightly for a while from $F–T_p$ to $F–T_f$, then increased from $F–T_f$ to $F–T_e$. The maximum soil moisture contents occurred during the $F–T_e$ period, decreased gradually from $F–T_m$ to $F–T_l$, and finally decreased during the six different periods according to the temperature. The soil moisture contents at depths of 0–1 cm, 1–5 cm, and 5–10 cm during $F–T_{po}$ were lower than those during $F–T_l$ but greater than that during $F–T_p$.

The soil moisture content increased and then decreased during the six time periods according to the temperature, and the fitted curves follow a Gaussian distribution.

The soil moisture content y follow a Gaussian distribution during the six freeze–thaw periods $(F–T)_n$ according to the temperature, as follows:

$$y = y_0 + \frac{A}{w\sqrt{\pi/2}} e^{-2\left(\frac{(F–T)_n - (F–T)_c}{w}\right)^2} \tag{1}$$

where y_0, A, w, and $(F–T)_c$ are parameters. $A > 0$; soil moisture content offset: y_0; the y_{max} corresponds to center: $(F–T)_c$; soil moisture content variation with time width: w; temperature area: A.

Table 3 shows the curve-fitting of the soil moisture content at distances of 10, 50, 100, and 150 m at depths of 0–1 cm, 1–5 cm, and 5–10 cm, respectively.

At the location 10 m away from the bottom of the 168 m long sloping black soil land sampling point, at depths of 0–1 cm, 1–5 cm, and 5–10 cm during the freeze–thaw period change in six different periods, the Gaussian distribution, when considering the reduced chi-squared statistic, is found to be with average standard deviations of 2.54, 2.31, and 17.79, respectively. The R^2 values of 0.905, 0.921, and 0.499 and Prob > F values of 0.0012, 0.001, and 0.0071, respectively, were observed. Prob > F means Probability, less than 0.05 means significant. Additionally, a similar Gaussian distribution was found at the locations 50, 100, and 150 m away from the bottom, at depths of 0–1 cm, 1–5 cm, and 5–10 cm during the freeze–thaw period change in six different periods.

Table 3 shows that the soil moisture contents in the plow layer of black soils at depths of 0–1 cm, 1–5 cm, and 5–10 cm, and at locations 10, 50, 100, and 150 m away from the bottom during the freeze–thaw period, change in six different periods ($F–T_p$, $F–T_f$, $F–T_e$, $F–T_m$, $F–T_l$, and $F–T_{po}$) and following a Gaussian distribution.

Often, only the conditions before and after the freeze–thaw cycles are considered [2–4]. As the data are not fine enough, the daily freeze–thaw cycle is missed every day during the daily data interval. However, the freeze–thaw period can be separated into six periods in time according to the temperature (air and soil): $F–T_p$, $F–T_f$, $F–T_e$, $F–T_m$, $F–T_l$, and $F–T_{po}$. It is necessary to understand how the soil moisture contents change during the periods between before and after freeze–thaw cycles. Obtaining the soil moisture contents during these periods and understanding the causes of their variations is necessary.

For the results of soil moisture content distribution during the freeze–thaw period in November 2011–April 2012, the soil sample test data of wild slope farmland during the freeze–thaw period from November 2013–April 2014 were used for verification. According to the freeze–thaw cycle, periods $F–T_p$, $F–T_f$, $F–T_e$, $F–T_m$, $F–T_l$, and $F–T_{po}$, according to the temperature, stratified sampling the soil moisture contents at different locations of 10, 50, 100, and 150 m away from the bottom, respectively. The result is that the soil moisture content at depths of 0–1 cm, 1–5 cm, and 5–10 cm is in accordance with a Gaussian function type distribution. The Gaussian R^2 values are 0.742 average soil moisture content at different locations and at different depths. Additionally, from November 2011 to April 2012 in freeze–thaw periods, according to the temperature monitoring of soil moisture sensor, results showed a similar distribution with the Gaussian, and R^2 values are 0.837 average of sun and shady soil moisture content at different depths.

Table 3. Curve-fitting parameters of the soil moisture content.

Model		Gaussian					
Equation		$y = y_0 + \frac{A}{w\sqrt{\pi/2}} e^{-2(\frac{(F-T)n - (F-T)c}{w})^2}$					
Location	Depth	0–1 cm		1–5 cm		5–10 cm	
10 m away from the bottom	Reduced Chi-Sqr	2.54		2.31		17.79	
	R^2	0.905		0.921		0.499	
	Prob > F	0.0012		0.001		0.0071	
	Parameters	Value	Standard Error	Value	Standard Error	Value	Standard Error
	y_0	33.57	0.92	34.3	1.12	35.56	3.12
	$(F-T)_c$	127.86	3.7	115.01	2.16	113.83	4.47
	W	10.6	5.87	23.41	3.42	24.38	9.37
	A	160.1	100.9	745.2	199.8	840.4	527.7
50 m away from the bottom	Reduced Chi-Sqr	5.38		18.15		17.21	
	R^2	0.905		0.612		0.703	
	Prob > F	0.0027		0.0092		0.0081	
	Parameters	Value	Standard Error	Value	Standard Error	Value	Standard Error
	y_0	23.49	2.32	29.49	3.23	26.11	4.14
	$(F-T)_c$	126.54	4.35	113.43	3.47	121.97	6.31
	W	55.04	9.94	25.95	8.64	46.72	14.33
	A	1286.6	326.8	1013.7	479.2	1118.1	497.2
100 m away from the bottom	Reduced Chi-Sqr	2.91		15.76		15.69	
	R^2	0.956		0.73		0.55	
	Prob > F	0.0017		0.0089		0.0087	
	Parameters	Value	Standard Error	Value	Standard Error	Value	Standard Error
	y_0	18.31	1.72	27.99	3.02	25.43	3.96
	$(F-T)_c$	118.3	2.39	120.6	32.1	125.61	9.92
	W	61.83	7.86	20.13	30.46	54.31	22.84
	A	1791.3	265.9	579.2	1837.6	923.5	547.1
150 m away from the bottom	Reduced Chi-Sqr	6.69		38.79		10.76	
	R^2	0.87		0.544		0.705	
	Prob > F	0.0036		0.0245		0.0063	
	Parameters	Value	Standard Error	Value	Standard Error	Value	Standard Error
	y_0	24.52	2.51	24.83	4.58	25.87	3.2
	$(F-T)_c$	115.75	2.7	120.73	49.38	118.72	5.25
	W	39.85	8.15	19.6	45.74	38.11	12.26
	A	1117.9	254.2	657.3	3268.8	784.4	334.2

In the early spring, when the temperature rises and the snow melts [11,12,20,23,27,30,42], water infiltrates into the soil moisture, the shallow soil moisture increases, the bottom soil freezes and does not melt, and the soil moisture in the freeze–thaw cycle shows a Gaussian distribution, which is consistent with the natural phenomenon of peak soil moisture content.

4.4. Soil Temperature Observed Value Associated with the CMADS-ST Analysis

Soil temperature data monitored at Site 1 is a group of every 1 h, divided into four layers (sunny slope) or three layers (shady slope), the time scale from November 2011 to mid-April 2012. We averaged the ground temperature of four daily time periods, namely, 02:00, 08:00, 14:00, and 20:00. The temperature of these four time periods was added and then divided by four. The average daily soil temperature for different layers can be derived. We selected CMADS-ST data of the top five layers (first layer: 0.007 m; second layer: 0.028 m; third layer: 0.062 m; fourth layer: 0.119 m; fifth layer: 0.212 m) for comparison, from the time scale from November 2011 to mid-April 2012, of daily time resolution, from the locations 139–207(45°59′07″ N 128°37′45″ W) and 137–205(45°19′09″ N 127°57′47.7″ W). Soil temperature at Site 1 was correlated with the daily average soil stratification temperature of locations 139–207 and 137–205 CMADS-ST near the observation point. The soil temperature observations from November 2011 to 20 April 2012 was compared with CMADS-ST, as shown in Figure 11.

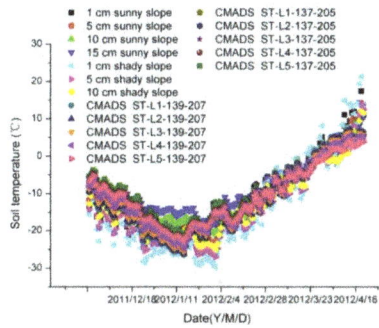

Figure 11. Comparison of observed soil temperature with the China meteorological assimilation driving datasets for the soil and water assessment tool (SWAT) model–soil temperature (CMADS-ST) (November 2011 to mid-April 2012).

The soil temperature (sunny slope and shady slope) monitoring data of Site 1 were used to analyze the CMADS-ST position points 139–207 and 137–205, respectively, and a linear correlation was found. Soil temperatures (sunny slope and shady slope) and locations 139–207 and 137–205 correlation coefficient R^2 value was calculated. The result showed that the sunny slope temperature R^2 is between 0.841–0.897, with an average value of 0.880. The mean value between 0.822–0.907 on the shady slope is 0.875; R^2 (CMADS-ST for locations 139–207) >R^2 (CMADS-ST for locations 137–205). It can be shown that position 139–207 is closer to Site 1 than 137–205. The results of CMADS-ST are in good agreement with the soil temperature measured by the fixed point of Site 1. The practicality of CMADS-ST in black soil slope farmland in the seasonal frozen soil area of the study area is very good. The variation of soil layer temperature in the seasonally frozen ground zone of the longer series CMADS-ST from 1 January 2009 to 31 December 2013 is shown in Figure 12.

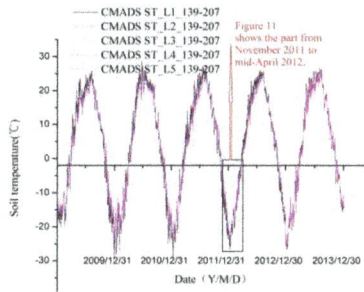

Figure 12. Seasonal variation of soil layer temperature with the CMADS-ST in seasonally frozen ground zone between 1 January 2009 and 31 December 2013 for CMADS-ST locations 139–207. The black box section shows some soil layer temperature data from November 2011 to mid-April 2012, as shown in Figure 11.

The soil temperature data of the sunny slope is more correlated with CMADS-ST, which coincides with the selection of the sunny slope in the wild slope farmland as the test Site 2. The monitored hourly data can be found in the freeze–thaw cycle with the time of day, melting during the day, and freezing at night (refer to Figure 3, the sunny slope soil experienced 39 freeze–thaw cycles, and the shady slope soil 47 cycles); however, the data period is short and the monitoring points are limited; CMADS-ST daily data can only see a large freeze–thaw cycle in the winter of the yearly cycle (refer to Figure 12); however, CMADS has a lot of spatiotemporal data, applied to a wide range of areas with a long series [11,31,33,38–54].Using fixed-point monitoring of refined soil temperature, soil moisture content, precipitation, temperature, nitrogen and phosphorus of nutrients, spatiotemporal CMADS data can be better promoted and applied.

5. Conclusions

This study, which investigated the distribution of soil moisture contents in the surface black soils of seasonal frozen grounds during freeze–thaw periods, lays a foundation for protecting the water resources downstream instead of water carrying soil into the water and the soil resources of sloping black soil farmlands. The soil moisture content in the black soil plow layer changed with temperature during a freeze–thaw period in an area with seasonally frozen grounds.

It was found that the soil moisture content and soil temperature fit line is consistent with Gaussian distribution during the freeze–thaw period. The soil moisture content and time with temperature are in accordance with Gaussian distribution during the freeze–thaw period. This is consistent with the natural phenomenon, that is, the surface soil moisture content in the early spring will peak, whereas in the seasonal frozen ground zone, the temperature increases, the snow melts, the soil surface melts, and the bottom layer freezes.

In this paper, the natural freeze–thaw cycle from late autumn to winter and early spring was divided into six periods according to the temperature: The pre-freezing period (F–T_p), freezing period (F–T_f), early freeze–thaw period (F–T_e), mid freeze–thaw period (F–T_m), late freeze–thaw period (F–T_l), and post-thawing period (F–T_{po}). According to the soil temperature in the six time periods during the different freeze–thaw periods, it was verified that the soil moisture contents followed a Gaussian distribution at different depths. The maximum surface soil moisture content was reached during the F–T_e period, due to air temperature being more than 0 °C at daytime, surface soil temperature more than 0 °C, and deep soil temperature less than 0 °C at noon, and air temperature less than 0 °C at night, surface soil temperature less than 0 °C in the morning and at night, start melting snow, which is consistent with the natural phenomenon of early spring peak soil moisture content under temperature rise and snow melt. The soil moisture contents gradually increased from the top to the bottom in sloping black soil farmland during the freeze–thaw period.

The used measurements and analytical methods are somewhat valuable as a result of difficult implementation due to the high latitude and freezing climate. These results suggest that further research is necessary on the use of spatiotemporal data CMADS-ST and precipitation datasets of the northeast black soil area for the freeze–thaw cycle prediction of soil moisture content change trend. Based on the relationship between soil temperature and soil moisture content, we can use CMADS-ST data to predict and analyze soil moisture changes in large areas (black soil areas) rather than at several points during freezing and thawing cycles. This study has important significance for decision-making for protecting water and soil environments in black soil slope farmland.

Author Contributions: X.Z. and S.X. conceived and designed the field experiments; X.Z. and T.L. performed the field experiments; P.Q. analyzed the soil sample; G.Q. analyzed the CMADS-ST; X.Z. contributed test site/monitoring instrument/reagents/materials/analysis tools; and X.Z. wrote the paper.

Funding: This study was supported by the National Natural Science Foundation of China (NSFC) (41201264, 51579157, 51779156), the National key research and development plan project of China, (2018YFC0507000), the Heilongjiang Province of China Youth Science Foundation (QC2010099), Heilongjiang Province of China Science and Technology Special Foundation (201606).

Conflicts of Interest: The authors declare no conflict of interest.

References

1. Kaighin, A.M.; Alemohammad, S.H.; Akbar, R.; Konings, A.G.; Yueh, S.; Entekhabi, D. The global distribution and dynamics of surface soil moisture. *Nat. Geosci.* **2017**, *10*, 100–104. [CrossRef]
2. Yang, M.; Yao, T.; Gou, X.; Koike, T.; He, Y. The soil moisture distribution, thawing-freezing processes and their effects on the seasonal transition on the Qinghai-Xizang (Tibetan) plateau. *J. Asian Earth Sci.* **2003**, *21*, 457–465. [CrossRef]

3. Li, T.; Shao, M.; Jia, X.; Huang, L. Profile distribution of soil moisture in the gully on the northern Loess Plateau, China. *Catena* **2018**, *171*, 460–468. [CrossRef]

4. Zhang, X.; Zhao, W.; Wang, L.; Liu, Y.; Liu, Y.; Feng, Q. Relationship between soil water content and soil particle size on typical slopes of the Loess Plateau during a drought year. *Sci. Total Environ.* **2019**, *648*, 943–954. [CrossRef] [PubMed]

5. Wei, W.; Feng, X.; Yang, L.; Chen, L.; Feng, T.; Chen, D. The effects of terracing and vegetation on soil moisture retention in a dry hilly catchment in China. *Sci. Total Environ.* **2019**, *647*, 1323–1332. [CrossRef]

6. Yang, L.; Chen, L.; Wei, W. Effects of vegetation restoration on the spatial distribution of soil moisture at the hillslope scale in semi-arid regions. *Catena* **2015**, *124*, 138–146. [CrossRef]

7. Yang, L.; Wei, W.; Chen, L.; Mo, B. Response of deep soil moisture to land use and afforestation in the semi-arid Loess Plateau, China. *J. Hydrol.* **2012**, *475*, 111–122. [CrossRef]

8. Wang, X.; Pan, Y.; Zhang, Y.; Dou, D.; Hu, R.; Zhang, H. Temporal stability analysis of surface and subsurface soil moisture for a transect in artificial revegetation desert area, China. *J. Hydrol.* **2013**, *507*, 100–109. [CrossRef]

9. Zhao, N.; Yu, F.; Li, C.; Wang, H.; Liu, J.; Mu, W. Investigation of Rainfall-Runoff Processes and Soil Moisture Dynamics in Grassland Plots under Simulated Rainfall Conditions. *Water* **2014**, *6*, 2671–2689. [CrossRef]

10. Lazzari, M.; Piccarreta, M.; Manfreda, S. The role of antecedent soil moisture conditions on rainfall-triggered shallow landslides. *Nat. Hazards Earth Syst. Sci. Discuss.* **2018**. [CrossRef]

11. Meng, X.; Dan, L.; Liu, Z. Energy balance-based SWAT model to simulate the mountain snowmelt and runoff—Taking the application in Juntanghu watershed (China) as an example. *J. Mt. Sci.* **2015**, *12*, 368–381. [CrossRef]

12. Wang, Y.; Meng, X. Snowmelt runoff analysis under generated climate change scenarios for the Juntanghu River basin in Xinjiang, China. *Tecnol. Cienc. Agua* **2016**, *7*, 41–54.

13. Ala, M.; Ya, L.; Anzhi, W.; Cunyang, N. Characteristics of soil freeze–thaw cycles and their effects on water enrichment in the rhizosphere. *Geoderma* **2016**, *264*, 132–139.

14. Chen, S.; Ouyang, W.; Hao, F.; Zhao, X. Combined impacts of freeze–thaw processes on paddy land and dry land in Northeast China. *Sci. Total Environ.* **2013**, *456*, 24–33. [CrossRef] [PubMed]

15. Kelln, C.; Barbour, S.L.; Qualizza, C. Controls on the spatial distribution of soil moisture and solute transport in a sloping reclamation cover. *Can. Geotech. J.* **2008**, *45*, 351–366. [CrossRef]

16. Messiga, A.J.; Ziadi, N.; Morel, C.; Parent, L.E. Soil phosphorus availability in no-till versus conventional tillage following freezing and thawing cycles. *Can. J. Soil Sci.* **2010**, *90*, 419–428. [CrossRef]

17. Lei, Z.; Shang, S.; Yang, S.; Wang, Y.; Zhao, D. Simulation on phreatic evaporation during soil freezing. *J. Hydraul. Eng.* **1999**, *6*, 8–12.

18. Harlan, R.L. Analysis of coupled heat-fluid transport in partially frozen soil. *Water Resour. Res.* **1973**, *9*, 1314–1323. [CrossRef]

19. Zhang, H.; Chen, X.; Hu, Y. Freeze under the conditions of unsaturated soil, water, heat, and solute coupled transport modeling. *Yangtze River* **2009**, *40*, 78–80.

20. Warrach, K.; Mengelkamp, H.T.; Raschke, E. Treatment of frozen soil and snow cover in the land surface model SEWAB. *Theor. Appl. Climatol.* **2001**, *69*, 23–37. [CrossRef]

21. Sinha, T.; Cherkauer, K.A. Time series analysis of soil freeze and thaw processes in Indiana. *J. Hydrometeorol.* **2008**, *9*, 936–950. [CrossRef]

22. Zhang, X.; Sun, S. The impact of soil freezing/thawing processes on water and energy balances. *Adv. Atmos. Sci.* **2011**, *28*, 169–177. [CrossRef]

23. Wei, D.; Chen, X.; Wang, T. Migration of different snow cover conditions, soil freezing and soil water. *J. Anhui Agric. Sci.* **2007**, *12*, 3570–3572.

24. Liu, T.; Huang, Y.; Zhang, R.; Li, J. Experiment study on the law of thermal motion of water of soil freeze-thaw on the black land. In *Proceedings of the Second International Conference on Mechanic Automation and Control Engineering Institute of Electrical and Electronics Engineers, Inner Mongolia, China, 15–17 July 2011*.

25. Li, R.; Shi, H.; Flerchinger, G.N.; Akae, T.; Wang, C. Simulation of freezing and thawing soils in inner Mongolia Hetao irrigation district, China. *Geoderma* **2012**, *173*, 28–33. [CrossRef]

26. French, H.K.; van der Zee, S.E. Improved management of winter operations to limit subsurface contamination with degradable deicing chemicals in cold regions. *Environ. Sci. Pollut. Res.* **2014**, *21*, 8897–8913. [CrossRef] [PubMed]

27. Zhao, X.; Xu, S.; Liu, Z. Study on Freeze-Thaw erosion lead to black soil and agricultural non-point source pollution. In *8th China Water Forum*; China Water & Power Press: Beijing, China, 2010; pp. 268–272.

28. Zhao, X.; Liu, T.; Xu, S.; Liu, Z. Freezing-thawing process and soil moisture migration within the black soil plow layer in seasonally frozen ground regions. *J. Glaciol. Geocryol.* **2015**, *37*, 233–240.

29. Zhao, X.; Liu, Z.; Xu, S.; Liu, T. Study of the black soil plow layer moisture changing with temperature in freeze-thaw cycle period in the seasonal frozen soil regions. *J. Glaciol. Geocryol.* **2015**, *37*, 931–939.

30. Zhao, X.; Xu, S.; Li, M. Freeze-Thaw lead to black soil erosion and non-point source pollution preferences: Alterable fuzzy optimum model of semi-structural decision applied. *World Hydrol.* **2015**, *272*, 14–18.

31. Meng, X.; Wang, H.; Wu, Y.; Long, A.; Wang, J.; Shi, C.; Ji, X. Investigating spatiotemporal changes of the land surface processes in Xinjiang using high-resolution CLM3.5 and CLDAS: Soil temperature. *Sci. Rep.* **2017**, *7*, 13286. [CrossRef] [PubMed]

32. Nayak, H.P.; Osuri, K.K.; Sinha, P.; Nadimpalli, R.; Mohanty, U.C.; Chen, F.; Rajeevan, M.; Niyogi, D. High-resolution gridded soil moisture and soil temperature datasets for the Indian monsoon region. *Sci. Data* **2018**, *5*, 180264. [CrossRef]

33. Yang, K.; Zhang, J. Evaluation of reanalysis datasets against observational soil temperature data over China. *Clim. Dyn.* **2018**, *50*, 317–337. [CrossRef]

34. Davtian, N.; Menot, G.; Bard, E.; Poulenard, J.; Podwojewski, P. Consideration of soil types for the calibration of molecular proxies for soil pH and temperature using global soil datasets and Vietnamese soil profiles. *Org. Geochem.* **2016**, *101*, 140–153. [CrossRef]

35. Gómez, I.; Caselles, V.; Estrela, M.J.; Niclòs, R. Impact of Initial Soil Temperature Derived from Remote Sensing and Numerical Weather Prediction Datasets on the Simulation of Extreme Heat Events. *Remote Sens.* **2016**, *8*, 589. [CrossRef]

36. Ambadan, J.T.; Berg, A.A.; Merryfield, W.J.; Lee, W. Influence of snowmelt on soil moisture and on near surface air temperature during winter–spring transition season. *Clim. Dyn.* **2018**, *51*, 1295–1309. [CrossRef]

37. Meng, X.; Wang, H.; Lei, X.; Cai, S.; Wu, H. Hydrological Modeling in the Manas River Basin Using Soil and Water Assessment Tool Driven by CMADS. *Teh. Vjesn.* **2017**, *24*, 525–534.

38. Meng, X.; Wang, H.; Cai, S.; Zhang, X.; Leng, G.; Lei, X.; Shi, C.; Liu, S.; Shang, Y. The China Meteorological Assimilation Driving Datasets for the SWAT Model (CMADS) Application in China: A Case Study in Heihe River Basin. *Preprints* **2016**, 120091. [CrossRef]

39. Shi, C.; Xie, Z.; Qian, H.; Liang, M.; Yang, X. China land soil moisture EnKF data assimilation based on satellite remote sensing data. *China Earth Sci.* **2011**, *54*, 1430. [CrossRef]

40. Stamnes, K.; Tsay, S.C.; Wiscombe, W.; Jayaweera, K. Numerically stable algorithm for discrete-ordinate method radiative transfer in multiple scattering and emitting layered media. *Appl. Opt.* **1988**, *27*, 2502–2509. [CrossRef] [PubMed]

41. Meng, X. Spring Flood Forecasting Based on the WRF-TSRM mode. *Teh. Vjesn.* **2018**, *25*, 27–37.

42. Zhao, F.; Wu, Y. Parameter Uncertainty Analysis of the SWAT Model in a Mountain Loess Transitional Watershed on the Chinese Loess Plateau. *Water* **2018**, *10*, 690. [CrossRef]

43. Liu, J.; Shanguan, D.; Liu, S.; Ding, Y. Evaluation and Hydrological Simulation of CMADS and CFSR Reanalysis Datasets in the Qinghai Tibet Plateau. *Water* **2018**, *10*, 513. [CrossRef]

44. Cao, Y.; Zhang, J.; Yang, M. Application of SWAT Model with CMADS Data to Estimate Hydrological Elements and Parameter Uncertainty Based on SUFI-2 Algorithm in the Lijiang River Basin, China. *Water* **2018**, *10*, 742. [CrossRef]

45. Shao, G.; Guan, Y.; Zhang, D.; Yu, B.; Zhu, J. The Impacts of Climate Variability and Land Use Change on Streamflow in the Hailiutu River Basin. *Water* **2018**, *10*, 814. [CrossRef]

46. Zhou, S.; Wang, Y.; Chang, J.; Guo, A.; Li, Z. Investigating the Dynamic Influence of Hydrological Model Parameters on Runoff Simulation Using Sequential Uncertainty Fitting-2-Based Multilevel-Factorial-Analysis Method. *Water* **2018**, *10*, 1177. [CrossRef]

47. Qin, G.; Liu, J.; Wang, T.; Xu, S.; Su, G. An Integrated Methodology to Analyze the Total Nitrogen Accumulation in a Drinking Water Reservoir Based on the SWAT Model Driven by CMADS: A Case Study of the Biliuhe Reservoir in Northeast China. *Water* **2018**, *10*, 1535. [CrossRef]

48. Guo, B.; Zhang, J.; Xu, T.; Croke, B.; Jakeman, A.; Song, Y.; Yang, Q.; Lei, X.; Liao, W. Applicability Assessment and Uncertainty Analysis of Multi-Precipitation Datasets for the Simulation of Hydrologic Models. *Water* **2018**, *10*, 1611. [CrossRef]

49. Dong, N.; Yang, M.; Meng, X.; Liu, X.; Wang, Z.; Wang, H.; Yang, C. CMADS-Driven Simulation and Analysis of Reservoir Impacts on the Streamflow with a Simple Statistical Approach. *Water* **2018**, *11*, 178. [CrossRef]

50. Vu, T.; Li, L.; Jun, K. Evaluation of Multi-Satellite Precipitation Products for Streamflow Simulations: A Case Study for the Han River Basin in the Korean Peninsula, East Asia. *Water* **2018**, *10*, 642. [CrossRef]

51. Meng, X.; Wang, H. Significance of the China Meteorological Assimilation Driving Data sets for the SWAT Model (CMADS) of East Asia. *Water* **2017**, *9*, 765. [CrossRef]

52. Meng, X.; Wang, H.; Shi, C.; Wu, Y.; Ji, X. Establishment and Evaluation of the China Meteorological Assimilation Driving Datasets for the SWAT Model (CMADS). *Water* **2018**, *10*, 1555. [CrossRef]

53. Meng, X. Simulation and spatiotemporal pattern of air temperature and precipitation in Eastern Central Asia using RegCM. *Sci. Rep.* **2018**, *8*, 3639. [CrossRef] [PubMed]

54. Tian, Y.; Zhang, K.; Xu, Y.; Gao, X.; Wang, J. Evaluation of Potential Evapo-transpiration Based on CMADS Reanalysis Dataset over China. *Water* **2018**, *10*, 1126. [CrossRef]

MDPI

St. Alban-Anlage 66

4052 Basel

Switzerland

Tel. +41 61 683 77 34

Fax +41 61 302 89 18

www.mdpi.com

Water Editorial Office

E-mail: water@mdpi.com

www.mdpi.com/journal/water

www.ingramcontent.com/pod-product-compliance
Lightning Source LLC
Chambersburg PA
CBHW051708210326
41597CB00032B/5409